Lecture Notes in Computer Science 14578

Advanced Research in Computing and Software Science
Subline of Lecture Notes in Computer Science

More information about this series at https://link.springer.com/bookseries/558

José A. Soto · Andreas Wiese

Editors

LATIN 2024: Theoretical Informatics

16th Latin American Symposium
Puerto Varas, Chile, March 18–22, 2024
Proceedings, Part I

 Springer

Editors
José A. Soto 🆔
DIM-CMM, Universidad de Chile
Santiago, Chile

Andreas Wiese 🆔
Technical University of Munich
Munich, Germany

ISSN 0302-9743   ISSN 1611-3349   (electronic)
Lecture Notes in Computer Science
ISBN 978-3-031-55597-8   ISBN 978-3-031-55598-5   (eBook)
https://doi.org/10.1007/978-3-031-55598-5

This Springer imprint is published by the registered company Springer Nature Switzerland AG
The registered company address is: Gewerbestrasse 11, 6330 Cham, Switzerland

Paper in this product is recyclable.

Preface

This volume contains the papers presented at the 16th Latin American Theoretical Informatics Symposium (LATIN 2024), held during March 18–22, 2024, in Puerto Varas, Chile. Previous editions of LATIN took place in São Paulo, Brazil (1992), Valparaíso, Chile (1995), Campinas, Brazil (1998), Punta del Este, Uruguay (2000), Cancún, Mexico (2002), Buenos Aires, Argentina (2004), Valdivia, Chile (2006), Búzios, Brazil (2008), Oaxaca, Mexico (2010), Arequipa, Peru (2012), Montevideo, Uruguay (2014), Ensenada, Mexico (2016), Buenos Aires, Argentina (2018), São Paulo, Brazil (2021) and Guanajuato, Mexico (2022). The symposium received 93 submissions from around the world. Each submission was double-blind reviewed by three or four program committee members, and carefully evaluated on quality, originality, and relevance to the conference. Committee members often reviewed the submissions with the help of additional external referees. Based on an extensive electronic discussion, the committee selected 44 papers. In addition to the accepted contributions, the symposium featured keynote talks by Pablo Barceló (Universidad Católica de Chile, Chile), Pierre Fraigniaud (Université Paris Cité and CNRS, France), Penny Haxell (University of Waterloo, Canada), Eunjung Kim (Korea Advanced Institute of Science and Technology, South Korea) and Jon Kleinberg (Cornell University, USA).

Also, LATIN 2024 featured two awards: the Imre Simon Test-of-Time Award and the Alejandro López-Ortiz Best Paper Award. In this edition, the Imre Simon Test-of-Time Award winners were Pierre Fraigniaud, Leszek Gąsieniec, Dariusz R. Kowalski and Andrzej Pelc for their paper "Collective Tree Exploration" which appeared in LATIN 2004. For the Alejandro López-Ortiz Best Paper Award, the program committee selected the paper "Faster Combinatorial k-Clique Algorithms" by Yarin Shechter, Amir Abboud and Nick Fischer. We thank our sponsor Springer for supporting both awards.

Our heartfelt thanks go to the authors for their excellent papers and cooperation, to the program committee members for their insightful discussions, to the subreferees for their careful reports, and to the steering committee for their valuable advice and feedback.

We would also like to recognize Nikhil Bansal, Conrado Martínez and Yoshiko Wakabayashi for their work on the 2024 Imre Simon Test-of-Time Award Selection Committee. Finally, the conference would not have been possible without our generous sponsors: ANID-Chile through grants BASAL FB210005 and ANILLO ACT210005, the Center for Mathematical Modelling (CMM), the research group Information and Computation in Market Design (ICMD) and Springer. We are also grateful for the facilities provided by EasyChair for the evaluation of the submitted papers and the discussions of the program committee.

January 2024

José A. Soto
Andreas Wiese

The Imre Simon Test-of-Time Award

The winner of the 2024 Imre Simon Test-of-Time Award, considering papers up to the 2014 edition of the Latin American Theoretical INformatics Symposium (LATIN), is

Collective Tree Exploration by Pierre Fraigniaud, Leszek Gąsieniec, Dariusz R. Kowalski and Andrzej Pelc, LATIN 2004, LNCS 2976, 141–151, 2004,

which later appeared as part of the journal article *Collective Tree Exploration*, by the same authors, and published in *Networks* 48(3): 166–177, 2006.

Collective Tree Exploration is an important milestone in the more general context of exploring an unknown environment. The problem arises in many applications, and over the years several researchers have studied efficient ways to explore different environments consisting of closed regions of the plane or graphs (typically used to model complex physical environments involving obstacles) under several computational/communication models.

For connected undirected graphs, it is well known that *depth-first search* (DFS) examines every single vertex and edge in an optimal way, provided that we can mark vertices and edges already explored. However, despite much interest, the amount of memory needed and the role of communication were much less understood. *Collective Tree Exploration* was one of the first contributions which addressed the issue of how much improvement can be obtained using $k \geq 2$ agents or *robots* instead of just $k = 1$, and which communication capabilities made a difference (and by how much).

In their work, Fraigniaud, Gąsieniec, Kowalski and Pelc consider an n-node tree that has to be explored by a team of k robots, starting from the root. All edges (and nodes) have to be visited by at least one robot, and this must be completed as quickly as possible. In each *synchronous* round, each of the k robots can stay where it is, or traverse one incident edge (either leading towards the root or towards some leaf). Two extreme communication scenarios are studied in the paper: in one, robots can share instantaneously all the information they have gathered until that moment (exploration with complete communication), in the other the robots share no information. A third scenario which is of interest is *exploration with write-read communication*, where robots can leave all the information that they have gathered for others to read. In the paper, an efficient algorithm is developed for the stronger scenario (complete communication), then the authors show that the same time complexity can be achieved in the less demanding and more practical model of write-read communication.

The first important contribution of the paper was to show that computing an optimal schedule for the exploration is NP-hard even if the full tree is known in advance; the proof was omitted in the conference extended abstract, but given in full detail in the journal version. The optimal algorithm with full knowledge of the tree is however an important piece of the investigation, as it sets the minimum exploration cost (= number of synchronous rounds) with which we can compare online exploration algorithms, that is, those that do not know the tree in advance and explore it in rounds.

The second fundamental result is the simple and elegant algorithm called "Collective Exploration" (CE) in the paper, which solves the problem on a tree of diameter D in $O(D + n/\log k)$ rounds. The *overhead* of CE is $O(k/\log k)$; it is the maximum competitive ratio between the exploration cost of the algorithm and that of the optimum, which is $\Theta(\max\{2n/k, D\})$, and taking the maximum over all possible trees of size n and all possible roots (starting points). This result was first established for the scenario with complete communication, then the authors showed how to simulate CE in the write-read communication scenario, while the time complexity remains the same.

The third main result of the work was to show that the overhead of any exploration algorithm (with no knowledge of the tree) is $\Omega(k)$ if there is no communication among the robots. Thus, combining these results, the authors give an interesting separation: without any form of communication k robots are essentially not better than one robot, on the other hand even a limited (and reasonable) amount of communication improves the exploration process, and allows us to take (some) advantage of having $k \geq 2$ robots.

Collaborative exploration of trees has been an important problem since its introduction in the ISTT 2024 awarded paper, cited many times by authors coming from different areas of Computer Science. *Collective Tree Exploration* has become a well-known and recognized reference by theoretical computer scientists working on exploration algorithms, but also for researchers in more practical areas like distributed robotics. Since the publication of the LATIN paper in 2004, and two years later of the journal version, many authors have studied variants, proposed new exploration algorithms for trees and other graphs, and analyzed the consequences of different communication capabilities, asynchronous settings and different ways to compare the collaborative online algorithms against the optimal algorithm that has complete knowledge of the tree. The area has been quite active during these years since the publication of *Collective Tree Exploration*, with the first improvement of the original overhead $O(k/\log k)$ obtained as recently as 2023.

The relevance of the problem addressed, the originality of the techniques used to solve it, the clarity of presentation and the continued and widespread recognition of this contribution throughout the years since its publication weighed heavily in the committee's choice.

The committee for the 2024 Imre Simon Test-of-Time Award,

Nikhil Bansal
Conrado Martínez
Yoshiko Wakabayashi

Organization

Program Committee Chairs

José A. Soto	Universidad de Chile, Chile
Andreas Wiese	Technical University of Munich, Germany

Steering Committee

Jacques Sakarovitch	CNRS and Télécom Paris, France
Armando Castañeda	Universidad Nacional Autónoma de México (UNAM), Mexico
Conrado Martínez	Universitat Politècnica de Catalunya, Spain
Flávio Keidi Miyazawa	Universidade Estadual de Campinas, Brazil
Cristina G. Fernandes	Universidade de São Paulo, Brazil
Michael A. Bender	Stony Brook University, USA

Program Committee

Shaull Almagor	Technion, Israel
Gabriela Araujo	Universidad Nacional Autónoma de México, Mexico
Flavia Bonomo	Universidad de Buenos Aires, Argentina
Fabio Botler	Universidade Federal do Rio de Janeiro, Brazil
Mario Bravo	Universidad Adolfo Ibáñez, Chile
Igor Carboni Oliveira	University of Warwick, UK
Timothy Chan	University of Illinois Urbana-Champaign, USA
Mark de Berg	TU Eindhoven, The Netherlands
Franziska Eberle	London School of Economics and Political Science, UK
Celina Figueiredo	Universidade Federal do Rio de Janeiro, Brazil
Johannes Fischer	TU Dortmund University, Germany
Emily Fox	University of Texas at Dallas, USA
Paweł Gawrychowski	University of Wrocław, Poland
Cristóbal Guzmán	Pontificia Universidad Católica de Chile, Chile
Christoph Haase	University of Oxford, UK
Adriana Hansberg	Universidad Nacional Autónoma de México, Mexico
Tobias Harks	University of Passau, Germany
Christoph Hertrich	London School of Economics and Political Science, UK
Martin Hoefer	Goethe University Frankfurt, Germany
Bart Jansen	TU Eindhoven, The Netherlands
Artur Jeż	University of Wrocław, Poland
Andrea Jiménez	Universidad de Valparaíso, Chile

Michael Kerber   Graz University of Technology, Austria
Thomas Kesselheim   University of Bonn, Germany
Arindam Khan   Indian Institute of Science, India
Stefan Kratsch   Humboldt University of Berlin, Germany
Jan Kretinsky   Technical University of Munich, Germany, and
                                  Masaryk University, Czech Republic
Ian Mertz   University of Warwick, UK
Pedro Montealegre   Universidad Adolfo Ibáñez, Chile
Ryuhei Mori   Nagoya University, Japan
Gonzalo Navarro   Universidad de Chile, Chile
Alantha Newman   Université Grenoble Alpes, France
Harumichi Nishimura   Nagoya University, Japan
André Nusser   University of Copenhagen, Denmark
Joël Ouaknine   Max Planck Institute for Software Systems, Germany
Dana Pizarro   Universidad de O'Higgins, Chile
Sergio Rajsbaum   Universidad Nacional Autónoma de México, Mexico
Andrea Richa   Arizona State University, USA
Saket Saurabh   Institute of Mathematical Sciences, India, and
                                  University of Bergen, Norway
Kevin Schewior   University of Southern Denmark, Denmark
Ildikó Schlotter   Centre for Economic and Regional Studies, Hungary
Sebastian Siebertz   University of Bremen, Germany
Jose A. Soto (Co-chair)   Universidad de Chile, Chile
Maya Stein   Universidad de Chile, Chile
Kavitha Telikepalli   Tata Institute of Fundamental Research, India
Roei Tell   Institute for Advanced Study, Princeton, USA, and
                                  Rutgers University, USA
Erik Jan van Leeuwen   Utrecht University, The Netherlands
Rob van Stee   University of Siegen, Germany
Jose Verschae   Pontificia Universidad Católica de Chile, Chile
Seeun William Umboh   University of Melbourne, Australia
Andreas Wiese (Co-chair)   Technical University of Munich, Germany

Organization Committee

Waldo Gálvez   Universidad de O'Higgins, Chile
José A. Soto   Universidad de Chile, Chile
Victor Verdugo   Universidad de O'Higgins, Chile
Andreas Wiese   Technical University of Munich, Germany

Additional Reviewers

Maximilian J. Stahlberg
Nicole Megow
Martín Ríos-Wilson
Lydia Mirabel Mendoza Cadena
Armando Castaneda
Kaustav Bose
Marta Grobelna
Pierre Vandenhove
Antonio Casares
Youssouf Oualhadj
Simon Weber
Sudebkumar Prasant Pal
Matt Gibson
Andrew Ryzhikov
Maël Le Treust
Stavros Kolliopoulos
Carolina Gonzalez
Luis Cunha
Lehilton L. C. Pedrosa
Abhinav Chakraborty
Lasse Wulf
André van Renssen
Leonidas Theocharous
Sanjana Dey
Tatsuya Gima
Bartlomiej Dudek
Bruno Netto
Yasuaki Kobayashi
Lucas De Meyer
Akira Suzuki
Alexandre Vigny
Torsten Mütze
Wanderson Lomenha
Jan Petr
Julien Portier
Sariel Har-Peled
Saladi Rahul
João Pedro de Souza Gomes da Costa
Aritra Banik
Anja Schedel
Raul Lopes
Tesshu Hanaka
François Dross

Hans Bodlaender
Augusto Modanese
Victor Larsen
Jens Schlöter
Matheus Pedrosa
Madhusudhan Reddy Pittu
Karol Pokorski
Sharma V. Thankachan
Julian Mestre
Alexander Braun
Maximilian Fichtl
Sugata Gangopadhyay
Eric Pérez
Ran Duan
Arturo Merino
K. Somasundaram
Asaf Yeshurun
Florian Dorfhuber
Vincent Froese
Andrei Draghici
Marc Vinyals
Torsten Ueckerdt
Elmar Langetepe
Martín Ríos-Wilson
Aditya Subramanian
Tobias Hofmann
Óscar C. Vásquez
Abdolhamid Ghodselahi
Jacob Calvert
Stefano Gogioso
Martin Koutecky
Syamantak Das
Sarita de Berg
Yuan Sha
Ge Xia
Shaily Verma
Andrea Marino
Neta Dafni
Venkatesh Raman
Benjamin Jauregui
Juan L. Reutter
Patrick Dinklage
Claudson Bornstein

Moses Ganardi
Jonas Ellert
Travis Gagie
Tomasz Kociumaka
Ernesto Araya Valdivia
Fahad Panolan
Valmir Barbosa
Stefan Schirra
Tassio Naia
Manuel Cáceres
Hadas Shachnai
Markus Bläser
Akanksha Agrawal
Mohammad Sadegh Mohagheghi
Eduardo Moreno
Giovanna Varricchio
Jamison Weber
Dolores Lara
César Hernández-Cruz
Vikash Tripathi
Ivan Bliznets
Pranabendu Misra
Ioan Todinca
Roohani Sharma
Patrick Eades
Alexandra Weinberger
Shaohua Li
Adam Kasperski
Nadia Brauner
Bertrand Simon
Łukasz Jeż
Ivan Rapaport
Daniel Rehfeldt
Andre Schidler
Anahi Gajardo
Pacôme Perrotin

Alberto Dennunzio
Mingyu Xiao
Benjamin Raichel
Britta Peis
Juan Pablo Contreras
Bart de Keijzer
Andrés Cristi
Michael Kaufmann
Philipp Kindermann
Juan Gutiérrez
Felix Schröder
Nicola Prezza
Jonas Ellert
Robert Bredereck
Claudio Telha Cornejo
Sung-Hwan Kim
Tomasz Kociumaka
Nikhil Balaji
Gerth Stølting Brodal
Jakub Łacki
Dominik Kempa
Adam Karczmarz
Nidhi Purohit
Kirill Simonov
Maximilian Prokop
Sven Jäger
Cristian Urbina
Bartlomiej Dudek
David Eppstein
Jean Cardinal
Mikkel Abrahamsen
Felipe A. Louza
Sabine Rieder
Zhouningxin Wang
Santiago Guzman Pro
Kathryn Nurse

Sponsors

ANID-Chile through grants BASAL FB210005 and ANILLO ACT210005
Center for Mathematical Modelling (CMM)
Information and Computation in Market Design (ICMD)
Springer

Contents – Part I

Approximation and Online Algorithms

Computational Geometry

Complexity Theory

Contents – Part II

Automata Theory and Formal Languages

Game Theory and Fairness

Algorithms and Data Structures

On 1-Bend Upward Point-Set Embeddings of st-Digraphs

Emilio Di Giacomo[1(✉)], Henry Förster[2], Daria Kokhovich[3],
Tamara Mchedlidze[3], Fabrizio Montecchiani[1], Antonios Symvonis[4],
and Anaïs Villedieu[5]

[1] University of Perugia, Perugia, Italy
{emilio.digiacomo,fabrizio.montecchiani}@unipg.it
[2] Universität Tübingen, Tübingen, Germany
henry.foerster@uni-tuebingen.de
[3] Utrecht University, Utrecht, The Netherlands
t.mtsentlintze@uu.nl
[4] National Technical University of Athens, Athens, Greece
symvonis@math.ntua.gr
[5] TU Wien, Vienna, Austria
avilledieu@ac.tuwien.ac.at

Abstract. We study the upward point-set embeddability of digraphs on one-sided convex point sets with at most 1 bend per edge. We provide an algorithm to compute a 1-bend upward point-set embedding of outerplanar st-digraphs on arbitrary one-sided convex point sets. We complement this result by proving that for every $n \geq 18$ there exists a 2-outerplanar st-digraph G with n vertices and a one-sided convex point set S so that G does not admit a 1-bend upward point-set embedding on S.

1 Introduction

A *point-set embedding (PSE)* of a planar graph $G = (V, E)$ on a given set of points S, with $|S| = |V|$, is a planar drawing Γ of G such that every vertex of G is represented by a point of S and each edge is drawn as a polyline connecting its end-vertices; if every edge has at most $b \geq 0$ bends, Γ is a b-bend PSE.

Gritzmann et al. [19] proved that the class of graphs that admit a PSE without bends along the edges on *every* set of points in general position coincides with the class of outerplanar graphs. Efficient algorithms to compute a PSE with no bends on any *given* set of points in general position exist for outerplanar graphs [8] and trees [9]. Cabello [10] proved that deciding whether a planar graph admits a PSE without bends on a *given* set of points is NP-complete. When bends are allowed, Kaufmann and Wiese [22] proved that every planar graph admits a PSE on every set of points with at most two bends per edge.

An *upward point-set embedding (UPSE)* of a directed graph $G = (V, E)$ on a given set of points S, with $|S| = |V|$, is a PSE with the additional property that each edge e is represented as a polyline monotonically increasing in the y-direction; also in this case we say that Γ is a b-bend UPSE if every edge has at most b bends. Clearly, for an UPSE to exist G must be an upward planar

© The Author(s), under exclusive license to Springer Nature Switzerland AG 2024
J. A. Soto and A. Wiese (Eds.): LATIN 2024, LNCS 14578, pp. 3–18, 2024.
https://doi.org/10.1007/978-3-031-55598-5_1

graph (and thus it must be a DAG). Different to the undirected case, a characterization of the upward planar digraphs that admit a UPSE without bends on *every* point set is still missing even for points in convex position. On the other hand, Binucci et al. [7] characterize DAGs that admit a 1-bend UPSE on every *upward one-sided convex* (UOSC) point set, i.e., a convex point set such that the bottommost point and the topmost point are adjacent in the convex hull of S; the same class has also been characterized by Heath and Pemmaraju [20] as the class of graphs that admit an upward 1-page book embedding. For points in convex position Binucci et al. [7] proved that there exist directed trees that do not admit an UPSE on every convex point set and many partial results exists about the embeddability of specific subclasses of directed trees on point sets with different properties [1,2,7,21]. Kaufmann et al. [21] studied the problem of deciding whether an upward planar graph admits an UPSE on a given set of points S and show that the problem can be solved in polynomial time for convex point sets, while it is NP-complete for point sets in general position. Arseneva et al. [2] proved that the problem remains NP-complete even for trees if one vertex is mapped to a specific point. As for the undirected case, two bends per edge suffice for UPSEs of upward planar graphs on any given set of points [18].

The results about (U)PSEs with zero and two bends naturally motivates the study of (U)PSEs with one bend. Testing whether a (upward) planar graph admits a 1-bend (U)PSE is NP-complete in both the upward and the non-upward variants. Indeed, it is easy to see that a 1-bend (U)PSE on a set of collinear points is, in fact, a 2-page (upward) book embedding and deciding whether a (upward) planar graph G admits a 2-page (upward) book embedding is NP-complete both in the non-upward [5] and in the upward case [4]. However, this relation between 1-bend (U)PSEs and 2-page (upward) book embeddings relies on the use of collinear points, and thus it does not hold for points in general or in convex position. The following problems are therefore open and worth to investigate.

Problem 1. Does every (upward) planar graph admit a 1-bend (U)PSE on every set of points in general or in convex position?

Problem 2. What is the complexity of testing whether a (upward) planar graph admits a 1-bend (U)PSE on a given set of points in general or in convex position?

We study the upward version of Problem 1 and our contribution is as follows.

- On the positive side, we show that every *st*-outerplanar graph (i.e., an outerplanar DAG with a single source and a single sink) admits a 1-bend UPSE on every UOSC point set (Theorem 1).
- We give a negative answer to the upward version of Problem 1 (Theorem 2). Namely, we prove that for every $n \geq 18$ there exists a 2-outerplanar *st*-digraph G with n vertices and an UOSC point set S such that G does not admit an UPSE on S with at most one bend per edge.

Concerning our second contribution, Di Giacomo et al. [12] proved that every two-terminal series-parallel digraph admits a 1-bend UPSE on any given set of

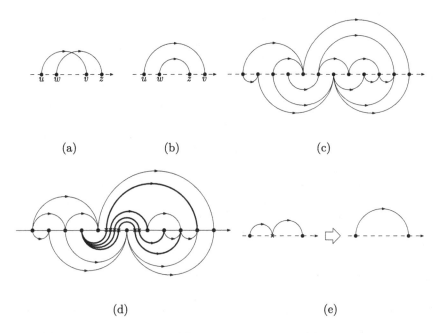

Fig. 1. (a) Two edges that cross; (b) two edges that nest; (c) an example of a 2UBE; (d) an example of a 2UTBE; the bold edges have spine crossings, shown with small crosses; (e) removal of unnecessary sub-edges.

points. This result has been extended by Mchedlidze and Symvonis [25] to the superclass of N-free graphs [1]. However, there exist st-outerplanar graphs that are not N-free digraphs (indeed, st-outerplanar graphs may contain the forbidden N-digraph), and vice-versa. We remark that the study of PSEs is a classical subject of investigation in the Graph Drawing and Computational Geometry literature where different (not necessarily upward) variants have been studied [3,11,14–16,24,27]. In particular, Everett et al. [17] and Löffler and Tóth [23] considered universal point sets for non-upward 1-bend drawings.

The paper is organized as follows. In Sect. 2 we give preliminary definitions. In Sect. 3 we prove necessary and sufficient conditions for the existence of a 1-bend UPSE. In Sect. 4 we describe the construction for outerplanar digraphs, while our negative example is described in Sect. 5. Open problems are in Sect. 6. Proofs marked with ($\star$) are sketched/removed and can be found in [13].

---

[1] The embedded N-graph is shaped like an N, i.e., it contains four vertices a, b, c, d and three edges (a, b), (c, b) and (c, d) such that (1) (a, b) enters b to the left of (c, b) and (2) (c, b) exists c to the left of (c, d). An embedded N-free graph does not contain the embedded N-graph as a subgraph.

2   Preliminaries

Let $G = (V, E)$ be an upward planar graph. A *2-page upward book embedding* (2UBE) of G consists of a total order $\prec$ of V, that is, a topological sorting of G, and of a partition of E into two sets, called *pages*, such that no two edges cross; two edges (u, v) with $u \prec v$ and (w, z) with $w \prec z$ *cross* if the two edges are in the same page and $u \prec w \prec v \prec z$ or $w \prec u \prec z \prec v$ (see Fig. 1(a)). Also, edges (u, v) and (w, z) *nest* if they are on the same page and $u \prec w \prec z \prec v$ or $w \prec u \prec v \prec z$ (see Fig. 1(b)). We write $u \preceq v$ if u precedes or coincides with v. A 2UBE can be visualized as an upward planar drawing such that all vertices of G lie along a horizontal line ℓ, called the *spine*, and each edge is represented as a semi-circle oriented in the direction of the spine and completely contained either above the spine (*top page*) or below the spine (*bottom page*). See Fig. 1(c) for an example of a 2UBE. A *2-page upward topological book embedding* (2UTBE) of G is a 2UBE of a subdivision of G. When considering a 2UTBE as a planar drawing, each subdivision vertex of an edge e can be regarded as a point where e crosses the spine, and therefore is also called a *spine crossing* (see Fig. 1(d)). Further, each of the "pieces" of an edge e defined by the subdivision vertices is called a *sub-edge* of e; specifically, the sub-edges that are in the top page are called *top sub-edges* and those that are in the bottom page are called *bottom sub-edges*. We write (sub-)edge to mean an element that is either an edge or a sub-edge. We assume that in a 2UTBE no spine crossing has two incident sub-edges that are in the same page; if so, the two sub-edges can be replaced by a single (sub-)edge (see Fig. 1(e)). A 2UTBE is a *single-top 2UTBE* if each edge has at most one top sub-edge (and hence at most two bottom sub-edges).

A set of points S is an *upward one-sided convex (UOSC) point set* if the points of S are in convex position and the lowest point of S is adjacent to the highest point of S in the convex hull. See Fig. 7(b) for an illustration. We denote by $CH(S)$ the convex hull of S. We always assume that all the points of S are to the left of the line passing through the topmost and the bottommost point.

3   Conditions for the Existence of a 1-Bend UPSE

We begin with a necessary condition for the existence of a 1-bend UPSE.

Lemma 1 ($\star$)**.** *Let $G = (V, E)$ be an upward planar graph. If G admits a 1-bend UPSE on an UOSC point set, then G admits a single-top 2UTBE.*

Proof (sketch). Let Γ be a 1-bend UPSE of G on an UOSC point set S. For each edge e of Γ, replace each intersection point between e and $CH(S)$ that is not an end-vertex of e, with a dummy vertex. We obtain a 1-bend upward planar drawing Γ' of a subdivision $G' = (V', E')$ of G, such that each edge is drawn completely outside $CH(S)$ or completely inside $CH(S)$. An edge of Γ' that is drawn completely outside $CH(S)$ has necessarily at least one bend, as edges with no bends necessarily lie inside $CH(S)$. Thus, an edge of Γ can be split by its intersection points with $CH(S)$ in at most three "pieces", at most one

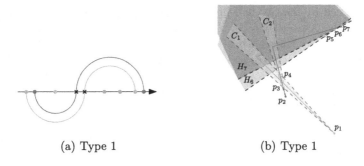

(a) Type 1   (b) Type 1

Fig. 2. (a) A Type 1 forbidden configuration; (b) A Type 1 impossible point set.

of which can be outside $CH(S)$. Each "piece" will be a sub-edge in the bottom page if it is inside $CH(S)$ and in the top page if it is outside $CH(S)$.   □

Given a 1-bend UPSE Γ on an UOSC point set S, we say that the 2UTBE γ that can be obtained as explained in the proof of Lemma 1 is *induced* by Γ.

We now give a sufficient condition for the existence of a 1-bend UPSE. We begin by introducing some additional definitions and technical lemmas. Let γ be a single-top 2UTBE of an upward planar graph G. A sub-edge (u, v) with $u \prec v$ *is nested inside* another sub-edge (w, z) with $w \prec z$ if the two sub-edges are in the same page and $w \preceq u \prec v \preceq z$. Notice that it cannot be that $w = u$ and $v = z$ at the same time. An (sub-)edge (w, z) with $w \prec z$ *covers* a vertex v if $w \prec v \prec z$. Let e_1 and e_2 be two edges of G. Edges e_1 and e_2 form a *forbidden configuration* in γ if the following three conditions hold simultaneously: (a) e_1 and e_2 both have a top sub-edge, one of the top sub-edges is nested inside the other, and the two sub-edges can possibly share a vertex; (b) e_1 and e_2 both have a bottom sub-edge, one of the bottom sub-edges is nested inside the other, and the sub-edges do not share a vertex; and (c) each bottom sub-edge covers at least one vertex and each top sub-edge covers at least two vertices. See Fig. 2(a) for an illustration of a forbidden configuration. We have four possible forbidden configurations: *Type 1 forbidden configuration* is such that the top sub-edges do not share a vertex and the two bottom sub-edges precede the two top sub-edges in the direction of the spine; *Type 2 forbidden configuration* is like the Type 1 forbidden configuration but with the top edges that share a vertex. *Type 3* and *Type 4 forbidden configurations* are like Type 1 and Type 2 respectively, but the top sub-edges precede the bottom sub-edges. We say that the 7 (or 6) vertices necessary to have a forbidden configuration are *the vertices that define the forbidden configuration*. These are the 4 (or 3) end-vertices of the two edges forming the forbidden configuration and the three vertices that are covered by their sub-edges. A single-top 2UTBE is *nice* if it has no forbidden configuration.

The next lemma shows that forbidden configurations are obstacles to the existence of a 1-bend UPSE for specific set of points. We describe 4 types of UOSC point sets, one for each type of forbidden configuration. Let $p_1, p_2, p_3, p_4, p_5, p_6, p_7$ be a set S of points ordered from bottom to top. Denote by C_i, with $i \in \{1, 2\}$

the cone defined by the two half-lines starting at p_i and passing through p_3 and p_4, respectively. Also, denote by H_i, with $i \in \{6, 7\}$ the half plane above the straight line passing through p_{i-1} and p_i. Finally, denote by T_1 the portion of C_1 that does not intersect H_6 and by T_2 the portion of C_2 that does not intersect H_7. We say that T_1 and T_2 cross each other if every segment connecting p_1 to the opposite side of T_1 crosses every segment that connects p_2 to the opposite side of T_2. If S is such that T_1 and T_2 cross, we say that S is a *Type 1 impossible point set* (see Fig. 2(b)). A *Type 2 impossible point set* is like a Type 1 impossible point set, but with p_6 and p_7 coincident – in this case the two half planes H_6 and H_7 are also coincident. *Type 3* and *Type 4 impossible points sets* are like Type 1 and Type 2 respectively, but mirrored vertically.

Lemma 2. *If a single-top 2UTBE γ contains a forbidden configuration of Type i, with $i \in \{1, 2, 3, 4\}$, then there does not exist a 1-bend UPSE whose induced 2UTBE is γ and such that the vertices that define the forbidden configuration are mapped to an impossible point set of Type i.*

Proof. Assume that γ has a Type 1 forbidden configuration (the other cases are analogous). Denote the two edges forming the forbidden configuration as e_1 and e_2, with the top sub-edge of e_1 nested inside the top sub-edge of e_2. Suppose that an UPSE exists whose induced 2UTBE is γ and such that the vertices of the forbidden configuration are mapped to the points of a Type 1 impossible point set. Then both e_1 and e_2 have one bend; the bend of e_1 is a point of $C_1 \cap H_6$, and the one of e_2 is a point of $C_2 \cap H_7$. This implies that the portion of e_1 drawn inside T_1 crosses the portion of e_2 drawn inside T_2 (see Fig. 2(a) and 2(b)). □

In the rest of this section, we prove that if G has a nice single-top 2UTBE then it admits a 1-bend UPSE on every UOSC point set. Let S be an UOSC point set of size n. Let γ be a 2UTBE of an n-vertex upward planar graph G and let $v_1, v_2, \ldots, v_{n'}$ be the sequence of vertices along the spine obtained by replacing each spine crossing with a dummy vertex. An *enrichment of S consistent with γ* is an UOSC point set S' such that: (i) $S \subset S'$; (ii) $|S'| = n'$; and (iii) if we denote by $p_1, p_2, \ldots, p_{n'}$ the points of S' in bottom-to-top order, then p_i is a dummy point if and only if v_i is a dummy vertex. See Fig. 3.

Let γ be a single-top 2UTBE of an upward planar graph G, and let γ' be the 2UBE obtained by replacing the spine crossings of γ with dummy vertices and let γ'_{top} be the 1-page book embedding obtained by γ' considering only the top page; we call γ'_{top} the *top-reduction* of γ. See Fig. 3(b). Let S be an n-point one-sided convex point set and let $S' = \langle p_1, p_2, \ldots, p_{n'} \rangle$ be an enrichment of S consistent with γ. We assign to each dummy vertex v_i in γ'_{top} a slope σ, which has to be used to draw the segment incident to the dummy vertex. If v_i is adjacent to a vertex v_j (real or dummy) with $j > i$, then σ is a slope of the II-IV quadrant defined by the Cartesian axes, while if v_i is adjacent to a vertex v_j (real or dummy) with $j < i$, then σ is a slope of the I-III quadrant. In either case the value of σ must be smaller, in absolute value, than the slope of any segment $\overline{p_k p_{k+1}}$ for $k = i, i+1, \ldots, j-1$ if $i < j$ or for $k = j, j+1, \ldots, i-1$ if $i > j$. Such a choice of slopes is called a *slope assignment* for γ'_{top}. Let e_1 and e_2 be two sub-edges with at least one dummy end-vertex each and such that e_2 is nested inside e_1. The slope assignment is *good for e_1 and e_2* if for any two slopes σ_1 assigned to e_1 and σ_2 assigned to e_2 in the same quadrant we have $|\sigma_1| < |\sigma_2|$. The slope assignment is *good* if it is good for every pair of nested sub-edges.

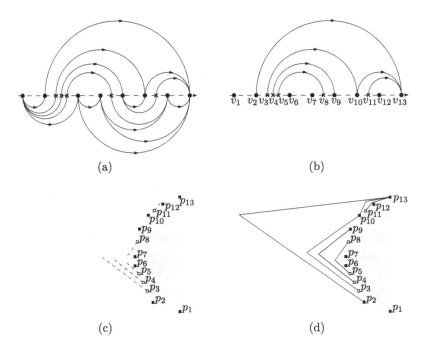

(a)   (b)

(c)   (d)

Fig. 3. (a) A single-top 2UTBE γ; (b) the top-reduction γ'_{top} of γ; (c) an enrichment of an UOSC point set S (black squares) consistent with γ with a good slope assignment. (d) A 1-bend UPSE of γ'_{top} on S' computed as in Lemma 3.

Lemma 3 ($\star$). *Let G be an n-vertex upward planar graph, let S be an UOSC point set. Let γ be a single-top 2UTBE of G and let γ'_{top} be the top-reduction of γ. If a good slope assignment is given, then γ'_{top} has a 1-bend UPSE on every enrichment S' of S consistent with γ such that all the edges are drawn outside $CH(S')$ and the segment incident to each dummy vertex is drawn with the assigned slope.*

Proof (sketch). Let $v_1, v_2, \ldots, v_{n'}$ be the vertices in γ'_{top} according to the spine order. Let $S' = \langle p_1, p_2, \ldots, p_{n'} \rangle$ be an enrichment of S consistent with γ; see Fig. 3. An edge is drawn only after all edges nested inside it are already drawn. Let $e = (v_i, v_j)$ be the current edge and suppose that $i < j$, i.e., that $v_i \prec v_j$. The edge e is drawn as the union of two segments: s_i incident to v_i and s_j incident to v_j. The segment s_i is drawn in the II quadrant, while segment s_j is drawn in the III quadrant. This guarantees that s_i and s_j meet at a bend point. If v_i (resp. v_j) is a dummy vertex, then s_i (resp. s_j) is drawn with the slope assigned to v_i (resp. v_j). Notice that the slope assigned to v_i (resp. v_j) is a slope of the II-IV quadrant (resp. I-III quadrant). If v_i (resp. v_j) is a real vertex, then s_i (resp. s_j) is drawn with a slope σ of the II-IV quadrant (resp. I-III quadrant); the absolute value of σ has to be smaller than the absolute value of any other slope used by the already drawn edges nested inside (v_i, v_j). If no edge is nested inside (v_i, v_j), then $|\sigma|$ has to be smaller than the absolute value of the slope of any segment $\overline{p_k p_{k+1}}$, for $k = i, i+1 \ldots, j-1$. It is easy to see that all the edges are drawn outside $CH(S')$ and no crossing exists.   □

Lemma 4 ($\star$). *Let G be an n-vertex upward planar graph, let γ be a single-top 2UTBE of G, and let e be a top sub-edge that covers exactly one vertex and that has no top sub-edge nested inside it. Let γ' be the 2UTBE obtained from γ by removing the edge e' containing the sub-edge e. Let Γ' be a 1-bend UPSE of $G \setminus \{e'\}$ on an UOSC point set S whose induced 2UTBE is γ'. Then it is possible to construct a 1-bend UPSE of G on S that has Γ' as a sub-drawing.*

Lemma 5 ($\star$). *Let $G = (V, E)$ be an n-vertex upward planar graph. If G admits a nice single-top 2UTBE, then G admits a 1-bend UPSE on every UOSC point set S of size n.*

Proof (sketch). If G admits a nice single-top 2UTBE γ, then we can compute a 1-bend UPSE on every one-sided convex point set S as follows. Let $S' = \langle p_1, p_2, \ldots, p_{n'} \rangle$ be an enrichment of S consistent with γ. We recursively remove all edges that have a top sub-edge covering only one vertex and no edges nested inside. These edges will be reinserted at the end in reverse order using Lemma 4.

Let γ' be the single-top 2UTBE resulting from the edge removal explained above and let G' be the corresponding graph. We now compute a 1-bend UPSE of G' on S'. We first map each vertex v_i to the point p_i ($i = 1, 2, \ldots, n'$). By the choice of the additional points, the dummy vertices are mapped to the dummy points. We then draw the bottom (sub-)edges as straight-line segments inside the convex hull $CH(S')$ of S'. Since the bottom-to-top order of the vertices along $CH(S')$ is the same as in γ', the (sub-)edges drawn inside $CH(S')$ do not cross.

Now, in order to draw the top (sub-)edges, we consider the top restriction of γ', and define a slope assignment, assigning to each dummy vertex d the slope of the segment incident to d that is in the bottom page (drawing the segment incident to d with this slope guarantee that no additional bend is created at d). We can prove that this slope assignment is good and thus by Lemma 3 all the top (sub-)edges can be drawn outside the convex hull respecting the slope assignment, which guarantees that each edge is drawn with one bend.   □

4   1-Bend UPSE of st-Outerplanar Graphs

A graph is *outerplanar* if it admits an *outerplanar drawing*, i.e., a planar drawing in which all vertices belong to the boundary of the outer face, which defines an *outerplanar embedding*. Unless otherwise specified, we will assume our graphs to have planar or outerplanar embeddings. An edge of an embedded planar graph G is *outer* if it belongs to the outer face, and it is *inner* otherwise. The *weak dual* $\overline{G}$ of G is the graph having a node for each inner face of G, and an edge between two nodes if and only if the two corresponding faces share an edge. If G is outerplanar, its weak dual $\overline{G}$ is a tree. If $\overline{G}$ is a path, G is an *outerpath*. A *fan* is an internally-triangulated outerpath whose inner edges all share an end-vertex.

An *st-digraph* is a directed acyclic graph with a single source s and a single sink t; an *st-outerplanar graph* (resp. *st-outerpath*) is an *st*-digraph whose underlying undirected graph is an outerplanar graph (resp. an outerpath). An *st-fan* is an *st*-digraph whose underlying graph is a fan and whose inner edges have s as an end-vertex. An *st*-outerplanar graph such that the edge (s,t) exists is *one-sided* if (s,t) is an outer edge, it is *two-sided* if (s,t) is an inner edge.

We recall a decomposition of *st*-outerpaths defined in [6]. The *extreme faces* of an *st*-outerpath G are the two faces that correspond to the two degree-one nodes of the weak dual $\overline{G}$. An *st*-outerpath G is *primary* if and only if one of its extreme faces is incident to s and the other one to t. Observe that this definition is stronger than the one used in [6], in the sense that a primary *st*-outerpath according to our definition is a primary *st*-outerpath also according to the definition in [6] (but the converse may not be true). Let G be a primary *st*-outerpath. Consider a subgraph F of G that is an xy-fan (for some vertices x, y of G). Let $\langle f_1, \ldots, f_h \rangle$ be the list of faces forming the path $\overline{G}$ ordered from s towards t. Note that the subgraph F of G is formed by a subset of faces that are consecutive in the path $\langle f_1, \ldots, f_h \rangle$. Let f_i be the face of F with the highest index, with $1 \leq i \leq h$. We say that F is *incrementally maximal* if $i = h$ or $F \cup f_{i+1}$ is not an xy-fan. For every face f_i we denote by $\mathrm{mid}(f_i)$ the unique vertex of f_i with one incoming edge and one outgoing edge in the boundary of f_i.

Definition 1. *An st-fan decomposition of a primary st-outerpath G is a sequence of $s_i t_i$-fans $F_i \subseteq G$, with $i = 1, \ldots, k$, such that: (i) F_i is incrementally maximal for each $i = 1, \ldots, k$; (ii) for any $1 \leq i < j \leq k$, F_i and F_j do not share any edge if $j > i + 1$, while F_i and F_{i+1} share a single edge, which we denote by e_i; (iii) $s_1 = s$; (iv) the tail of e_i is s_{i+1} for each $i = 1, \ldots, k-1$; (v) $e_i \neq (s_i, t_i)$ for each $i = 1, \ldots, k-1$; and (vi) $\bigcup_{i=1}^{k} F_i = G$. Refer to Fig. 4a-b.*

Lemma 6 ([6]). *Every primary st-outerpath G admits an st-fan decomposition.*

Let G be an *st*-outerplanar graph and let $\overline{P}$ be a path in the weak dual $\overline{G}$ of G whose two endpoints are such that one corresponds to a face containing s and the other one to a face containing t. Observe that the primal graph G_{core} of $\overline{P}$ is a primary *st*-outerpath by construction, we call it the *core* of G. On the other hand, if an outer edge (u, v) of G_{core} is not an outer edge of G, then it corresponds to a separation pair in G. In particular, (u, v) belongs to G_{core} and to another subgraph A_{uv} of G which is a one-sided uv-outerplanar graph; we call A_{uv} an *appendage* of G attached to (u, v); refer to Fig. 4c.

Property 1. *Let G be an st-outerplanar graph and let G_{core} be the core of G. The following properties hold:*

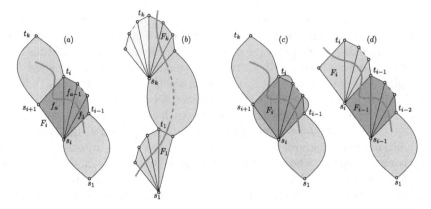

Fig. 4. (a) An st-fan in the middle of the st-fan decomposition. (b) The first and last st-fans of the st-fan decomposition. Since the outerpath is primary, the last fan can always be chosen to be one-sided with edge (s_k, t_k) regarded as a possible attachment edge of an appendage (light blue). (c) G_{core} is blue and gray, while green subgraphs represent appendages of G_{core}. (d) Illustration of Property 1b. (Color figure online)

(a) Every outer edge of G_{core} is potentially an attachment edge of an appendage.
(b) Let $F_1, \ldots, F_k$ be an st-fan decomposition of G_{core} and let P be its dual path. Let s_i, t_i denote the source and the sink of F_i. The fans F_{i-1} and F_i share the edge (s_i, t_{i-1}); See Fig. 4d (Stronger version of Lemma 3 in [6]).
(c) Path P enters F_i, $i = 2, \ldots, k$ through the edge (s_i, t_{i-1}) and leaves F_i, $i = 1, \ldots, k-1$ through the edge (s_{i+1}, t_i).
(d) Let F_i be a two-sided st-outerpath and let $f_1, \ldots, f_a$ be the faces of F_i as visited by P. Faces $f_1, \ldots, f_{a-1}$, $a \geq 2$, lie on one side of (s_i, t_i) and only the face f_a lies on the other side of (s_i, t_i). Refer to Fig. 4a.

Property (a) holds by definition. If Properties (b) and (c) do not hold, then G_{core} has either more than one sink or more than one source. Finally, assuming that Property (d) does not hold, implies that G_{core} is not an outerpath.

In this section we utilize a tool, called *Hamiltonian completion*, that is another way to look at 2UTBEs. An upward planar graph G has a 2UBE if and only if it is *subhamiltonian*, i.e., it is a spanning subgraph of an upward planar st-digraph $\tilde{G}$ that has a directed Hamiltonian st-path [26]. More generally, there is an analogy between upward topological book embeddings and a more general form of subhamiltonicity. Let G be an upward planar graph and $\tilde{G} = (V, \tilde{E})$ be an embedded st-digraph such that: (1) $G = (V, E)$ is a spanning subgraph of $\tilde{G}$, (2) $\tilde{G}$ has a directed Hamiltonian st-path H, and (3) each edge in E is crossed by at most one edge of $\tilde{E} \setminus E$. We say that H is a *subhamiltonian path* of G and $\tilde{G}$ is an *HP-completion* of G. See Fig. 5 for an example of subhamiltonian paths.

Lemma 7 ([26]). *An upward planar graph has a 2UTBE with at most one spine-crossing per edge if and only if it has an HP-completion. The order of the vertices along the spine in the 2UTBE is the same as in the subhamiltonian path.*

The subhamiltonian path crosses some edges of G by splitting them into *sub-edges*. We inherit the definition of nesting (sub-)edges from 2UTBE to HP-completion. Thus, the (sub-)edges (u, v), (w, z) *nest* in $\tilde{G}$ if in the embedding of $\tilde{G}$ they are on the same side of the path H and $u \prec w \prec z \prec v$ or $w \prec u \prec v \prec z$. Since the order of the vertices on the spine of the book and along the Hamiltonian path coincide, two (sub-)edges nest in $\tilde{G}$ if and only if they nest in the corresponding 2UTBE. We now prove the key result of this section.

Lemma 8 ($\star$). *Every primary st-outerpath has an HP-completion without nesting sub-edges.*

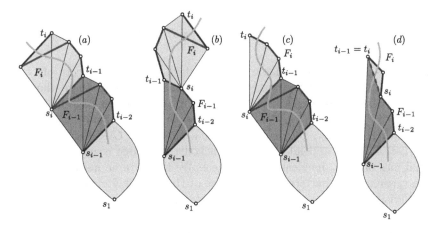

Fig. 5. Proof of Lemma 8: (a) Case 1. (b–d) Case 2, 3.a, and 3.b. The subhamiltonian path H_i is drawn in dark red. (Color figure online)

Proof (sketch). Let G be a primary st-outerpath and $F_1, \ldots, F_k$ be its st-fan decomposition. Let $\overline{P}$ be the dual path of G. Let G_i be the subgraph of G composed by $F_1, \ldots, F_i$, $i = 1, \ldots, k$, therefore $G = G_k$. We construct the subhamiltonian path H_i in G_i by induction on i, assuming the next invariants for H_{i-1} in G_{i-1}:

$\mathcal{I}1$ Subhamiltonian path H_{i-1} in G_{i-1} terminates with the edge (s_i, t_{i-1}).

$\mathcal{I}2$ Path H_{i-1} crosses the edge (s_{i-1}, t_{i-1}) (in a point referred to as p_{i-1}) if and only if F_{i-1} is two-sided. No other edge of F_{i-1} is crossed by H_{i-1}.

$\mathcal{I}3$ H_{i-1} does not create nesting sub-edges in G_{i-1}.

We show how to construct H_i so to maintain the invariants. We have three cases (two are omitted) based on whether F_{i-1} and F_i are two-sided or not.

Case 1: **both F_{i-1} and F_i are two-sided.** Refer to Fig. 5a. Consider the dual path $\overline{P}$ in F_i and let $f_1, \ldots, f_a$, be the faces of F_i as visited by $\overline{P}$. By Property 1(d), since F_i is two-sided, faces $f_1, \ldots, f_{a-1}$, $a \geq 2$, lie on one side of (s_i, t_i) and only the face f_a lies on the other side of (s_i, t_i). Note that, by Properties 1(b) and 1(c), F_{i-1} and F_i share (s_i, t_{i-1}) and P enters F_i through (s_i, t_{i-1}); it follows that mid$(f_1) = t_{i-1}$. By induction hypothesis H_{i-1} terminates at (s_i, t_{i-1}). Therefore, we can set path H_i to be H_{i-1} concatenated with mid$(f_1), \ldots,$ mid$(f_a), t_i$. Note that mid$(f_a) = s_{i+1}$, thus Invariant $\mathcal{I}1$ holds. Also, H_i crosses (s_i, t_i) and no other edge of F_i, hence Invariant $\mathcal{I}2$ holds as well. Finally, concerning the only two edges of F_{i-1} and F_i that are crossed by H_i, the order in which their end-vertices s_i, t_i, s_{i-1}, and t_{i-1} and their crossing points p_{i-1} and p_i are visited is $s_{i-1}, p_{i-1}, s_i, t_{i-1}, p_i, t_i$, which implies that their sub-edges do not nest. No other sub-edge is created by H_i, thus $\mathcal{I}3$ holds.   □

Lemma 9. *Every st-outerplanar graph has an HP-completion without nesting sub-edges.*

Proof. Let G be an st-outerplanar graph and let G_{core} be the core of G. By Lemma 8, G_{core} has an HP-completion with subhamiltonian path H' that does not create nesting sub-edges. By Property 1(a), every outer edge of G_{core} is potentially an attachment edge of an appendage of G. We expand the subhamiltonian path H' of G_{core} to a subhamiltonian path H in G as follows, refer to Fig. 6. Let A be an appendage of P attached to an edge e and let f be the internal face of G_{core} incident to the edge e. We flip A to lie inside f. We visit all the vertices of A that are not the source or the sink of A either immediately after H' visits the source of A (blue appendage in Fig. 6.c) or immediately before it visits the sink of A (pink appendage in Fig. 6.c), or both things at the same time (green appendages in Fig. 6c). After this procedure the edges crossed by H are exactly the edges of G_{core} crossed by H', i.e., no new sub-edge is created. Further, the vertices of G_{core} are visited in the same order by H and by H'. Hence, since H' did not create nesting sub-edges in G_{core}, so does H in G.   □

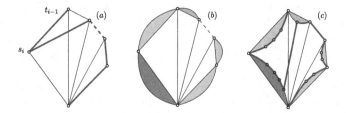

Fig. 6. Augmenting the subhamitonian path to visit appendages. (Color figure online)

By Lemmas 7 and 9 every st-outerplanar graph has a nice 2UTBE with at most one spine-crossing per edge. By Lemma 5 we have the following.

Theorem 1. *Every st-outerplanar graph admits a 1-bend UPSE on every UOSC point set.*

5   1-Bend UPSE Are Not Always Possible

We describe a 2-outerplanar st-digraph G and an UOSC point set S such that G does not admit a 1-bend UPSE on S. An st-digraph is 2-outerplanar if removing all vertices of the outer face yields an outerplanar digraph.

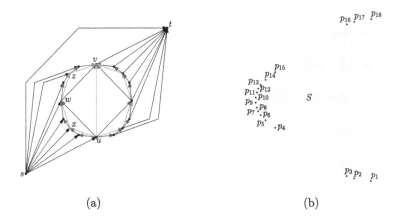

(a)   (b)

Fig. 7. (a) An st-digraph G and (b) an UOSC point set S for the proof of Lemma 10 (Color figure online)

Lemma 10 ($\star$). *There exists a 2-outerplanar st-digraph G and an UOSC point set S such that G does not admit a 1-bend UPSE on S.*

Proof (sketch). Let G be the st-digraph of Fig. 7(a) and let S be the point set of Fig. 7(b) By Lemma 1, if G has a 1-bend UPSE Γ on S, then Γ induces a single-top 2UTBE. We show that every single-top 2UTBE γ of G has a forbidden configuration of Type i, for some $i \in \{1, 2, 3, 4\}$, that is necessarily mapped to a Type i impossible point subset of S. By Lemma 2 a 1-bend UPSE cannot exist. Let $p_1, p_2, \ldots, p_{18}$ be the points of S in bottom-to-top order. Let π_l be the path from u to v to the left of (u, v) (red in Fig. 7(a)) and let π_r be the path from u to v to the right of (u, v) (blue in Fig. 7(a)). The edge (u, v) (yellow in Fig. 7(a)) has vertices on both sides. Thus, in every 2UTBE it crosses the spine either once or twice and the vertices of π_l must appear along the spine in the order they appear along π_l; the same holds for π_r. We have different cases. In each case we denote by $v_1, v_2, \ldots, v_n$ the sequence of vertices along the spine (thus vertex v_i is mapped to point p_i). In all cases u is mapped to p_2 and v is mapped to p_{17}.
Case 1: Edge (u, v) crosses the spine once (see Fig. 8(a)). The first sub-edge of (u, v) is either a bottom or a top sub-edge. Assume the first case (the other one is symmetric) case the vertex w coincides with v_6 and the edges (w, v) and (u, v) form a Type 2 forbidden configuration with the spine crossings between v_9 and v_{10}. Since, $p_2, p_6, p_9, p_{10}, p_{16}, p_{17}$ form a Type 2 impossible point set (see Fig. 8(b)), by Lemma 2 a 1-bend UPSE cannot exist in this case.

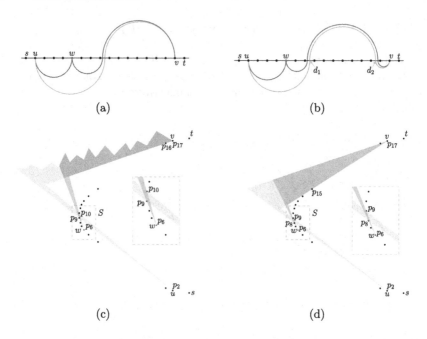

Fig. 8. Theorem 2: (a)–(c) Case 1. (b)–(d) Case 2.A.

Case 2: Edge (u, v) crosses the spine twice. In this case (u, v) consists of three sub-edges (u, d_1), (d_1, d_2), and (d_2, v), where d_1 and d_2 are spine crossings. Only (d_1, d_2) is a top sub-edge. Thus, the vertices of π_l have to be distributed in the two intervals defined by (u, d_1) and (d_2, v). We distinguish six sub-cases (five are omitted) depending on the distribution of the vertices of π_l.

Case 2.A: w is between u and d_1 with a single vertex of π_l between d_2 and v (see Fig. 8(b)). In this case w coincides with v_6 and both (w, v) and (u, v) cross the spine between v_8 and v_9 and between v_{15} and v_{16}. The edges (w, v) and (u, v) form a Type 2 forbidden configuration. Since $p_2, p_6, p_8, p_9, p_{15}, p_{17}$ form a Type 2 impossible point set (see Fig. 8(d)), by Lemma 2 a 1-bend UPSE cannot exist.   □

The following theorem is easily derived from Lemma 10 by suitably adding, for every $n \geq 18$, $n - 18$ vertices to G and $n - 18$ points to S.

Theorem 2. *For every $n \geq 18$ there exists an n-vertex 2-outerplanar st-digraph G and an UOSC point set S such that G does not admit a 1-bend UPSE on S.*

6 Open Problems

Various questions remain open related to Problem 1 and 2 of Sect. 1, such as: (i) Investigate the non-upward version of Problem 1. (ii) Study Problem 2. In particular, characterize the digraphs admitting a 1-bend UPSE on every UOSC point set.

References

1. Angelini, P., Frati, F., Geyer, M., Kaufmann, M., Mchedlidze, T., Symvonis, A.: Upward geometric graph embeddings into point sets. In: Brandes, U., Cornelsen, S. (eds.) GD 2010. LNCS, vol. 6502, pp. 25–37. Springer, Heidelberg (2010). https://doi.org/10.1007/978-3-642-18469-7_3
2. Arseneva, E., et al.: Upward point set embeddings of paths and trees. In: Uehara, R., Hong, S.H., Nandy, S.C. (eds.) WALCOM 2021. LNCS, vol. 12635, pp. 234–246. Springer, Cham (2021). https://doi.org/10.1007/978-3-030-68211-8_19
3. Badent, M., Di Giacomo, E., Liotta, G.: Drawing colored graphs on colored points. Theor. Comput. Sci. **408**(2–3), 129–142 (2008)
4. Bekos, M.A., Da Lozzo, G., Frati, F., Gronemann, M., Mchedlidze, T., Raftopoulou, C.N.: Recognizing DAGs with page-number 2 is NP-complete. Theor. Comput. Sci. **946**, 113689 (2023)
5. Bernhart, F., Kainen, P.C.: The book thickness of a graph. J. Comb. Theory Ser. B **27**(3), 320–331 (1979)
6. Bhore, S., Da Lozzo, G., Montecchiani, F., Nöllenburg, M.: On the upward book thickness problem: combinatorial and complexity results. Eur. J. Comb. **110**, 103662 (2023)
7. Binucci, C., et al.: Upward straight-line embeddings of directed graphs into point sets. Comput. Geom. **43**(2), 219–232 (2010)
8. Bose, P.: On embedding an outer-planar graph on a point set. Comput. Geom. Theory Appl. **23**, 303–312 (2002)
9. Bose, P., McAllister, M., Snoeyink, J.: Optimal algorithms to embed trees in a point set. J. Graph Algorithms Appl. **2**(1), 1–15 (1997)
10. Cabello, S.: Planar embeddability of the vertices of a graph using a fixed point set is NP-hard. J. Graph Algorithms Appl. **10**(2), 353–363 (2006)
11. Di Giacomo, E., Didimo, W., Liotta, G., Meijer, H., Wismath, S.K.: Point-set embeddings of trees with given partial drawings. Comput. Geom. **42**(6–7), 664–676 (2009)
12. Di Giacomo, E., Didimo, W., Liotta, G., Wismath, S.K.: Book embeddability of series-parallel digraphs. Algorithmica **45**(4), 531–547 (2006). https://doi.org/10.1007/s00453-005-1185-7
13. Di Giacomo, E., et al.: On 1-bend upward point-set embeddings of *st*-digraphs. CoRR 2401.03226 (2024). http://arxiv.org/2401.03226
14. Di Giacomo, E., Gasieniec, L., Liotta, G., Navarra, A.: On the curve complexity of 3-colored point-set embeddings. Theor. Comput. Sci. **846**, 114–140 (2020)
15. Di Giacomo, E., Liotta, G., Trotta, F.: Drawing colored graphs with constrained vertex positions and few bends per edge. Algorithmica **57**(4), 796–818 (2010). https://doi.org/10.1007/s00453-008-9255-2
16. Dujmović, V., et al.: On point-sets that support planar graphs. Comput. Geom. **46**(1), 29–50 (2013)
17. Everett, H., Lazard, S., Liotta, G., Wismath, S.K.: Universal sets of *n* points for one-bend drawings of planar graphs with *n* vertices. Discrete Comput. Geom. **43**(2), 272–288 (2010). https://doi.org/10.1007/s00454-009-9149-3
18. Giordano, F., Liotta, G., Mchedlidze, T., Symvonis, A., Whitesides, S.: Computing upward topological book embeddings of upward planar digraphs. J. Discrete Algorithms **30**, 45–69 (2015)
19. Gritzmann, P., Mohar, B., Pach, J., Pollack, R.: Embedding a planar triangulation with vertices at specified points. Am. Math. Monthly **98**(2), 165–166 (1991)

20. Heath, L.S., Pemmaraju, S.V.: Stack and queue layouts of directed acyclic graphs: part II. SIAM J. Comput. **28**(5), 1588–1626 (1999)
21. Kaufmann, M., Mchedlidze, T., Symvonis, A.: On upward point set embeddability. Comput. Geom. **46**(6), 774–804 (2013)
22. Kaufmann, M., Wiese, R.: Embedding vertices at points: few bends suffice for planar graphs. J. Graph Algorithms Appl. **6**(1), 115–129 (2002)
23. Löffler, M., Tóth, C.D.: Linear-size universal point sets for one-bend drawings. In: Di Giacomo, E., Lubiw, A. (eds.) GD 2015. LNCS, vol. 9411, pp. 423–429. Springer, Cham (2015). https://doi.org/10.1007/978-3-319-27261-0_35
24. Mchedlidze, T.: Upward planar embedding of an n-vertex oriented path on $O(n^2)$ points. Comput. Geom. **46**(8), 1003–1008 (2013)
25. Mchedlidze, T., Symvonis, A.: Crossing-free acyclic hamiltonian path completion for planar st-digraphs. In: Dong, Y., Du, D.Z., Ibarra, O. (eds.) ISAAC 2009. LNCS, vol. 5878, pp. 882–891. Springer, Heidelberg (2009). https://doi.org/10.1007/978-3-642-10631-6_89
26. Mchedlidze, T., Symvonis, A.: Crossing-optimal acyclic hamiltonian path completion and its application to upward topological book embeddings. In: Das, S., Uehara, R. (eds.) WALCOM 2009. LNCS, vol. 5431, pp. 250–261. Springer, Heidelberg (2009). https://doi.org/10.1007/978-3-642-00202-1_22
27. Pach, J., Wenger, R.: Embedding planar graphs at fixed vertex locations. Graphs Comb. **17**(4), 717–728 (2001). https://doi.org/10.1007/PL00007258

Decoding Tree Decompositions
from Permutations

Samuel Eduardo da Silva and Uéverton S. Souza$^{(\boxtimes)}$ (iD)

Instituto de Computação, Universidade Federal Fluminense, Niterói, Brazil
samueleduardo@id.uff.br, ueverton@ic.uff.br

Abstract. Most algorithmic strategies for solving problems considering treewidth parameterization require that a tree decomposition is given. Given a graph $G = (V, E)$ and denoting by $\mathcal{C}_G$ the family of chordal graphs (triangulations) G' such that $V(G) = V(G')$ and $E(G) \subseteq E(G')$, the treewidth of a graph G can be defined alternatively as the size of the smallest maximum clique of a graph in $\mathcal{C}_G$, minus one. In addition, any tree decomposition $\mathcal{T}$ of a graph $G' \in \mathcal{C}_G$ is also a tree decomposition of G. In this paper, we are interested in the main subproblem to be solved by the most popular heuristics for treewidth computation, called TREE DECOMPOSITION DECODING. In such a problem, we are given a graph $G = (V, E)$ and a permutation ρ of $V(G)$ and asked to determine the width of the tree decomposition $\mathcal{T}$ of G that is an optimum tree decomposition of the minimal triangulation $G' \in \mathcal{C}_G$ having ρ as perfect elimination ordering. From (G, ρ), it is easy to find the solution to the problem by first constructing the triangulation G' arising from ρ. However, in the worst case, such constructions of G' require $\Theta(|V(G)|^2)$ space. In this work, we propose two algorithms for solving the problem; both avoid the construction of triangulations G'. The first performers in $\mathcal{O}(|V(G)| \cdot \ell)$ space and $\mathcal{O}(|V(G)|^2 \cdot \ell)$ time, where ℓ is the number of leaves of the tree decomposition encoded by ρ. The second is faster in practice and achieves a different trade-off, solving the problem within $\mathcal{O}(|E(G)| + |V(G)|)$ space and $\mathcal{O}(|E(G)| \cdot \log |V(G)|)$ time.

Keywords: Tree decomposition · Treewidth · Perfect elimination order · Decoder

1 Introduction

A simple graph G is called *chordal* when every induced cycle of G has exactly three vertices, i.e., every cycle of size at least four of G has at least one chord. The chordal graph class is one of the main and most important graph classes, used in several theoretical frameworks in Algorithmic Graph Theory and having many practical applications.

Given a graph G, a vertex $v \in V(G)$ is called *simplicial* if its neighborhood induces a clique. In 1961, Dirac [6] proved that any chordal graph is either

This research has received funding from Rio de Janeiro Research Support Foundation (FAPERJ) and National Council for Scientific and Technological Development (CNPq).

J. A. Soto and A. Wiese (Eds.): LATIN 2024, LNCS 14578, pp. 19–34, 2024.
https://doi.org/10.1007/978-3-031-55598-5_2

complete or has at least two non-adjacent simplicial vertices. Based on that, Fulkerson and Gross [7], in 1965, presented a polynomial-time algorithm to recognize chordal graphs, which finds and removes simplicial vertices of the input graph until there are no more simplicial vertices.

A permutation $\rho = v_1, v_2, \ldots, v_n$ of the n vertices of a graph G is called a *perfect elimination ordering* of G if for all $1 \leq i \leq n$, the vertex v_i is a simplicial vertex in $G[\{v_i, \ldots, v_n\}]$ (subgraph of G induced by $\{v_i, \ldots, v_n\}$). At this point, by Dirac's characterization of chordal graphs [6], it is easy to see that a graph is chordal if and only if it admits a perfect elimination ordering of its vertex set. Also, the Fulkerson and Gross algorithm [7] asserts that the input graph G is chordal if and only if it finds a perfect elimination ordering of G.

Note that any perfect elimination ordering ρ of a tree T can be obtained from the successive removal of simplicial vertices with degree one (leaves) in the current graph until getting a single vertex. This notion of perfect elimination ordering by removing simplicial vertices with bounded degree can be generalized to define a family of subclasses of chordal graphs called *k-trees*. Given an integer $k \geq 1$, the class of k-trees is defined recursively as follows: a complete graph with $k + 1$ vertices is a k-tree; a k-tree T with $n + 1$ vertices where $n \geq k + 1$ can be constructed from a k-tree H with n vertices by adding a new vertex adjacent to exactly k vertices that form a k-clique in H. Also, any perfect elimination ordering of a k-tree can be obtained from the successive removal of simplicial vertices with degree k in the current graph until a clique of size k is obtained. Trees are precisely the class of 1-trees.

A graph G is a *partial k-tree* if G is a subgraph of a k-tree. The minimum integer k such that a graph G is a partial k-tree is a width parameter that aims to measure some "distance" from G to a tree. Such a width parameter is equivalent to the notion of *treewidth* commonly defined under *tree decompositions*.

A *tree decomposition* of a graph G is a pair $\mathcal{T} = (T, \{X_t\}_{t \in V(T)})$, where T is a tree in which every node t is associated with a subset $X_t \subseteq V(G)$, called a *bag*, satisfying the following conditions (c.f. [4]):

- $\cup_{t \in V(T)} X_t = V(G)$, i.e., every vertex is in at least one bag;
- for every edge $uv \in E(G)$, there is a node t of T where the bag X_t contains both u and v;
- for all $u \in V(G)$, the set $T_u = \{t \in V(T) : u \in X_t\}$ (set of bags containing u) is a subtree of T.

The *width* of a tree decomposition $\mathcal{T}$ is $\max\{|X_t| - 1\}$ for $t \in V(T)$ (the size of the largest bag minus one). The *treewidth* of a graph, denoted by $tw(G)$, is the smallest width among all its possible tree decompositions. Nowadays, it is well known that a graph G is a partial k-tree if and only if $tw(G) \leq k$.

Tree decompositions are widely used in designing dynamic programming algorithms for graph problems on instances with bounded treewidth (see [4]), and the treewidth of a graph is one of the most important and useful graph parameters. Several classes of graphs have treewidth bounded by a constant, and many NP-hard problems can be solved in polynomial time in classes of graphs with bounded treewidth (see [4]). Most algorithmic strategies for solving problems using treewidth and tree decompositions require that a tree decomposition with a "sufficiently small" width is previously computed. However, computing the treewidth of a graph (thus, finding a tree decomposition with minimum width) is an NP-hard problem, as demonstrated in [1]. From these facts, the need to compute tree-decompositions that do not necessarily have the minimum width but have a small enough width for applications emerges.

Since k-trees are chordal graphs, any graph G with treewidth at most k is a subgraph of a chordal graph having maximum clique of size at most $k + 1$. Furthermore, Dirac [6] showed that a chordal graph G with perfect elimination ordering $\rho = v_1, \ldots, v_n$ has treewidth $\max\{|N(v_i) \cap \{v_i, \ldots, v_n\}|\}$, where $N(v_i)$ is the neighborhood of v_i in G. Thus, if G' is a chordal supergraph of a graph G and G' has maximum clique size at most $k+1$, then G is a partial k-tree because G' is also a partial k-tree. Therefore, given a graph $G = (V, E)$ and denoting by $\mathcal{C}_G$ the family of chordal graphs G' such that $V(G) = V(G')$ and $E(G) \subseteq E(G')$, the treewidth of a graph G can be defined alternatively as the cardinality of the smallest maximum clique size of a graph in $\mathcal{C}_G$, minus one.

From a chordal graph G' whose maximum clique size is at most $k+1$, one can successively remove simplicial vertices whose neighborhood is a clique of size at most k. From this perfect elimination ordering $\rho = v_1, v_2, \ldots, v_n$, a tree decomposition $\mathcal{T}$ can be built where each bag is formed by the closed neighborhood of the simplicial vertices during its removal/elimination concerning ρ in G'; and, letting X_i be the bag formed by the closed neighborhood of v_i during its elimination concerning ρ in G', if $|X_i| \geq 2$ the parent of the bag X_i correspondent to the least $j > i$ such that $v_j \in X_i$; otherwise, $X_i = \{v_i\}$ and its parent is the root of $\mathcal{T}$ whose bag is empty. We say that such a tree decomposition $\mathcal{T}$ *arises from* ρ. Since $\mathcal{T}$ has width k, such a tree decomposition that arises from ρ is an optimum tree decomposition of G'. Figure 1 illustrates a chordal graph G' and its tree decomposition $\mathcal{T}$ that arises from a given perfect elimination ordering ρ.

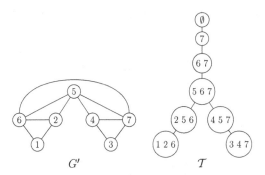

Fig. 1. Chordal graph G' with $V(G') = \{1, 2, 3, 4, 5, 6, 7\}$ and its tree decomposition $\mathcal{T}$ that arises from the perfect elimination ordering $\rho = 1, 2, 3, 4, 5, 6, 7$. The vertices highlighted in red inside the bags are the simplicial vertices that generate such a bag when eliminated from the current graph. (Color figure online)

Also, it is possible to transform any simple graph G into a chordal graph by adding some edges; the resulting graph is called a *triangulation* of G. A process for obtaining a triangulation of a graph G can be described as follows:

   (*i*) create a copy of the original graph, named G^*;
   (*ii*) choose any vertex x from G^*, add edges between vertices in the neighborhood of x until they form a clique;
  (*iii*) remove x from G^*;
   (*iv*) add to the original graph all the edges added in the previous step;
   (*v*) repeat step (*ii*) if there are still vertices in G^*.

At the end of this process, the order in which the vertices were removed from the copy of the original graph provides a perfect elimination ordering ρ of the resulting graph G'. We also say that G' *arises from* ρ. Thus, a typical strategy of heuristics to compute a tree decomposition of the graph G is: (*a*) compute, according to some criteria, a permutation ρ of the vertices of G representing a perfect elimination ordering of a triangulation G' of G; (*b*) compute an optimal tree decomposition $\mathcal{T}$ of G'; (*c*) return $\mathcal{T}$ as a tree decomposition of G.

Note that, with respect to polynomial-time computations, the criteria to obtain the permutation ρ must be heuristic (non-exact) because finding the perfect elimination ordering that results in a triangulation with the smallest treewidth is NP-hard since it is equivalent to computing the treewidth of G.

At this point, we can observe that any permutation of the vertices of G represents a perfect elimination ordering of a triangulation of G and encodes a tree decomposition for G. Therefore, in this paper, we are interested in the problem that we formalize as follows:

TREE DECOMPOSITION DECODING

Instance: A graph $G = (V, E)$, and a permutation ρ of $V(G)$.

Goal: Determine the width of the tree decomposition $\mathcal{T}$ of G that arises from ρ; i.e., the width of the optimum tree decomposition of the triangulation G' of G that arises from ρ.

TREE DECOMPOSITION DECODING is the main subproblem to be solved by the most popular heuristics for the problem of computing the treewidth of a graph. For solving TREE DECOMPOSITION DECODING, one can use the information that the treewidth of a triangulation G' is the size of its largest clique minus one. Thus, from (G, ρ), by constructing the triangulation G' that arises from ρ, we can obtain in $\mathcal{O}(|V(G')| + |E(G')|)$ time the solution for the problem. However, constructions of G' take $\Theta(n^2)$ space in the worst case where $n = |V(G)|$ because for each simplicial vertex, one must transform its current neighborhood into a clique. Therefore, the main issue of such constructions is that $\Theta(n)$ width requires $\Theta(n^2)$ space, which might be too much for large instances. However, the triangulation does not necessarily need to be computed to evaluate a solution encoded by ρ (determine its width). Since it is the main drawback of algorithms for TREE DECOMPOSITION DECODING, one can ask the following algorithmic questions:

Question 1. Is it possible to obtain the solution T for TREE DECOMPOSITION DECODING without constructing the triangulation G' arising from ρ?

Avoiding typical auxiliary structures is challenging but can be the key property of some algorithmic breakthroughs. Besides, it is worth asking the following:

Question 2. Can TREE DECOMPOSITION DECODING be solved in $o(n^2)$ space and $\mathcal{O}(n^2)$ time?

Question 3. Can TREE DECOMPOSITION DECODING be solved in $o(n^2)$ time?

This paper has answered the first question and presented two approaches. The algorithm *Leaf decoder* avoids the construction of the triangulation G' and solves the problem in $\mathcal{O}(n \cdot \ell)$ space and $\mathcal{O}(n^2 \cdot \ell)$ time, where n is the number of vertices of the input graph G, and ℓ is the number of leaves of the tree decomposition encoded by the permutation ρ. The algorithm *Set decoder*, using a different strategy, solves the problem in $\mathcal{O}(m \cdot \log n)$ time, where $m = |E(G)|$. However, it may consume more auxiliary memory ($\mathcal{O}(m + n)$), which can be an issue in huge instances.

Since the triangulation G' that arises from ρ is a chordal graph, the number of leaves in the tree decomposition T encoded by $\rho = v_1, v_2, \ldots, v_n$ is precisely the number of vertices v_i having no neighbor in $G'[\{v_1, \ldots, v_i\}]$. Therefore, we partially answer the second question by presenting an algorithm that performs in $o(n^2)$ space when the tree decomposition encoded by ρ has $o(n)$ leaves. In particular, if the underlying triangulation that arises from ρ has a bounded clique cover number, then the solution is decoded by our algorithm in $\mathcal{O}(n)$ space and $\mathcal{O}(n^2)$ time. At the same time, the best decoders based on computing triangulations would use $\Theta(n^2)$ space and $\Theta(n^2)$ time. A similar phenomenon occurs if ρ encodes a path decomposition with a large width. Thus, avoiding the construction of G', in $\mathcal{O}(n)$ space, we can solve some dense instances of the problem that tend to require $\Omega(n^2)$ space if the construction of G' is required. Computational experiments show that in addition to saving space, the proposed algorithm is also competitive in terms of running time, decoding the solution of some instances faster than the method based on computing triangulations. We also partially answer the third question with the *Set decoder* algorithm, which performs in $o(n^2)$ time when $m = o(\frac{n^2}{\log n})$. In addition, it is the fastest decoder in practice, according to our experiments.

It is worth mentioning that even when G is a graph having bounded treewidth, depending on the permutation ρ, the associated graph G' may have $\Theta(n^2)$ edges. The extreme case would be when the first element of the permutation is a universal vertex of G, forcing the triangulation G' to be a complete graph. Also, the reader may wonder why ρ could be any permutation rather than one obtained by an appropriate heuristic. One of the motivations is the engineering of heuristic algorithms, such as genetic algorithms, that are based on generating a population of solutions encoded through permutations. Although constructive heuristics generate some members of the population, due to the population's need for diversity, some permutations are commonly generated randomly and must be evaluated by some decoder. In addition, it is worth mentioning that, in some scenarios, it may be desirable to find a feasible solution in a short time. Thus, running more elaborate constructive heuristics may be unfeasible in the context of huge graphs. Our computational experiments point to real benchmark instances where this occurs. However, taking a random permutation and using an efficient decoding algorithm would still be possible in such a case. We remark that besides the theoretical contribution regarding the state of practice, by taking a random permutation of the vertices and using our decoding algorithms, we can obtain feasible solutions for large instances that triangulation-based decoders and classical constructive heuristics are not able to get within the same time. Finally, space constraints impose a rule, and therefore, our contribution complements state-of-the-art by offering a solution for situations in which the construction of triangulations is not pertinent.

Due to space constraints, our computational experiments are omitted.

2   Related Works

Most studies regarding heuristics for the treewidth computation of a graph G are based on finding perfect elimination orderings of triangulations of G, where vertices are selected following some criteria to build a permutation ρ of $V(G)$. Thus, the feasible solution to be returned is the tree decomposition of the minimal triangulation that arises from ρ. The most prominent constructive heuristics in the literature are *Min-Degree* and *Fill-in* (c.f. [9]). The Min-Degree heuristic selects the vertex with a minimum degree in the current graph successively, while in the Fill-in heuristic, the vertex to be chosen is the one whose neighborhood is the closest to becoming a clique through additions of edges (completions). Both criteria aim to minimize the size of the largest clique of the chordal graph to be produced. The *Min-Degree* heuristic is faster than the *Fill-in* heuristic, but *Fill-in* typically finds a better solution in practice. In 2019, Gaspers et al. [9] proposed turbocharging both heuristics. Besides, exact and heuristic implementations for the treewidth computation were proposed in PACE 2017 [5]. The goal of this challenge was to investigate the applicability of algorithmic ideas studied in the vast literature of the treewidth computation, and it has inspired some interesting works such as [2,11].

Besides, many approaches to solving optimization problems consist of studying how to encode feasible solutions as permutations. Bean [3] showed that any optimization problems in which their solutions can be encoded as permutations can be used in metaheuristic algorithms to manipulate these permutations to obtain better solutions. After manipulating encoded solutions, it is necessary to decode them so that we can infer a solution value (value encoding). A deterministic algorithm called decoder takes a solution encoded as a permutation and associates it with a value (its encoded value). Decoding the solution can be computationally expensive and a bottleneck for metaheuristics since it is a recurrent subroutine of the entire process. Regarding TREE DECOMPOSITION DECODING, in 1976, Rose, Tarjan, and Lueker [10] presented an algorithm with $\mathcal{O}(n + m')$ running time that computes the chordal supergraph G' of a graph G that arises from a given permutation ρ of $V(G)$ representing the elimination ordering of G', where $n = |V(G)|$ and $m' = |E(G')|$. In addition, from G', its corresponding tree decomposition can be obtained in polynomial time [8]. Therefore, TREE DECOMPOSITION DECODING can be solved in $\mathcal{O}(n^2)$ space and $\mathcal{O}(n^2)$ time. In this paper, we present the algorithms *leaf decoder* that solves TREE DECOMPO-SITION DECODING in $\mathcal{O}(n \cdot \ell)$ space and $\mathcal{O}(n^2 \cdot \ell)$ time and the *set decoder* that solves it in $\mathcal{O}(m + n)$ space and $\mathcal{O}(m \cdot log n)$ time. To the best of our knowledge, these are the first decoders that neither compute triangulations of the input graph nor perform any completion operation (addition of edges).

3   Leaf Decoder Algorithm

As explained in the Introduction, any permutation of the vertices of a graph G can be seen as a perfect elimination ordering of a triangulation of G. During the triangulation process, edges are added to transform the original graph into a chordal graph. Now, we propose an algorithm to compute the width of the tree decomposition that arises from the permutation ρ of $V(G)$ (representing a perfect elimination ordering of the triangulation of G), but without adding edges to the original graph G and avoiding constructing the triangulation.

Definition 1. *Given a graph G, a permutation ρ of $V(G)$, and the tree decom-position $\mathcal{T}$ of G that arises from ρ, a vertex $v \in V(G)$ is a leaf vertex according to ρ if v is contained in exactly one bag of $\mathcal{T}$.*

Observation 1. *Given a graph G, a permutation ρ of $V(G)$, and the tree decom-position $\mathcal{T}$ of G that arises from ρ, the number of leaf vertices according to ρ is exactly the number of leaf nodes of $\mathcal{T}$.*

Proof. By construction, except the root bag, each bag is formed by the closed neighborhood of a vertex v during their elimination concerning ρ in G'. Thus, each leaf node t of $\mathcal{T}$ has exactly one leaf vertex v in X_t. Also, by construction, the bag of a node t with child t' is formed by the closed neighborhood of a vertex $v \in X_{t'}$. Thus, there is no non-leaf node t having leaf vertices.   □

Our algorithm is divided into two procedures. The first procedure aims to find the set of leaf vertices according to ρ. These vertices can be seen as those that dictate which branch of the tree decomposition each vertex will be on.

Observation 2. *Given a graph G, a permutation $\rho = v_1, v_2, \ldots, v_n$ of $V(G)$, and the tree decomposition $\mathcal{T}$ of G that arises from ρ, a vertex $v_i \in V(G)$ is a leaf vertex if and only if v_i is an isolated vertex in $G[\{v_1, v_2, \ldots, v_i\}]$.*

Proof. Let v_i be an isolated vertex in $G[\{v_1, v_2, \ldots, v_i\}]$, and let $\rho_i = v_1, v_2, \ldots, v_i$. Clearly, the triangulation of $G[\{v_1, v_2, \ldots, v_i\}]$ that arises from

ρ_i has v_i as isolated vertex. Thus, v_i is a leaf vertex according to ρ. Now, suppose that v_i is a leaf vertex according to ρ. Let G' be the triangulation of G that arises from ρ. By definition, v_i has no neighbor in $\{v_1, v_2, \ldots, v_{i-1}\}$ in the graph G'. Since G' is a supergraph of G then v_i is an isolated vertex in $G[\{v_1, v_2, \ldots, v_i\}]$.   □

3.1   Determining Descendant Leaves for Each Node

For computing the leaf vertices, we use as data structure a list F indexed by the vertices of the graph, where each position stores an ordered set/list of linked elements (for better performance). The elements p to be stored for a vertex v of G are pairs (f, c) of integers f and c, where the first integer (f) of the pair p is denoted by $p.f$ and indicates a leaf vertex that v reaches, and the second integer (c) of p denoted by $p.c$ indicates through which vertex of the graph such a leaf f reaches v (i.e., the first neighbor of v in the path between f and v). Thus, the goal of Algorithm 1 is not just determining the set of leaf vertices of G according to ρ, but also determining for each vertex v_i of G a subset of leaf vertices that safely are descendant from node i (bag X_i) according to the tree decomposition $\mathcal{T}$ that arises from ρ. Recall that the node of the tree decomposition $\mathcal{T}$ formed during the elimination of v_i (according to ρ) is denoted by node i (its bag is X_i).

Algorithm 1: FindLeaves(G, ρ)

---

1   Let $\rho = 1, 2, \ldots, n$;
2   **for** x *from* 1 *to* n **do**
3      $F[x] \leftarrow \emptyset$;
4      $Leaves[x] \leftarrow 1$;

5   **for** x *from* 1 *to* n **do**
6   **if** $Leaves[x] = 1$ **then**
7   add $\{x, x\}$ to $F[x]$
8   **foreach** $w \in N(x)$ **do**
9   **if** $w > x$ **then**
10            $Leaves[w] \leftarrow 0$;
11   **else**
12   **foreach** p *in* $F[w]$ **do**
13   **if** *the leaf* $p.f$ *is in some element of* $F[x]$ **then**
14   Let p' be the element of $F[x]$ such that $p'.f = p.f$;
15   replace p' by $(p.f, \min\{w, p'.c\})$ in $F[x]$;
16   **else**
17   add to $F[x]$ the pair $(p.f, w)$;

18   **return** (F);

The Algorithm 1 describes the process of finding the leaf vertices and initializing the leaf set of each vertex v_i with some leaf vertices which are descendants

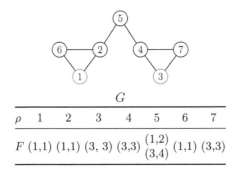

ρ	1	2	3	4	5	6	7
F	(1,1)	(1,1)	(3, 3)	(3,3)	(1,2)(3,4)	(1,1)	(3,3)

Fig. 2. A graph G, a permutation $\rho = 1,2,3,4,5,6,7$ of the vertices of G, and the structure F after the computation of Algorithm 1 on (G, ρ). Highlighted in red are the leaf vertices of the tree decomposition $\mathcal{T}$ that arises from ρ. The decomposition $\mathcal{T}$ and the triangulation G' arising from ρ are illustrated in Fig. 1. (Color figure online)

of the node corresponding to v_i in the tree decomposition $\mathcal{T}$ that arises from ρ. The algorithm takes as input a graph G and a permutation of the vertices ρ. Let x be a vertex of G, $F[x]$ denotes the list F at position x and represents the descendant leaves of x in $\mathcal{T}$, and $N(x)$ denotes the neighborhood of x into G. The variable *Leaves* is a binary vector indexed by the vertices of G, such that *Leaves*$[i]= 1$ means that v_i is a leaf vertex. Initially, *Leaves*$[i]= 1$ for each i, and the structure will be updated as the algorithm progresses, according to ρ. Figure 2 illustrates the structure F after the computation of Algorithm 1.

Time and Space Complexity. This procedure traverses all vertices and edges of G, taking $\Omega(n + m)$ time, where for each edge, the descendant leaf sets of its endpoints are parsed and updated if necessary. Considering that such sets of descendant leaves are stored in ascending order in a linked list for each edge, this step can be performed in $\mathcal{O}(\ell)$ time. Therefore, the procedure *FindLeaves* can be performed in $\mathcal{O}((n + m) \cdot \ell)$ time and $\mathcal{O}(n \cdot \ell)$ space, where n is the number of vertices, m the number of edges of the input graph, and ℓ is the maximum number of descendant leaves of a vertex.

3.2 Determining the Width of the Encoded Tree Decomposition

The second procedure of the algorithm computes the bags of the tree decomposition or, alternatively, the size of each bag if the objective is to evaluate the permutation by determining the width of its tree decomposition rather than returning the decomposition. The main goal is to identify for each vertex which successor vertices according to ρ should be in its bag. For this, the current vertex v_i and its successor v_j must have descendant leaves in common in their leaf sets; that is, they are vertices that are part of the same branch in the tree decomposition, but not only that, it is also necessary that the vertex that takes the successor v_j to the leaf must be in a position less than or equal to the position of the current vertex v_i in ρ because the current vertex bag must be a minimal separator between the "already-forgotten" vertices and the "not-yet-forgotten" vertices in the path from the leaf node to the root node in tree decomposition that arises from the permutation ρ. In addition, by determining that v_j should be in the bag of v_i, new descendant leaves for v_j can be recognized.

Algorithm 2: FillBags(F, ρ)

---

1 $width \leftarrow 0$;
2 **foreach** $x \in \rho$ **do**
3       $T[x].bag \leftarrow \{x\}$;
4       $T[x].size \leftarrow 1$;
5       $T[x].parent \leftarrow n + 1$ (representing the root node);
6   **foreach** $y \in \rho$ *and* $y > x$ **do**
7           $inter \leftarrow \emptyset$;
8   **if** *there are* $p \in F[x]$, $p' \in F[y]$ *with* $p.f = p'.f$ *and* $p.c' \leq x$ **then**
9   **foreach** $p \in F[x]$ *such that there is* $p' \in F[y]$ *with* $p.f = p'.f$ **do**
10   add p to $inter$;
11              $T[x].bag \leftarrow T[x].bag \cup \{y\}$;
12              $T[x].size \leftarrow T[x].size + 1$;
13              $T[x].parent \leftarrow \min\{y, T[x].parent\}$
14   **if** $inter \neq \emptyset$ **then**
15              $notinter \leftarrow F[x] \setminus inter$;
16   **foreach** $p \in inter$ **do**
17   **foreach** p' *in* $F[y]$ *such that* $p'.f = p.f$ **do**
18   replace p' by $(p.f, p.f)$ in $F[y]$;
19   **foreach** $p \in nointer$ **do**
20   add $(p.f, p.f)$ to $F[y]$;
21      $width \leftarrow \max\{width, T[x].size - 1\}$;
22 **return** $(T, width)$;

Algorithm 2 summarizes the entire process necessary in the second procedure. It receives as a parameter the structure of sets F generated by the Algorithm 1 and the permutation of the vertices ρ. As a data structure to store the tree decomposition or the size of its bags, we use a list T with n positions where each position is associated with a node of the tree decomposition ordered according to ρ. Each position in the list T is a record with three fields, *bag*, *size*, and *parent*, storing the vertices of the bag of node i, its size, and the index of its parent node, respectively. We are implicitly assuming that there is a root node with an empty bag whose position is $n + 1$. We remark that to save space and determine just the width of the decomposition, it is enough to omit the field "bag" of the structure. As previously discussed, our primary goal is saving space; however, we consider the field "bag" to simplify the argumentation about correctness.

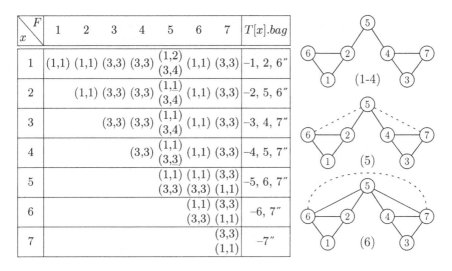

x \ F	1	2	3	4	5	6	7	$T[x].bag$
1	(1,1)	(1,1)	(3,3)	(3,3)	(1,2) (3,4)	(1,1)	(3,3)	–1, 2, 6″
2		(1,1)	(3,3)	(3,3)	(1,1) (3,4)	(1,1)	(3,3)	–2, 5, 6″
3			(3,3)	(3,3)	(1,1) (3,4)	(1,1)	(3,3)	–3, 4, 7″
4				(3,3)	(1,1) (3,3)	(1,1)	(3,3)	–4, 5, 7″
5					(1,1) (1,1) (3,3) (3,3) (3,3) (1,1)			–5, 6, 7″
6					(1,1) (3,3) (3,3) (1,1)			–6, 7″
7					(3,3) (1,1)			–7″

Fig. 3. Evolution of the data structures throughout the execution of Alg. 2. The first element of a bag represents the current simplicial vertex and the label of the correspondent node on $\mathcal{T}$. The second element of a bag (if any) determines the parent's label of the current node. The root node is the parent of node 7 in $\mathcal{T}$. Highlighted in red are the sets of pairs to be traversed at each step. (Color figure online)

Figure 3 describes the execution of Algorithm 2 according to a perfect elimination ordering ρ. Initially, the first vertex in ρ is the vertex 1 (i.e., $x = 1$), and it has $\{(1,1)\}$ as set of leaves. The possible vertices to be added to its bag are vertices 2, 5, and 6, as they have leaf 1 in their leaf sets. However, only vertices 2 and 6 are introduced in its bag because vertex 5, despite having leaf 1 in $F[5]$, is accessed by vertex 2, which has not yet been forgotten. Therefore, vertex 5 is not necessary in the bag of node 1, while the other two are accessed by 1 itself, which represents precisely the edges between these vertices.

When $x = 2$, vertex 2 is analyzed. It has a leaf set $\{(1,1)\}$, so the possible vertices to be introduced in its bag are vertices 5 and 6. Both are introduced in its bag, as they have leaf 1, and these are accessed by forgotten vertices or by 2 itself. Hence, the pair $p' = (1,2)$ containing the leaf vertex 1 in $F[5]$ is then updated to $(1,1)$ (underlined), as 2 will be now forgotten. This update on $F[5]$ is because if vertex 5 is forgotten before vertices that still have 1 in their leaf set, it should get those vertices into your bag (each non-leaf bag is a separator).

When $x = 3$, the vertex to be considered is the vertex 3, which has the element $(3,3)$ in $F[3]$, implying the vertex 3 is a leaf vertex. The possible vertices to be introduced in its bag are 4, 5, and 7, but only 4 and 7 are introduced because 5 reaches the leaf vertex 3 through vertex 4, and it has not yet been forgotten.

The next vertex is 4, that is $x = 4$. The possible vertices to be introduced into its bag are 5 and 7, and both are added. Thus, the pair $(3,4)$ in $F[5]$ is then updated to $(3,3)$ (underlined), because vertex 4 is being forgotten.

Next, we consider vertex 5 having leaf set $F[5]$ containing $(1,1)$ and $(3,3)$. When $x = 5$, the importance of the updates made in the previous steps can be noted since the possible vertices to be introduced are 6 and 7, which had as leaf sets, respectively $\{(1,1)\}$ and $\{(3,3)\}$, implying that they must be introduced in the bag of node 5. This is analogous to introducing the edges $(5,6)$ and $(5,7)$ in G'. Also, note that originally $F[5] = \{(1,2),(3,4)\}$, so the updates are fundamental for correctness. After that, an important step is that as vertex 5 is being forgotten, and it has introduced 6 and 7 into its bag, all its pairs in $F[5]$ are "inherited" to the leaf sets $F[6]$ and $F[7]$.

When $x = 6$, we consider the vertex 6 as simplicial. At this moment, it has $F[6] = \{(1,1),(3,3)\}$. The only possible vertex to be introduced into its bag is 7, which is introduced due to the previous step. Here, the importance of the previous update on $F[6]$ and $F[7]$ becomes evident, since originally $F[6] \cap F[7] = \emptyset$. This is analogous to creating an edge between 6 and 7 in G'.

Finally, when $x = 7$, only vertex 7 remains, implying that its bag is $\{7\}$ and its parent should be the root node with an empty bag. Thus, the algorithm ends.

The tree decomposition $\mathcal{T}$ for the instance (G, ρ) of Fig. 3 is presented in Fig. 1 together with the corresponding supergraph G' that arises from (G, ρ).

Time and Space Complexity. This second procedure traverses all vertices, and for each vertex v_x, it analyzes all the vertices v_y not yet removed according to ρ $(y > x)$, taking $\Omega(n^2)$ time, where given v_x, v_y the current leaf set of both vertices is compared. Again, by considering that such sets of descendant leaves are stored in ascending order in a linked list, the steps from lines 6–18 can be performed in $\mathcal{O}(\ell)$ time for each pair v_x, v_y of vertices, where ℓ is the maximum number of descendant leaves of a vertex (which is upper bounded by the number of leaves of the tree decomposition that arises from the permutation ρ). Therefore, the second procedure can be performed in $\mathcal{O}(n^2 \cdot \ell)$ time.

Regarding space complexity, the version considering the field "bag" is performed in $\mathcal{O}(n \cdot (k+\ell))$ space, where k is the width of the decomposition. However, if the goal is to evaluate the width of the encoded tree decomposition rather than computing it, then one can omit the field "bag" and the algorithm would run in $\mathcal{O}(n \cdot \ell)$ space.

By adding both procedures together, the complete algorithm to evaluate the solution encoded by ρ has $\mathcal{O}(n^2 \cdot \ell + (n + m) \cdot \ell) = \mathcal{O}(n^2 \cdot \ell)$ running time and $\mathcal{O}(n \cdot \ell)$ running space.

In practice, it was observed that the applicability of the new decoder becomes noticeable in relatively large and dense instances. This is because the value ℓ tends to be small for dense instances while its width tends to be large, making our algorithm more practical in this scenario. In particular, G (or G') having bounded clique cover number implies that the tree decomposition has a small number of leaf vertices. For instance, if G is co-bipartite, then the tree decomposition arising from ρ has at most 2 leaves.

4 Optimizing Running Time: The Set Decoder Algorithm

Algorithm 3 is developed to obtain better performance concerning running time. Let v_1, ..., v_n be the elimination ordering of G and let $P_i = \{v_1, \ldots, v_i\}$ denote the prefix of eliminated vertices at point i. Let $C_i \subseteq P_i$ be the connected component of $G[P_i]$ that contains v_i. It can be observed that the number of forward-edges of vertex v_i in the elimination order is equal to $|N(C_i)|$. Therefore, we can solve the problem by keeping track of the connected components of $G[P_i]$ and their neighborhoods. Note that the sum of the sizes of the neighborhoods over all components is always at most m. We can explicitly keep track of the components and their neighborhoods by storing them in sets represented by balanced binary trees and doing small-to-large merging when components merge.

Algorithm 3: SetDecoder(ρ, G)

---

1 $width \leftarrow 0$;
2 $C \leftarrow \emptyset$;
3 $U \leftarrow \emptyset$;
4 **foreach** $x \in \rho$ **do**
5      $C_x \leftarrow$ FindDisjointSet(C, x);
6      $U_{C_x} \leftarrow$ FindJointSet(U, C_x);
7   **foreach** $y \in N(x)$ **do**
8          $C_y \leftarrow$ FindDisjointSet(C, y);
9   **if** $y < x$ **then**
10   **if** $C_x \neq C_y$ **then**
11   UnionDisjointSet(C, C_x, C_y);
12                 $U_{C_y} \leftarrow$ FindJointSet(U, C_y);
13                 $U_{C_x} \leftarrow$ UnionJointSet (U, U_{C_x}, U_{C_y});
14   **else**
15             $U_{C_x} \leftarrow U_{C_x} \cup y$;
16     $U_{C_x} \leftarrow U_{C_x} \setminus x$;
17     $width \leftarrow \max\{width, |U_{C_x}|\}$;
18 **return** $width$;

The algorithm identifies the disjoint set containing x or creates it if it does not exist yet and stores this set in a variable named C_x (representing the current connected component of x) with the function FindDisjointSet, keeping the current set of connected components updated and stored in C. Now, on line 6, it obtains the neighbors of the connected component C_x denoted as U_{C_x} with the function FindJointSet, keeping the current set of connected components' neighborhoods updated and stored in U. Continuing to line 7, inside an inner loop, the algorithm iterates through each vertex y in the neighbors of x, represented by $N(x)$. Line 8 entails finding the disjoint set to which y belongs, or creating if it does not exist yet, and storing it in the variable C_y (representing the current

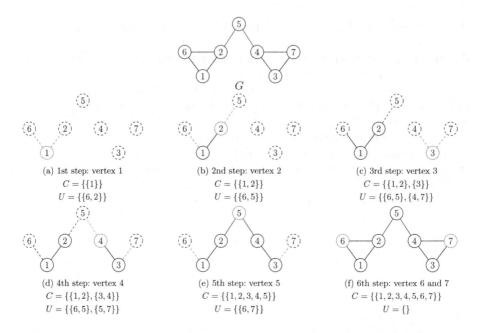

Fig. 4. Step-by-step of Algorithm 3 on a graph G. (Color figure online)

connected component of y) with the function FindDisjointSet. On line 9, the algorithm checks whether y is less than x. If this condition does not hold, in line 15, y is merged in U_{C_x}; otherwise, in line 10, it verifies whether the disjoint sets C_x and C_y are different. If they are indeed different, in line 11, it is done the union of the disjoint sets C_x and C_y with the function UnionDisjointSet, which also keeps C updated. On line 12, it obtains the neighbors of the connected component C_y denoted by U_{C_y}, with the function FindJointSet. Finally, line 13 updates the neighbors of C_x, merging them with the neighbors of C_y using the UnionJointSet function (it also keeps U updated). In line 16, x is removed from U_{C_x} and in line 17, $witdh$ is updated with the maximum value between $width$ and $|U_{C_x}|$. In line 18, the $width$ is returned.

The Fig. 4 shows the step-by-step of the Algorithm 3 on the graph G and $\rho = 1, 2, 3, 4, 5, 6, 7$. The process begins with vertex 1 (a), initializing sets: C with only $\{1\}$ for connected components and U with its neighbors $\{6, 2\}$. Subsequently, vertex 2 (b) leads to the merging of components into C $\{1, 2\}$ and these neighborhoods resulting U $\{6, 5\}$. Moving on to vertex 3 (c), a new component forms in C $\{1, 2\}$ and $\{3\}$, with U containing $\{6, 5\}$ and $\{4, 7\}$, the neighborhoods, respectively. In the fourth step, vertex 4 (d) results in C $\{1, 2\}$ and $\{3, 4\}$, while U holds $\{6, 5\}$ and $\{5, 7\}$ neighborhoods. Vertex 5 (e) leads to a single merged connected component, C $\{1, 2, 3, 4, 5\}$, with U containing $\{6, 7\}$ neighbors. Finally, vertices 6 and 7 (f) are considered together; both do not have any neighbors that have not yet been eliminated, resulting in the entire graph being a single connected component, C $\{1, 2, 3, 4, 5, 6, 7\}$, and U becoming empty, signifying the completion of the algorithm's execution. The largest size of some subset of U (neighborhood of some connected component, edges in red in each step) formed during the process was 2, which is the algorithm's response.

Time and Space Complexity. In lines 2 and 5 of Algorithm 3, we traverse all edges of G. Consequently, during this process, we incur a time complexity of $\Omega(m)$. For each edge, it is necessary to determine if the sets C_x and C_y are identical. This can be accomplished in $\mathcal{O}(\log n)$ time by implementing the FindDisjointSet and UnionDisjointSet operations using an efficient UnionFind algorithm with path compression techniques [12].

It is important to note that the number of sets of the connected components neighborhoods U is limited to n. This limitation reflects the maximum number of connected components in graph G. Consequently, for any given vertex, it will belong to at most $\mathcal{O}(n)$ sets and go through a maximum of $\mathcal{O}(n)$ merge operations. Consequently, if the UnionJointSet operation is implemented similarly to UnionFind but permitting non-disjoint sets, the resulting tree structure formed by these sets will have a size $\mathcal{O}(n^2)$, with its height limited to $\mathcal{O}(\log n)$. As a result, each merge/union operation costs $\mathcal{O}(\log n)$. In summary, Algorithm 3 performs in $\mathcal{O}(m \cdot \log n)$ runtime.

For each iteration, any edge of G contributes with at most one element in some subset of U. Thus, in the worst case, the sum of the sizes of the subsets in U results in an auxiliary memory of size $\mathcal{O}(m)$. The sum of the sizes of subsets within C is constrained by n since they are disjoint sets, resulting in a space requirement of $\Omega(n)$. Therefore, Algorithm 3 requires $\mathcal{O}(m+n)$ auxiliary space.

References

1. Arnborg, S., Corneil, D.G., Proskurowski, A.: Complexity of finding embeddings in a k-tree. SIAM J. Algebraic Discrete Methods **8**(2), 277–284 (1987)
2. Bannach, M., Berndt, S., Ehlers, T.: Jdrasil: a modular library for computing tree decompositions. In: 16th International Symposium on Experimental Algorithms (SEA 2017). Schloss Dagstuhl-Leibniz-Zentrum fuer Informatik (2017)
3. Bean, J.C.: Genetic algorithms and random keys for sequencing and optimization. ORSA J. Comput. **6**(2), 154–160 (1994)
4. Cygan, M., et al.: Parameterized Algorithms. Springer, Cham (2015). https://doi.org/10.1007/978-3-319-21275-3
5. Dell, H., Komusiewicz, C., Talmon, N., Weller, M.: The PACE 2017 parameterized algorithms and computational experiments challenge: the second iteration. In: Lokshtanov, D., Nishimura, N. (eds.) 12th International Symposium on Parameterized and Exact Computation (IPEC 2017), Volume 89 of Leibniz International Proceedings in Informatics (LIPIcs), pp. 30:1–30:12. Dagstuhl (2018)
6. Dirac, G.A.: On rigid circuit graphs. Abh. Math. Semin. Univ. Hambg. **25**, 71–76 (1961). https://doi.org/10.1007/BF02992776
7. Fulkerson, D., Gross, O.: Incidence matrices and interval graphs. Pac. J. Math. **15**(3), 835–855 (1965)

8. Galinier, P., Habib, M., Paul, C.: Chordal graphs and their clique graphs. In: Nagl, M. (ed.) WG 1995. LNCS, vol. 1017, pp. 358–371. Springer, Heidelberg (1995). https://doi.org/10.1007/3-540-60618-1_88

9. Gaspers, S., Gudmundsson, J., Jones, M., Mestre, J., Rümmele, S.: Turbocharging treewidth heuristics. Algorithmica **81**(2), 439–475 (2019). https://doi.org/10.1007/s00453-018-0499-1

10. Rose, D.J., Tarjan, R.E., Lueker, G.S.: Algorithmic aspects of vertex elimination on graphs. SIAM J. Comput. **5**(2), 266–283 (1976)

11. Tamaki, H.: Positive-instance driven dynamic programming for treewidth. J. Comb. Optim. **37**(4), 1283–1311 (2019). https://doi.org/10.1007/s10878-018-0353-z

12. Wu, K., Otoo, E.: A simpler proof of the average case complexity of union-find withpath compression. Technical report, Lawrence Berkeley National Lab. (LBNL), Berkeley, CA, United States (2005)

Enumerating m-Length Walks in Directed Graphs with Constant Delay

Duncan Adamson[1(✉)], Paweł Gawrychowski[2], and Florin Manea[3]

[1] Materials Innovation Factory, University of Liverpool, Liverpool, UK
D.A.Adamson@Liverpool.ac.uk
[2] Institute of Computer Science, University of Wrocław, Wrocław, Poland
[3] Department of Computer Science, University of Göttingen, Göttingen, Germany

Abstract. In this paper, we provide a novel enumeration algorithm for the set of all walks of a given length within a directed graph. Our algorithm has worst-case constant delay between outputting succinct representations of such walks, after a preprocessing step requiring linear time relative to the size of the graph. We apply these results to the problem of enumerating succinct representations of the strings of a given length from a prefix-closed regular language (languages accepted by a finite automaton which has final states only).

1 Introduction

Enumerating all members of a given class of combinatorial objects is one of the fundamental problems in computer science. Enumeration problems take a description of the class of objects and produce every object satisfying this description. Often, the number of objects in each class is of exponential size relative to the size of the description: for example, the set of walks of length m in the complete graph K_n with n vertices has size n^m. Due to the large size of these classes, the usual goal of enumeration algorithms is to reduce the delay between outputting consecutive objects, either in terms of the worst case or the average case. Various enumeration problems appear in diverse contexts with a wide range of applications. Comprehensive surveys of enumeration problems and their connections to various areas of computer science and mathematics have been provided by Segoufin [28] (with a focus on logic), Wasa [32] (which provides a list of enumeration problems from multiple areas, ranging from graph theory to computational geometry or to combinatorics on words and automata), and Uno [30] (focused on the amortized analysis of enumeration algorithms). Interesting applications of enumeration algorithms include, among others, database theory [4,5,26,27], combinatorics and algorithms on strings and the study of formal languages [1,2,5,15,29].

This paper is primarily motivated by the problem of enumerating the set of all *crystal structures* of a given size. This problem originates in chemistry, with the problem of *crystal structure prediction*. In one dimension, the crystal structure prediction problem asks, given an alphabet of "blocks" (3-dimensional

J. A. Soto and A. Wiese (Eds.): LATIN 2024, LNCS 14578, pp. 35–50, 2024.
https://doi.org/10.1007/978-3-031-55598-5_3

structures), what is the optimal way to arrange these objects to minimise some pairwise objective function [11]. Currently, this problem is solved via heuristic techniques [21], leaving the possibility of missing the optimal solution in numerous instances, or by costly integer programming-based algorithms [16] that often correspond to checking every possible solution. At the same time, existing knowledge from chemistry allows certain solutions to be ruled out without the costly process of simulating the predicted structure [21], simply based on the presence of some bad (or forbidden) block combinations. By considering the set of blocks as an alphabet, we can canonically represent the set of various arrangements of blocks as strings over this alphabet. In this way, the set of valid arrangements of blocks forms as a *prefix-closed* regular language, i.e., a set of strings where every prefix of every string is also in the set. The class of prefix-closed regular languages has nice language theoretic properties [10,19], and, interestingly w.r.t. our motivation, includes the class of languages of strings that avoid a given set of forbidden factors (corresponding to the bad combinations of blocks). Thus, in this framework, we may solve the problem of enumerating the set of valid crystal structures, for a given set of blocks, by solving the problem of enumerating all strings in a given prefix-closed regular language.

Rather than restricting ourselves purely to prefix-closed regular languages (or crystal structures), and since deterministic finite automata (which are natural ways to specify such languages) can be canonically represented as directed, labelled multi-graphs, we look at the problem of enumerating (succinct representations of) all walks of a given length in a directed graph (potentially parameterized by their starting vertex or set of starting vertices). This seems to lift the language-enumeration problem discussed above to a more abstract setting. Therefore, we will first consider Problem 1, and then see how we can use the obtained results to solve the more concrete problem introduced above.

Problem 1. Given directed graph G, and integer $m > 0$, enumerate efficiently succinct representations of all walks of length m in G.

To fully specify this problem, one needs to define exactly what is the output of the enumeration. One possibility would be to output each walk of length m explicitly (as a sequence of edges); this would inherently lead to $O(m)$-time delay between the consecutive walks (as we first need to finish outputting the first walk, before starting the next one). To achieve $O(1)$-delay, an implicit, succinct representation of the output is needed. However, such a representation has to be meaningful: one should be able to canonically and efficiently retrieve the explicit list of enumerated walks from the list of implicitly represented walks. Ideally, an algorithm solving Problem 1 would also permit explicitly outputting, on demand, the current walk of the enumeration, at any step of the computation. Also, ideally, the preprocessing done by the algorithm would take linear time in the size of G and would not depend on m, allowing the constructed data structures to be reused to enumerate walks of other lengths. Worth noting, straightforward implicit representations of the enumerated walks (e.g., outputting in step i of the enumeration the implicit description "the ith walk of G in lexicographic order (as induced by a total order on edges/vertices)") are usually not meaningful (or do not fulfil our other requirements), as obtaining the walks explicitly would usually require non-trivial work (in the enumeration or in the preprocessing phase).

Our Contributions. We solve Problem 1 by providing an enumeration algorithm with worst-case constant delay after a preprocessing phase running in linear time in the total size of the graph G, while being independent of m.

Our *first contribution*, essential in achieving this result, is to introduce the notion of *default walks* in the graph G. More precisely, each vertex of G is associated with a *default edge*, corresponding to the first edge on the longest (possibly infinite) walk in the graph starting at that vertex. These *default edges* allow the definition of *default walks* in the graph (walks consisting only of default edges). In this framework, arbitrary walks in the graph can be represented implicitly as the concatenation of multiple default walks (specified by starting vertex and length) and the non-default edges which connect these walks.

Our *second contribution*, is an enumeration algorithm for the walks of length m within a graph G, which crucially uses the notions of default edges and walks both for the sake of efficiency of enumeration and as the basis for the output of the enumerated walks. The main idea behind this algorithm is to maintain internally, while going through the walks we want to enumerate, the implicit representation of the current walk as a list of default walks in the graph. By augmenting the graph of default edges with a series of non-trivial data structures, the implicit representation of the enumerated walks allows us to efficiently compute the representation of the next walk in the enumeration and present it in a succinct (yet, meaningful, in the sense mentioned above) way, by simply referencing (by length) the "prefix" shared by the current walk and the next one, and then extending it by appending a non-default edge and a single, succinctly represented default walk. This implicit representation also allows us to output explicitly, on-demand, at every point of the computation, the current walk in linear time w.r.t. its length. Finally, we show that only constant time is needed in our enumeration algorithms to move between consecutive walks and output their representations. Our main results are, as such, efficient solutions to Problem 1.

Result 1 (Theorem 2). *Given directed graph $G = (V, E)$, with n vertices, and integer m, we can enumerate succinct representations of all walks of length m in G, with $O(1)$-delay, after $O(|E|)$-time preprocessing.*

The above result can be extended to enumerate with $O(1)$-delay all the walks in a graph, whose length is between two given integers ℓ and m. More interestingly, these results can be directly applied to languages accepted by a prefix closed automaton (PCA for short, a deterministic finite automaton with final states only), by noting that there is a bijective correspondence between the walks starting in the initial state of a PCA and the strings of the language.

Result 2 (Theorem 3). *Given integer m and PCA $\mathcal{A}$, we can enumerate succinct representations of all strings of length m accepted by $\mathcal{A}$, with $O(1)$ delay after a linear time preprocessing w.r.t. the size of $\mathcal{A}$.*

We also extend our results to obtain efficient algorithms for the problems of ranking and unranking strings in prefix closed regular languages w.r.t. the order in which they are output by the enumeration algorithm from Theorem 3.

Result 3 (Theorem 5). *Given PCA A, with n states, accepting the prefix-closed regular language $L(A)$ over the alphabet Σ with σ letters, and string $w \in L(A)$, with $|w| = m$, we can compute in polynomial time the number of strings of length m output in our enumeration of $L(A)$ before outputting w. Moreover, given integer i, we can compute in polynomial time (relative to the output) the ith string of length m output in our enumeration algorithm of $L(A)$.*

Related Work. The problem of enumerating walks and paths (walks with no repeated vertex) within graphs is highly studied with a wealth of existing results. Wasa lists a series of such enumeration tasks and the complexity of their solutions in [32]. A notable work regarding the enumeration of paths within a graph is the backtracking technique by Read and Tarjan [22], proving a delay between outputting paths of at most $O(|E|)$. We note that in the general case, where each path must be explicitly output and there is no upper bound on the length of the path, this is optimal. However, when the length of the paths is bounded by some $m < |E|$, the potentially significant cost of backtracking limits the efficiency of this algorithm. Other initial work on enumerating paths focused on matrix-based approaches, requiring exponential time and space for precomputation, without intermediary output. Danielson [12] (later strengthened by Rubin [23]) provided such an approach for enumerating all simple paths (paths that do not visit any vertex more than once). Kamae [18] used this approach to output all cycles and paths in a directed graph; their approach was strengthened by Mateti and Deo [20], Wild [33], and Birmelé et al. [9]. We note that these algorithms are not focused, as ours are, on outputting (succinct representations of) each path (or cycle) sequentially with minimal delay, but rather on collecting all the paths, and thus optimise the time taken to compute the paths efficiently, rather than focusing on the output. Additional work has focused on specific classes of paths, such as Hamiltonian paths [34], ST-paths [9,14], and chordless paths [31].

Regarding the application of our results on walk-enumeration to the enumeration of strings from prefix closed languages, we also recall the rich literature regarding the enumeration of strings. We again point to the survey of Wasa for a series of classical results [32], as well as to the surveys by Gruber et al. [15] and Shallit [29]. Several works have focused on more general classes of languages, at the cost of allowing enumeration with non-constant delay, relative to the length of the strings [1,2,26,27]. Of particular interest to us is the recent work by Amarilli and Monet [6], who have provided an algorithm for outputting all strings recognised by a given regular language with bounded delay (which depends on the size of the automaton). This is achieved by partitioning the input language into *orderable regular languages*, i.e., a language whose strings can be ordered (in a potentially infinite sequence) in such a way that the edit distance between the ith and the $i + 1^{th}$ members of the sequence is bounded. In the other direction, there have been several algorithms for enumerating strings within specific subclasses of prefix closed languages. Quite close to our work are the results of Ruskey and Sawada from [24], where they give a relatively straightforward algorithm for enumerating the strings of length n, over an alphabet with σ letters, which do not contain a given factor f of length m with constant amortized delay, after a preprocessing taking $O(m\sigma)$ time. We note that our solution outperforms the one from [24] for this specific class by reducing the preprocessing time to $O(m)$, and the delay to constant in the worst-case. Similarly, constant amortised delay enumeration algorithms have since been provided for a large number of classes of cyclic strings, most relevantly necklaces and bracelets with a forbidden factor [24, 25].

For the interested reader, a full version of this paper containing complete proofs a pseudo code can be found in arXiv [3].

2 Preliminaries: Definitions and Sketch of the Algorithm

The computational model we use in this paper is the RAM with logarithmic word size relative to the size of the input graph or automaton.

Let $\mathbb{N} = \{1, 2, \ldots\}$ be the set of strictly positive integers and let $[n] = \{1, \ldots, n\}$ for $n \in \mathbb{N}$. Let $G = (V, U)$ be a directed graph (multi-graph) with the set of vertices V and the set of edges $U \subseteq V \times V$ (respectively, U is a multiset of pairs from $V \times V$); the direction of an edge (v, u) is from v to u. The directed (multi-)graph G is labelled, with labels over an alphabet $\Sigma = [\sigma]$, if there exists a function $\mathcal{L} : U \to \Sigma$ which labels each edge of G with a letter from Σ. An example relevant to our paper is that of finite automata, which are directed multi-graphs whose edges, called transitions in that context, are labelled by letters from an input alphabet. We assume that the sets Σ and V are totally ordered.

A walk of length k in G is a sequence $\pi = ((v_1, v_2), (v_2, v_3), \ldots, (v_{k-1}, v_k))$ such that $(v_i, v_{i+1}) \in U$, for all $i \in [k-1]$; the length k of π is denoted by $|\pi|$, and $v_1, \ldots, v_k$ are the vertices on the walk π. Given a walk π, we refer to its first (respectively, last) ℓ edges as the *prefix* (respectively, *suffix*) of length ℓ of π.

In this paper, we first develop an algorithm for the enumeration of walks of length m within a directed graph G, and then apply these results to enumerate strings accepted by a specific class of deterministic finite automaton, called prefix closed automata (PCA, for short). For space reasons, we refer the reader to [17] for definitions regarding strings and automata.

Algorithm Sketch. The key idea behind our approach is to enumerate the set of walks via an *implicit* yet *meaningful* representation of the walks within G, i.e. a representation that does not require the walk to be explicitly output but allows the explicit representation to be retrieved in linear time. We do so by creating a pseudoforest $D(G)$ from G, which has the same vertices as G. Moreover, each vertex has at most one outgoing edge, called *default edge*, corresponding to the first edge on (one of) the longest walk(s) leaving that vertex in G. We refer to $D(G)$ as the *default graph* and note that there is a unique (potentially infinite) walk leaving the vertex v in $D(G)$. The walks in $D(G)$ are called *default walks*. Using the default graph, we can succinctly represent default walks as tuples: the unique default walk of length ℓ starting with v is represented as (v, ℓ).

When solving Problem 1, we maintain the most recently enumerated walk π (which is a walk in G) as a list $(v_1, \ell_1)(u_1, v_2)(v_2, \ell_2) \ldots, (u_{k-1}, v_k)(v_k, \ell_k)$ where (v_i, ℓ_i) is the default walk starting in v of length ℓ_i, ending at u_i, and (u_i, v_{i+1}) is a non-default edge of G (so, not an edge of $D(G)$). To modify this walk π, and continue the enumeration, we find the last vertex v' in the current walk π that has a *branch*, i.e., a non-default edge, (v', u'), which has not been considered yet as a continuation for the prefix of π which connects v_1 to v', and which starts at least one walk as long as the suffix of the current walk π which connects v' to u_k. Once v' has been identified, a new walk is constructed, represented by $(v_1, \ell_1)(u_1, v_2)(v_2, \ell_2) \ldots, (u_{i-1}, v_i)(v_i, \ell_i')(v', u')(u', \ell_{i+1}')$, i.e. a walk sharing the first $m - (\ell_{i+1}' + 1)$ edges with π, followed by the non-default edge from v' to u', then the default walk from u' of corresponding length. Importantly, this new walk can be enumerated by only changing at most three entries in the list representing π: first, updating the tuple (v_i, ℓ_i) to (v_i, ℓ_i'); second, adding the edge (v', u'); third, adding the tuple (u', ℓ_{i+1}'). In this way, we require only a constant number of steps to update the walk.

The main challenge of this algorithm is determining these branches in constant time while keeping the preprocessing linear. In Sect. 3 we define the data structures used to achieve this result and show that these can be built in $O(|E|)$ time, where $|E|$ is the number of edges in the input graph. Section 4 formalises the algorithms, and explains why the worst-case delay between outputs is $O(1)$.

3   Toolbox: Default Graphs and Data Structures

Default edges, default walks, default graphs: definitions and basic facts. For the remainder of this paper, we consider the directed graph $G = (V, E)$, where $V = \{v_1, v_2, \ldots, v_n\}$ is a set of n vertices, and $E \subseteq V \times V$ is a set of directed edges represented by ordered pairs of vertices (v, u). We assume that we store a list of all outgoing and incoming edges for every vertex $v \in V$. We now introduce the primary data structures that are used for our enumeration algorithm.

Firstly, observe that the longest walk starting with vertex v either has length at most n or infinite length. Hence, we compute, for each vertex $v \in V$, the length $\pi(v) \in [n] \cup \{\infty\}$ of the longest walk starting at v. Moreover, we compute and store an ordered list L_v for each vertex $v \in V$, containing the pairs $((v, u), \ell)$, where $\ell \in [n] \cup \{\infty\}$ is the length of the longest walk from v starting with the edge (v, u). The list L_v is ordered in decreasing order of the length-component of its elements, with ties broken according to the ordering of the target vertices of the edge-component of these elements, as induced by the ordering on V. We can show the following lemma.

Lemma 1. *Given a directed graph $G = (V, E)$, the lengths $\pi(v)$ and the lists L_v, for all $v \in V$, can be computed in $O(|E|)$ time.*

Now, for each vertex v, the first edge (v, u) of the list L_v is the first edge on (one of) the longest walk(s) starting with v. This first edge of the list L_v is called in the following *default edge* of v (note that, by the definition of L_v, the notion of default edge is unambiguous, although there might be more than one longest walk starting with v). At this point, it is also important to note that, for each edge (v, u), the longest walk from v starting with (v, u) has a length equal to the length of the longest walk starting with u plus 1. Consequently, there is a walk from v starting with the edge (v, u) whose length is maximal among all walks starting with (v, u) and whose second edge is the default edge of u.

Further, we define the *default graph* $D(G) = (V, D(E))$. $D(G)$ has the same vertex set V and the edge set $D(E)$ containing exactly the default edges of the vertices from V. By Lemma 1, as the default edges of G (that is, the edges of $D(G)$) can be retrieved by simply taking the first edges of each of the lists L_v, with $v \in V$, we get that $D(G)$ can be constructed efficiently, in $O(|V|)$ time. Moreover, because each vertex v of the default graph $D(G)$ has at most a single outgoing edge (the default edge of v), this graph is a pseudoforest, consisting of a collection of *default components*: disjoint cycles, trees whose roots are on cycles, and, respectively, independent trees, which are not connected to any cycle. In the trees of this collection, the orientation of the edges is induced by the orientation of the edges in G, i.e., from children to parents, so from the leaves towards the root. For an example, see Fig. 1.

The walks of the graph $D(G)$ are called *default walks*. We use the notation (v, ℓ) to represent the unique default walk of length ℓ starting from v.

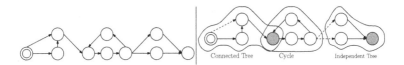

Fig. 1. The different classes of default components: On the left we have a directed graph; on the right, for that graph, default edges are shown as solid lines, and non-default edges are dashed. From left to right, we have a tree that is connected to a cycle, a cycle, and an independent tree. The tree-roots are grey.

We use the default walks as a tool to represent all walks in G. More precisely, given a walk $\pi = ((v_1, v_2), (v_2, v_3), \ldots, (v_{k-1}, v_k))$, we represent the walk π as a sequence $(v_{i_1}, \ell_{i_1})(v_{t_1}, v_{i_2})(v_{i_2}, \ell_{i_2})(v_{t_2}, v_{i_3}) \ldots (v_{t_{r-1}}, v_{i_r})(v_{i_r}, \ell_r)$, where default walks and non-default edges alternate:

- $v_{i_1} = v_1$ and v_k is the final vertex of the default walk (v_{i_r}, ℓ_r).
- For $1 \leq x \leq r$, (v_{i_x}, ℓ_{i_x}), with $\ell_x \geq 0$, is the longest default walk which is a prefix of the suffix $(v_{i_x}, v_{i_x+1}, \ldots, v_k)$ of π. The vertex v_{t_x} is the last vertex on the default walk (v_{i_x}, ℓ_{i_x}).
- For $1 \leq x \leq r - 1$, $(v_{t_x}, v_{i_{x+1}})$ is a non-default edge and $i_{x+1} = t_x + 1$.

Alternatively, to obtain this representation of π, we could first select all the non-default edges of π. These edges are connected, along π, by default walks (of length greater or equal to 0). This yields the aforementioned representation.

Before defining the data structures allowing us to process efficiently default graphs, we make one more observation. Consider a vertex v and the first edge

(v, u) of L_v (i.e., the default edge of v). Now, the first edge (v, u'), with $u \neq u'$, on the longest walk starting from v with any edge other than (v, u), is given by $((v, u'), \ell')$, the second element of L_u. So, these lists enable us to decide in constant time if, for a given length $\ell \leq m$ and vertex v, there exists some walk from v of length at least ℓ, other than the default walk: we get this information by looking at the second element of L_v. This will become crucial in finding branches, as described in the sketch of our algorithm, during the enumeration.

Efficient algorithms and data structures for the default graph. To work efficiently with the representations of arbitrary walks based on default walks, we need to be able to efficiently process the default graph. To this end, we now present a set of combinatorial lemmas, tools, and data structures providing a deeper understanding of the default graph. We begin by showing that the components of $D(G)$ can be computed efficiently.

Lemma 2. *Given default graph $D(G)$, we can compute in $O(|E|)$ time all its default components, and store for each vertex v the default component containing it.*

Trees appearing as default components of $D(G)$ (independent or attached to a cycle) can be represented as the root, followed by a list of children for each vertex; additionally, the default edge of each vertex points to its parent in the tree. Each cycle α is represented by its length $|\alpha|$ and an array containing $2|\alpha|$ elements. More precisely, for each cycle α, we have an initial vertex r, which is the first element in the array corresponding to α. We then traverse through α twice, while writing in the array the vertices of α in the order we meet them in this traversal. So, each element of α appears exactly twice in the array associated with the cycle, with exactly $|\alpha| - 1$ positions between its two occurrences. For each vertex v of α, it is enough to store its first occurrence i_v in the corresponding array. While a bit cumbersome at first view, this representation of cycles makes a bit simpler the usage of default graphs in the enumeration.

If vertex v is on a cycle α of length $|\alpha|$, the default walk (v, ℓ) corresponds to following α starting at v for ℓ edges. Letting v be the ith vertex on α (the initial vertex being the first), then the ending vertex of the default walk (v, ℓ) is the vertex found on the $((i + \ell) \mod |\alpha|)^{th}$ position of the cycle.

If vertex v is in an independent tree rooted at r, the default walk (v, ℓ) goes towards the root of the tree, traversing ℓ edges. We define the root as being on level 0, and the children of a vertex on level i are on level $i + 1$. Therefore, if v is on level h in its tree, the ending vertex of the default walk (v, ℓ) is the ancestor of v on level $h - \ell$ of this tree. Thus, to be able to retrieve the ending vertices of default walks in trees quickly, we build, for all our trees (both independent and attached to cycles), *level ancestor* data structures [8]. For a tree of size τ, these data structures can be computed in $O(\tau)$ time and enable us to answer in $O(1)$ queries $LA(v, j)$: return the ancestor of v which is on level j of the tree.

If v is in a tree whose root r is on a cycle, the walk (v, ℓ) goes towards the root of the tree and potentially also goes around the cycle, traversing ℓ edges in total. Following the ideas already described above, to retrieve the ending vertex of a default walk (v, ℓ) in a tree with root r attached to a cycle, we will first check if the walk ends inside the tree or enters the cycle. This can be done by verifying if the level h of v inside its tree is greater or equal to ℓ. If yes, we can compute again the level ancestor of v, which is on level $h - \ell$. If not, then the ending vertex of the walk (v, ℓ) is on the cycle. The number of edges traversed in the cycle by this walk is $\ell - h$, and to find the ending vertex of (v, ℓ) it is enough to find the ending vertex of the default walk $(r, \ell - h)$, which is a walk on a cycle, and can be treated as above.

According to the above, it is important to store, for the vertices of the tree components of $D(G)$, their level. To allow a uniform treatment of all the vertices, we define the *depth of vertex v in its default component*, denoted d_v^t, if v is in a tree, or, respectively, d_v^c, if v is on a cycle. This is defined in one of three ways, depending on which kind of component contains v. Before giving the definition, we note that the vertices contained in trees whose root is on a cycle will have two such depths, one w.r.t. the tree and one w.r.t. the cycle.

- If v is in a tree, then d_v^t is simply the level of v in the respective tree.
- For a cycle α, we define the depth of each position $i \leq 2|\alpha|$ of the array associated to α as $d_i^c = 2|\alpha| - i + 1$. If v is a vertex on α, we define $d_v^c = d_{i_v}^c$, where i_v is the first (i.e., leftmost) occurrence of v in the array associated to α.
- If v is in a tree with root r connected to a cycle α, then we associate to v a second value d_v^c (the depth of v w.r.t. the cycle α), defined as $d_v^c = d_v^t + d_r^c$.

Using the default graph for enumeration. The algorithm presented in Sect. 4 enumerates all walks of length m in G starting at a given vertex v_0. We note that the preprocessing does not depend on the choice of the vertex v_0, it can be done once for all vertices. Our approach can be then extended by selecting (at the time when the values $\pi(v)$ are computed) a list of all vertices with a default walk of length at least m, and then repeating the enumeration for each vertex of that list.

An important primitive of this algorithm, allowing us to move in our enumeration from one walk π_1 to the next one π_2 (both starting with v_0), is to check whether there exists a vertex v on some default walk (s, ℓ) (which is part of the representation of the first walk π_1) from which we can follow a non-default edge, instead of the default edge which we have followed in the walk π_1, and obtain a new walk starting in v_0 of length m. Such a vertex v is called, for simplicity of exposure, a *branching vertex* w.r.t. the walk π_1 (notice, though, that some vertices might be branching w.r.t. some walks π_1 and not branching w.r.t. others, depending on the position in which they appear on these walks; however, we will only use this name when there is no danger of confusion). Thus, we need to check the existence of a vertex v on (s, ℓ), such that, if the walk from v_0 to such a vertex v along π_1 has length ℓ_v, then there is a walk starting with a non-default edge of v, with length at least $m - \ell_v$.

We achieve this by finding the vertex v of the default walk (s, ℓ) maximising the sum between the length of the walk from v_0 to v along π_1 and the length of the longest walk starting in v with a non-default edge, and seeing if this sum is greater or equal to m. If this sum is not greater or equal to m, then there is no vertex with the desired properties on (s, ℓ). If the sum is greater or equal to m, the vertex v has the desired properties, and thus we can use it next in our enumeration. To identify this vertex v, we note that v is exactly that vertex for which the sum of the length of the walk from s to v and the length of the longest walk starting in v with a non-default edge is maximum (as all walks from v_0 to vertices on the default walk (s, ℓ), along π_1, share the prefix of π_1 connecting v_0 to s, the starting vertex of that default walk (s, ℓ)).

We define for each vertex v a weight w_v, corresponding to the length of the longest walk from v which starts with a non-default edge, i.e. the length of the walk starting with the second edge stored in L_v, and then continuing with the longest default walk starting with the end vertex of that edge. After the preprocessing of Lemma 1, w_v can be retrieved in $O(1)$ time, for each v.

Let WalkMaxNode(s, ℓ) denote, for each default walk (s, ℓ) of $D(G)$, the pair (v, d) where v is the vertex of this walk such that the sum of w_v and the distance d between s and v along the default walk (s, ℓ) is maximum. In Lemma 3, we show that WalkMaxNode queries can be answered in $O(1)$-time, after linear time preprocessing. Indeed, building on the data structures introduced above and taking into account the particular structure of the default graph, these queries reduce to either computing walk minimum queries in trees, which can be handled efficiently [13], or to range minimum queries in arrays corresponding to the cycles, which again can be computed efficiently [7].

Lemma 3. *We can build in $O(|E|)$ time data structures, allowing us to answer* WalkMaxNode(s, ℓ) *queries in $O(1)$ time, for each default walk (s, ℓ) of $D(G)$.*

4   Enumeration

The Main Algorithm. The basic idea of our enumeration algorithm is to represent and compute the walks of length m of G starting at v_0 as sequences of default walks and the single non-default edges connecting them. This is achieved using a recursive procedure, Enumerate. In each call, Enumerate takes two parameters: an edge $(v, u) \in E$, and a number ℓ. In the call Enumerate$((v, u), \ell)$ we go through all walks of length ℓ starting from u; each such walk is the suffix of a walk of length m starting in v_0, going through v, which will be output.

Before making the first call, we preprocess the graph G in $O(|E|)$ time as described in Sect. 3: building the default components and associated data structures which allow their efficient processing. In particular, we construct in $O(|E|)$ time the data structures needed to answer in $O(1)$-time WalkMaxNode-queries. Note that the space used by the respective structures is also linear, $O(|E|)$.

During the enumeration, done via calls to the recursive procedure Enumerate, we maintain as global variables two stacks $\mathcal{C}$ and $\mathcal{S}$ (of size at most m), containing tuples $((v, u), \ell)$, with $(v, u) \in E, \ell \in [n]$. Intuitively, if the content of the stack $\mathcal{S}$ is, at some step of the computation, the sequence $\langle (\uparrow, v_0, \ell_0), ((u_0, v_1), \ell_1), \ldots,$

$((u_{t-1}, v_t), \ell_t))$ (where the top of the stack is to the right of this sequence), then the currently enumerated walk π is the concatenation of p_0, (u_0, v_1), p_1, (u_1, v_2), ..., p_{t-1}, (u_{t-1}, v_t), p_t, where, for $i \geq 0$, p_i is the default walk of length ℓ_i starting with the vertex v_i and ending with some vertex u_i, and $(u_i, v_{i+1}) \in E$, for $0 \leq i \leq t - 1$. This allows the walk π to be retrieved explicitly from its representation on the stack $\mathcal{S}$ in $O(m)$ time. With respect to the execution of our algorithm, $\mathcal{S}$ corresponds to the stack of currently active recursive calls. The usage of $\mathcal{C}$ is more subtle. Intuitively, at every step of the computation, $\mathcal{C}$ contains (bottom to top, in the same order as in $\mathcal{S}$) exactly those tuples $((v, u), \ell)$ of $\mathcal{S}$ for which the default walk of length ℓ starting in v still contains branching vertices leading to walks of length m which have not yet been enumerated. As such, $\mathcal{C}$ facilitates the quick identification of the next walk to output in our enumeration: such a walk should go through one of the branching points of the default walk found on top of this stack $\mathcal{C}$. With respect to the execution of our algorithm, $\mathcal{C}$ corresponds to the currently active recursive calls which were not tail calls. We assume, w.l.o.g., that in the computational model we use, tail calls are implemented so that no new stack frame is added to the call stack. Hence, $\mathcal{C}$ corresponds, at each moment of our algorithm's execution, to the current call stack.

The enumeration starts with the call Enumerate$((\uparrow, v_0), m)$, which outputs first the tuple (v_0, m), corresponding to the first m edges of the default walk starting at v_0. Note that this first object in our enumeration is output in $O(1)$, after the preprocessing. In a general step, we call Enumerate$((v, u), \ell)$. Due to space constraints, we give only a high-level description of our algorithm.

In general, Enumerate$((v, u), \ell)$ first outputs the triple $(m - \ell - 1, (v, u), \ell)$. That is, the current walk in the enumeration, denoted by π in the following, is obtained from the previous one by keeping its prefix of length $m - \ell - 1$ edges (which ends in some vertex v), to which we append the edge (v, u), and the default walk of length ℓ starting in u. We also store the tuple $((v, u), \ell)$ in the two stacks.

Next, we need to see if there are more walks of length ℓ starting with u, other than the default walk. If no, we simply remove the triple $((u, v), \ell)$ from both stacks and return. If yes, we need to discover each of them, and this will be done by further recursive calls to Enumerate; all the walks produced in this way will share at least the first $m - \ell$ edges with π. The efficient identification of these walks relies on the data structures constructed in the preprocessing phase for the default components of G. However, to see how exactly they are identified, we need to understand what we are looking for. Basically, each such walk follows a (potentially empty) prefix of the default walk of length ℓ starting with u, which ends with a branching vertex w.r.t. the walk π starting in v_0. So, we only need to discover the respective branching vertices and process them one by one, by recursive calls. We can find the branching vertices by WalkMaxNode-queries and, by Lemma 3, such queries can be answered in $O(1)$-time due to the structure of the default graph (composed of trees and cycles).

This is done as follows. Let r be the ending vertex of the default walk (u, ℓ). We first retrieve $(v', d) = \text{WalkMaxNode}(u, \ell)$. If $d + w_{v'} \geq \ell$, then this vertex is one of the branching vertices we were looking for. Indeed, the above inequality means that the walk starting from v' with the edge given by the second element of $L_{v'}$ is at least as long as the walk from v' to r along the default walk (u, ℓ). Thus, we can go through the non-default edges leaving v' (as given by $L_{v'}$), as long as they start walks which are at least as long as the default walk from v' to r, and start a corresponding recursive call of Enumerate for each of them. Once we are done with all these edges, and, accordingly, with the vertex v', we might still have other branching vertices on the default walk (u, ℓ). Such a vertex is either on the segment of this walk between u and the predecessor of v' or on the segment of this walk between the successor of v' and r, and can be identified using WalkMaxNode-queries, as above. In general, if the WalkMaxNode-query on such a segment of the default walk (u, ℓ) (delimited by vertices v_1 and v_2) returns a vertex v'' from which we can follow a long-enough walk starting with a non-default edge, we process v'' exactly as we processed vertex v' above, then search for more branching vertices in the two segments delimited, resp., by v_1 and the predecessor of v'' and the successor of v'' and v_2. If the WalkMaxNode-query on a segment of the default walk (u, ℓ) returns a vertex v'' from which no long-enough walk starting with a non-default edge can be found, we stop processing that entire segment: there are no more relevant branching vertices to be found there.

From the above, we guarantee that we only call Enumerate when we are certain that it will lead to the discovery of at least one walk which was not yet enumerated. The total number of WalkMaxNode-queries which are answered during the call $\text{Enumerate}((v, u), \ell)$ is proportional to the number of branching vertices on the default walk (u, ℓ). So, the overall time complexity of the execution of $\text{Enumerate}((v, u), \ell)$ is proportional to the number of walks starting from u of length ℓ (each corresponding to a walk of length m we need to enumerate).

To obtain enumeration with constant delay in the worst case, there is one more additional subtlety. Namely, we need to make sure that as soon as a call to Enumerate is finished, we will directly consider a default walk on which we certainly find branching vertices. This is ensured by the usage of tail calls and of the stack $\mathcal{C}$. Indeed, we need to make sure that the last step of the procedure Enumerate is a recursive call (and this is not hard to do); before executing this call, we remove the tuple $((v, u), \ell)$ from $\mathcal{C}$ (as it will also be removed from the call stack). Thus, in our computational model, after the respective tail recursive call is finished, we will return to a previous call which still has at least one walk to explore (so, one branching vertex that was not completely processed); the respective previous call corresponds to the tuple found now on top of $\mathcal{C}$, and, as such, we can check where the respective tuple is found in $\mathcal{S}$ and set it to be the current top of the respective stack too (what came above it is no longer important for the rest of the computation, as it does not lead to new walks).

The correctness of our approach is proven by induction, by showing that every walk is output exactly once via the tree of recursive calls whose root is

Enumerate($\uparrow, q_0, n$) (in this tree, the vertices are instances of Enumerate called in our algorithm, and their children the calls initiated in the respective instances).

Lemma 4. *The call* Enumerate$((\uparrow, v_0), m)$ *outputs a representation of every walk of length m starting at v_0 in the graph G exactly once.*

Theorem 1. *Given integer m, directed graph $G = (V, E)$, and vertex v_0, we can enumerate, without repetitions, succinct representations of all walks of length m starting from v_0 in G with $O(1)$-delay, after an $O(|E|)$-time preprocessing.*

It is worth noting that the space used by our algorithm (on top of the $O(|E|)$-space used by the data structures produced during the preprocessing) is upper bounded, at any point, by the length of the currently enumerated walk. We can immediately extend Theorem 1 to output every walk of length m in G: in the preprocessing phase, we collect every vertex with a default walk of length at least m, and then use the algorithm of Theorem 1 for each such vertex.

Theorem 2. *Given integer m and directed graph $G = (V, E)$, we can enumerate, without repetitions, succinct representations of all walks of length m in G with $O(1)$-delay, after an $O(|E|)$-time preprocessing.*

We can also enumerate for integers $\ell \leq m$, after exactly the same preprocessing (which is independent of the length of the enumerated walks), representations of the walks of G with length between ℓ and m.

Applications for Automata. The result of Theorem 1 can be immediately applied to prefix closed regular languages (PCLs), given by the prefix closed automata (PCA, incomplete deterministic finite automata with final states only) accepting them. For this, we represent the input PCA A as a directed, labelled multi-graph, $G(A)$, and enumerate all walks of length m starting at the vertex v_0 corresponding to the initial state in A. Our algorithm still works without any change because between two vertices of $G(A)$ we have at most one edge with a certain label (although we might have multiple edges), and all our data structures can be extended canonically to this setting. As there is a bijective correspondence between the walks in $G(A)$ and the strings of $L(A)$, the following theorem follows.

Theorem 3. *Given integer m and PCA A, we can enumerate, without repetitions, succinct representations of all strings of length m of $L(A)$ with $O(1)$-delay after an $O(|A|)$-time preprocessing.*

Among PCLs, the class of languages $L_{\mathcal{F}}$, of the strings over Σ that do not contain any *forbidden factor* from a finite set of strings $\mathcal{F}$, is of particular interest. However, when $\mathcal{F} = \{f\}$ the following result holds, outperforming [24].

Theorem 4. *Given integer m and string $f \in \Sigma^*$, we can enumerate, without repetitions, succinct representations of the strings of length m over Σ which do not contain f as a factor, with $O(1)$-delay, after an $O(|f|)$-time preprocessing.*

The final problems considered here are those of *ranking* and *unranking* strings in prefix-closed regular languages. The *rank* of a string $w \in \Sigma^m$ in a language is the number of strings smaller than w in the language under some ordering. The ranking problem requires computing the rank of a given string w. The unranking operation (problem) takes a number i as the input and asks for the string of rank i. In both cases, we refer to the ordering induced by the enumeration algorithm of Sect. 4. Both ranking and unranking require identifying a walk in the tree of recursive calls with root Enumerate($\uparrow, q_0, m$). For ranking, we identify the walk corresponding to w, and the branching vertices occurring on it, then count the total number of walks of length m corresponding to the leaves of subtrees of recursive calls occurring prior to this walk (assuming that the recursive calls made by an instance are ordered in the tree left to right according to their call-order). This can be done by running Enumerate($\uparrow, q_0, m$) and simply performing only the recursive calls that correspond to branching vertices on the walk labelled with w, and retrieving the number of induced walks for those that should have been called before them. For unranking, we run again Enumerate($\uparrow, q_0, m$) and we make only those recursive calls which lead to the ith walk of length m, in the order of our enumeration. The following two results are obtained, w denotes here the size of the memory word in our model).

Theorem 5. Ranking. *Given PCA A and string $w \in L(A)$ of length m, we can compute the number of strings accepted by A which are output before w in our enumeration algorithm in $O\left(\frac{n \log \sigma}{\mathtt{w}}(n\sigma + m(n^\omega + m + \sigma))\right)$-time, where $2 \leq \omega \leq 3$ is the exponent for matrix multiplication.* **U.** *Given integers i and m, and PCA A, we can compute the ith string w of length m output in our enumeration algorithm in $O\left(\frac{n \log \sigma}{\mathtt{w}}(n\sigma + m(n^\omega + m + \sigma))\right)$-time.*

Acknowledgements. The authors thank the reviewers for their helpful comments. Duncan Adamson was supported by the Leverhulme Trust via the Leverhulme Centre for Functional Material Design and by DFG Heisenberg-project number 389613931. Florin Manea was supported by DFG Heisenberg-project number 466789228.

References

1. Ackerman, M., Mäkinen, E.: Three new algorithms for regular language enumeration. In: Ngo, H.Q. (ed.) COCOON 2009. LNCS, vol. 5609, pp. 178–191. Springer, Heidelberg (2009). https://doi.org/10.1007/978-3-642-02882-3_19
2. Ackerman, M., Shallit, J.: Efficient enumeration of words in regular languages. Theoret. Comput. Sci. **410**(37), 3461–3470 (2009)
3. Adamson, D., Gawrychowski, P., Manea, F.: Enumerating m-length walks in directed graphs with constant delay (2024)
4. Amarilli, A., Bourhis, P., Mengel, S., Niewerth, M.: Constant-delay enumeration for nondeterministic document spanners. ACM Trans. Database Syst. **46**(1), 2:1–2:30 (2021)
5. Amarilli, A., Jachiet, L., Muñoz, M., Riveros, C.: Efficient enumeration for annotated grammars. In: PODS 2022, pp. 291–300. ACM (2022)

6. Amarilli, A., Monet, M.: Enumerating regular languages with bounded delay. In: STACS 2023. LIPIcs, vol. 254, pp. 8:1–8:18. Schloss Dagstuhl - Leibniz-Zentrum für Informatik (2023)

7. Bender, M.A., Farach-Colton, M.: The LCA problem revisited. In: Gonnet, G.H., Viola, A. (eds.) LATIN 2000. LNCS, vol. 1776, pp. 88–94. Springer, Heidelberg (2000). https://doi.org/10.1007/10719839_9

8. Bender, M.A., Farach-Colton, M.: The level ancestor problem simplified. Theoret. Comput. Sci. **321**(1), 5–12 (2004)

9. Birmelé, E., et al.: Optimal listing of cycles and st-Paths in undirected graphs. In: SODA 2013, pp. 1884–1896. SIAM (2013)

10. Cevorová, K., Jirásková, G., Mlynárcik, P., Palmovský, M., Sebej, J.: Operations on automata with all states final. In: AFL 2014. EPTCS, vol. 151, pp. 201–215 (2014)

11. Collins, C., et al.: Accelerated discovery of two crystal structure types in a complex inorganic phase field. Nature **546**(7657), 280–284 (2017)

12. Danielson, G.: On finding the simple paths and circuits in a graph. IEEE Trans. Circuit Theory **15**(3), 294–295 (1968)

13. Demaine, E.D., Landau, G.M., Weimann, O.: On Cartesian trees and range minimum queries. Algorithmica **68**(3), 610–625 (2014)

14. Grossi, R., Marino, A., Versari, L.: Efficient algorithms for listing k disjoint st-Paths in graphs. In: Bender, M.A., Farach-Colton, M., Mosteiro, M.A. (eds.) LATIN 2018. LNCS, vol. 10807, pp. 544–557. Springer, Cham (2018). https://doi.org/10.1007/978-3-319-77404-6_40

15. Gruber, H., Lee, J., Shallit, J.O.: Enumerating regular expressions and their languages. In: Pin, J. (ed.) Handbook of Automata Theory, pp. 459–491. European Mathematical Society Publishing House, Zürich, Switzerland (2021)

16. Gusev, V.V., et al.: Optimality guarantees for crystal structure prediction. Nature **619**(7968), 68–72 (2023)

17. Hopcroft, J.E., Ullman, J.D.: Introduction to Automata Theory, Languages and Computation. Addison-Wesley, London (1979)

18. Kamae, T.: A systematic method of finding all directed circuits and enumerating all directed paths. IEEE Trans. Circuit Theory **14**(2), 166–171 (1967)

19. Kao, J., Rampersad, N., Shallit, J.O.: On NFAs where all states are final, initial, or both. Theoret. Comput. Sci. **410**(47–49), 5010–5021 (2009)

20. Mateti, P., Deo, N.: On algorithms for enumerating all circuits of a graph. SIAM J. Comput. **5**(1), 90–99 (1976)

21. Oganov, A.R.: Crystal structure prediction: reflections on present status and challenges. Faraday Discuss. **211**, 643–660 (2018)

22. Read, R.C., Tarjan, R.E.: Bounds on backtrack algorithms for listing cycles, paths, and spanning trees. Networks **5**(3), 237–252 (1975)

23. Rubin, F.: Enumerating all simple paths in a graph. IEEE Trans. Circuits Syst. **25**(8), 641–642 (1978)

24. Ruskey, F., Sawada, J.: Generating necklaces and strings with forbidden substrings. In: Du, D.-Z.-Z., Eades, P., Estivill-Castro, V., Lin, X., Sharma, A. (eds.) COCOON 2000. LNCS, vol. 1858, pp. 330–339. Springer, Heidelberg (2000). https://doi.org/10.1007/3-540-44968-X_33

25. Sawada, J.: Generating bracelets in constant amortized time. SIAM J. Comput. **31**(1), 259–268 (2001)

26. Schmid, M.L., Schweikardt, N.: Spanner evaluation over SLP-compressed documents. In: PODS 2021, pp. 153–165. ACM (2021)

27. Schmid, M.L., Schweikardt, N.: Query evaluation over SLP-represented document databases with complex document editing. In: PODS 2022, pp. 79–89. ACM (2022)
28. Segoufin, L.: Enumerating with constant delay the answers to a query. In: ICDT 2013, pp. 10–20. ACM (2013)
29. Shallit, J.: Decidability and enumeration for automatic sequences: a survey. In: Bulatov, A.A., Shur, A.M. (eds.) CSR 2013. LNCS, vol. 7913, pp. 49–63. Springer, Heidelberg (2013). https://doi.org/10.1007/978-3-642-38536-0_5
30. Uno, T.: Amortized analysis on enumeration algorithms. In: Kao, M.Y. (ed.) Encyclopedia of Algorithms. Springer, New York (2016). https://doi.org/10.1007/978-1-4939-2864-4_730
31. Uno, T., Satoh, H.: An efficient algorithm for enumerating chordless cycles and chordless paths. In: Džeroski, S., Panov, P., Kocev, D., Todorovski, L. (eds.) DS 2014. LNCS (LNAI), vol. 8777, pp. 313–324. Springer, Cham (2014). https://doi.org/10.1007/978-3-319-11812-3_27
32. Wasa, K.: Enumeration of enumeration algorithms. CoRR abs/1605.05102 (2016)
33. Wild, M.: Generating all cycles, chordless cycles, and Hamiltonian cycles with the principle of exclusion. J. Discrete Algorithms 6(1), 93–102 (2008)
34. Yau, S.: Generation of all Hamiltonian circuits, paths, and centers of a graph, and related problems. IEEE Trans. Circuit Theory 14(1), 79–81 (1967)

The Hardness of Local Certification
of Finite-State Dynamics

Diego Maldonado[1], Pedro Montealegre[2(✉)], and Martín Ríos-Wilson[2]

[1] Facultad de Ingeniería, Universidad Católica de la Santísima Concepción,
Concepción, Chile
dmaldonado@ucsc.cl
[2] Facultad de Ingeniería y Ciencias, Universidad Adolfo Ibáñez, Peñalolén, Chile
{p.montealegre,martin.rios}@uai.cl

Abstract. Finite-State Dynamics (FSD) is one of the simplest and constrained distributed systems. An FSD is defined by an n-node network, with each node maintaining an internal state selected from a finite set. At each time-step, these nodes synchronously update their internal states based solely on the states of their neighboring nodes.

Rather than focusing on specific types of local functions, in this article, our primary focus is on the problem of determining the maximum time required for an FSD to reach a stable global state. This global state can be seen as the acceptance state or as the output of a distributed computation. For fixed k and q, we define the problem convergence(k,q), which consists of deciding if a q-state FSD converges in at most k time-steps.

Our main focus is to study the problem convergence from the perspective of distributed certification, with a focus on the model of proof-labeling schemes (PLS). First, we study the problem convergence on arbitrary graphs and show that every PLS has certificates of size $\Theta(n^2)$ (up to logarithmic factors). Then, we turn to the restriction of the problem on graphs of maximum degree Δ. Roughly, we show that the problem admits a PLS with certificates of size Δ^{k+1}, while every PLS requires certificates of size at least $2^{k/6} \cdot 6/k$ on graphs of maximum degree 3.

Keywords: Local Certification · Proof Labeling Schemes · Finite State Dynamics

1 Introduction

Networks serve as the backbone of numerous scientific domains, ranging from the social sciences, where they represent human connections, to logistics, as seen in

This research was supported by Centro de Modelamiento Matemático (CMM), FB210005, BASAL funds for centers of excellence from ANID-Chile (P.M.), FONDE-CYT 1230599 (P.M.), Programa Regional STIC-AMSUD (CAMA) cod. 22-STIC-02 (P.M., M.R.-W.), ECOS project C19E02 (M.R.-W.), ANID FONDECYT Postdoctorado 3220205 (M.R.-W.) and Fondo Especial de Actividades Académicas proyecto ingeniería 2030 ING222010004 (D.M.).

J. A. Soto and A. Wiese (Eds.): LATIN 2024, LNCS 14578, pp. 51–65, 2024.
https://doi.org/10.1007/978-3-031-55598-5_4

traffic patterns, and even electrical engineering, as in circuitry. Distributed computing explores the capabilities and constraints of algorithms that operate across these networks. Given the pervasive influence of the Internet, most contemporary models aim to understand devices capable of accessing it. Typically, distributed computing employs the message-passing model, where nodes can send extensive messages to neighboring nodes and perform local computations. However, some networks, such as emerging wireless networks like ad-hoc or sensor networks, do not perfectly conform to this traditional model. Their underlying devices have constraints that do not align with the classical message-passing model.

A recent trend involves using distributed computing techniques, especially the message-passing model, for sub-microprocessor networks like those in biological cells or nano-mechanical devices [10]. However, the fundamental differences in capability between biological or nano nodes and silicon-based devices necessitate a distinct network model, one designed for nodes inherently more limited than Turing machines. A natural model to consider in that context is the one of finite-state dynamics [21,27]. Within this framework, numerous agents are interconnected through an undirected graph known as the *interaction graph*. Each agent communicates solely with its immediate neighbors. Every node on this graph has an inherent local state, with the collective states of all nodes forming a *configuration* or global state for the dynamic system. This internal state gets updated, relying on its prior state and those of its neighbors. This updating is governed by a *local rule*. Moreover, the function that transforms one configuration into another using local rules is termed the *global rule*. As this rule is repeatedly applied, a series of global configurations ensues. Given that there is a finite number of configurations, the dynamics will eventually lead to a repetitive sequence of node states. If, after successive applications of the global rule, the system reaches a stable configuration where the internal state of each node remains unchanged, we say that the system *converges*.

An interesting feature of the finite-state dynamics formalism is its capacity to describe complex behavior while utilizing a constant-size information representation through the states of the nodes [3]. In the realm of theoretical computer science, the study of finite-state dynamics revolves around several key topics. One significant question involves investigating whether the structure of the underlying interaction graph can influence its dynamics [22–24]. Another well-explored approach is the study of decision problems linked to the dynamics, often motivated by specific questions pertaining to particular models. In this context, a central question centers on determining the point at which the complexity of the dynamics, viewed as the complexity of a particular decision problem, becomes related to the dynamical behavior.

In the field of distributed systems, numerous models exhibit similarities or share characteristics with finite-state dynamics. An example of a model in which the messages distributed by the nodes in the network are bounded is the case of the beeping model [8,17]. In this model, nodes can either beep or stay silent, and they can only distinguish between two situations: when all their neighbors are silent or when at least one neighbor is beeping. In addition, they can either beep or listen to their neighbors, but they cannot do both at the same time. A crucial difference between the FSD model and the beeping model is that in the latter, the local computation is performed by a Turing machine (so the set of states is not finite). This model has interesting applications as the beeping process can be interpreted as communication between simple machines, such as sensors, or it can be seen as a way to communicate messages in a biological network [1,2].

Another example of a model related to finite-state dynamics is the model called networked state machines (nFSM) introduced in [10]. In this case, the local computation is done using a finite set of states, but the essential difference between this model and finite-state dynamics is that nFSM operates asynchronously and incorporates elements of randomness. In the previously cited papers, the dynamics induced by these models are interpreted as the distributed computing of coloring or a maximal independent set [1,2,10].

Generally speaking, the main measure of performance in distributed systems is the number of rounds required to output a solution in the worst case. In this sense, studying the time required for the system to compute a solution, the convergence time, is crucial. However, to our knowledge, this has not been addressed from a more dynamics-centered approach or a local certification approach.

Existing literature has established a connection between the computational complexity of these problems and the dynamical properties of the system [5,25,34]. Nevertheless, to our knowledge, these problems have not been explored from a distributed standpoint.

Recently, a novel approach based on distributed algorithms and local decisions has been introduced [31]. This approach has been applied to study opinion dynamics in social systems. In this work, we extend this approach to examine a broader framework, specifically focusing on distributed certification for determining whether the dynamics converge.

The Model. We now define the model more formally, while we refer to the preliminaries section for further details. We consider simple finite undirected graphs G. Each node has one over a finite set of states Q. A node u of G also knows a deterministic local function f_u, that specifies how the node updates its state, computed according to the states on its closed neighborhood. In order to simplify the model, we consider only synchronous dynamics, meaning that all nodes update their states at the same time on each time-step. The synchronous application of the local functions on every node produces a *global function F* over G, which establishes how the configuration of states of every node evolves over time. The pair (G, F) is called a *finite-state dynamic system*.

Since the possible state configurations of the system is finite, and the local functions are deterministic, we have that every sequence of state configurations of a finite-state dynamics system is eventually periodic. A state configuration is called a *fixed point* if the dynamics does not change under the application of the global function. A finite-state dynamics system is *convergent* if every state configuration reaches a fixed point after a certain number of times-steps. The *convergence time* of a system is the maximum number of time-steps required to reach a fixed point from any initial configuration. Obviously, the *convergence time* of a system is finite only if the system is convergent.

The Convergensen Problem. In the study of finite-state dynamics, there are multiple possible research questions. Most of these questions are related to the study of one particular dynamical behavior. However, instead of focusing on a specific system, the main aim of this work is to focus on identifying properties of an arbitrary dynamics that can be studied using distributed methods. A natural task in this context consists of predicting the long-term behavior of the system. For example, it might be interesting to ask if observing the dynamics for a given number of time-steps is enough for the system to attain a particular state (such as an acceptance state, for instance), or ascertain whether a computation processed by the network will eventually converge.

Within distributed systems characterized by finite states, the properties based on localized knowledge of the update rules constitute a key element to consider. In fact, given the inherent nature of distributed systems, it becomes crucial to understand the evolution of the dynamics and the long-term behavior of the system based solely on this local data.

The main research question of this article is: Using local knowledge, can we predict if a dynamic system will converge within a given number of time-steps? More precisely, given a pair of positive integers k and q, we define the problem CONVERGENCE(k, q) which consists of, given a q-state finite-state dynamic system, determining if the dynamics converges in at most k time-steps.

This problem is not solvable by a local algorithm, as the state evolution of a node might be affected by the state configuration in remote locations of the network. For that reason, we tackle the CONVERGENCE(k, q) problem from the perspective of local certification, more specifically, *proof-labeling schemes*.

Local Certification. A local certification algorithm for a distributed decision problem is a prover-verifier pair where the prover is an untrustworthy oracle assigning certificates to the nodes, and the verifier is a distributed algorithm enabling the nodes to check the correctness of the certificates by a certain number of communication rounds with their neighbors. Note that the certificates may not depend on the instance G only, but also on the identifiers id assigned to the nodes. Proof-labeling schemes (PLSs) are a specific type of local certification algorithms, where the information exchanged between the nodes during the verification phase is limited to only one round, and the contents are limited to the certificates. The prover-verifier pair must satisfy the following two properties.

Completeness: On yes-instances, the untrustworthy prover can assign certificates to the nodes such that the verifier accepts at all nodes;

Soundness: On no-instances, for every certificate assignment to the nodes by the untrustworthy prover, the verifier rejects in at least one node.

The main complexity measure for proof-labeling schemes is the size of the certificates assigned to the nodes by the prover.

1.1   Our Results

First, we observe that for each pair of positive integers k and q, the problem CONVERGENCE(k, q) can be verified on n-node graphs with certificates of size

$$b_{\max}(k + 1) \cdot \log(\mathsf{id}_{\max}) \cdot |f_{\max}(n, q)|,$$

where $b_{\max}(k+1)$ is the maximum number of edges having an endpoint in a node at distance at most $k + 1$, the identifiers are assigned with values in $[\mathsf{id}_{\max}]$ and $|f_{\max}(n, q)|$ is the maximum number of bits required to encode a local function. This is achieved by providing each node with the necessary part of the network to simulate $k+1$ time-steps of any configuration. In our first main result, we show that such a simple algorithm is in fact the best one possible in general, up to logarithmic factors. More precisely, we show that every Proof-Labeling Scheme (PLS) for CONVERGENCE$(2, 4)$ requires certificates of size $\Omega(n^2/(\log n))$.

Observe that even if we consider a finite number of states, the encoding of a local function can be exponentially large relative to the number of neighbors of a node. Therefore, we consider local rules that admit *succinct representations*, which can be encoded using a logarithmic number of bits per neighbor. Examples of such local rules are the ones found in artificial neural networks, the modeling of opinion dynamics, or different biological processes. Our lower-bound does not hide the complexity in the encoding of the local rules, as it holds even when the problem is restricted to local functions that admit a succinct representation.

Later, we focus on graphs with bounded degree. Restricted to this case, the straightforward approach gives a PLS for CONVERGENCE(k, q) with certificates of size at most

$$\Delta((\Delta - 1)^{k+1} - 1) \cdot q^{\Delta+1} \log(q) \cdot \log(\mathsf{id}_{\max})$$

where $\Delta \geq 2$ is the maximum degree of the input graph. In our second main result, we show that the exponential dependency on k cannot be avoided in the bounded degree case. More precisely, we show that every PLS for problem CONVERGENCE$(k, 3)$ requires certificates of size at least $2^{k/6} \cdot 6/k$ even when the problem is restricted to input graphs of maximum degree 3.

1.2   Related Work

Finite-State Dynamics and Majority Voting Dynamics. An interesting distributed problem is the one of predicting the result of a two-candidate election. This process is typically modeled as a majority voting dynamics. The model can be seen as a group of n agents represented by nodes in a network who are surveyed about their preferred candidate in an upcoming election with two choices. Over a series of time steps, each agent adjusts their vote based on the majority opinion of their network neighbors, ultimately determining the leading candidate after T time steps. The challenge lies in predicting the leading candidate which in general is a very hard task.

In [31] the authors study the problem ELECTION-PREDICTION, consisting of predicting the leading candidate after a certain number of time-steps from the perspective of local certification. In particular, they show that graphs with sub-exponential growth admit a proof labeling scheme of size $\mathcal{O}(\log n)$ for problem ELECTION-PREDICTION. Additionally, they deduce upper bounds for graphs with bounded degree, where certificate sizes are sub-linear in n. Furthermore, they explore lower bounds for the unbounded degree case, establishing that the local certification of ELECTION-PREDICTION on arbitrary n-node graphs require certificates of at least $\Omega(n)$ bits. Interestingly, the authors show that the upper bounds are tight, even for graphs with constant growth.

In terms of the techniques used in this paper, the authors present an interesting approach for deriving an upper bound based on the analysis of the maximum number of time steps in which an individual may change their opinion during the majority dynamics. In particular, they show for different families of graphs (one of them being the graphs with sub-exponential growth) that this number is bounded when the dynamics are observed at every two time steps. In addition, the lower bounds are deduced via a reduction to the disjointedness problem in non-deterministic communication complexity.

Local Certification. Since the introduction of PLSs [29], various variants have been introduced. As we mentioned, a stronger form of PLS are locally checkable proofs [26], where each node can send not only its certificates but also its state and look around within a given radius. Other stronger forms of local certifications are d-PLS [14], where nodes perform communication at a distance of $d \geq 1$ before deciding. Authors have studied many other variants of PLSs, such as randomized PLSs [20], quantum PLSs [18], interactive protocols [9,28,33], zero-knowledge distributed certification [6], and certain PLSs that use global certificates in addition to the local ones [16], among others. On the other hand, some trade-offs between the size of the certificates and the number of rounds of the verification protocol have been exhibited [14]. Also, several hierarchies of certification mechanisms have been introduced, including games between a prover and a disprover [4,13].

PLSs have been shown to be effective for recognizing many graph classes. For example, there are compact PLSs (i.e., with logarithmic size certificates) for the recognition of acyclic graphs [29], planar graphs [15], graphs with bounded genus [11], and H-minor-free graphs, provided that H contains no more than four vertices [7]. In a recent breakthrough, Bousquet et al. [12] proved a 'meta-theorem', stating that there exists a PLS for deciding any monadic second-order logic property with $O(\log n)$-bit certificates on graphs of bounded *tree-depth*. This result has been extended by Fraigniaud et al [19] to the larger class of graphs with bounded *tree-width*, using certificates on $O(\log^2 n)$ bits.

2   Preliminaries

In this article, we denote by $[m, n]$ the set of integers greater than or equal to m and less than or equal to n. We also denote $[n]$ as the interval $[1, n]$.

Let $G = (V, E)$ be a graph. We denote by $N_G(v)$ the *neighborhood of v in G*, defined by $N_G(v) = \{u \in V : \{u, v\} \in E\}$. The *degree* of v, denoted $d_G(v)$, is the cardinality of $N_G(v)$. The *maximum degree* of G, denoted Δ_G, is the maximum value of $d_G(v)$ taken over all $v \in V$. We denote by $N_G[v]$ the set $N_G(v) \cup \{v\}$ and call it the *close neighborhood* of v. We say that two nodes $u, v \in V$ are *connected* if there exists a path in G joining them. In the following, we only consider connected graphs. The *distance* between u, v, denoted $d_G(u, v)$, is the minimum length (number of edges) of a path connecting them. For a graph G, $v \in V(G)$, and $p \geq 0$, we denote $B_v(p)$ as the set of all edges where one of the endpoints is a node at distance at most p from v in G. We denote by $b_v(p)$ the cardinality of $B_v(p)$, and denote by $b_{\max}(p)$ the maximum $b_v(p)$ over $v \in V(G)$. In the following, we omit the sub-indices when they are obvious by the context.

Finite State Dynamics. Let $G = (V, E)$ be a graph, Q a finite set. A *finite state dynamic system over G* is a function $F : Q^V \to Q^V$ such that, for each $v \in V$ there exists a function $f_v : Q^{N[v]} \to Q$ satisfying $F(x)_v = f_v(x|_{N[v]})$. The functions $\{f_v\}_{v \in V}$ are called the *local functions* of F and the elements of Q are the *states* of F. The elements of Q^V are called *configurations* of G.

We consider the model where F is distributed over the network in a way that each node v receives its local function $f_v : Q^{N[v]} \to Q$ as input. A vertex v identifies $d(v)$ ports that enumerate its incident edges, where $d(v)$ is the size of the neighborhood of v. Each input variable of f_v is identified with one of the ports, except one that is identified with v. Therefore, f_v can be encoded in at most $|Q|^{d(v)+1} \log |Q| + (d(v) + 1) \log(d(v) + 1)$ bits.

In some cases we are interested in the local functions that can be encoded with much fewer bits. For instance, consider the dynamics with states $Q = \{-1, 1\}$ and where each node takes the majority state over its neighbors, which can be encoded with a constant number of bits per node. We say that $F : Q^V \to Q^V$ is *succinct* if, for each $v \in V$ we have that f_v can be encoded with $\mathcal{O}(d(v) \log(d(v)))$ bits.

Local Decision. Let $G = (V, E)$ be a simple connected n-node graph. A *distributed language* $\mathcal{L}$ is a Turing-decidable collection of tuples $(G, \mathsf{id}, \mathsf{In})$, called *network configurations*, where $\mathsf{In} : V \to \{0, 1\}^*$ is called an *input function* and $\mathsf{id} : V \to [n^c]$ is an injective function that assigns to each vertex a unique identifier in $[n^c]$ with $c > 1$. In this article, all our distributed languages are independent of the id assignments. In other words, if $(G, \mathsf{id}, \mathsf{In}) \in \mathcal{L}$ for some id, then $(G, \mathsf{id}', \mathsf{In}) \in \mathcal{L}$ for every other id'.

Given $d > 0$, a *local decision algorithm* for a distributed language $\mathcal{L}$ is an algorithm on instance $(G, \mathsf{id}, \mathsf{In})$, where each node v in $V(G)$ receives the sub-graph induced by all nodes within a distance of at most d from v, including their identifiers and inputs. The integer $d > 0$ depends only on the algorithm, not on the size of the input. Each node performs unbounded computation on the information received, and decides whether to accept or reject, with the following requirements:

– When $(G, \mathsf{id}, \mathsf{In}) \in \mathcal{L}$, then every node accepts.
– When $(G, \mathsf{id}, \mathsf{In}) \notin \mathcal{L}$, there is at least one vertex that rejects.

Distributed Languages for Finite-State Dynamics. Consider a graph G, a finite-state dynamic (FSD) F over G, and a configuration x. The *orbit* of x is the sequence of configurations $\{x^t\}_{t>0}$ such that $x^0 = x$ and for every $t > 0$, $x^t = F(x^{t-1})$. We say that the dynamics of x *converge* in at most $k \geq 0$ time-steps if x^k is a fixed point, i.e., $x^k = F(x^k)$. We denote by CONVERGENCE(k, q) the set of pairs (G, F) satisfying that every configuration x converges in at most k time-steps. Formally,

$$
\text{CONVERGENCE}(k, q) = \left\{ (G, F) : \begin{array}{l} F : Q^{V(G)} \to Q^{V(G)} \text{ is a FSD over } G, \\ |Q| \leq q, \text{ and} \\ F(x^k) = x^k \text{ for every } x \in Q^{V(G)} \end{array} \right\}
$$

It is easy to see that there are no local decision algorithms for problem CONVERGENCE(k, q). That is, there are no algorithms in which every node of a network exchanges information solely with nodes in its vicinity and decides whether the dynamics converge within a limited number of time-steps.

2.1 Communication Complexity

Given a Boolean function $f : X \times Y \to \{0, 1\}$, where Alice is given an input $x \in X$ and Bob is given an input $y \in Y$, the deterministic communication complexity of f is the minimum number of bits Alice and Bob need to exchange to compute $f(x, y)$, over all possible deterministic communication protocols. In the non-deterministic version of communication complexity, a third party, called the prover, is allowed to send a message (called a certificate) to one or both of the communicating parties to assist in computing the function. The challenge is to determine the minimal size of such a hint that would enable the parties to compute the function with the least amount of communication between them. In this article, we prove our lower-bounds by reducing CONVERGENCE to a problem DISJOINTNESS in communication complexity. This problem corresponds to the function DISJOINTNESS$_n : 2^{[n]} \times 2^{[n]} \to \{0, 1\}$ such that

$$\text{DISJOINTNESS}_n(A, B) = 1 \text{ if and only if } A \cap B = \emptyset.$$

The following result is given in [30].

Proposition 1. $N^{cc}(\text{DISJOINTNESS}_n) = n$.

3   Finite-State Dynamics on Arbitrary Graphs

In this section, we tackle problem CONVERGENCE on arbitrary graphs. We begin by giving an upper bound on the certification size. For $q > 0$, we denote by $|f_{\max}(n, q)|$ the maximum number of bits required to encode a local function of finite-state dynamics over an n-node graph on q states.

Theorem 1. *For each $q > 1$, CONVERGENCE(k,q) admits a Proof-Labeling Scheme (PLS) with certificates of size*

$$b_{\max}(k+1) \cdot \log(id_{\max}) \cdot |f_{\max}(n,q)|,$$

on graphs with identifiers in $[id_{\max}]$.

Proof. Let G, F be an instance of CONVERGENCE(k,q). The certification algorithm provides each node $v \in V(G)$ with the following information:

- The set $B_v^v(k+1)$ representing $B_v(k+1)$.
- The set of all local functions f_w^v of nodes w that are endpoints of edges in $B_v^v(k+1)$.

Observe that the certificates can be encoded in at most $b_{\max}(k+1)\cdot\log(id_{\max})\cdot |f_{\max}(n,q)|$ bits. Given the certificates, v can reconstruct all the neighborhoods and local functions of nodes at distance at most $k+1$ from v. In particular, v can determine all the nodes up to distance $k+2$ and all the local functions of nodes up to distance $k+1$.

In the verification algorithm, each node v checks the consistency of the information provided to its neighbors and verifies that v converges to a fixed-point in k time-steps for every configuration assigned to nodes with endpoints in $B_v(k)$. Formally, v checks the following conditions for each node $u \in N(v)$:

a. All the edges in $B_u^u(k)$ belong to $B_v^v(k+1)$.
b. If w is an endpoint of an edge in $B_u(k+1) \cap B_v(k+1)$ then $f_w^v = f_w^u$. In particular, $f_v^v = f_v^u$.
c. For every configuration of the nodes with endpoints in $B_v^v(k+1)$, v simulates $k+1$ time-steps and checks that the configuration reached by v after k time-steps is a fixed point.

Node v accepts only if all conditions are satisfied. Let us analyze now the soundness and completeness.

Completeness: Let us suppose that (G, F) is a *yes-instance*. Clearly, if every node v receives the certificates as they are defined (i.e., $B_v^v(k+1) = B_v(k+1)$, $f_w^v = f_w$ for every w that is an endpoint of an edge in $B_v(k+1)$), then every node accepts.

Soundness: Let us suppose that (G, F) is a *No-instance*. In other words, there exists a configuration x of G for which $F(x^k) \neq x^k$. Let v be a node such that $F(x^k)_v \neq x_v^k$. Assuming that every node accepts conditions a. and b., and since these conditions are satisfied for every node, we have that $B_v^v(k+1) = B_v(k+1)$ and $f_w^v = f_w$ for every w that is an endpoint of an edge in $B_v(k+1)$. Observe that x_v^{k+1} only depends on the initial configuration of nodes at distance at most $k+2$ from v, and the local functions of nodes at distance at most $k+1$ from v. Therefore, v rejects the condition c of the verification algorithm. □

Notice that when $b_{max}(k+1) = \mathcal{O}(n^2)$ (for instance, on dense graphs of small diameter), the upper bound above is $\mathcal{O}(n^2 \log(\text{id}_{max})|f_{max}|)$, which is greater than the trivial upper bound $\mathcal{O}(n^2 \log(\text{id}_{max}) + n|f_{max}|)$ that involves providing each node with all the edges of the graph and all the local functions. In the next theorem, we show that, up to logarithmic factors, there is also a quadratic lower bound for the convergence in at most two time-steps, and specifically for dynamics on four states that admit a succinct representation.

Theorem 2. *Every Proof Labeling Scheme for* CONVERGENCE$(2, 4)$ *has certificates of size* $\Omega(n^2/(\log n))$. *This holds even when the problem is restricted to local functions that admit a succinct representation.*

Proof. We reduce CONVERGENCE$(2, 4)$ to DISJOINTNESS in two-player communication complexity. Given an instance of DISJOINTNESS, we build an instance of problem CONVERGENCE$(2, 4)$, which is composed of a *lower-bound graph* and a dynamic picked from a family of *lower-bound dynamics*.

We begin by giving a high-level description of the construction. Let n be a positive integer, and let $\mathbf{A}, \mathbf{B} \subseteq \binom{[n]}{2}$ be a pair of sets, interpreted as an instance of DISJOINTNESS. The set $\mathbf{A}$ is associated with the graph $G_{\mathbf{A}}$ with the vertex set $\{v_1, \ldots, v_n\}$ such that, for each $i, j \in [n]$, node v_i is adjacent to v_j if and only if $\{i, j\} \in \mathbf{A}$. The graph $G_{\mathbf{B}}$ is defined analogously using the set $\mathbf{B}$ instead of $\mathbf{A}$.

The lower-bound graph has vertices containing disjoint copies of $G_{\mathbf{A}}$ and $G_{\mathbf{B}}$, as well as a *bit gadget*, which is connected to every other vertex. A configuration satisfies an *admissibility condition* when the state of the bit gadget encodes the binary representation of an element of $[n] \times [n]$, and this element represents an edge that is both in $G_{\mathbf{A}}$ and $G_{\mathbf{B}}$ (i.e., when $\mathbf{A} \cap \mathbf{B} \neq \emptyset$). The dynamic of the system is designed to oscillate in a limit-cycle of period two only when the admissibility condition is satisfied. When the admissibility condition is not satisfied, the dynamic reaches a fixed point in at most two time-steps.

Lower-bound graph. For each $n > 0$, the *lower-bound graph* $G(\mathbf{A}, \mathbf{B})$ is a graph of size $2n + 4\lceil \log n \rceil$. The vertex set of $G(\mathbf{A}, \mathbf{B})$ is partitioned into four subsets, namely, $V_A = \{v_1^A, \ldots, v_n^A\}$, $V_B = \{v_1^B, \ldots, v_n^B\}$, $D_A = \{d_1^A, \ldots, d_\ell^A\}$ and $D_B = \{d_1^B, \ldots, d_\ell^B\}$, where $\ell = 2\lceil \log n \rceil$. The edge set of $G_{\mathbf{A}, \mathbf{B}} = (V, E)$ contains all the edges with one endpoint in V_A and the other in D_A, all edges between nodes in V_B and D_B, and all edges with endpoints in $D_A \cup D_B$. It also contains, for each $i, j \in [n]$ the edge $\{v_i^A, v_j^A\}$ (respectively, $\{v_i^B, v_j^B\}$) if node v_i is adjacent to v_j in $G_{\mathbf{A}}$ (respectively, $G_{\mathbf{B}}$).

Nodes in $D_A \cup D_B$ are called the *bit gadget* and the nodes of $V_A \cup V_B$ are the *set gadget*.

States of the system. For all nodes the set of states is $Q = \{0, 1\} \times \{0, 1\}$. Given a node u in state $s(u) = (m(u), c(u)) \in Q$ we say that $m(u)$ is the *mark* of u and $c(u)$ is the *clock* of u.

Given the set of states Q, we define now lower-bound dynamics given by lower-bound local functions, that we call $F(\mathbf{A}, \mathbf{B})$. The local functions of nodes in $D_A \cup D_B$ will be independent on the input of the players, while the ones nodes of V_A (respectively V_B) depend only on **A** (respectively **B**). We describe first the local functions of nodes in $D_A \cup D_B$.

Local functions of nodes of the bit gadget. First, we define the following *admissibility conditions* for a given configuration, which is checked for every node in D_A (respectively D_B):

d1) Every node in $D_A \cup D_B$ has the same clock.
d2) There are exactly two nodes in V_A (resp. V_B) with mark 1.
d3) For each $s \in \{1,\ldots,\ell\}$, the mark of d_s^A equals the one of d_s^B.

The dynamic of the nodes u in D_A and D_B is then defined as follows: if the admissibility conditions are satisfied, then the mark of u remains unchanged, and the clock of u switches to 0 if the clock was 1 and vice-versa. If the admissibility condition is not satisfied, the whole state of u (mark and clock) remains unchanged. In any case, the marks in the nodes of the bit gadget do not change under any circumstances.

For each $s \in [\ell]$, the local function of d_s^A (respectively d_s^B) is succinct. Indeed, to define the function, we simply need to indicate which neighbors of d_s^A (resp. d_s^B) belong to D_A and which to V_A (resp. D_B and V_B), and which node is d_s^B (resp. d_s^A). This can be encoded using $\mathcal{O}(1)$ bit per neighbor.

Observe that while the admissibility condition is verified, the clocks in the nodes of $D_A \cup D_B$ continuously switch between 0 and 1. Suppose that in the configuration on a given time-step t, there are nodes $d_1, d_2 \in D_A \cup D_B$ such that the admissibility condition is satisfied for d_1 but not for d_2. Then, in time-step t, we have that $c(d_1) = c(d_2)$, but in $t+1$, $c(d_1) \neq c(d_2)$. This implies that in time-step $t+1$, the admissibility condition (d1) is not satisfied by any node in $D_A \cup D_B$, which results in every node in that set being fixed in its state forever.

Local functions of nodes of the set gadgets. Let $i \in [n]$. The following admissibility conditions are considered for node v_i^A:

v1) The mark of v_i^A is 1.
v2) Exactly one neighbor of v_i^A in V_A has mark 1.
v3) The mark of $d_1^A,\ldots,d_{\ell/2}^A$ or $d_{\ell/2+1}^A,\ldots,d_\ell^A$ represents the binary representation of i.

The local function of v_i^A indicates that the mark of node v_i^A is 1 if the admissibility conditions are satisfied, and 0 otherwise. The clock of v_i^A does not change under any circumstance.

The local function of v_i^A is succinct for every $i \in [n]$. To define the function, it is necessary to simply indicate which neighbors of v_i^A belong to D_A and which to V_A. This can be encoded using $\mathcal{O}(1)$ bit per neighbor. Additionally, specifying the index i requires $\mathcal{O}(\log n)$ bits. Analogous admissibility conditions and local functions are defined for node v_i^B by switching the subindices A by B.

Independently of the initial configuration, the admissibility condition is satisfied for at most two nodes in V_A, as the marks in D_A can represent at most two indices in $[n]$. Then, in the first time-step, the mark of every node in V_A is 0, except for at most two nodes. The mark of the remaining nodes, namely v_i^A and v_j^A, is 1 only if the admissibility condition is satisfied for two nodes. In particular, condition (v2) implies that v_i^A is adjacent to v_j^A. Hence $\{i, j\} \in \mathbf{A}$. In any case, the state of every node of $V_A \cup V_B$ remains unchanged after the first time-step.

The Reduction. We now show that $(G(\mathbf{A}, \mathbf{B}), F(\mathbf{A}, \mathbf{B}))$ converges in at most two time-steps if and only if $\mathbf{A} \cap \mathbf{B} = \emptyset$. Let us suppose that $\mathbf{A} \cap \mathbf{B} \neq \emptyset$, and let $\{i, j\} \in \mathbf{A} \cap \mathbf{B}$. We define the following initial configuration x.

- For every $s \in [n] \setminus \{i, j\}$, nodes v_s^A and v_s^B have initial configuration $(0, 0)$.
- The initial configuration of v_i^A, v_j^A, v_i^B, and v_j^B is $(1, 0)$.
- For each $s \in [\ell/2]$, the initial configuration of d_s^A and d_s^B is $(b, 0)$, where b is the s-th bit in the binary representation of i.
- For each $s \in [\ell/2 + 1, \ell]$, the initial configuration of d_s^A and d_s^B is $(b, 0)$, where b is the s-th bit in the binary representation of j.

Observe that in x all the admissibility conditions are satisfied. Moreover, in the next time-steps, the mark of every node in $G(\mathbf{A}, \mathbf{B})$ remains unchanged, and the clocks of every node in $D_A \cup D_B$ switch between 0 and 1 back and forth. Therefore, the dynamic $(G(\mathbf{A}, \mathbf{B}), F(\mathbf{A}, \mathbf{B}))$ does not converge.

Now let us suppose that $\mathbf{A} \cap \mathbf{B} = \emptyset$. Let x be any initial configuration, and let y be the configuration obtained in the first time-step. Observe that the state of every node in $V_A \cup V_B$ is fixed in the state of y in the next time-steps. If fewer than two nodes in V_A (respectively V_B) have mark 1 in y, all nodes in D_A (respectively D_B) are fixed in their state forever. Then, in the third time-step, all nodes in D_B (respectively D_A) are also fixed. Suppose then that in y, exactly two nodes $v_i^A, v_j^A \in V_A$ and two nodes $v_p^B, v_q^B \in V_B$ have mark 1. By condition (v3), in y the marks of the nodes in D_A are the binary representation of i and j. By condition (v2), v_i^A and v_j^A are adjacent. Hence $\{i, j\} \in \mathbf{A}$. Similarly, the marks on D_B are the binary representations of p and q, and v_p^B is adjacent to v_q^B. Hence $\{p, q\} \in \mathbf{B}$. Since $\mathbf{A} \cap \mathbf{B} = \emptyset$, condition (d3) is not satisfied in y, implying that y is a fixed point.

We deduce that $(G, F(\mathbf{A}, \mathbf{B})) \in \text{CONVERGENCE}(2, 4)$ if and only if $\mathbf{A} \cap \mathbf{B} = \emptyset$. Let π be a PLS for CONVERGENCE$(2, 4)$. We define the following two-player protocol $\mathcal{P}$ for DISJOINTNESS. On instance $\mathbf{A}$, Alice computes nondeterministically the certificates that π would give on all nodes in $V_A \cup D_A \cup D_B$, and simulates the verification protocol of π on all nodes in $V_A \cup D_A$. Alice communicates a single bit to Bob indicating if every node in $V_A \cup D_A$ has accepted, as well as all the certificates of $D_A \cup D_B$. Analogously, Bob computes the certificates of $D_A \cup D_B \cup V_B$ and simulates the verification protocol of π on all nodes in $V_B \cup D_B$. Bob accepts if all nodes in V_B accept and his certificates for $D_A \cup D_B$ are the same as those generated by Alice. The correctness of $\mathcal{P}$ follows directly from the soundness and completeness of π. Let $C(n)$ be the maximum size of a certificate produced by π on graphs of size n. According to Proposition 1, it follows that $C(2n + 2\lceil \log n \rceil) \cdot \lceil \log n \rceil = \Omega(n^2)$, implying that $C(n) = \Omega(n^2 / \log n)$. □

4   Finite-State Dynamics in Graphs of Bounded Degree

Given that the problem is hard in general graphs, we focus our study on finite-state dynamics over graphs of bounded degree. The full proofs of this section are given in the full version of this article [32].

The local functions defined on graphs of bounded degree are all succinct. Moreover, an analysis of the bound given by Theorem 1 provides a non-trivial upper-bound for the certificate size.

Corollary 1. *For each $q > 1$ and for each $\Delta > 2$, problem* CONVERGENCE(k, q) *admits a Proof-Labeling Scheme with certificates of size at most*

$$\Delta((\Delta - 1)^{k+1} - 1) \cdot q^{\Delta+1} \log(q) \cdot \log(id_{\max})$$

on graphs of maximum degree Δ with assignments of identifiers in $[id_{\max}]$.

In the following result, we state that the exponential dependency on k is necessary even on graphs of bounded degree. Our proof follows similar ideas to the proof of Theorem 2. However, we need to introduce a series of gadgets that allows us to control the maximum degree of the lower-bound graph. One of our gadgets simulates a *binary decoder*. A Boolean circuit C is called a *binary decoder* if there exists a positive integer t such that C has t inputs and 2^t outputs, named $\{v_1, \ldots, v_{2^t}\}$. This Boolean circuit satisfies that for each $i \in [2^t]$, the output value of v_i is True if and only if the truth values of the inputs (mapped to 0/1 values) correspond to the binary representation of i. In the full version of this article [32], we provide a construction of a binary decoder with small depth, as well as a finite-state dynamic that simulates its behavior.

Theorem 3. *Let $k \geq 2$. Every PLS for* CONVERGENCE$(k, 3)$ *requires certificates of size at least $2^{k/6} \cdot 6/k$, even when the problem is restricted to input graphs of degree at most 3.*

5   Discussion

In this paper, we study local certification of the problem CONVERGENCE(t, q) which asks whether a finite-state dynamics with q states converges in at most t time steps. We show that in general there is an upper bound of $b_{\max} \cdot \log(id_{\max}) \cdot |f_{\max}|$, and we show as a lower bound that certificates of size $\Omega(n^2/\log n)$ are required even for $t = 2$ and $q = 4$. In both cases, the size of the representation of the function F plays an important role. We present the following open questions:

First, which specific families of functions might exhibit a succinct representation? An interesting case study might be studying if the lower bound construction is still valid in the context of set-valued functions, such as neural networks, which depend only on the subset of states in the neighborhood.

Second, how does the difficulty of solving CONVERGENCE compare to other decision problems commonly studied for finite-state dynamics, such as the *reachability problem* (determining whether a configuration y is reachable from a starting configuration x under the dynamics) and the *prediction problem* (determining if the state of a node will change after T time steps, given a node, time T, and an initial configuration)?

Finally, what is the certification cost for other global properties of F, such as reversibility, injectivity, or nilpotency (where a function is nilpotent if there is only one possible fixed point)? How are these problems related to problem CONVERGENCE and at which point are they comparable? Do they require certificates of bigger size compared to CONVERGENCE?

References

1. Afek, Y., Alon, N., Bar-Joseph, Z., Cornejo, A., Haeupler, B., Kuhn, F.: Beeping a maximal independent set. Distrib. Comput. **26**(4), 195–208 (2013)
2. Afek, Y., Alon, N., Barad, O., Hornstein, E., Barkai, N., Bar-Joseph, Z.: A biological solution to a fundamental distributed computing problem. Science **331**(6014), 183–185 (2011)
3. Atlan, H.: Self-organizing networks: weak, strong and intentional, the role of their underdetermination. In: Carsetti, A. (ed.) Functional Models of Cognition. Theory and Decision Library, vol. 27, pp. 127–142. Springer, Dordrecht (1999). https://doi.org/10.1007/978-94-015-9620-6_9
4. Balliu, A., D'Angelo, G., Fraigniaud, P., Olivetti, D.: What can be verified locally? J. Comput. Syst. Sci. **97**, 106–120 (2018)
5. Barrett, C.L., Hunt, H.B., III., Marathe, M.V., Ravi, S., Rosenkrantz, D.J., Stearns, R.E.: Complexity of reachability problems for finite discrete dynamical systems. J. Comput. Syst. Sci. **72**(8), 1317–1345 (2006)
6. Bick, A., Kol, G., Oshman, R.: Distributed zero-knowledge proofs over networks. In: 33rd ACM-SIAM Symposium on Discrete Algorithms (SODA), pp. 2426–2458 (2022)
7. Bousquet, N., Feuilloley, L., Pierron, T.: Local certification of graph decompositions and applications to minor-free classes. In: 25th International Conference on Principles of Distributed Systems (OPODIS). LIPIcs, vol. 217, pp. 22:1–22:17. Schloss Dagstuhl - Leibniz-Zentrum für Informatik (2021)
8. Cornejo, A., Kuhn, F.: Deploying wireless networks with beeps. In: Lynch, N.A., Shvartsman, A.A. (eds.) Distributed Computing, pp. 148–162. Springer, Heidelberg (2010). https://doi.org/10.1007/978-3-642-15763-9_15
9. Crescenzi, P., Fraigniaud, P., Paz, A.: Trade-offs in distributed interactive proofs. In: 33rd International Symposium on Distributed Computing (DISC). LIPIcs, vol. 146, pp. 13:1–13:17. Schloss Dagstuhl - Leibniz-Zentrum für Informatik (2019)
10. Emek, Y., Wattenhofer, R.: Stone age distributed computing. In: Proceedings of the 2013 ACM Symposium on Principles of Distributed Computing, pp. 137–146 (2013)
11. Esperet, L., Lévêque, B.: Local certification of graphs on surfaces. Theor. Comput. Sci. **909**, 68–75 (2022)
12. Feuilloley, L., Bousquet, N., Pierron, T.: What can be certified compactly? compact local certification of mso properties in tree-like graphs. In: Proceedings of the 2022 ACM Symposium on Principles of Distributed Computing, pp. 131–140 (2022)
13. Feuilloley, L., Fraigniaud, P., Hirvonen, J.: A hierarchy of local decision. Theor. Comput. Sci. **856**, 51–67 (2021)
14. Feuilloley, L., Fraigniaud, P., Hirvonen, J., Paz, A., Perry, M.: Redundancy in distributed proofs. Distrib. Comput. **34**(2), 113–132 (2021)

15. Feuilloley, L., Fraigniaud, P., Montealegre, P., Rapaport, I., Rémila, É., Todinca, I.: Compact distributed certification of planar graphs. In: Algorithmica, pp. 1–30 (2021)
16. Feuilloley, L., Hirvonen, J.: Local verification of global proofs. In: 32nd International Symposium on Distributed Computing. LIPIcs, vol. 121, pp. 25:1–25:17. Schloss Dagstuhl - Leibniz-Zentrum für Informatik (2018)
17. Flury, R., Wattenhofer, R.: Slotted programming for sensor networks. In: Proceedings of the 9th ACM/IEEE International Conference on Information Processing in Sensor Networks, pp. 24–34 (2010)
18. Fraigniaud, P., Gall, F.L., Nishimura, H., Paz, A.: Distributed quantum proofs for replicated data. In: 12th Innovations in Theoretical Computer Science Conference (ITCS). LIPIcs, vol. 185, pp. 28:1–28:20. Schloss Dagstuhl - Leibniz-Zentrum für Informatik (2021)
19. Fraigniaud, P., Montealegre, P., Rapaport, I., Todinca, I. (2022). A meta-theorem for distributed certification. In: Parter, M. (ed.) Structural Information and Communication Complexity. SIROCCO 2022. LNCS, vol. 13298, pp. 116–134. Springer, Cham (2022). https://doi.org/10.1007/978-3-031-09993-9_7
20. Fraigniaud, P., Patt-Shamir, B., Perry, M.: Randomized proof-labeling schemes. Distrib. Comput. **32**(3), 217–234 (2019)
21. D Frischknecht, S., Keller, B., Wattenhofer, R.: Convergence in (social) influence networks. In: Proceedings of the Distributed Computing: 27th International Symposium, DISC 2013, Jerusalem, 14–18 October 2013, vol. 27. pp. 433–446. Springer (2013)
22. Gadouleau, M.: On the stability and instability of finite dynamical systems with prescribed interaction graphs. Electron. J. Combinator. P3–32 (2019)
23. Gadouleau, M.: On the influence of the interaction graph on a finite dynamical system. Nat. Comput. **19**(1), 15–28 (2020)
24. Gadouleau, M., Richard, A.: Simple dynamics on graphs. Theoret. Comput. Sci. **628**, 62–77 (2016)
25. Goles, E., Montealegre, P.: Computational complexity of threshold automata networks under different updating schemes. Theoret. Comput. Sci. **559**, 3–19 (2014)
26. Göös, M., Suomela, J.: Locally checkable proofs in distributed computing. Theory Comput. **12**(1), 1–33 (2016)
27. Guseo, R., Guidolin, M.: Modelling a dynamic market potential: a class of automata networks for diffusion of innovations. Technol. Forecast. Soc. Change **76**(6), 806–820 (2009). https://doi.org/10.1016/j.techfore.2008.10.005
28. Kol, G., Oshman, R., Saxena, R.R.: Interactive distributed proofs. In: ACM Symposium on Principles of Distributed Computing, pp. 255–264. ACM (2018)
29. Korman, A., Kutten, S., Peleg, D.: Proof labeling schemes. Distrib. Comput. **22**(4), 215–233 (2010)
30. Kushilevitz, E.: Communication complexity. In: Advances in Computers, vol. 44. Elsevier (1997)
31. Maldonado, D., Montealegre, P., Ríos-Wilson, M., Theyssier, G.: Local certification of majority dynamics. arXiv preprint arXiv:2309.01852 (2023)
32. Maldonado, D., Montealegre, P., Ríos-Wilson, M.: The hardness of local certification of finite-state dynamics (2023)
33. Naor, M., Parter, M., Yogev, E.: The power of distributed verifiers in interactive proofs. In: 31st ACM-SIAM Symposium on Discrete Algorithms (SODA), pp. 1096–115. SIAM (2020)
34. Ríos-Wilson, M., Theyssier, G.: Intrinsic Simulations and Universality in Automata Networks (2022). https://hal.science/hal-03779748

Iterated Straight-Line Programs

Gonzalo Navarro[1,2]([✉]) and Cristian Urbina[1,2]([✉])

[1] CeBiB – Centre for Biotechnology and Bioengineering, Santiago, Chile
[2] Department of Computer Science, University of Chile, Santiago, Chile
{gnavarro,crurbina}@dcc.uchile.cl

Abstract. We explore an extension to straight-line programs (SLPs) that outperforms, for some text families, the measure δ based on substring complexity, a lower bound for most measures and compressors exploiting repetitiveness (which are crucial in areas like Bioinformatics). The extension, called iterated SLPs (ISLPs), allows rules of the form $A \rightarrow \Pi_{i=k_1}^{k_2} B_1^{i^{c_1}} \cdots B_t^{i^{c_t}}$, for which we show how to extract any substring of length λ, from the represented text $T[1..n]$, in time $O(\lambda + \log^2 n \log \log n)$. This is the first compressed representation for repetitive texts breaking δ while, at the same time, supporting direct access to arbitrary text symbols in polylogarithmic time. As a byproduct, we extend Ganardi et al.'s technique to balance any SLP (so it has a derivation tree of logarithmic height) to a wide generalization of SLPs, including ISLPs.

Keywords: Grammar compression · Substring complexity · Repetitiveness measures

1 Introduction

Motivated by the data deluge, and by the observed phenomenon that many of the fastest-growing text collections are highly repetitive, recent years have witnessed an increasing interest in (1) defining measures of compressibility that are useful for highly repetitive texts, (2) develop compressed text representations whose size can be bounded in terms of those measures, and (3) provide efficient (i.e., polylogarithmic time) access methods to those compressed texts, so that algorithms can be run on them without ever decompressing the texts [23,24]. We call *lower-bounding measures* those satisfying (1), *reachable measures* those (asymptotically) reached by the size of a compressed representation (2), and *accessible measures* those reached by the size of representations satisfying (3).

For example, the size γ of the smallest "string attractor" of a text T is a lower-bounding measure, unknown to be reachable [14], and smaller than the size reached by known compressors. The size b of the smallest "bidirectional macro scheme" of T [31], and the size z of the Lempel-Ziv parse of T [20], are reachable measures. The size g of the smallest context-free grammar generating (only) T [5] is an accessible measure [3]. It holds $\gamma \leq b \leq z \leq g$ for every text.

Funded with Basal Funds FB0001, ANID, Chile; and ANID-Subdirección de Capital Humano/Doctorado Nacional/2021-21210580.

J. A. Soto and A. Wiese (Eds.): LATIN 2024, LNCS 14578, pp. 66–80, 2024.
https://doi.org/10.1007/978-3-031-55598-5_5

One of the most attractive lower-bounding measures devised so far is δ [6,29]. Let $T[1..n]$ be a text over alphabet $[1..\sigma]$, and T_k be the number of distinct substrings of length k in T, which define its so-called substring complexity. Then the measure is $\delta(T) = \max_k T_k/k$. This measure has several attractive properties: it can be computed in linear time and lower-bounds all previous measures of compressibility, including γ, for every text. While δ is known to be unreachable, the measure $\delta' = \delta \log \frac{n \log \sigma}{\delta \log n}$ has all the desired properties: $\Omega(\delta')$ is the space needed to represent some text family for each n, σ, and δ; within $O(\delta')$ space it is possible to represent every text T and access any length-λ substring of T in time $O(\lambda + \log n)$ [18], together with more powerful operations [13,17,18].

As for g, a *straight-line program (SLP)* is a context-free grammar that generates (only) T, and has size-2 rules of the form $A \to BC$, where B and C are nonterminals, and size-1 rules $A \to a$, where a is a terminal symbol. The SLP size is the sum of all its rule sizes. A *run-length SLP (RLSLP)* may contain, in addition, size-2 rules of the form $A \to B^k$, representing k repetitions of nonterminal B. A RLSLP of size g_{rl} can be represented in $O(g_{rl})$ space, and within that space we can offer fast string access and other operations [6, Appendix A]. It holds $\delta \le g_{rl} = O(\delta')$, where g_{rl} is the smallest RLSLP that generates T [18,23] (the size g of the smallest grammar or SLP, instead, is not always $O(\delta')$).

While δ lower-bounds all previous measures on every text, δ' is not the smallest accessible measure. In particular, g_{rl} is always $O(\delta')$, and it can be smaller by up to a logarithmic factor. Indeed, g_{rl} is a minimal accessible measure as far as we know. It is asymptotically between z and g [23]. An incomparable accessible measure is $z_{end} \ge z$, the size of the LZ-End parse of the text [15,19].

The belief that δ is a lower bound to every reachable measure was disproved by the recently proposed L-systems [26,27]. L-systems are like SLPs where all the symbols are nonterminals and the derivation ends at a specified depth in the derivation tree. The size ℓ of the smallest L-system generating $T[1..n]$ is a reachable measure of repetitiveness and was shown to be as small as $O(\delta/\sqrt{n})$ on some text families, thereby sharply breaking δ as a lower bound. Measure ℓ, however, is unknown to be accessible, and thus one may wonder whether there exist accessible text representations that are smaller than δ.

In this paper we devise such a representation, which we call *iterated SLP (ISLP)*. ISLPs extend SLPs (and RLSLPs) by allowing a more complex version of the rule $A \to B^k$, namely $A \to \Pi_{i=k_1}^{k_2} B_1^{i^{c_1}} \cdots B_t^{i^{c_t}}$ of size $2 + 2t$. We show how to extract a substring of length λ from the ISLP of a text T in time $O(\lambda + \log^2 n \log \log n)$ provided the ISLP is balanced, that is, its derivation tree is of height $O(\log n)$.

Just like SLPs and RLSLPs can be balanced [10,28] while retaining their asymptotic size, we show how to balance a more general class of SLP extensions we call generalized SLPs (GSLPs). GSLPs, which include ISLPs, allow rules of the form $A \to x$, where x is a *program* that outputs the right-hand side of the rule. We show that, if every nonterminal appearing in x's output does it at least twice, then the GSLP can be balanced in the same way as SLPs. This byproduct of our results can be of independent interest to provide polylogarithmic-time access to other extensions of context-free grammars.

2   Preliminaries

We explain some concepts and notation used in the rest of the paper.

Strings. Let $\Sigma = [1 .. \sigma]$ be an *alphabet*. A *string* $T[1 .. n]$ of length n is a finite sequence $T[1] T[2] \ldots T[n]$ of n symbols in Σ. We denote by ε the unique string of length 0. We denote by Σ^* the set of all finite strings with symbols in Σ. The i-th symbol of T is denoted by $T[i]$, and the sequence $T[i] \ldots T[j]$ is denoted by $T[i .. j]$. The *concatenation* of $X[1 .. n]$ and $Y[1 .. m]$ is defined as $X \cdot Y = X[1] \ldots X[n] Y[1] \ldots Y[m]$ (we omit the dot when there is no ambiguity). If $T = XYZ$, then X (resp. Y, resp. Z) is a *prefix* (resp. *substring*, resp. *suffix*) of T. A *power* T^k stands for k consecutive concatenations of the string T. We denote by $|T|_a$ the number of occurrences of the symbol a in T. A *string morphism* is a function $\varphi : \Sigma^* \to \Sigma^*$ such that $\varphi(xy) = \varphi(x) \cdot \varphi(y)$ for any strings x and y.

Straight-Line Programs. A *straight-line program* (SLP) is a context-free grammar [30] that contains only terminal rules of the form $A \to \mathtt{a}$ with $\mathtt{a} \in \Sigma$, and binary rules of the form $A \to BC$ for variables B and C whose derivations cannot reach again A. These restrictions ensure that each variable of the SLP generates a unique string, defined as $\exp(A) = \mathtt{a}$ for a rule $A \to \mathtt{a}$, and as $\exp(A) = \exp(B) \cdot \exp(C)$ for a rule $A \to BC$. A *run-length straight-line program* (RLSLP) is an SLP that also admits run-length rules of the form $A \to B^k$ for some $k \geq 3$, with their expansion defined as $\exp(A) = \exp(B)^k$. The *size* of an SLP is the sum of the lengths of the right-hand sides of its rules; the size of an RLSLP is defined similarly, assuming that rules $A \to B^k$ are of size 2 (i.e., two integers to represent B and k).

The *derivation tree* of an SLP is an ordinal tree where the nodes are the variables, the root is the initial variable, and the leaves are the terminal variables. The children of a node are the variables appearing in the right-hand side of its rule (in left-to-right order). The *height* of an SLP is the length of the longest path from the root to a leaf node in the derivation tree. The height of an RLSLP is obtained by *unfolding* its run-length rules, that is, writing a rule B^k as $BB \ldots B$ where B appears k times, to obtain an equivalent SLP (actually, a slight extension where the right-hand sides can feature more than two variables).

SLPs and RLSLPs yield measures of repetitiveness g and g_{rl}, defined as the size of the smallest SLP and RLSLP generating the text, respectively. Clearly, it holds that $g_{rl} \leq g$. It also has been proven that g is NP-hard to compute [5].

Other Repetitiveness Measures. For self-containedness, we describe the most important repetitiveness measures and relate them with the accessible measures g and g_{rl}; for more details see a survey [23].

Burrows-Wheeler Transform. The *Burrows-Wheeler Transform* (BWT) [4] is a reversible permutation of T, which we denote by $\mathtt{bwt}(T)$. It is obtained by sorting lexicographically all the rotations of the string T and concatenating their last symbols, which can be done in $O(n)$ time. The measure r is defined as the size of the *run-length encoding* of $\mathtt{bwt}(T)$. Usually, T is assumed to be appended with a sentinel symbol $\$$ strictly smaller than any other symbol in T, and then we call $r_\$$ the size of the run-length encoding of $\mathtt{bwt}(T\$)$. This measure is then reachable, and fully-functional indexes of size $O(r_\$)$ exist [8], but interestingly, it is unknown to be accessible. While this measure is generally larger than others, it can be upper-bounded by $r_\$ = O(\delta \log \delta \log \frac{n}{\delta})$ [16].

Lempel-Ziv Parsing. The *Lempel-Ziv parsing* (LZ) [20] of a text $T[1..n]$ is a *factorization* into non-empty *phrases* $T = X_1 X_2 \ldots X_z$ where each X_i is either the first occurrence of a symbol or the longest prefix of $X_i \ldots X_z$ with a copy in T starting at a position in $[1..|X_1 \ldots X_{i-1}|]$. LZ is called a *left-to-right* parsing because each phrase has its *source* starting to the left, and it is optimal among all parsings satisfying this condition. It can be constructed greedily from left to right in $O(n)$ time. The measure z is defined as the number of phrases in the LZ parsing of the text, and it has been proved that $z \leq g_{rl}$ [25]. While z is obviously reachable, it is unknown to be accessible. A close variant $z_{end} \geq z$ [19] that forces phrase sources to be end-aligned with a preceding phrase, has been shown to be accessible [15].

Bidirectional Macro Schemes. A *bidirectional macro scheme* (BMS) [31] is a factorization of a text $T[1..n]$ where each phrase can have its source starting either to the left or to the right. The only requirement is that by following the pointers from phrases to sources, we should eventually be able to fully decode the text. The measure b is defined as the size of the smallest BMS representing the text. Clearly, b is reachable, but it is unknown to be accessible. It holds that $b \leq z$, and it was proved that $b \leq r_\$$ [25]. Computing b is NP-hard [9].

String Attractors. A *string attractor* for a text $T[1..n]$ is a set of positions $\Gamma \subseteq [1..n]$ such that any substring of $T[i..j]$ has an occurrence $T[i'..j']$ crossing at least one of the positions in Γ (i.e., there exist $k \in \Gamma$ such that $i' \leq k \leq j'$). The measure γ is defined as the size of the smallest string attractor for the string T, and it is NP-hard to compute [14]. It holds that γ lower bounds the size b of the smallest bidirectional macro scheme and can sometimes be asymptotically smaller [2]. On the other hand, it is unknown if γ is reachable.

Substring Complexity. Let $T[1..n]$ be a text and T_k be the number of distinct substrings of length k in T, which define its so-called substring complexity. Then the measure is $\delta = \max_k T_k/k$ [6,29]. This measure can be computed in $O(n)$ time and lower-bounds γ, and thus all previous measures of compressibility, for every text. On the other hand, it is known to be unreachable [18]. The related measure $\delta' = \delta \log \frac{n \log \sigma}{\delta \log n}$ is reachable and accessible, and still lower-bounds b and all other reachable measures on some text family for every n, σ, and δ [18]. Besides, g_{rl} (and thus z, b, and γ, but not g) are upper-bounded by $O(\delta \log \frac{n \log \sigma}{\delta \log n})$.

L-systems. An *L-system* (for compression) is a tuple $L = (V, \varphi, \tau, S, d, n)$ extending a Lindenmayer system [21,22], where V is the set of variables (which are also considered as terminal symbols), $\varphi : V \to V^+$ is the set of rules (and also a morphism of strings), $\tau : V \to V$ is a coding, $S \in V$ the initial variable, and d and n are integers. The string generated by the system is $\tau(\varphi^d(S))[1 \mathinner{..} n]$. The measure ℓ is defined as the size of the smallest L-system generating the string. It has been proven that ℓ is incomparable to δ (ℓ can be smaller by a $\sqrt{n}$ factor) and almost any other repetitiveness measure considered in the literature [26,27].

3   Iterated Straight-Line Programs

We now define iterated SLPs and show that they can be much smaller than δ. Some proofs in this section are omitted due to space constraints.

Definition 1. *An* iterated straight-line program *of* degree d *(d-ISLP) is an SLP that allows in addition iteration rules of the form* $A \to \prod_{i=k_1}^{k_2} B_1^{i^{c_1}} \cdots B_t^{i^{c_t}}$, *where* $1 \le k_1, k_2$, $0 \le c_1, \ldots, c_t \le d$ *are integers and* $B_1 \ldots B_t$ *are variables that cannot reach A (so the ISLP generates a unique string). Iteration rules have size* $2 + 2t = O(t)$ *and expand to* $\exp(A) = \prod_{i=k_1}^{k_2} \exp(B_1)^{i^{c_1}} \cdots \exp(B_t)^{i^{c_t}}$, *where if* $k_1 > k_2$ *the iteration goes from* $i = k_1$ *downwards to* $i = k_2$. *The size* size(G) *of a d-ISLP G is the sum of the sizes of all of its rules.*

Definition 2. *The measure* $g_{it(d)}(T)$ *is defined as the size of the smallest d-ISLP that generates T, whereas* $g_{it}(T) = \min_{d \ge 0} g_{it(d)}(T)$.

The following observations show that ISLPs subsume RLSLPs, and thus, can be smaller than the smallest L-system.

Proposition 1. *For any $d \ge 0$, it always holds that* $g_{it(d)} \le g_{rl}$.

Proof. Just note that a rule $A \to \prod_{i=1}^{k} B^{i^0}$ from an ISLP simulates a rule $A \to B^k$ from a RLSLP. In particular, 0-ISLPs are equivalent to RLSLPs.   □

Proposition 2. *For any $d \ge 0$, there exists a string family where* $g_{it(d)} = o(\ell)$.

Proof. Navarro and Urbina show a string family where $g_{rl} = o(\ell)$ [27]. Hence, $g_{it(d)}$ is also $o(\ell)$ in this family.   □

We now show that $d = 1$ suffices to obtain ISLPs that are significantly smaller than δ for some string families.

Lemma 1. *Let $d \ge 1$. There exists a string family with* $g_{it(d)} = O(1)$ *and* $\delta = \Omega(\sqrt{n})$.

Proof. Such a family is formed by the strings $s_k = \prod_{i=1}^{k} \mathtt{a}^i \mathtt{b}$. The 1-ISLPs with initial rule $S_k \to \prod_{i=1}^{k} A^i B$, and rules $A \to \mathtt{a}$, $B \to \mathtt{b}$, generate each string s_k in the family using $O(1)$ space. On the other hand, it has been proven that $\delta = \Omega(\sqrt{n})$ in the family cs_k [27]. As δ can only decrease by 1 after the deletion of a character [1], $\delta = \Omega(\sqrt{n})$ in the family s_k too.   □

On the other hand, ISLPs can perform worse than other compressed representations; recall that $\delta \leq \gamma \leq b \leq r_\$$.

Lemma 2. *Let $\mu \in \{r, r_\$, \ell\}$. There exists a string family with $g_{it(d)} = \Omega(\log n)$ and $\mu = O(1)$.*

Lemma 3. *There exists a string family satisfying that $z = O(\log n)$ and $g_{it(d)} = \Omega(\log^2 n / \log \log n)$.*

One thing that makes ISLPs robust is that they are not very sensitive to reversals, morphism application, or edit operations (insertions, deletions, and substitutions of a single character). This makes $g_{it(d)}$ more robust than measures like r and $r_\$$, which are sensitive to all these transformations [1,7,11,12].

Lemma 4. *Let G be a d-ISLP generating T. Then there exists a d-ISLP of size $|G|$ generating the reversed text T^R. Let φ be a morphism. Then there exists a d-ISLP of size $|G| + c_\varphi$ generating the text $\varphi(T)$, where c_φ is a constant depending only on φ. Moreover, there exists a d-ISLP of size at most $O(|G|)$ generating T' where T and T' differ by one edit operation.*

4 Accessing ISLPs

We have shown that $g_{it(d)}$ breaks the lower bound δ already for $d \geq 1$. We now show that the measure is accessible. Concretely, we will show that any substring of length λ can be extracted in time $O(\lambda + (h + \log n) \log n \log \log n)$, where h is the height of the grammar tree, and in Sect. 5 we show that ISLPs can be balanced so they have $h = O(\log n)$. In total, we obtain the following result.

Theorem 1. *Let $T[1..n]$ be represented by a d-ISLP of size g_{it}. Then, there exists a data structure of size $O(g_{it})$ that extracts any substring of T of length λ in time $O(\lambda + \log^2 n \log \log n)$ on a RAM machine of $\Theta(\log n)$ bits, using $O(\log^2 n \log \log n)$ additional words of working space.*

In fact, our extraction time is $O(\lambda + d \log d \log n + d^2 \log d)$ using $O(d^2 \log d)$ working space, which reduces to $O(\lambda + \log n)$ time and $O(1)$ working space for $d = O(1)$ (recall that 1-ISLPs already break the δ lower-bound), and yields the result in the theorem if $d = O(\log n)$. For larger d, we start with a technical result that shows that we can always force d to be $O(\log n)$ without asymptotically increasing the size. From now on in the paper, we will disregard for simplicity the case $k_1 > k_2$ in the rules $A \rightarrow \Pi_{i=k_1}^{k_2} B_1^{i^{c_1}} \cdots B_t^{i^{c_t}}$, as their treatment is analogous to that of the case $k_1 \leq k_2$.

Lemma 5. *If a d-ISLP G generates $T[1..n]$, then there is also a d'-ISLP G' of the same size that generates T, for some $d' \leq \log_2 n$.*

Proof. For any rule $A = \prod_{i=k_1}^{k_2} B_1^{i^{c_1}} \cdots B_t^{i^{c_t}}$, any $i \in [k_1..k_2]$, and any c_j, it holds that $n \geq |\exp(A)| \geq i^{c_j}$, and therefore $c_j \leq \log_i n$, which is bounded by $\log_2 n$ for $i \geq 2$. Therefore, if $k_2 \geq 2$, all the values c_j can be bounded by some $d' \leq \log_2 n$. A rule with $k_1 = k_2 = 1$ is the same as $A \rightarrow B_1 \cdots B_t$, so all values c_j can be set to 0 without changing the size of the rule at all. $\square$

4.1   Data Structures

We define some data structures that extend ISLPs allowing us to efficiently navigate it within $O(g_{it})$ space. Per Lemma 5, we assume $d = O(\log n)$.

Consider a rule $A \to \prod_{i=k_1}^{k_2} B_1^{i^{c_1}} \ldots B_t^{i^{c_t}}$. Though t can be large, there are only $d + 1$ distinct values c_j. We will make use of auxiliary polynomials

$$f_r(i) = \sum_{j=1}^{r} |\exp(B_j)| \cdot i^{c_j},$$

for $r \in [1, t]$, to navigate within the "blocks" i: $f_r(i)$ computes cumulative lengths inside the product expression $B_1^{i^{c_1}} \ldots B_t^{i^{c_t}}$, up to the variable B_r, for a given i.

We now show how to compute any $f_r(i)$ in time $O(d)$ using $O(t)$ space for each A. An array $S_A[1..t]$ stores cumulative length information, as follows $S_A[r] = \sum_{1 \leq j \leq r, c_j = c_r} |\exp(B_j)|$. That is, $S_A[r]$ adds up the lengths of the symbol expansions up to B_r that must be multiplied by i^{c_r}. A second array, $C_A[1..t]$, stores the values $c_1, \ldots, c_t$. We preprocess C_A to solve predecessor queries of the form $pred(A, r, c) = \max\{j \leq r, C_A[j] = c\}$, that is, the latest occurrence of c in C_A to the left of position r, for every $c = 0, \ldots, d$. This query can be answered in $O(d)$ time because the elements in C_A are also in $\{0, \ldots, d\}$: cut C_A into chunks of length $d + 1$, and for each chunk $C_A[(d + 1) \cdot j + 1 .. (d + 1) \cdot (j + 1)]$ store precomputed values $pred(A, (d + 1) \cdot j, c)$ for all $c \in \{0, \ldots, d\}$. This requires $O(t)$ space. To compute the values $r_c = pred(A, r, c)$ for all c, find the chunk $j = \lceil r/(d + 1) \rceil - 1$ where r belongs, initialize every $r_c = pred(A, (d + 1) \cdot j, c)$ for every c (which is stored with the chunk j), and then scan the chunk prefix $C_A[(d + 1) \cdot j + 1 .. r]$ left to right, correcting every $r_c \leftarrow k$ if $c = C_A[k]$, for $k = (d + 1) \cdot j + 1 .. r$. We can then evaluate $f_r(i)$ in $O(d)$ time by computing all values r_c as explained (i.e., the last position to the left of r where the exponent is c), and adding up $S_A[r_c] \cdot i^c$ (because $S_A[r_c]$ adds up all $|\exp(B_j)|$ that must be multiplied by i^c in $f_r(i)$). We also define the polynomial

$$f^+(k) = \sum_{i=k_1}^{k} f_t(i)$$

to select a "block": $f^+(k)$ computes the cumulative sum of the length of the whole expressions $B_1^{i^{c_1}} \cdots B_t^{i^{c_t}}$ until $i = k$. Note we cannot afford to store all the $k_2 - k_1 + 1$ values $f^+(k)$, but we can exploit the fact that the polynomials $f_t(i)$ have degree at most d, and thus $f^+(k)$ is a polynomial on k of degree at most $d + 1$. Storing f^+ as a polynomial, then, requires only $O(d)$ space, instead of the $O(k)$ space needed to store all of its values. This can still be excessive, however, as it blows the space by an $O(\log n)$ factor in a rule like $A \to \Pi_{i=k_1}^{k_2} B^{i^d}$, which is of size 4 but f^+ is of degree $d + 1$. We will instead compute $f^+(k)$ in $O(d)$ arithmetic operations by reusing the same data structures we store for $f_r(i)$: for each $c = 0, \ldots, d$, we compute $t_c = pred(A, t, c)$ and $s_c = S_A[t_c]$. Instead of accumulating $s_c \cdot i^c$, however, we accumulate $s_c \cdot \sum_{i=k_1}^{k} i^c = s_c \cdot (p_c(k) - p_c(k_1 - 1))$, where $p_c(k) = \sum_{i=1}^{k} i^c$.

We cannot afford storing all the $O(kd)$ values $p_c(k)$, but since there are only $d + 1 = O(\log n)$ functions p_c and each one is a polynomial of degree $c+1 = O(\log n)$, they can be represented as polynomials using $O(\log^2 n)$ integers. Further, they can be computed at query time[1], before anything else, in $O(d^2)$ arithmetic operations using, for each c, the formula[2]

$$p_c(k) \;=\; k^c + \frac{1}{c+1} \cdot \sum_{j=0}^{c} \binom{c+1}{j} B_j \cdot k^{c+1-j},$$

which is a polynomial on k of degree at most $d + 1$. The formula requires $O(c)$ arithmetic operations once the numbers B_j are computed. Those B_j are the Bernoulli (rational) numbers. All the Bernoulli numbers from B_0 to B_d can be computed in $O(d^2)$ arithmetic operations using the recurrence $\sum_{j=0}^{d} \binom{d+1}{j} B_j = 0$, from $B_0 = 1$. The numerators and denominators of the rationals B_j fit in $O(j \log j) = O(d \log d) = O(\log n \log d)$ bits,[3] so they can be operated in $O(\log d)$ time in a RAM machine with word size $\Theta(\log n)$. Therefore, the total preprocessing time to later compute any $f^+(k)$ is $O(d^2 \log d)$. We note, however, that due to the length of the numerators and denominators of the fractional Bernoulli numbers, the time to compute any $f^+(k)$ is $O(d \log d)$.

Example 1. Consider the ISLP of Proposition 2, defined by the rules $S \rightarrow \prod_{i=1}^{k_2} A^i B$, $A \rightarrow$ a, and $B \rightarrow$ b. The polynomials associated with the representation of the rule S are $i^{c_1} = i$ and $i^{c_2} = 1$. Then, we construct the auxiliary polynomials $f_1(i) = |\exp(A)|i^{c_1} = i$ and $f_2(i) = |\exp(A)|i^{c_1} + |\exp(B)|i^{c_2} = i+1$. Finally, we construct the auxiliary polynomial $f^+(k) = \sum_{i=1}^{k} f_2(i) = \sum_{i=1}^{k}(i + 1) = \frac{1}{2}k^2 + \frac{3}{2}k$. Figure 1 shows a more complex example to illustrate C_A and S_A.

4.2   Direct Access in Time $O((h + \log n)\, d \log d)$

We start with the simplest query: given the data structures of size $O(g_{it})$ defined in the previous sections, return the symbol $T[l]$ given an index l.

For SLPs with derivation tree of height h, the problem is easily solved in $O(h)$ time by storing the expansion size of every nonterminal, and descending from the root to the corresponding leaf using $|\exp(B)|$ to determine whether to descend to the left or to the right of every rule $A \rightarrow BC$. This is easy to generalize in RLSLP rules $A \rightarrow B^k$, because every repetition corresponds to the same string, of length $|\exp(B)|$. The general idea for d-ISLPs is similar, but now determining which child to follow in repetition rules is more complex.

To access the l-th character of the expansion of $A \rightarrow \prod_{i=k_1}^{k_2} B_1^{i^{c_1}} \cdots B_t^{i^{c_t}}$ we first find the value i such that $f^+(i - 1) < l \leq f^+(i)$ by using binary search.

---

[1] Indeed, the polynomials $p_c(k)$ are independent of the grammar, so they can be computed once for all queries and for all grammars.

[2] See Wolfram Mathworld's https://mathworld.wolfram.com/BernoulliNumber.html, Eqs. (34) and (47).

[3] See https://www.bernoulli.org, sections "Structure of the denominator", "Structure of the nominator", and "Asymptotic formulas".

	1	2	3	4	5	6	7	8
S_A	2	3	6	7	14	13	5	3
C_A	1	2	1	0	0	1	2	3

$f_8(i) = 3i^3 + 5i^2 + 13i + 14$

$f^+(k) = \frac{9}{12}k^4 + \frac{38}{12}k^3 + \frac{117}{12}k^2 + \frac{256}{12}k$

Fig. 1. Data structures built for the ISLP rule $A \to \prod_{i=1}^{5} B^i C^{i^2} D^i E E E^i B^{i^2} C^{i^3}$, with $|\exp(B)| = 2$, $|\exp(C)| = 3$, $|\exp(D)| = 4$, and $|\exp(E)| = 7$. We show some of the polynomials to be simulated with these data structures.

Then, we find the value r such that $f_{r-1}(i) < l - f^+(i-1) \leq f_r(i)$ by using binary search in the subindex of the polynomials. We then know that the search follows by B_r, with offset $l - f^+(i-1) - f_{r-1}(i)$ inside $|\exp(B_r)|^{i^{c_r}}$. The offset within B_r is then easily computed with a modulus, as in RLSLPs. Algorithm 1 gives the details.

We carry out the first binary search so that, for every i we try, if $f^+(i) < l$ we immediately answer $i + 1$ if $l \leq f^+(i+1)$; instead, if $l \leq f^+(i)$, we immediately answer i if $f^+(i-1) < l$. As a result, the search area is initially of length $|\exp(A)|$ and, if the answer is i, the search has finished by the time the search area is of length $\leq f^+(i) - f^+(i-1) = f_t(i)$. Thus, there are $O(1 + \log(|\exp(A)|/f_t(i)))$ binary search steps. The second binary search is modified analogously so that it carries out $O(1 + \log(f_t(i)/(i^{c_r}|\exp(B_r)|)))$ steps, for a total of at most $O(1 + \log(|\exp(A)|/|\exp(B_r)|))$ steps. As the search continues by B_r, the sum of binary search steps telescopes to $O(h + \log n)$ on an ISLP of height h, and the total time is $O((h + \log n) d \log d) = O((h + \log n) \log n \log \log n)$.

Example 2. We show how to access the b at position 14 of the string $T = \prod_{i=1}^{5} a^i b$. Consider the ISLP G and its auxiliary polynomials computed in Example 1. We start by computing $f^+(2) = 5$. As $l > 5$, we go right in the binary search and compute $f^+(4) = 14$. As $l \leq 14$ we go left, compute $f^+(3) = 9$ and find that $i = 4$. Hence, $T[l]$ lies in the expansion of $A^i B = A^4 B$ at position $l_1 = l - f^+(i-1) = 5$. Then, we compute $f_1(4) = 4$. As $l_1 > 4$, we turn right and compute $f_2(4) = 5$, finding that $r = 2$. Hence, $T[l]$ lies in the expansion of $B^{i^0} = B^1$ at position $l_2 = l_1 - f_{r-1}(i) = 1$.

4.3   Extracting Substrings

Once we have accessed $T[l]$, it is possible to output the substring $T[l .. l + \lambda - 1]$ in $O(\lambda + h)$ additional time, as we return from the recursion in Algorithm 1. We carry the parameter λ of the number of symbols (yet) to output, which is first decremented when we finally arrive at line 3 and find the first symbol, $T[l]$, which we now output immediately. From that point, as we return from the recursion, instead of returning the symbol $T[l]$, we return the number λ of symbols yet to output, doing some extra work until $\lambda = 0$.

Algorithm 1. Direct access for ISLPs in $O((h + \log n)\, d \log d)$ time

---

Input: An ISLP G of height h, a variable A of G, and a position $l \in [1, |\mathbf{exp}(A)|]$.
Output: The character $\mathbf{exp}(A)[l]$ at position l in $\mathbf{exp}(A)$.

```
 1: function ACCESS(G, A, l)
 2:     if A → a then
 3:         return a
 4:     if A → BC then
 5:         if l ≤ |exp(B)| then
 6:             return ACCESS(G, B, l)
 7:         else
 8:             return ACCESS(G, C, l − |exp(B)|)
```
9:     **if** $A \to \prod_{i=k_1}^{k_2} B_1^{i^{c_1}} \ldots B_t^{i^{c_t}}$ **then**
10:        $i \leftarrow \texttt{arg\_successor}([f^+(k_1) \ldots f^+(k_2)], l)$
11:        $l \leftarrow l - f^+(i-1)$
12:        $r \leftarrow \texttt{arg\_successor}([f_1(i) \ldots f_t(i)], l)$
13:        $l \leftarrow l - f_{r-1}(i)$
14:        **return** ACCESS$(G, B_r, l \bmod |\mathbf{exp}(B_r)|)$

---

1. If we return from line 5, we output $\min(\lambda, |\mathbf{exp}(C)|)$ symbols from nonterminal C, by invoking a new procedure REPORT(G, C, λ), which returns the new number λ of symbols yet to report; this number is then returned by ACCESS.
2. If we return from line 7, we just return the current value of λ to the caller.
3. If we return from line 13, we must report:
   (a) $i^{c_r} - \lceil l / |\mathbf{exp}(B_r)| \rceil$ further copies of $\mathbf{exp}(B_r)$.
   (b) i^{c_s} copies of $\mathbf{exp}(B_s)$, for $s = r + 1, \ldots, t$.
   (c) the expansions $\mathbf{exp}(B_1)^{j^{c_1}} \cdots \mathbf{exp}(B_t)^{j^{c_t}}$, for $j = i + 1, \ldots, k_2$.
   For each expansion $\mathbf{exp}(C)$ to report, we invoke REPORT(G, C, λ) and update λ to the new number of symbols yet to report. We stop if $\lambda = 0$.

Procedure REPORT(G, C, λ) outputs $\mathbf{exp}(C)$ in $O(|\mathbf{exp}(C)|)$ time if $\lambda \geq |\mathbf{exp}(C)|$, as it simply traverses the leaves of a tree without unary paths. In this case it returns $\lambda - |\mathbf{exp}(C)|$. Otherwise, it traverses only the first λ leaves of the derivation tree of C, in time $O(\lambda + h)$, and returns zero. Once a call to REPORT returns zero, it is never called again; therefore the total time we spend is $O(\lambda + h)$.

5   Balancing ISLPs

We show that any d-ISLP can be balanced so that its derivation tree is of height $O(\log n)$. Actually, we introduce a new type of SLP, which allows us to prove a more general balancing result that subsumes ISLPs.

Definition 3. *A* generalized straight-line program *(GSLP) is an SLP that allows special rules of the form $A \to x$, where x is a* program *(in any Turing-complete language) of length $|x|$ whose output $\mathtt{OUT}(x)$ is a nonempty sequence of variables, none of which can reach A. The rule $A \to x$ contributes $|x|$ to the* *size of the GSLP; the standard SLP rules contribute as usual. If it holds for all special rules that no variable appears exactly once inside* OUT(x), *then the GSLP is said to be* balanceable.

We can choose any desired language to describe the programs x. Though in principle $|x|$ can be taken as the Kolmogorov complexity of OUT(x), we will focus on very simple programs and on the asymptotic value of x. In particular, RLSLPs allow rules of the form $A \rightarrow B^k$ of size 2, and we can have a program of size $O(1)$ that outputs k copies of B; ISLPs allow rules of the form $\prod_{k_1}^{k_2} B_1^{i_1^c} \cdots B_t^{i_t^c}$ of size $2 + 2t$, and we can have a program of size $O(t)$ that writes the corresponding $f^+(k_2)$ symbols. Note that in both cases the GSLP is balanceable as long as special rules satisfy $k > 1$ (for RLSLPs), or if $k_1 \neq k_2$ (for ISLPs); otherwise they can be replaced by alternative rules of the same asymptotic size.

We will prove that any balanceable GSLP can be balanced without increasing its asymptotic size. Our proof generalizes that of Ganardi et al. [10, Theorem 1.2] for SLPs in a similar way to how it was extended to balance RLSLPs [28]. Just as Ganardi et al., in this section we will allow SLPs to have rules of the form $A \rightarrow B_1 \cdots B_t$, of size t, where each B_j is a terminal or a nonterminal; this can be converted into a strict SLP of the same asymptotic size.

A *directed acyclic graph* (DAG) is a directed multigraph $D = (V, E)$ without cycles (nor loops). We denote by $|D|$ the number of edges in this DAG. For our purposes, we assume that any DAG has a distinguished node r called the *root*, satisfying that any other node can be reached from r and r has no incoming edges. We also assume that if a node has k outgoing edges, they are numbered from 1 to k, so edges are of the form (u, i, v). The *sink nodes* of a DAG are the nodes without outgoing edges. The set of sink nodes of D is denoted by W. We denote the number of paths from u to v as $\pi(u, v)$, and $\pi(u, V) = \sum_{v \in V} \pi(u, v)$ for a set V of nodes. The number of paths from the root to the sink nodes is $n(D) = \pi(r, W)$.

One can interpret an SLP G generating a string T as a DAG D: There is a node for each variable in the SLP, the root node is the initial variable, variables of the form $A \rightarrow a$ are the sink nodes, and a variable with rule $A \rightarrow B_1 B_2 \ldots B_t$ has outgoing edges (A, i, B_i) for $i \in [1, t]$. Note that if D is a DAG representing G, then $n(D) = |\exp(G)| = |T|$.

Definition 4. *(Ganardi et al. [10, p. 5]) Let D be a DAG, and define the pairs $\lambda(v) = (\lfloor \log_2 \pi(r, v) \rfloor, \lfloor \log_2 \pi(v, W) \rfloor)$. The* symmetric centroid decomposition *(SC-decomposition) of a DAG D produces a set of edges between nodes with the same λ pairs defined as $E_{scd}(D) = \{(u, i, v) \mid \lambda(u) = \lambda(v)\}$, partitioning D into disjoint paths called* SC-paths *(some of them possibly of length 0).*

The set E_{scd} can be computed in $O(|D|)$ time. If D is the DAG of an SLP G, then $|D|$ is $O(|G|)$. The following lemma justifies the name "SC-paths".

Lemma 6. *(Ganardi et al. [10, Lemma 2.1]) Let $D = (V, E)$ be a DAG. Then every node has at most one outgoing and at most one incoming edge from $E_{scd}(D)$. Furthermore, every path from the root r to a sink node contains at most $2 \log_2 n(D)$ edges that do not belong to $E_{scd}(D)$.*

Note that the sum of the lengths of all SC-paths is at most the number of nodes of the DAG, or equivalently, the number of variables of the SLP.

The following definition and technical lemma are needed to construct the building blocks of our balanced GSLPs.

Definition 5. *(Ganardi et al. [10, p. 7]) A weighted string is a string $T \in \Sigma^*$ equipped with a weight function $||\cdot|| : \Sigma \to \mathbb{N}\backslash\{0\}$, which is extended homomorphically. If A is a variable in an SLP G, then we write $||A||$ for the weight of the string $\exp(A)$ derived from A.*

Lemma 7. *(Ganardi et al. [10, Proposition 2.2]) For every non-empty weighted string T of length n one can construct in linear time an SLP G generating T with the following properties:*

- *G contains at most $3n$ variables*
- *All right-hand sides of G have length at most 4*
- *G contains suffix variables $S_1, ..., S_n$ producing all non-trivial suffixes of T*
- *every path from S_i to some terminal symbol a in the derivation tree of G has length at most $3 + 2(\log_2 ||S_i|| - \log_2 ||a||)$*

Theorem 2. *Given a balanceable GSLP G generating a string T, it is possible to construct an equivalent GSLP G' of size $O(|G|)$ and height $O(\log n)$.*

Proof. Transform the GSLP G into an SLP H by replacing their special rules $A \to x$ by $A \to \text{OUT}(x)$, and then obtain the SC-decomposition $E_{scd}(D)$ of the DAG D of H. Observe that the SC-paths of H use the same variables of G, so it holds that the sum of the lengths of all the SC-paths of H is less than the number of variables of G. Also, note that any special variable $A \to x$ of G is necessarily the endpoint (i.e., the last node of a directed path) of an SC-path in D. To see this note that $\lambda(A) \neq \lambda(B)$ for any B that appears in $\text{OUT}(x)$, because $\log_2 \pi(A, W) \geq \log_2(|\text{OUT}(x)|_B \cdot \pi(B, W)) \geq 1 + \log_2 \pi(B, W)$ where $|\text{OUT}(x)|_B \geq 2$ because G is balanceable. This implies that the balancing procedure of Ganardi et al. on H, which transforms the rules of variables that are not the endpoint of an SC-path in the DAG D, will not touch variables that were originally special variables in G.

Let $\rho = (A_0, d_0, A_1), (A_1, d_1, A_2), \ldots, (A_{p-1}, d_{p-1}, A_p)$ be an SC-path of D. It holds that for each A_i with $i \in [0..p-1]$, in the SLP H its rule goes to two distinct variables, one to the left and one to the right. Thus, for each variable A_i, with $i \in [0..p-1]$, there is a variable A'_{i+1} that is not part of the path. Let $A'_1 A'_2 \ldots A'_p$ be the sequence of these variables. Let $L = L_1 L_2 \ldots L_s$ be the subsequence of left variables of the previous sequence. Then construct an SLP of size $O(s) \subseteq O(p)$ for the sequence L (seen as a string) as in Lemma 7, using $|\exp(L_i)|$ in H as the weight function. In this SLP, any path from the suffix nonterminal S_i to a variable L_j has length at most $3 + 2(\log_2 ||S_i|| - \log_2 ||L_j||)$. Similarly, construct an SLP of size $O(t) \subseteq O(p)$ for the sequence $R = R_1 R_2 \ldots R_t$ of right symbols in reverse order, as in Lemma 7, but with prefix variables P_i instead of suffix variables. Each variable A_i, with $i \in [0..p-1]$, derives the same string as $w_l A_p w_r$, for some suffix w_l of L and some prefix w_r of R. We can find rules deriving these prefixes and suffixes in the SLPs produced in the previous step, so for any variable A_i, we construct an equivalent rule of length at most 3. Add these equivalent rules, and the left and right SLP rules to a new GSLP G'. Do this for all SC-paths. Finally, add the original terminal variables and special variables (which are left unmodified) of the GSLP G, so G' is equivalent to G.

The SLP constructed for L has all its rules of length at most 4, and $3s \leq 3p$ variables. The same happens with R. The other constructed rules also have a length of at most 3, and there are p of them. Summing over all SC-paths, we have $O(|G|)$ size. The special variables cannot sum up to more than $O(|G|)$ size. Thus, the GSLP G' has size $O(|G|)$.

Any path in the derivation tree of G' is of length $O(\log n)$. To see why, let $A_0, \ldots, A_p$ be an SC-path. Consider a path from a variable A_i to an occurrence of a variable that is in the right-hand side of A_p in G'. Clearly, this path has length at most 2. Now consider a path from A_i to a variable A'_j in L with $i < j \leq p$. By construction this path is of the form $A_i \rightarrow S_k \rightarrow^* A'_j$ for some suffix variable S_k (if the occurrence of A'_j is a left symbol), and its length is at most $1 + 3 + 2(\log_2 ||S_k|| - \log_2 ||A'_j||) \leq 4 + 2 \log_2 ||A_i|| - 2 \log_2 ||A'_j||$. Analogously, if A'_j is a right variable, the length of the path is bounded by $1+3+2(\log_2 ||P_k|| - \log_2 ||A'_j||) \leq 4+2 \log_2 ||A_i|| - 2 \log_2 ||A'_j||$. Finally, consider a maximal path to a leaf in the derivation tree of G'. Factorize it as $A_0 \rightarrow^* A_1 \rightarrow^* \cdots \rightarrow^* A_k$ where each A_i is a variable of H (and also of G). Paths $A_i \rightarrow^* A_{i+1}$ are like those defined in the paragraph above, satisfying that their length is bounded by $4 + 2 \log_2 ||A_i|| - 2 \log_2 ||A_{i+1}||$. Observe that between each A_i and A_{i+1}, in the DAG D there is almost an SC-path, except that the last edge is not in E_{scd}. The length of this path is at most

$$\sum_{i=0}^{k-1} (4 + 2 \log_2 ||A_i|| - 2 \log_2 ||A_{i+1}||) \leq 4k + 2 \log_2 ||A_0|| - 2 \log_2 ||A_k||$$

By Lemma 6, $k \leq 2 \log_2 n$, which yields the upper bound $O(\log n)$.

The resulting GSLP can be modified to contain standard SLP rules of size at most two, with only a constant increase in size and depth.   □

By the above theorem, Lemma 5, and because ISLPs can be made balanceable, we obtain the following.

Corollary 1. *Given a d-ISLP G generating a string T, there is an equivalent d'-ISLP G' of size $O(|G|)$, with $d' \leq d$, $d' = O(\log n)$, and height $h' = O(\log n)$.*

6   Conclusions

We have introduced a new extension to straight-line programs (SLPs) and run-length SLPs (RLSLPs) called *iterated SLPs (ISLPs)*. ISLPs permit so-called *iteration rules* of the form $A \rightarrow \Pi_{i=k_1}^{k_2} B_1^{i^{c_1}} \cdots B_t^{i^{c_t}}$, of size $O(t)$. While it had already been shown that the lower-bound (and unreachable) measure δ, which was text-wise smaller than every preceding measure of repetitiveness, could be outperformed by a *reachable* measure (L-systems) on some text families [26, 27], the size g_{it} of the smallest ISLP generating a text is the first *accessible* measure that also outperforms δ (by the same margin, $O(\delta/\sqrt{n})$ on a text of length n).

With SLPs or RLSLPs representing a text $T[1..n]$, an arbitrary symbol of T can be accessed in $O(\log n)$ time. We have shown that, just as SLPs and RLSLPs [10, 28], ISLPs can be balanced without asymptotically increasing their space, and used it to devise an algorithm to access any arbitrary text position in time $O(\log^2 n \log\log n)$ within $O(g_{it})$ space. They are also similarly resistant to edits and other text manipulations.

References

1. Akagi, T., Funakoshi, M., Inenaga, S.: Sensitivity of string compressors and repetitiveness measures. Inf. Comput. **291**, 104999 (2023)
2. Bannai, H., Funakoshi, M., I, T., Köppl, D., Mieno, T., Nishimoto, T.: A separation of γ and b via Thue–Morse words. In: Lecroq, T., Touzet, H. (eds.) SPIRE 2021. LNCS, vol. 12944, pp. 167–178. Springer, Cham (2021). https://doi.org/10.1007/978-3-030-86692-1_14
3. Bille, P., Landau, G.M., Raman, R., Sadakane, K., Rao, S.S., Weimann, O.: Random access to grammar-compressed strings and trees. SIAM J. Comput. **44**(3), 513–539 (2015)
4. Burrows, M., Wheeler, D.: A block sorting lossless data compression algorithm. Technical report 124, Digital Equipment Corporation (1994)
5. Charikar, M., et al.: The smallest grammar problem. IEEE Trans. Inf. Theory **51**(7), 2554–2576 (2005)
6. Christiansen, A.R., Ettienne, M.B., Kociumaka, T., Navarro, G., Prezza, N.: Optimal-time dictionary-compressed indexes. ACM Trans. Algorithms **17**(1), Article 8 (2020)
7. Fici, G., Romana, G., Sciortino, M., Urbina, C.: On the impact of morphisms on BWT-Runs. In: Bulteau, L., Lipták, Z. (eds.) 34th Annual Symposium on Combinatorial Pattern Matching (CPM 2023). Leibniz International Proceedings in Informatics (LIPIcs), vol. 259, pp. 10:1–10:18. Schloss Dagstuhl - Leibniz-Zentrum für Informatik, Dagstuhl, Germany (2023)
8. Gagie, T., Navarro, G., Prezza, N.: Fully-functional suffix trees and optimal text searching in BWT-runs bounded space. J. ACM **67**(1), Article 2 (2020)
9. Gallant, J.K.: String compression algorithms. Ph.D. thesis, Princeton University (1982)
10. Ganardi, M., Jeż, A., Lohrey, M.: Balancing straight-line programs. J. ACM **68**(4), 1–40 (2021)
11. Giuliani, S., Inenaga, S., Lipták, Z., Prezza, N., Sciortino, M., Toffanello, A.: Novel results on the number of runs of the burrows-wheeler-transform. In: Bureš, T., et al. (eds.) SOFSEM 2021: Theory and Practice of Computer Science, pp. 249–262. Springer, Cham (2021)
12. Giuliani, S., Inenaga, S., Lipták, Z., Romana, G., Sciortino, M., Urbina, C.: Bit catastrophes for the burrows-wheeler transform. In: Drewes, F., Volkov, M. (eds.) Developments in Language Theory, pp. 86–99. Springer, Cham (2023)
13. Kempa, D., Kociumaka, T.: Collapsing the hierarchy of compressed data structures: suffix arrays in optimal compressed space. CoRR 2308.03635 (2023)

14. Kempa, D., Prezza, N.: At the roots of dictionary compression: string attractors. In: Proceedings of 50th Annual ACM Symposium on the Theory of Computing (STOC), pp. 827–840 (2018)

15. Kempa, D., Saha, B.: An upper bound and linear-space queries on the LZ-End parsing. In: Proceedings of ACM-SIAM Symposium on Discrete Algorithms (SODA), pp. 2847–2866 (2022)

16. Kempa, D., Kociumaka, T.: Resolution of the burrows-wheeler transform conjecture. In: Proceedings of 61st IEEE Annual Symposium on Foundations of Computer Science (FOCS), pp. 1002–1013 (2020)

17. Kociumaka, T., Navarro, G., Olivares, F.: Near-optimal search time in δ-optimal space, and vice versa. CoRR 2206.00781 (2023)

18. Kociumaka, T., Navarro, G., Prezza, N.: Towards a definitive compressibility measure for repetitive sequences. IEEE Trans. Inf. Theory **69**(4), 2074–2092 (2023)

19. Kreft, S., Navarro, G.: On compressing and indexing repetitive sequences. Theoret. Comput. Sci. **483**, 115–133 (2013)

20. Lempel, A., Ziv, J.: On the complexity of finite sequences. IEEE Trans. Inf. Theory **22**(1), 75–81 (1976)

21. Lindenmayer, A.: Mathematical models for cellular interactions in development I. Filaments with one-sided inputs. J. Theoret. Biol. **18**(3), 280–299 (1968)

22. Lindenmayer, A.: Mathematical models for cellular interactions in development II. Simple and branching filaments with two-sided inputs. J. Theoret. Biol. **18**(3), 300–315 (1968)

23. Navarro, G.: Indexing highly repetitive string collections, Part I: repetitiveness measures. ACM Comput. Surv. **54**(2), Article 29 (2021)

24. Navarro, G.: Indexing highly repetitive string collections, Part II: compressed indexes. ACM Comput. Surv. **54**(2), Article 26 (2021)

25. Navarro, G., Ochoa, C., Prezza, N.: On the approximation ratio of ordered parsings. IEEE Trans. Inf. Theory **67**(2), 1008–1026 (2021)

26. Navarro, G., Urbina, C.: On stricter reachable repetitiveness measures. In: Proceedings of 28th International Symposium on String Processing and Information Retrieval (SPIRE), pp. 193–206 (2021)

27. Navarro, G., Urbina, C.: L-systems for measuring repetitiveness. In: Proceedings of 34th Annual Symposium on Combinatorial Pattern Matching (CPM), p. article 14 (2023)

28. Navarro, G., Olivares, F., Urbina, C.: Balancing run-length straight-line programs. In: Proceedings of 29th International Symposium on String Processing and Information Retrieval (SPIRE), pp. 117–131 (2022)

29. Raskhodnikova, S., Ron, D., Rubinfeld, R., Smith, A.D.: Sublinear algorithms for approximating string compressibility. Algorithmica **65**(3), 685–709 (2013)

30. Sipser, M.: Introduction to the Theory of Computation. Cengage Learning, London (2012)

31. Storer, J.A., Szymanski, T.G.: Data compression via textual substitution. J. ACM **29**(4), 928–951 (1982)

Computing Largest Minimum Color-Spanning Intervals of Imprecise Points

Ankush Acharyya[1]([✉]), Vahideh Keikha[2], Maria Saumell[3],
and Rodrigo I. Silveira[4]

[1] Department of Computer Science and Engineering,
National Institute of Technology, Durgapur, India
`aacharyya.cse@nitdgp.ac.in`
[2] Institute of Computer Science, The Czech Academy of Sciences, Prague,
Czech Republic
`keikha@cs.cas.cz`
[3] Department of Theoretical Computer Science, Faculty of Information Technology,
Czech Technical University in Prague, Prague, Czech Republic
`maria.saumell@fit.cvut.cz`
[4] Department de Matemàtiques, Universitat Politècnica de Catalunya,
Barcelona, Spain
`rodrigo.silveira@upc.edu`

Abstract. We study a geometric facility location problem under imprecision. Given n unit intervals in the real line, each with one of k colors, the goal is to place one point in each interval such that the resulting *minimum color-spanning interval* is as large as possible. A minimum color-spanning interval is an interval of minimum size that contains at least one point from a given interval of each color. We prove that if the input intervals are pairwise disjoint, the problem can be solved in $O(n)$ time, even for intervals of arbitrary length. For overlapping intervals, the problem becomes much more difficult. Nevertheless, we show that it can be solved in $O(n^2 \log n)$ time when $k = 2$, by exploiting several structural properties of candidate solutions, combined with a number of advanced algorithmic techniques. Interestingly, this shows a sharp contrast with the 2-dimensional version of the problem, recently shown to be NP-hard.

Keywords: Color-spanning interval · Imprecise points · Algorithms

1 Introduction

Color-spanning problems arise naturally in certain facility location problems where the goal is to find a "good" location for a facility (with respect to a set of

A. Acharyya was supported by the DST-SERB grant number SRG/2022/002277. V. Keikha was supported by the CAS PPPLZ grant L100302301, and the institutional support RVO: 67985807. M. Saumell was supported by the Czech Science Foundation, grant number 23-04949X. R. Silveira was partially supported by grant PID2019-104129GB-I00/MCIN/AEI/10.13039/501100011033.

J. A. Soto and A. Wiese (Eds.): LATIN 2024, LNCS 14578, pp. 81–96, 2024.
https://doi.org/10.1007/978-3-031-55598-5_6

sites), and where, in addition, each site has a category (or *color*). In such settings, the goal is often to consider only one site of each category. For instance, one may be interested in a location such that the maximum distance to reach one site of each color is as small as possible. Then, from a geometric point of view, if sites are points in the plane, and the distance used is the Euclidean distance, one is looking for a smallest circle that contains at least one point from each color. This is known as a *minimum color-spanning circle*.

Plenty of variants of color-spanning objects have been studied. Typically, the object sought is a two-dimensional region of some type, such a circle, a square, or a strip. Then one can aim at finding the smallest, largest, narrowest, etc., color-spanning object of such a type (see, for example, [1,5]).

However, it is well-known that data used in real-world instances is not 100% accurate. This is especially true for geometric data, which—in most applications—originates from inaccurate measuring devices, such as GPS receivers o laser scanners. This motivated a flurry of research on uncertainty models for geometric algorithms, where the imprecision in the data is modeled explicitly. One of the simplest and most studied models for geometric uncertainty is based on regions: instead of assuming that the exact location of each site is known, one assumes that the site lies within a region (e.g., a disk). In principle, any location within the site's region is possible. Choosing one location inside each site's region results in a *realization* of the imprecise sites. Since many different realizations are possible, natural optimization problems arise. Most typically, one is interested in understanding extreme realizations: those that give the best possible situation, or the worst one. For instance, in the context of color-spanning circles, if each site is modeled by, say, a disk, then one can wonder how to place one point inside each disk, so that the resulting set of points gives the smallest or largest color-spanning circles.

Many problems in computational geometry have been studied for the region-based imprecision model (see, e.g., [10,11]). Depending on the region (e.g., a line segment, a square, a disk) and the actual problem (e.g., convex hull, triangulation, etc.), some problems become very difficult already under very simple imprecision models, while some others can still be solved efficiently.

In this work, we study the problem of finding a minimum color-spanning circle of largest size, for imprecise points modeled as 1D intervals on the real line. Our motivation stems from recent work by Acharyya et al. [2], where the problem was studied for regions consisting of disks in 2D. While the authors of [2] managed to find efficient algorithms to find a minimum color-spanning circle of *smallest* size, the maximization versions resulted more difficult. In fact, they proved that the problem of placing one point in each disk, such that the minimum color-spanning circle has largest possible size, is NP-hard, even for unit input disks and only two colors. Given this somewhat surprising negative result, in this paper, we study the same problem, one dimension lower, where disks become intervals on the real line, and the minimum color-spanning circle becomes the *minimum color-spanning interval*. More formally, the problem we study is defined as follows. See Fig. 1 for an example with $k = 3$.

Fig. 1. Example of unit intervals with $k = 3$ colors, with an optimal realization that results in four *L-MCSI*s (indicated in gray). Note that the three leftmost representatives also form a color-spanning interval, but it is not minimum.

Largest Minimum Color-Spanning Interval (*L-MCSI*): Given n unit-length closed intervals on the real line $\mathcal{I} = \{I_1, I_2, \ldots, I_n\}$, each with one of k colors, specified in sorted order with respect to their left endpoints, find a realization of $\mathcal{I}$ such that the length of the minimum color-spanning interval(s) (*MCSI*) of the realization is as large as possible.

Contributions. We first show that, if the input intervals are pairwise disjoint, the *L-MCSI* problem can be solved in $O(n)$ time, even if intervals have arbitrary lengths. It turns out that the main difficulty of the problem originates from intervals that overlap. Intuitively, when two intervals of different color overlap, the points in each interval can be placed in any of two orders. This, repeated for all pairs of intersecting intervals of different color, results in a combinatorial explosion of possible orderings, already for two colors (i.e., $k = 2$). Thus most of this work is devoted to efficiently solving the problem for $k = 2$. As is usual in this type of optimization setting, we focus on the decision version of the problem: Given n colored unit-intervals, is there a realization with a minimum color-spanning interval of length at least q? We show that this problem has a rich structure that allows us to compute solutions efficiently. Indeed, a key contribution is a detailed analysis of the structure of certain canonical (sub)solutions, which we call *leftmost*, and their decomposition into so-called *tabular subsolutions*. The careful combination of leftmost (sub)solutions leads to an $O(n^3)$-time algorithm to solve the decision version of the problem.

By applying a number of advanced algorithmic techniques, we show how to decrease this running time substantially, from $O(n^3)$ to $O(n \log n)$. This allows to solve the *L-MCSI* problem in $O(n^2 \log n)$ time. Further, we also present a $(1 - \varepsilon)$-approximation algorithm that runs in $O(n \log n \log \frac{1}{\varepsilon})$ time.

Related Work. An extensive literature exists devoted to color-spanning objects. For a set of n points of k colors in the plane, the problem of identifying a set of k points of disjoint colors such that the chosen points have smallest possible diameter is NP-hard [7]. Polynomial algorithms are known for computing the smallest color-spanning circle or square [1], strip or rectangle [5], among others.

Several strategies have been proposed in the literature to deal with imprecision in geometric data. Here we only mention a few relevant results for the region-based model, which is the one adopted in this article. In this model, finding a placement of points within a set of disks that maximizes or minimizes the radius of the smallest enclosing circle of the points can be solved in $O(n)$ time [11]. Finding a placement of points within a set of line segments or squares that maximizes or minimizes the area or the perimeter of the convex hull can be solved with algorithms with running times ranging from $O(n)$ to $O(n^{13})$, while some variants are NP-hard [10]. Other objective functions and/or regions lead to many other variants that have also been studied in the literature. Imprecision problems have also been previously studied for one-dimensional points, represented by intervals, as we do in this work. For instance, in the 1D k-center problem on imprecise points, one is given n intervals, and the goal is to find a point in each interval that maximizes or minimizes the distances to the resulting k-centers; something that can be done in $O(n)$ time [8].

The problem of finding the largest minimum color-spanning interval is also related to the 1D *dispersion problem*: given n (uncolored) intervals, choose one point from each interval such that the minimum distance between any pair of consecutive points is maximized. Multiple variants of this problem exist, depending on the exact objective and constraints taken into account. See, e.g., [6,12] and references therein. If the intervals are given in sorted order which has to be preserved for the representatives, the problem can be solved in $O(n)$ time [9].

As already mentioned, the related work most relevant to ours is that of Acharyya et al. [2], who studied algorithms for the minimum color-spanning disk problem for imprecise points in 2D. Given n colored disks, they consider finding one point in each disk such that the minimum color-spanning circle of the selected points has maximum radius. They show this problem is NP-hard even for unit input disks and only two colors. In contrast, the minimization version of the problem, for k colors, can be solved in $O(nk \log n)$ time [2].

1.1   Preliminaries

The input to the problem is a set of closed intervals $\mathcal{I} = \{I_1, I_2, \ldots, I_n\}$ of unit length on the real line, where each interval is colored with one of k colors. We will sometimes refer to the intervals also as *segments*. We assume that no two intervals of the same color are at the same position. The intervals are given sorted from left to right, breaking ties arbitrarily if needed. For each i, we use x_i to denote the left endpoint of I_i. We assume that $x_1 = 0$. Given a realization (also called a *representation*) $\mathcal{P}$ of $\mathcal{I}$, we call the chosen point of I_i the *representative* of I_i. We denote it by r_i (with some abuse of notation, r_i is sometimes used to denote the coordinate of the representative). Sometimes, *realization* applies to a proper subset of $\mathcal{I}$ rather than the entire $\mathcal{I}$. In this case, we say that I_i is *represented* in the realization, if the realization contains a representative of I_i. When $k = 2$, we use c or c' to denote one of the two colors. For a fixed c, we use $\bar{c}$ to denote the color that is not c. We use c_i to denote the color of I_i.

2   Disjoint Case

In this section, input segments are disjoint. We use the concept of *minimal color-spanning interval* (*mCSI*), defined as a color-spanning interval not properly contained within any other color-spanning interval. A minimum color-spanning interval is also a minimal color-spanning interval, with minimum length.

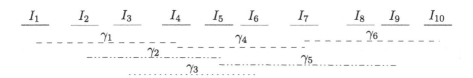

Fig. 2. An example where the chains of $mCSI$s are $\{\gamma_1, \gamma_4, \gamma_6\}$, $\{\gamma_2, \gamma_5\}$, $\{\gamma_3\}$.

In general, $mCSI$s depend on the realization. However, if intervals are disjoint, the only thing that varies is the location of the first and last representatives. All other intervals represented in the $mCSI$ are fully contained in the $mCSI$, thus the positions of their representatives are irrelevant. Hence, from a combinatorial point of view, an $mCSI$ is a sequence of consecutive intervals. It follows that the possible combinatorial $mCSI$s are determined by the intervals alone:

Lemma 1. *If the input segments are disjoint, every realization leads to the same set of at most $n - k + 1$ combinatorial $mCSI$s.*

The set of combinatorial $mCSI$s can be computed in $O(n)$ time [4]. Let $\gamma = \{\gamma_1, \gamma_2, \ldots, \gamma_\ell\}$ be this set, sorted by the leftmost segment represented. We divide γ into subsets called *chains of $mCSI$s* as follows: Start with γ_1, and let the rightmost segment represented in γ_1 be I_β. If there is an $mCSI$ γ_i starting at I_β, add it to the subset of γ_1, and repeat with γ_i; otherwise, stop. At the end of the process, remove from γ all the γ_i in the same subset as γ_1, and construct a new subset starting with the remaining element in γ with the smallest index. Repeat until γ is empty. See Fig. 2 for an example.

Next, we process each chain of $mCSI$s separately as follows: Let $\mathcal{I}'$ be the set of segments obtained from taking the leftmost and rightmost segments represented in each γ_i contained in the chain of $mCSI$s. Among all $mCSI$s contained in the chain of $mCSI$s, the minimum one is achieved by the pair of representatives of consecutive segments of $\mathcal{I}'$ at minimum distance. The position of the representatives of the segments not in $\mathcal{I}'$ is irrelevant. Since we want to maximize the length of the $MCSI$s, we compute a realization of $\mathcal{I}'$ maximizing the minimum distance between consecutive representatives in time $O(|\mathcal{I}'|)$ [9].

After repeating this procedure for all chains of $mCSI$s, we obtain a realization for all those segments that are the leftmost or rightmost of some γ_i. For all the other segments, we choose a representative arbitrarily. We output the obtained representation as the solution to the problem. Note that the algorithm does not require segments to be unit-length. Since the input segments are sorted, we have:

Theorem 1. *If the input intervals have arbitrary lengths and are pairwise disjoint, the L-MCSI problem can be solved in $O(n)$ time.*

Finally, we observe that the previous result can also be extended to the case where there is no pair of intersecting intervals of *distinct* color.

3   Case $k = 2$: Decision Problem

In the rest of the paper, we focus on the case where $k = 2$ and there is at least one pair of intersecting intervals of distinct color. We assume the two colors are red and blue, denoting the set of red and blue intervals by $\mathcal{R}$ and $\mathcal{B}$, respectively. Let q^* be the length of an *MCSI* in a realization solving the *L-MCSI* problem.

Lemma 2. *If at least two intervals of different colors intersect, then $\frac{1}{2} \leq q^* \leq 2$.*

Hence, the decision problem only makes sense for this range. For $k = 2$, it can be rephrased as: For a given value q, does there exist a realization of $\mathcal{R} \cup \mathcal{B}$ such that the distance between any pair of representatives of distinct color is at least q? Such a realization is said to satisfy the *separation property*.

In the subsequent sections, we assume that no two consecutive intervals leave a gap of length q or greater between them. Indeed, if the opposite happens, we can divide the problem into two independent subproblems.

All of our algorithms for the decision problem use the following concept:

Definition 1. *A* leftmost solution *is a solution where each representative is either at the left endpoint of its interval or at a distance q from a representative of distinct color located to its left.*

Observe that any solution can be easily transformed into a leftmost solution. Thus, our algorithms try to compute a leftmost solution. The main difficulty is that, if two intervals of distinct color overlap by q or more, there are two possible orderings of the representatives that satisfy the separation property. We divide $(\frac{1}{2}, 2]$ into three subranges and present a different algorithm for each of them.

3.1   Decision Problem for $q \in (1, 2]$

This is an easy case due to the following observation:

Lemma 3. *In the decision problem for $q \in (1, 2]$, it is enough to consider realizations where $r_i \leq r_{i+1}$, for all $i = 1, 2, \ldots, n - 1$.*

Hence, we can process the segments from left to right and try to compute a leftmost solution. The algorithm stops if, for some segment I_i, there is no placement for r_i in I_i satisfying the separation property (with respect to the representatives that have already been placed).

Proposition 1. *For $k = 2$, the decision problem for $q \in (1, 2]$ can be solved in $O(n)$ time.*

3.2 Decision Problem for $q \in \left(\frac{3}{4}, 1\right]$

In this case, there might be more left-to-right orderings for the representatives leading to affirmative answers for the problem. To deal with them, we introduce a bipartite graph H with vertices $\mathcal{R} \cup \mathcal{B}$ where a red and a blue segment are adjacent if and only if the length of their intersection is at least q. Let C_i be a connected component of H and $V(C_i)$ be its set of vertices.

We define $\mathcal{J}_{C_i}$ as the interval spanned by $V(C_i)$, that is, its leftmost (resp., rightmost) point is the leftmost (resp., rightmost) point of the leftmost (resp., rightmost) segment(s) in $V(C_i)$.

Lemma 4. *For any $I_j \in \mathcal{R} \cup \mathcal{B}$, if $I_j \subseteq \mathcal{J}_{C_i}$, then $I_j \in V(C_i)$.*

Corollary 1. *For any two intervals $\mathcal{J}_{C_i}$ and $\mathcal{J}_{C_j}$ with $i \neq j$, none of them is contained in the other.*

By Corollary 1, given two intervals $\mathcal{J}_{C_i}$ and $\mathcal{J}_{C_j}$ with $i \neq j$, their left endpoints do not coincide. Thus, we can uniquely sort the intervals $\mathcal{J}_{C_i}$ from left to right according to their left endpoints. We rename the components C_i so that this left-to-right order is $\mathcal{J}_{C_1}, \mathcal{J}_{C_2} \cdots$.

Lemma 5. *In the decision problem for $q \in \left(\frac{3}{4}, 1\right]$, it is enough to consider realizations where, for any $I_{i'} \in C_i$ and $I_{j'} \in C_j$ with $i < j$, $r_{i'} \leq r_{j'}$.*

Proof. Let the realization $\{r_1, r_2, \ldots, r_n\}$ be a yes-certificate for the problem.

If $\mathcal{J}_{C_i} \cap \mathcal{J}_{C_j} = \emptyset$, we are done. Otherwise, we denote the positions of the left and right endpoints of $\mathcal{J}_{C_i}$ by $\ell(\mathcal{J}_{C_i})$ and $r(\mathcal{J}_{C_i})$, respectively. We show that $x_{i'} \leq x_{j'}$: If $x_{j'} > r(\mathcal{J}_{C_i})$, we have $x_{i'} < r(\mathcal{J}_{C_i}) < x_{j'}$. If $\ell(\mathcal{J}_{C_i}) \leq x_{j'} \leq r(\mathcal{J}_{C_i})$, Lemma 4 implies that $x_{j'} + 1 > r(\mathcal{J}_{C_i})$. Since $x_{i'} + 1 \leq r(\mathcal{J}_{C_i})$, the conclusion follows. Finally, $x_{j'} < \ell(\mathcal{J}_{C_i})$ is not possible (it would imply $x_{j'} < \ell(\mathcal{J}_{C_j})$).

If $c_{i'} \neq c_{j'}$, $r_{i'} \leq r_{j'}$ is implied by $x_{i'} \leq x_{j'}$ and $|I_{i'} \cap I_{j'}| < q$. If $c_{i'} = c_{j'}$ and $r_{i'} > r_{j'}$, the fact that $x_{i'} \leq x_{j'}$ allows us to reassign the representatives ($r_{i'}$ becomes the representative of $I_{j'}$ and vice versa), and this change does not decrease the length of $MCSI(s)$.   □

The final key lemma is the following:

Lemma 6. *In any realization of $V(C_i)$ satisfying the separation property, there is exactly one alternation between red and blue representatives.*

We can now sketch our algorithm: By Lemma 5, we can process $C_1, C_2 \ldots$ in this order. By Lemma 6, in the relevant realizations of $V(C_i)$ either we have first all red representatives and then all the blue ones, or vice versa. We compute one leftmost representations of each type as follows: We take each of the two computed leftmost representations of $V(C_{i-1})$, and we try to compute a leftmost representation of $V(C_i)$ of the specific type with the condition that the separation property is satisfied in the final representation of $V(C_{i-1}) \cup V(C_i)$. Among the two computed representations of $V(C_i)$ of that type, we keep the one where the rightmost representative is as much to the left as possible.

Theorem 2. *For $k = 2$, the decision problem for $q \in \left(\frac{3}{4}, 1\right]$ can be solved in $O(n)$ time.*

Fig. 3. Left: A leftmost solution. Right: A sequence of four tabular subsolutions $\mathcal{T}_1, \mathcal{T}_4, \mathcal{T}_7, \mathcal{T}_8$ that is almost identical to the leftmost solution.

4   Case $k = 2$: Decision Problem for $q \in \left(\frac{1}{2}, \frac{3}{4}\right]$

This case requires a more involved algorithm and demands the use of more sophisticated algorithmic machinery to achieve an efficient running time. Here we present a summarized version with all the key ingredients.

4.1   Basic Algorithm

Suppose that, for some I_i and $x \in I_i$, there exists an interval I_j of the opposite color with leftmost point in the interval $(x - q, x + q - 1)$. Then, choosing x as a representative for I_i makes it impossible to find a representative for I_j not violating the separation property. Thus, we say that x is a *forbidden* position for the representative of I_i.

Definition 2. *A representation of a subset of intervals is* valid *if it satisfies the separation property and no representative is placed at a forbidden position.*

Observe the leftmost solution illustrated in Fig. 3 (left). The section that starts at r_4 is formed by a red representative placed at the left endpoint of its interval (r_4), followed by blue representatives r_2, r_5 at distance q, in turn, followed by red representative r_6 at distance q, and finishing before r_7, where a new representative is placed at the left endpoint of its interval. In order to describe these groups of representatives, we introduce the following definition:

Definition 3. *A tabular subsolution $\mathcal{T}_i$ starting at I_i is a valid representation of a subset of intervals including I_i such that: (i) $r_i = x_i$; (ii) $r_j \geq r_i$ for all representatives r_j of $\mathcal{T}_i$; (iii) all representatives with $r_j > r_i$ are placed at a distance q to the right of a representative of opposite color; (iv) if for two intervals of the same color I_k, I_j we have that $\mathcal{T}_i$ contains a representative r_k for I_k and $r_k \in I_j$, then $\mathcal{T}_i$ contains a representative r_j for I_j and $r_j = r_k$.*

Figure 3 (right) shows a sequence of tabular subsolutions that is almost identical to the leftmost solution from Fig. 3 (left). They only differ at r_3, which in the tabular subsolution is forced to be at the same position as r_4 (by Definition 3), but this position is not leftmost. Clearly, transforming the sequence of tabular subsolutions into the leftmost solution is trivial, so in our algorithm, we work with sequences of tabular subsolutions.

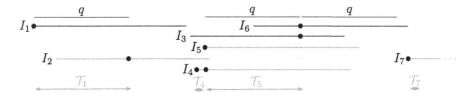

Fig. 4. A valid sequence of tabular subsolutions.

The strategy of our algorithm is to try to construct a sequence of tabular subsolutions such that: (i) the separation property is satisfied; (ii) the sequence contains a representative for every interval. Such a sequence gives a solution to the decision problem.

The first key question about such a sequence is under which conditions some T_i can be concatenated with some T_k, with $k < i$.

Definition 4. *Let $k < i$. A tabular subsolution T_i is compatible with a tabular subsolution T_k if the rightmost representative of T_k is to the left of x_i and every pair of representatives of different colors, one of which is in T_i and the other in T_k, are at distance at least q.*

Denote by $\tilde{L}[k]$ the set of representatives in the sequence to the left of x_i. It is less clear which elements must belong to $\tilde{L}[k]$. Requiring that $\tilde{L}[k]$ contains representatives for all I_j with $j < i$ is not the right answer, as witnessed by the example in Fig. 4: $r_3 \notin \tilde{L}[1]$ (here we are looking at $i = 4$, $k = 1$). It is not even necessary to require that this r_j is either in $\tilde{L}[k]$ or in T_i, since in Fig. 4, r_3 is actually in T_5 (not in T_4). Interestingly, we can show that a set of necessary conditions on $\tilde{L}[k]$ is also sufficient.

Let us explain these conditions. We denote by $\mathcal{I}_i^{s,<}$ the set of intervals I_j such that $c_j = c_i$ and $x_j < x_i - 1$. Since I_j lies completely to the left of I_i, any sequence where $r_j \notin \tilde{L}[k]$ does not contain any representative for I_j. We thus obtain the first necessary condition (C1) (see below).

Let us next denote by $\mathcal{I}_i^{o,<<}$ (resp., $\mathcal{I}_i^{o,<}$) the set of intervals I_j such that $c_j \neq c_i$ and $x_j \leq x_i - (1+q)$ (resp., $x_j \in (x_i - (1+q)), x_i - q])$. If $I_j \in \mathcal{I}_i^{o,<<}$, it lies to the left of I_i, so again we need $r_j \in \tilde{L}[k]$ (condition (C2); see below). Furthermore, I_j leaves a gap with I_i greater than or equal to q, which implies that any choice of r_j is compatible with $r_i = x_i$. On the other hand, if $I_j \in \mathcal{I}_i^{o,<}$, it has a portion of length smaller than q to the right of x_i. Thus, if $r_i = x_i$, r_j can only be placed to the left of x_i with $r_j \leq x_i - q$ (condition (C3); see below).

Let us next examine the leftmost tabular subsolution T_l of a sequence. As before, the choice $r_l = x_l$ forces any interval in $\mathcal{I}_l^{s,<} \cup \mathcal{I}_l^{o,<<} \cup \mathcal{I}_l^{o,<}$ to have its representative to the left of x_l. Since there is nothing to the left of x_l, we obtain a new condition (C0). All in all, the list of necessary conditions is as follows:

(C0) $\mathcal{I}_l^{s,<} = \mathcal{I}_l^{o,<<} = \mathcal{I}_l^{o,<} = \emptyset$.

(C1) For all $I_j \in \mathcal{I}_i^{s,<}$, $r_j \in \tilde{L}[k]$.

(C2) For all $I_j \in \mathcal{I}_i^{o,<<}$, $r_j \in \tilde{L}[k]$.

(C3) For all $I_j \in \mathcal{I}_i^{o,<}$, $r_j \in \tilde{L}[k]$ and $r_j \leq x_i - q$.

We now have all the ingredients to define our main object of interest:

Definition 5. *A sequence $\mathcal{S}$ of tabular subsolutions is* valid *if it satisfies the following: (i) the leftmost tabular subsolution of $\mathcal{S}$ satisfies condition (C0); (ii) for every pair of consecutive subsolutions $\mathcal{T}_k, \mathcal{T}_i$, with $k < i$, $\mathcal{T}_k$ and $\mathcal{T}_i$ are compatible and conditions (C1)–(C3) are satisfied.*

Lemma 7. *Let $\mathcal{S}$ be a valid sequence of tabular subsolutions. Suppose that the rightmost representative of $\mathcal{S}$ is at position x and has color c. Then, $\mathcal{S}$ contains representatives exactly for all intervals I_j of color c such that $x_j \leq x$ and for all intervals I_b of color $\bar{c}$ such that $x_b < x + q - 1$.*

Notice that $\mathcal{S}$ might contain more than one representative for some intervals (see, e.g., I_4 in Fig. 4).

Lemma 7 leads to our desired conclusion, stated in Lemma 8. Before, we introduce the following concept:

Definition 6. *The* maximal tabular subsolution *starting at I_i, denoted $M[i]$, is the tabular subsolution starting at I_i containing the largest number of representatives of intervals.*

In Fig. 4, $M[5] = \mathcal{T}_5$ because no red interval contains the position $r_5 + 2q$. Notice that we might have $M[i] = \emptyset$.

Lemma 8. *Let $\mathcal{S}$ be a valid sequence of tabular subsolutions such that the rightmost tabular subsolution is a maximal tabular subsolution containing a representative for I_n. Then, $\mathcal{S}$ satisfies the separation property and it contains at least one representative (but possibly more) for every interval.*

Proof. Within a tabular subsolution the separation property is satisfied. Additionally, the fact that consecutive subsolutions are compatible guarantees that the separation property is satisfied between pairs of consecutive subsolutions. This is enough to guarantee that the property is satisfied in the entire sequence.

Let us next prove that $\mathcal{S}$ contains representatives for all intervals. Let $\mathcal{T}_i$ be the rightmost tabular subsolution of $\mathcal{S}$, and x be the position of the representative for I_n in $\mathcal{T}_i$. Suppose that I_n has color c. By Lemma 7, if $\mathcal{S}$ contains a representative for an interval, it also contains representatives for all intervals of the same color to its left. Thus, $\mathcal{S}$ contains representatives for all intervals of color c. To prove that it also contains representatives for all intervals of color $\bar{c}$, we consider two cases, where we denote by $I_{n'}$ the rightmost interval of color $\bar{c}$.

Suppose first that x is the position of the rightmost representative of $\mathcal{S}$. By Lemma 7, $\mathcal{S}$ contains representatives for all intervals I_b of color $\bar{c}$ such that $x_b < x + q - 1$. Thus, if $x_{n'} < x + q - 1$, we are done. Notice that $x_{n'} > x + q$ is not possible, because we would have $x_n \leq x < x + q < x_{n'}$, which contradicts the fact that I_n is a rightmost interval from the set. Hence, the remaining case is $x_{n'} \in [x + q - 1, x + q]$. Then $I_{n'}$ contains $x + q$. Since x is the rightmost position of $\mathcal{T}_i$ and $\mathcal{T}_i$ is maximal, $\mathcal{T}_i$ does not contain representatives at $x + q$ because this

Algorithm 1. Decision problem for $q \in (1/2, 3/4]$ (simplified)

Input: $\mathcal{I}$
Output: a solution, if it exists; otherwise, "there is no solution"
1: **for** $i = 1, 2, \ldots, n$ **do**
2:     compute $M[i]$
3:     $appendable[i] := F$ ▷ F stands for false, and T stands for true
4:     $p[i] := \emptyset$ ▷ $p[i]$ stores an index k such that the predecessor of $M[i]$ in a valid sequence of tabular subsolutions is a tabular subsolution starting at I_k
5: **for** $i = 1, 2, \ldots, n$ such that $M[i] \neq \emptyset$ **do**
6:     **if** $\mathcal{I}_i^{s,<} = \mathcal{I}_i^{o,<<} = \mathcal{I}_i^{o,<} = \emptyset$ **then**
7:         $appendable[i] := T$
8:     **else**
9:         **if** $\exists k < i$ such that $appendable[k] = T$ and some portion $\mathcal{T}_k$ of $M[k]$ is compatible with $M[i]$ and is s.t. conditions (C1)–(C3) are satisfied **then**
10:             $appendable[i] := T$
11:             $p[i] := k$
12:     **if** $appendable[i] = T$ and $M[i]$ contains a representative for I_n **then**
13:         reconstruct and return a solution based on the reverse of the sequence $i, p[i], p[p[i]], \ldots$
14:         finish
15: **return** "there is no solution"

---

is a forbidden position. Hence, there exists an interval of color c with leftmost point in $(x, x + 2q - 1)$. This contradicts the fact that I_n is a rightmost interval.

Finally, suppose that x is not the position of the rightmost representative of S. Such rightmost representative cannot be at $x + 2q, x + 3q \ldots$, because then I_n would not be a rightmost interval from the set. Thus, the rightmost representative of S is at $x + q$. Since all intervals I_b of color $\bar{c}$ satisfy $x_b \leq x_n \leq x$ and at least one of them contains $x + q$, we derive that $I_{n'}$ contains $x + q$. Thus, S contains a representative for $I_{n'}$ and, by Lemma 7, also for all intervals of color $\bar{c}$ to its left. Hence, S contains representatives for all intervals of color $\bar{c}$. □

The goal of the algorithm is to find a sequence S satisfying the hypothesis of Lemma 8.

Definition 7. *We say that $M[i] \neq \emptyset$ is appendable if there exists a valid sequence of tabular subsolutions with $M[i]$ as the rightmost tabular subsolution.*

For each i with $M[i] \neq \emptyset$, the algorithm (sketched in Algorithm 1) determines if $M[i]$ is appendable. To this end, there are two options: If $\mathcal{I}_i^{s,<} = \mathcal{I}_i^{o,<<} = \mathcal{I}_i^{o,<} = \emptyset$ (condition (C0)), $M[i]$ is appendable because on its own it forms a valid sequence of tabular subsolutions. Otherwise, the algorithm looks for some $k < i$ such that $M[k]$ is appendable, and some portion $\mathcal{T}_k$ of $M[k]$ is compatible with $M[i]$ and is such that conditions (C1)–(C3) are satisfied.

Lemma 9. *Algorithm 1 is correct and runs in $O(n^3)$ time.*

Proof (sketch). The first observation is that, if the algorithm sets *appendable*[i] = T, then $M[i] \neq \emptyset$ and there exists a valid sequence of tabular subsolutions finishing with $M[i]$. This happens because the algorithm checks precisely the conditions defining valid sequences. Thus, if the algorithm answers "yes", it finds a sequence satisfying the conditions of Lemma 8. Such a sequence can be trivially transformed into a solution to the problem.

Conversely, if there exists a leftmost solution where the indices of the intervals having a representative at the leftmost endpoint are $i_1, i_2, \ldots, i_m$, it is easy to show by induction that, for each of these indices, the algorithm sets *appendable*[·] := T. Since the leftmost solution contains a representative for I_n, one of these indices will then satisfy both conditions of line 12 and the algorithm will answer "yes".

Let us next analyze the running time. Each $M[i]$ can be computed in $O(n)$ time by scanning the set of intervals from left to right and placing the representatives of $M[i]$ also from left to right.

In the main loop, let $k < i$ be such that *appendable*[k] = T. To check conditions (C1)–(C3), a possible approach is as follows: suppose that the associated valid sequence of tabular subsolutions finishing with $M[k]$ is stored. Then, in this sequence we replace $M[k]$ by the longest $\mathcal{T}_k$ such that $\mathcal{T}_k$ and $M[i]$ are compatible. To check (C1)–(C3), we traverse the obtained sequence and check that the relevant intervals are indeed represented there; in the case of (C3), we also check that the position of the representative is $\leq x_i - q$. Thus, the test for a particular k takes $O(n)$ time and the main loop takes $O(n^3)$ time. □

4.2   Speed Up of the Algorithm

The goal of this subsection is to describe a more sophisticated version of the algorithm running in $O(n \log n)$ time. Due to lack of space, we focus on the two main parts of the algorithm.

Computing $M[i]$. If x_i is a forbidden position for r_i, $M[i] = \emptyset$. This can be tested in $O(\log n)$ time. In the following, let us assume that $M[i] \neq \emptyset$.

In the new algorithm, it is enough to be able to compute the position of the rightmost representatives of $M[i]$. Suppose that they have color c and they are at $x_i + kq$. We can observe that either: (i) there is no interval of color $\bar{c}$ that contains the point $x_i + (k+1)q$; or (ii) there is such an interval, but the position $x_i + (k+1)q$ is forbidden for that interval.

To take care of (i), we describe a preprocessing of the intervals that allows to detect in $O(\log n)$ time the first time that there are no intervals of the appropriate color containing some point of the form $x_i + (k+1)q$.

Suppose that we are dealing with red color. We take the minimum segment spanning all the intervals, and we divide it into the *red region*, containing all points of the segment covered by at least one of the red intervals, and the *white region*, containing the rest. Afterwards, we "cut" the segment into portions of length $2q$. Our assumptions imply that the number of such portions is $O(n)$. We then translate these portions of length $2q$ in the plane as follows: All portions

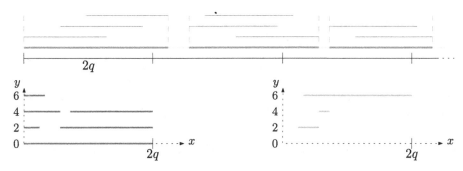

Fig. 5. Construction of T^r. Top: red input intervals (and the red region). Bottom: left, portions of length $2q$ have been cut and translated; right: final set of segments T^r. (Color figure online)

have their left endpoint on the line $x = 0$. Additionally, the portion $[0, 2q)$ is placed at $y = 0$, the portion $[2q, 4q)$ is placed at $y = 2$, the portion $[4q, 6q)$ is placed at $y = 4$, and so on (see Fig. 5 for an example). Finally, we make the red regions disappear and the white regions become a set of (open) segments in the plane, called T^r. We then preprocess T^r in $O(n \log n)$ time and $O(n)$ space to answer ray-shooting queries into upwards direction in $O(\log n)$ time [3].

Let $2jq + \delta$, where j is a non-negative integer and $\delta \in [0, 2q)$, be the position of the leftmost red representative of $M[i]$. Then the red representatives of $M[i]$ are at $2jq + \delta$, $2(j+1)q + \delta$, $2(j+2)q + \delta \ldots$ Consider the following query: If we shoot a vertical ray with origin at $(\delta, 2j+1)$ in upwards direction, when is the first time that it intersects a segment of T^r? If the answer to this query is the point $(\delta, 2k)$ $(k > j)$, the first time that we cannot place the next red representative as there is no red interval covering that position happens at $2kq + \delta$.

We take care of (ii) similarly: In the preprocessing phase, we mark all portions corresponding to forbidden positions for one of the colors. For example, if I_j is blue, all points in $(x_j + (1 - q), x_j + q)$ are forbidden for the red representatives. The union of all such segments gives the *dark red* region. As in the previous case, we cut the spanning interval into portions of length $2q$, and we distribute the portions into a vertical column of segments of length $2q$. The starting point of a maximal tabular solution $M[i]$ tells us from which position we shoot a vertical ray upwards and compute the first intersection with a dark red region.

Naturally, we create analogous data structures to deal with color blue. All in all, the total cost of computing all the $M[i]$ becomes $O(n \log n)$.

Concatenating Two Subsolutions. The other bottleneck of the algorithm is the test in line 9.

The first observation is that, by Lemma 7, for each of the sets $\mathcal{I}_i^{s,<}$, $\mathcal{I}_i^{o,<<}$ and $\mathcal{I}_i^{o,<}$, it is enough to check that the *rightmost* interval in the set has representative in $\tilde{L}[k]$. We use the notation, $\overrightarrow{\mathcal{I}}_i^{\cdots}$ to denote such rightmost intervals.

Let us first suppose that $\mathcal{I}_i^{o,<} = \emptyset$. Then, it is enough to determine if there exists $k < i$ such that $appendable[k] = T$ and some portion T_k of $M[k]$ is compatible with $M[i]$ and is such that $\tilde{L}[k]$ contains representatives for $\overrightarrow{\mathcal{I}}_i^{s,<}$ and $\overrightarrow{\mathcal{I}}_i^{o,<<}$. With some preprocessing, this query can be answered fast: Every time that $appendable[k]$ is set to T, we add to a data structure S the point $(last^r[k], last^b[k])$, where $last^r[k]$ and $last^b[k]$ store the index of the rightmost red and blue interval with representative in the valid sequence of tabular subsolutions (found by the algorithm) finishing with $M[k]$. Then, if say, $\overrightarrow{\mathcal{I}}_i^{s,<}$ is red and $\overrightarrow{\mathcal{I}}_i^{o,<<}$ is blue, we need to determine if there is some point $(x, y) \in S$ such that $x \geq \mathrm{ind}(\overrightarrow{\mathcal{I}}_i^{s,<})$ and $y \geq \mathrm{ind}(\overrightarrow{\mathcal{I}}_i^{o,<<})$ (where $\mathrm{ind}(I_j)$ denotes the index of interval I_j). This is a 2-sided orthogonal range query for points in the plane, which can be answered in $O(\log n)$ time after the appropriate preprocessing.

Next, let us suppose that $\mathcal{I}_i^{o,<} \neq \emptyset$. In this case, we can prove that checking (C3) is enough:

Lemma 10. *Let S be a valid sequence of tabular subsolutions. Let $M[k]$ be the rightmost subsolution of S, I_i be an interval of color c with $i > k$, and T_k be the longest tabular subsolution starting at I_k that is compatible with $M[i]$. Suppose that $M[i] \neq \emptyset$, $\mathcal{I}_i^{o,<} \neq \emptyset$, and $T_k \neq \emptyset$. If T_k and $M[i]$ satisfy condition (C3), then T_k and $M[i]$ also satisfy conditions (C1) and (C2).*

To check (C3), we use a preprocessing similar to that for computing $M[i]$. Let us consider a red interval I_i such that $M[i] \neq \emptyset$ and $\mathcal{I}_i^{o,<} \neq \emptyset$. Then $\overrightarrow{\mathcal{I}}_i^{o,<}$ is a blue interval, and the positions for its representative relevant for (C3) are those in the so-called *good segment* of $\overrightarrow{\mathcal{I}}_i^{o,<}$: $\overrightarrow{\mathcal{I}}_i^{o,<} \cap (-\infty, x_i - q]$. We take the collection S of all such good segments. Exactly as before, we cut the minimum segment spanning all the intervals into portions of size $2q$, and we place the portions in a vertical column of segments. The set S is cut consistently and thus becomes a collection of horizontal segments at different heights. We call it G^b.

Let us next consider some k such that $k < i$ and $appendable[k] := T$. The positions of the blue representatives of $M[k]$ (if any) are $2jq + \delta, 2(j + 1)q + \delta, \ldots, 2j'q + \delta$, where j, j' are non-negative integers, $j' \geq j$ and $\delta \in [0, 2q)$. If the vertical ray starting at $(\delta, 2j)$ and going upwards intersects the good segment of $\overrightarrow{\mathcal{I}}_i^{o,<}$ within the portion up to $(\delta, 2j')$, we derive that $M[k]$ contains a representative for $\overrightarrow{\mathcal{I}}_i^{o,<}$ at a position $\leq x_i - q$, and the pair satisfies (C3). Hence, we perform such ray shooting query repeatedly until we do not find any new good segment intersected by the ray. Every time that a good segment of some $\overrightarrow{\mathcal{I}}_i^{o,<}$ is intersected by the vertical ray, we remove it from G^b and we store $p[i] := k$. For all i such that $\mathcal{I}_i^{o,<} \neq \emptyset$, the test in line 9 simply becomes checking whether $p[i]$ has already been assigned a value.

All in all, the total cost of all the involved operations is $O(n \log n)$.

Observe that segments in S may overlap, which is an issue for standard ray-shooting data structures. Hence, the described general approach is slightly modified to handle this issue.

Theorem 3. *For $k = 2$, the decision problem for $q \in \left(\frac{1}{2}, \frac{3}{4}\right]$ can be solved in $O(n \log n)$ time.*

5 Case $k = 2$: Optimization Problem

In this section we present an algorithm for the optimization problem. Recall that q^* is the length of an *MCSI* in a realization solving the problem.

Lemma 11. *There exist i, j with $i < j$ such that $q^* = \frac{x_j + 1 - x_i}{m}$, where $m \in \mathbb{N}$.*

How many values of type $(x_j + 1 - x_i)/m$ are there? Since $q^* \leq 2$, we can assume that no two consecutive intervals leave a gap of length 2 or greater. Together with $x_1 = 0$, this implies that $x_n + 1 < 3n$. Recall also that $\frac{1}{2} \leq q^*$. For fixed i, j with $i < j$, we have that $\frac{1}{2} \leq \frac{x_j + 1 - x_i}{m} \leq 2$, with $m \in \mathbb{N}$, if and only if $\frac{x_j + 1 - x_i}{2} \leq m \leq 2(x_j + 1 - x_i)$. Since $x_j + 1 - x_i < 3n$, there are less than $6n$ values of $m \in \mathbb{N}$ satisfying the equations. Thus, the total number of candidates for q^* is $O(n^3)$. To avoid listing all of them, we proceed as follows.

We first run a binary search on the values of type $\frac{x_n + 1 - x_1}{m} \in \left[\frac{1}{2}, 2\right]$. In $O(n \log^2 n)$ time we find some $m_0 \in \mathbb{N}$ such that $\frac{x_n + 1 - x_1}{m_0 + 1} \leq q^* < \frac{x_n + 1 - x_1}{m_0}$. For any other i, j with $i < j$, we are then only interested in the values of the form $\frac{x_n + 1 - x_1}{m_0 + 1} < \frac{x_j + 1 - x_i}{m} < \frac{x_n + 1 - x_1}{m_0}$. Hence, we have $\frac{x_j + 1 - x_i}{x_n + 1 - x_1} m_0 < m < \frac{x_j + 1 - x_i}{x_n + 1 - x_1}(m_0 + 1)$. Since $x_j - x_i < x_n - x_1$, this equation has at most one solution. Thus, each pair i, j with $i < j$ gives at most one candidate and we can run a second binary search in this set of $O(n^2)$ candidates to find q^*. In conclusion:

Theorem 4. *For $k = 2$, the L-MCSI problem can be solved in $O(n^2 \log n)$ time.*

Additionally, given some $\varepsilon > 0$, by using the bisection method on $\left[\frac{1}{2}, 2\right]$, we can obtain some value $\hat{q}$ such that $q^* \geq \hat{q} > q^* - \varepsilon/2$ in $O(n \log n \log \frac{1}{\varepsilon})$ time. Since $\hat{q} > q^* - \varepsilon/2 \geq q^* - \varepsilon q^* = q^*(1 - \varepsilon)$, we obtain

Corollary 2. *For $k = 2$, we can compute a $(1 - \varepsilon)$-approximation for the L-MCSI problem in $O(n \log n \log \frac{1}{\varepsilon})$ time.*

References

1. Abellanas, M., et al.: Smallest color-spanning objects. In: auf der Heide, F.M. (ed.) ESA 2001. LNCS, vol. 2161, pp. 278–289. Springer, Heidelberg (2001). https://doi.org/10.1007/3-540-44676-1_23
2. Acharyya, A., Jallu, R.K., Keikha, V., Löffler, M., Saumell, M.: Minimum color spanning circle of imprecise points. Theor. Comput. Sci. **930**, 116–127 (2022)
3. de Berg, M. (ed.): Ray shooting into a fixed direction. In: Ray Shooting, Depth Orders and Hidden Surface Removal. LNCS, vol. 703, pp. 67–84. Springer, Heidelberg (1993). https://doi.org/10.1007/BFb0029819
4. Chen, D.Z., Misiołek, E.: Algorithms for interval structures with applications. Theor. Comput. Sci. **508**, 41–53 (2013)

5. Das, S., Goswami, P.P., Nandy, S.C.: Smallest color-spanning object revisited. Int. J. Comput. Geom. Appl. **19**(5), 457–478 (2009)
6. Fiala, J., Kratochvíl, J., Proskurowski, A.: Systems of distant representatives. Discret. Appl. Math. **145**(2), 306–316 (2005)
7. Fleischer, R., Xu, X.: Computing minimum diameter color spanning sets is hard. Inf. Process. Lett. **111**(21–22), 1054–1056 (2011)
8. Hu, R., Zhang, J.: Computing k-centers of uncertain points on a real line. Oper. Res. Lett. **50**(3), 310–314 (2022)
9. Li, S., Wang, H.: Dispersing points on intervals. Discret. Appl. Math. **239**, 106–118 (2018)
10. Löffler, M., van Kreveld, M.: Largest and smallest convex hulls for imprecise points. Algorithmica **56**(2), 235–269 (2010)
11. Löffler, M., van Kreveld, M.: Largest bounding box, smallest diameter, and related problems on imprecise points. Comput. Geom. **43**(4), 419–433 (2010)
12. Naredla, A.M.: Algorithms for Geometric Facility Location: Centers in a Polygon and Dispersion on a Line. Ph.D. thesis, University of Waterloo (2023)

Total Domination, Separated-Cluster, CD-Coloring: Algorithms and Hardness

Dhanyamol Antony[1]([envelope]), L. Sunil Chandran[1]([envelope]), Ankit Gayen[2]([envelope]),
Shirish Gosavi[1]([envelope]), and Dalu Jacob[1]([envelope])

[1] Indian Institute of Science, Bengaluru, India
{dhanyamola,sunil,shirishgp,dalujacob}@iisc.ac.in
[2] Chennai Mathematical Institute, Chennai, India
ankitg@cmi.ac.in

Abstract. Domination and coloring are two classic problems in graph theory. In this paper, our major focus is on the CD-COLORING problem, which incorporates the flavors of both domination and coloring in it. Let G be an undirected graph. A proper vertex coloring of G is said to be a *cd-coloring*, if each color class has a dominating vertex in G. The minimum integer k for which there exists a *cd-coloring* of G using k colors is called the *cd-chromatic number* of G, denoted as $\chi_{cd}(G)$. A set $S \subseteq V(G)$ is said to be a *total dominating set*, if any vertex in G has a neighbor in S. The *total domination number* of G, denoted as $\gamma_t(G)$, is defined to be the minimum integer k such that G has a total dominating set of size k. A set $S \subseteq V(G)$ is said to be a *separated-cluster* (also known as *sub-clique*) if no two vertices in S lie at a distance exactly 2 in G. The *separated-cluster number* of G, denoted as $\omega_s(G)$, is defined to be the maximum integer k such that G has a separated-cluster of size k.

In this paper, we contribute to the literature connecting CD-COLORING with the problems, TOTAL DOMINATION and SEPARATED-CLUSTER. For any graph G, we have $\chi_{cd}(G) \geq \gamma_t(G)$ and $\chi_{cd}(G) \geq \omega_s(G)$. First, we explore the connection of CD-COLORING problem to the well-known problem TOTAL DOMINATION. Note that TOTAL DOMINATION is known to be NP-Complete for *triangle-free 3-regular graphs*. We generalize this result by proving that both the problems CD-COLORING and TOTAL DOMINATION are NP-Complete, and do not admit any subexponential-time algorithms on *triangle-free d-regular graphs, for each fixed integer $d \geq 3$*, assuming the Exponential Time Hypothesis. We also study the relationship between the parameters $\chi_{cd}(G)$ and $\omega_s(G)$. Analogous to the well-known notion of 'perfectness', here we introduce the notion of 'cd-perfectness'. We prove a sufficient condition for a graph G to be *cd-perfect* (i.e. $\chi_{cd}(H) = \omega_s(H)$, for any induced subgraph H of G). Our sufficient condition is also necessary for certain graph classes (like *triangle-free* graphs). This unified approach of 'cd-perfectness' has several exciting consequences. In particular, it is interesting to note that the same framework can be used as a tool to derive both positive and negative results concerning the algorithmic complexity of CD-COLORING and SEPARATED-CLUSTER.

J. A. Soto and A. Wiese (Eds.): LATIN 2024, LNCS 14578, pp. 97–113, 2024.
https://doi.org/10.1007/978-3-031-55598-5_7

Keywords: Total Domination · CD-coloring · Separated-Cluster ·
CD-perfectness · Triangle-free d-regular graphs

1   Introduction

In graph theory, one of the prominent directions of research is to study the
relationship between any two correlated graph parameters and explore the algorithmic consequences on the graphs for which these two parameters coincide. For
instance, for a given graph, the *chromatic number* χ and the *clique number* ω,
are two such classic parameters in the literature. It is a well-known fact that for
any graph G, $\chi(G) \geq \omega(G)$. Perfect graphs are exactly the graphs G having the
property that for any induced subgraph H of G, $\chi(H) = \omega(H)$. The notion of
perfectness thereby unifies the results concerning *colorings* and *cliques* for many
important graph classes. The celebrated *'Strong Perfect Graph Theorem'* [1]
gives a different perspective on perfect graphs by characterizing them by their
structure instead of parameters. Along these lines of research, here we explore
the interconnections between a few related graph parameters. Domination and
coloring are two important and well-motivated problems in graph theory. The
central problem, *'cd-coloring,'* in this paper incorporates the flavors of both domination and coloring. Even though the other two problems, *'total domination'*
and *'separated-cluster'* have their own significance and are of independent interest, they share an interesting relationship with the *'cd-coloring'* problem. In
this paper, we explore these relationships in detail and obtain several exciting
algorithmic consequences.

Let G be an undirected graph and let $c : V(G) \longrightarrow [k]$ be a proper vertex coloring of G that partitions the vertex set into k color classes, say $C_1, C_2, \ldots, C_k$.
Then c is said to be a *cd-coloring* of G if, for each $j \in \{1, 2, \ldots, k\}$, there exists
a vertex $v_j \in V(G)$ such that $C_j \subseteq N[v_j]$. Note that each non-trivial color
class C_j in G has to be dominated by a vertex $v_j \in V(G) \setminus C_j$. Hence the
name *class-domination* coloring, which we shortly call *cd*-coloring. The minimum integer k for which there exists a *cd-coloring* of G using k colors is called
the *cd-chromatic number* of G, denoted as $\chi_{cd}(G)$. Given an input graph G,
the problem CD-COLORING seeks to find the *cd-chromatic number* of G. CD-
COLORING is known to be NP-Complete for several special classes of graphs,
including bipartite graphs [2] and chordal graphs [3] and polynomial-time solvable for graph classes like trees [4], co-bipartite graphs [4], split graphs [2], and
claw-free graphs [3]. Shalu et al. [3] obtained a complexity dichotomy for CD-
COLORING for H-free graphs. CD-COLORING is also studied in the paradigm
of parameterized complexity [5,6] and approximation complexity [7]. In addition to its theoretical significance, CD-COLORING has a wide range of practical
applications in social networks [7] and genetic networks [8].

A set $S \subseteq V(G)$ is said to be a *total dominating set* if any vertex in G has a
neighbor in S. The *total domination number* of G, denoted as $\gamma_t(G)$, is defined
to be the minimum integer k such that G has a total dominating set of size k.
Given an input graph G, in the problem TOTAL DOMINATION, we intend to find

the total domination number of G. Total domination is one of the most popular variants of domination, and there are hundreds of research papers dedicated to this notion in the literature [2, 9–11].

A set $S \subseteq V(G)$ is said to be a *separated-cluster* (also known as *sub-clique*) if no two vertices in S lie at a distance exactly 2 in G. The *separated-cluster number* of G, denoted as $\omega_s(G)$, is defined to be the maximum integer k such that G has a separated-cluster of size k. Given an input graph G, in the problem SEPARATED-CLUSTER, our goal is to find the *separated-cluster number* of G. Note that a disjoint union of cliques in a graph G is called a *cluster* in G. We introduce this name, *'separated-cluster'* (instead of the existing name, *subclique*) because if no two vertices in S lie at a distance exactly 2 in G, then S is, in fact, a disjoint union of cliques, where any pair of cliques are separated by a distance greater than 2 in G (i.e. no two vertices belonging to two distinct cliques in S can be adjacent in G or have a common neighbor in G). Using this perspective, we can infer that SEPARATED-CLUSTER is a restricted version of the well-known problem, CLUSTER VERTEX DELETION, where we intend to find minimum number of vertices whose deletion results in a disjoint union of cliques [12] (or, alternatively, to find maximum number of vertices that can be partitioned into disjoint cliques). SEPARATED-CLUSTER is known to be NP-Complete for graph classes like bipartite graphs, chordal graphs, $3K_1$-free graphs [13], and polynomial-time solvable for trees, co-bipartite graphs, cographs, split graphs [13].

1.1   Total Domination and *cd*-Coloring

Domination and graph coloring problems are often related. There exist variants of graph coloring problems in the literature that concern domination. For instance, see [4, 13–15]. The close relationship between *CD-coloring* and TOTAL DOMINATION was established by Merouane et al. [2]. They proved that for any triangle-free graph G, we have $\gamma_t(G) = \chi_{cd}(G)$. It is also known that TOTAL DOMINATION is NP-Complete for bipartite graphs with bounded degree 3 and triangle-free cubic graphs [11]. Even though several bounds for γ_t are known for regular graphs [9], to the best of our knowledge, the complexity of TOTAL DOMINATION remains unknown for triangle-free d-regular graphs for each fixed integer $d \geq 4$.

Our Results: The following theorem is one of the major results in this section.

Theorem 1. CD-COLORING *is NP-Complete on triangle-free d-regular graphs, for each fixed integer $d \geq 3$. Further, the problem cannot be solved in time $2^{o(|V(G)|)}$, unless the Exponential-Time Hypothesis fails.*

It is fascinating to observe how a relatively new graph parameter, like *cd-coloring*, sheds light on a classic problem of total domination. For instance, the following corollary is an easy consequence of Theorem 1, and the fact that for any triangle-free graph G, $\gamma_t(G) = \chi_{cd}(G)$.

Corollary 1. *Total domination on triangle-free d-regular graphs is NP-Complete, for any constant $d \geq 3$. Further, the problem cannot be solved in time $2^{o(|V(G)|)}$, unless the Exponential-Time Hypothesis fails.*

1.2 Separated-Cluster and cd-Coloring

Interestingly, we can see that the notions of cd-chromatic number χ_{cd} and separated-cluster number ω_s also follow a similar relationship as their classic counterparts of chromatic number χ and clique number ω. Since any two vertices in a separated-cluster of a graph G cannot lie at a distance exactly 2 in G, we have $\chi_{cd}(G) \geq \omega_s(G)$. Moreover, using the same idea of 'Mycielskian' construction [16] (in classic coloring), we can have a family of graphs for which $\omega_s = 2$ but χ_{cd} is arbitrarily large.

By the definition of separated-cluster, a natural way to look at this problem is by considering the following auxiliary graph.

Definition 1 (Auxiliary Graph G^* [13]). *Given a graph G, the auxiliary graph G^* is the graph having $V(G^*) = V(G)$ and $E(G^*) = \{uv : u, v \in V(G)$ and $d_G(u, v) = 2\}$, where $d_G(u, v)$ denote the distance between the vertices u and v in G.*

It is easy to infer that $\omega_s(G) = \alpha(G^*)$, where $\alpha(G^*)$ is the independence number (the size of maximum cardinality independent set) of G^*. For a graph G, we observe that $\chi_{cd}(G) \geq k(G^*)$, where $k(G^*)$ is the clique cover number (the size of a minimum number of cliques needed to partition the vertex set) of G^*. Note that the parameters α and k are well-studied in the literature, particularly in connection with perfect graphs. By an equivalent definition of perfect graphs, we have that a graph G is perfect if for any induced subgraph H of G, $\alpha(H) = k(H)$.

Our Results: Motivated by the close relationship between the parameters $\chi_{cd}(G)$ and $\omega_s(G)$, respectively with the parameters $\alpha(G^*)$ and $k(G^*)$ of a graph G, here we initiate the study of 'cd-perfectness'. We say that a graph G is cd-perfect if for any induced subgraph H of G, we have $\chi_{cd}(H) = \omega_s(H)$. In some of the earlier works on cd-coloring, the researchers have observed that for certain classes of graphs, say $\mathcal{C}$, we have $\chi_{cd}(G) = \omega_s(G)$, for each graph $G \in \mathcal{C}$ [4,13]. But to the best of our knowledge, no one has tried to study the structural properties of the graphs for which these two parameters coincide. This general notion of cd-perfectness allows us to unify several (existing) results concerning cd-coloring and separated-clusters for several graph classes.

As an attempt to understand the structure of cd-perfect graphs, first, in Theorem 5, we prove a sufficient condition for a graph G to satisfy the equality, $\chi_{cd}(G) = k(G^*)$. This further leads us to a sufficient condition for a graph to be cd-perfect (Corollary 2). We then use this to bring several graph classes, like co-bipartite graphs, chordal bipartite graphs etc. under the common umbrella of cd-perfect graphs. Even though these results look structural in nature, they have several exciting algorithmic consequences. It is interesting to note that the same framework can be used as a tool to derive both positive and negative results concerning the algorithmic complexity of CD-COLORING and SEPARATED-CLUSTER. For instance,

- We use Theorem 5 to provide a unified approach (mostly with an improvement) for finding polynomial-time algorithms for CD-COLORING on graph classes including proper interval graphs, $3K_1$-free graphs, etc.

- A beautiful interplay between the problems, TOTAL DOMINATION, CD-COLORING, and SEPARATED-CLUSTER can be witnessed in Theorem 10, where we use Corollary 2, Theorem 9, and the fact that $\gamma_t(G) = \chi_{cd}(G)$ for triangle free-graphs G, to prove that the three problems mentioned above are equivalent and solvable in $O(n^2)$ time for chordal bipartite graphs. Consequently, this result improves and generalizes the existing $O(n^3)$-time algorithm for SEPARATED-CLUSTER on the class of P_6-free chordal bipartite graphs [17].

- We apply Theorem 5 and Proposition 3 to derive hardness results for both CD-COLORING and SEPARATED-CLUSTER for C_6-free bipartite graphs.

2   Preliminaries and Notations

All graphs considered in this paper are undirected, simple, and finite. For standard graph-theoretic definitions, notations, and terminologies, we refer to the book "Introduction to graph theory" by West [18]. However, here we present some expressions that are specific to this article. Let G be a graph. For a vertex $v \in G$, we denote the *open neighborhood* of v in G by $N_G(v)$ and the *closed neighborhood* of v in G by $N_G[v] = N_G(v) \cup \{v\}$ (sometimes, we omit the subscript, if the graph G is clear from the context). The degree of a vertex v in G is denoted as $d_G(v)$ and the maximum degree of a graph G by $\Delta(G)$. We denote by $d_G(u,v)$, the *distance* between two vertices u and v in G. The *square* of a graph G, denoted by G^2, is the graph having $V(G^2) = V(G)$, and $E(G^2) = \{uv : d_G(u,v) \leq 2\}$. We denote by $G - H$, the graph obtained from G by removing the vertices in H, i.e. $G - H = G[V(G) \setminus V(H)]$.

A graph G is said to be H-*free*, if it does not contain H as an induced subgraph. A graph G is said to be *d-regular* if every vertex of G has degree d. We denote the *star graph* on $t + 1$ vertices by $K_{1,t}$. A *claw*, $K_{1,3}$, is a star on 4 vertices. A *chordal bipartite graph* is a bipartite graph that does not contain any induced cycle of length k, where $k \geq 6$.

Even though the parameters χ_{cd} and ω_s have a strong similarity to their classic counterparts, χ, and ω, here we note an important property of χ_{cd} and ω_s, which is fundamentally different from χ and ω.

Note: For any induced subgraph H of a graph G, we have $\chi(H) \leq \chi(G)$ and $\omega(H) \leq \omega(G)$. But it can be seen that this is *not true* in general for the parameters χ_{cd} and ω_s. i.e. it is possible for G to have an induced subgraph H such that $\chi_{cd}(H) > \chi_{cd}(G)$ or $\omega_s(H) > \omega_s(G)$. For instance, let $G = K_{1,3}$, the star graph, with central vertex, say, u, and leaf vertices, say, v_1, v_2, v_3. It is easy to verify that $\chi_{cd}(G) = \omega_s(G) = 2$. On the other hand, consider the subgraph induced by leaf vertices, say, $H = G[\{v_1, v_2, v_3\}]$. We then have $\chi_{cd}(H) = \omega_s(H) = 3 > 2 = \omega_s(G) = \chi_{cd}(G)$.

In the following observation, we have a natural sufficient condition for an induced subgraph H of a graph G to have $\chi_{cd}(G) \geq \chi_{cd}(H)$. We use this observation later.

Observation 2. *Let H be an induced subgraph of the graph G. Suppose there exist no independent set $I \subseteq V(H)$ of vertices such that I has a dominating vertex in G but not in H, then $\chi_{cd}(G) \geq \chi_{cd}(H)$.*

We also make use of the following known results.

Proposition 1 [11]. *Total domination in bipartite graphs with bounded degree 3 is NP-Complete. Further, the problem cannot be solved in time $2^{o(|V(G)|)}$, unless the Exponential-Time Hypothesis fails.*

Proposition 2 [2]. *Let G be a triangle-free graph. Then $\chi_{cd}(G) = \gamma_t(G)$.*

3   Total Domination and *cd*-Coloring in Triangle-Free *d*-Regular Graphs

In this section, we prove Theorem 1, i.e. we derive the hardness result for CD-COLORING in triangle-free d-regular graphs for each fixed integer $d \geq 3$. As a corollary of this result, we obtain the hardness result for TOTAL DOMINATION in triangle-free d-regular graphs for each fixed integer $d \geq 3$. Note that the reduction for the hardness given in [11] for triangle-free cubic graphs does not seem to have an easy generalization. Hence, first, we propose a simple linear reduction for triangle-free cubic graphs. Then, we generalize this reduction for triangle-free d-regular graphs for each fixed integer $d \geq 3$, by using two linear reductions; one for odd values of d and the other for even values.

The Exponential-Time Hypothesis (ETH) along with the Sparsification Lemma implies that 3-SAT cannot be solved in subexponential-time, i.e. in time $2^{o(n+m)}$, where n is the number of variables and m is the number of clauses in the input formula. To prove that a problem does not admit a subexponential-time algorithm, it is sufficient to obtain a linear reduction from a problem which does not admit a subexponential time algorithm, where a linear reduction is a polynomial-time reduction in which the size of the resultant instance is linear in the size of the input instance. We refer to Chap. 14 of the book [19] for more details about these concepts. All reductions mentioned in this section are trivially linear.

To prove Theorem 1, we use three constructions. Construction 1 is used for a reduction from TOTAL DOMINATION on bipartite graphs with bounded degree 3 to CD-COLORING on triangle-free cubic graphs. Construction 2 is used for a reduction from CD-COLORING on triangle-free $(d-2)$-regular graphs to CD-COLORING on triangle-free d-regular graphs, for each odd integer $d \geq 5$, whereas Construction 3 is used for a reduction from CD-COLORING on triangle-free $(d-1)$-regular graphs to CD-COLORING on triangle-free d-regular graphs, for each even integer $d \geq 4$.

Construction 1. *Let G be a bipartite graph with bounded degree 3. We construct a gadget W, as shown in Fig. 1, in the following way:*

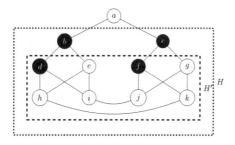

Fig. 1. The gadget W used in Construction 1. The shaded vertices, d, b, f and c, respectively dominates the color classes, $\{b, h, i\}, \{a, d, e\}, \{c, j, k\}$, and $\{f, g\}$, in a valid *cd*-coloring of W.

- *Introduce a star graph $K_{1,2}$, which we denote by S, having a as the root vertex, and b, c as the leaves. Note that we also call 'a' as the root vertex of W.*
- *Then we introduce two copies of the complete bipartite graph $K_{2,2}$, denoted by $C_1 = (A_1, B_1)$ and $C_2 = (A_2, B_2)$, respectively. Now the leaf b (resp. c) of S is adjacent to each vertex of the partition A_1 (resp. A_2) of C_1 (resp. C_2). Furthermore, each vertex of the partition B_1 of C_1 is adjacent to exactly one vertex of the partition B_2 of C_2 as shown in Fig. 1. Let $H = W - \{a\}$ and $H' = H - \{b, c\}$.*
- *Thereafter we construct a graph G_c from G using the gadget W as follows: for every vertex u of G with degree 2, introduce a gadget W such that the root vertex a of W is adjacent to u in G. Similarly, for each vertex v in G of degree 1, we introduce two copies of W such that the root vertices of both the gadgets are adjacent to v in G.*

The graph $G_c - G$ contains $11x + 22y$ vertices, where x and y denote the number of vertices in G having degree 2 and degree 1, respectively.

An example of the Construction 1 corresponding to the bipartite graph G with bounded degree 3 given in Fig. 2a is as shown in Fig. 2b. By Observation 2, it is clear that the number of colors needed for $V(H')$ in any *cd*-coloring of H is at least 4. Now it is easy to see that any 4-*cd*-coloring of a cycle C_6 in each H' can be extended to a 4-*cd*-coloring of the corresponding gadget W (by using the leaves of star graph S, and one of the vertices from the partition A_i of each C_i, for $i \in \{1, 2\}$, as the dominating vertices of the color classes). Note that none of the colors used for the vertices in the gadgets of G_c obtained in Construction 1 are used to color any vertex in G. Thus, we have the following lemma, based on the Construction 1, and as a consequence, we have Theorem 3.

Lemma 1. *Let G be a bipartite graph with bounded degree 3 and G_c be a graph obtained from Construction 1. Then $\gamma_t(G) = k$ if and only if $\chi_{cd}(G_c) = k + 4x + 8y$, where x and y denote the number of vertices in G having degree 2 and degree 1, respectively.*

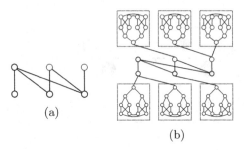

Fig. 2. (a) is a bipartite graph G with bounded degree 3, and (b) is the resultant graph G_c obtained from Construction 1 corresponding to the bipartite graph G in (a). The dotted rectangles represent the gadget W used for the construction.

Theorem 3. CD-COLORING *is NP-Complete on triangle free cubic graphs. Further, the problem cannot be solved in time* $2^{o(|V(G)|)}$*, unless the ETH fails.*

Proof. Notice that G_c has $|V(G)| + 11(x + 2y)$ vertices, where x and y are the number of vertices in G having degree 2 and degree 1, respectively. We know that there is a linear reduction from TOTAL DOMINATION on bipartite graphs with bounded degree 3 to CD-COLORING on triangle-free cubic graphs due to Lemma 1. Thus, by using Proposition 1, we are done.

Now, we generalize Construction 1 to prove the hardness of CD-COLORING on triangle-free d-regular graphs, for any constant $d \geq 4$. The following construction is used to prove the hardness of CD-COLORING on triangle-free d-regular graphs, for any odd integer $d \geq 5$.

Construction 2. *Let G be a triangle-free $(d-2)$-regular graph, for any odd integer $d \geq 5$. Let W be a gadget which is constructed in the following way.*

- *Introduce a star graph $K_{1,d-1}$, which we denote by S, having the vertex a as the root vertex. Note that we also call 'a' as the root vertex of W.*
- *Further introduce $2(d-1)^2$ vertices which induces $d-1$ disjoint copies of complete bipartite graphs $K_{d-1,d-1}$, namely $C_1, C_2 \ldots, C_{d-1}$, respectively. The adjacency between these complete bipartite graphs $C_i = (A_i, B_i)$, for $1 \leq i \leq (d-1)$, and the star graph S is in such a way that the vertices of A_i of the complete bipartite graph C_i is adjacent to the leaf vertex v_i of S. Furthermore, for each odd integer $i \leq (d-2)$, each vertex in the partite set B_i of C_i is adjacent to exactly one vertex of the partite set B_{i+1} of C_{i+1} (for instance refer Fig. 3, when $d = 5$). Now, by H_i' we denote the subgraph of W induced by the vertices in the two complete bipartite graphs C_i and C_{i+1}, for an odd integer $i \leq d - 2$. Then, by H_i we denote the subgraph of W induced by the vertices in H_i' and the two leaves v_i and v_{i+1} of the star graph S. Note that $H_i' = H_i - \{v_i, v_{i+1}\}$ and the gadget W contains $(2d^2 - 3d + 2)$ vertices.*
- *Now the graph G_c is constructed from G using W in such a way that for each vertex u in G, introduce two copies of W such that root vertices of both the copies of W is attached to u.*

Note that $G_c - G$ contains $2n(2d^2 - 3d + 2)$ vertices, where $n = |V(G)|$.

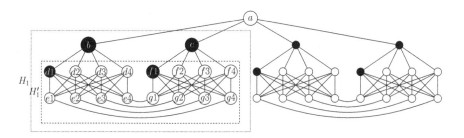

Fig. 3. An example of the gadget W used in Construction 2, for $d = 5$ such that the shaded vertices correspond to the dominating vertices of the color classes in a valid *cd*-coloring of W.

An example of the gadget W in Construction 2 is shown in Fig. 3. By Observation 2, it is clear that the number of colors needed for $V(H_i')$ in any *cd*-coloring of H_i is at least 4. Now it is easy to see that any 4-*cd*-coloring of a C_6 in each H_i' can be extended to a $2(d-1)$-*cd*-coloring of the gadget W (by using the leaves of star graph S, and one of the vertices from the partition A_i of each C_i, for $1 \leq i \leq d-1$, as the dominating vertices of the color classes). Note that none of the colors used for the vertices in the gadgets of G_c obtained in Construction 2 are used to color any vertex in G. Thus, we have the following lemma, based on the Construction 2.

Lemma 2. *For each odd integer $d \geq 5$, let G be a triangle-free $(d-2)$-regular graph, having n vertices. Then $\chi_{cd}(G) = k$ if and only if $\chi_{cd}(G_c) = k+4n(d-1)$.*

Now, the following construction is used to prove the hardness of CD-COLORING on triangle-free d-regular graphs, for any even integer $d \geq 4$.

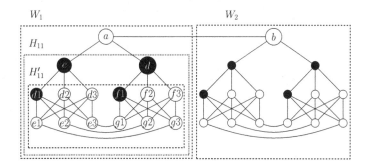

Fig. 4. An example of the gadget W used in Construction 3, for $d = 4$, such that shaded vertices correspond to the dominating vertices of the color classes in a valid *cd*-coloring of W.

Construction 3. *Let G be a triangle-free $(d-1)$-regular graph, for any even integer $d \geq 4$. Let W be a gadget which is constructed in the following way.*

- *Introduce two sets of $2d^2 - 5d + 3$ vertices which induces two subgraphs W_1 and W_2 in W.*
- *The adjacency among the vertices in W_1 (resp. W_2) is in such a way that the $d-1$ vertices of W_1 (resp. W_2) induces a star graph $K_{1,d-2}$, which we denote by S_1 (resp. S_2), having the vertex a in W_1 (resp. the vertex b in W_2) as the root vertex. Note that the vertices a and b are adjacent.*
- *Further in each set W_i, for $i \in \{1,2\}$, the remaining $(2d^2 - 6d + 4)$ vertices induces $d-2$ disjoint copies of complete bipartite graphs $K_{d-1,d-1}$, namely $C_{i1}, C_{i2} \ldots, C_{i(d-2)}$ respectively. The adjacency between these complete bipartite graphs C_{ij}, for $i \in \{1,2\}$ and $1 \leq j \leq d-2$ and the star graph S_i, is in such a way that the vertices of one of the partition A_{ij} of a complete bipartite graph C_{ij} is adjacent to the leaf v_j of S_i. Furthermore, for each odd integer $j \leq d-3$, each vertex of the partite set B_{ij} of C_{ij} is adjacent to exactly one vertex of $B_{i(j+1)}$ of $C_{i(j+1)}$ (for instance, refer Fig. 4, when $d = 4$). Now, by H'_{ij} we denote the subgraph of W_i induced by the vertices in the two complete bipartite graphs C_{ij} and $C_{i(j+1)}$, for an odd integer $j \leq d-3$. Then, by H_{ij} we denote the subgraph of W_i induced by the vertices in H'_{ij} and the two leaves v_j and v_{j+1} of the star graph S_i which are adjacent to the partitions A_{ij} and $A_{i(j+1)}$ of C_{ij} and $C_{i(j+1)}$, respectively. Note that $H'_{ij} = H_{ij} - \{v_j, v_{j+1}\}$.*
- *Now the graph G_c is constructed from G using W in the following way: consider an arbitrary pair-wise ordering $(v_1, v_2), (v_3, v_4) \ldots, (v_{n-1}, v_n)$ of vertices, in G. Note that such a pairing is possible, since n is even (as $d-1$ is odd). For a pair of vertices (v_j, v_{j+1}), for odd integer $j \leq n-1$ in this ordering, introduce a gadget W such that the vertex a (resp. b) of W is adjacent to v_j (resp. v_{j+1}) of G.*

The graph $G_c - G$ contains $n(2d^2 - 5d + 3)$ vertices, where $n = |V(G)|$.

An example of the gadget W in Construction 3 is shown in Fig. 4. By Observation 2, it is clear that the number of colors needed for $V(H'_{ij})$, for $i \in \{1,2\}$ and $1 \leq j \leq (d-2)/2$, in any cd-coloring of H_{ij} is at least 4. Now it is easy to see that a 4-cd-coloring of a C_6 in each H'_{ij} can be extended to a $4(d-2)$-cd-coloring of the gadget W (by using the leaves of star graphs S_i, for $i \in \{1,2\}$, and one of the vertices from the partition A_{ij} of each C_{ij}, for $1 \leq j \leq d-2$, as the dominating vertices of the color classes). It is easy to see that none of the colors used for the vertices in the gadgets of G_c obtained in Construction 3 are used to color any vertex in G. Thus we have the following lemma, based on the Construction 3.

Lemma 3. *For even integer $d \geq 4$, let G be a triangle-free $(d-1)$-regular graph having n vertices. Then $\chi_{cd}(G) = k$ if and only if $\chi_{cd}(G_c) = k + 2n(d-2)$.*

Now, we are ready to prove Theorem 1.

Proof of Theorem 1: Note that $|V(G_{c1})| = 2n(2d^2 - 3d + 2) + n$, where G_{c1} is obtained as per construction 2. Similarly, $|V(G_{c2})| = n(2d^2 - 5d + 3) + n$, where

G_{c2} is obtained as per construction 3. Hence, both the constructions are linear with respect to their size of inputs and hence, their associated reductions are also linear. Thus, we know that there is a linear reduction from triangle-free $(d-2)$-regular graphs to triangle-free d-regular graphs for odd integer $d \geq 5$ due to Lemma 2. We also know that there is a linear reduction from triangle-free $(d-1)$-regular graphs to triangle-free d-regular graphs for even integer $d \geq 4$ due to Lemma 3. Therefore, we are done using Proposition 1 and Theorem 3.

4   Separated-Cluster and *cd*-Coloring: *cd*-Perfectness

Here, we introduce the notion of *cd-perfect graphs*, which is defined below.

Definition 2 (*cd*-Perfect Graphs). *An undirected graph G is said to be cd-perfect if for any induced subgraph H of G, we have $\chi_{cd}(H) = \omega_s(H)$.*

In this section, we study $\chi_{cd}(G)$ and $\omega_s(G)$ of a graph G, by relating them to some well-known graph parameters of the auxiliary graph G^*. This simple reduction has various interesting consequences. Recall from Definition 1 that given a graph G, the auxiliary graph G^* is the graph having $V(G^*) = V(G)$ and $E(G^*) = E(G^2) \setminus E(G)$. i.e. $E(G^*) = \{uv : u, v \in V(G) \text{ and } d_G(u,v) = 2\}$. The following proposition trivially follows from the definition of G^*.

Proposition 3. *For any graph G, we have $\omega_s(G) = \alpha(G^*)$.*

We then note the observation below.

Observation 4. *For any graph G, we have $\chi_{cd}(G) \geq k(G^*)$*

Proof sketch: Let $\chi_{cd}(G) = l$, where $I_1, I_2, \ldots, I_l$ denote the corresponding color classes of G. Then, as each color class has a dominating vertex, it can be seen that $\{I_1, I_2, \ldots, I_l\}$ forms a clique cover of G^*.

(a) C_6 (b) C_6^1 (c) C_6^2 (d) C_6^3

Fig. 5. Set of graphs in $\mathcal{H}$

Note that the reverse inequality of Observation 4 is not necessarily true. For instance, consider the graph C_6 (an induced cycle on 6 vertices). It is not difficult to verify that $\chi_{cd}(C_6) = 4$. But as C_6^* is a disjoint union of two triangles, we have that $k(C_6^*) = 2 < \chi_{cd}(C_6)$. In Theorem 5, we prove a sufficient condition for a graph to satisfy the equality in Observation 4. First, we define certain graphs. Consider the graph C_6. Since C_6 is bipartite, observe that $V(C_6)$ can be partitioned into two independent sets, say A and B, where $|A| = |B| = 3$. We denote by C_6^1, C_6^2, and C_6^3, the graphs obtained by respectively adding *1 edge, 2 edges, and 3 edges* to exactly one of the partite sets A or B of C_6 (see Fig. 5). Let $\mathcal{H} = \{C_6, C_6^1, C_6^2, \text{ and } C_6^3\}$. We then have the following theorem.

Theorem 5. *Let G be an $\mathcal{H}$-free graph and G^* its corresponding auxiliary graph. Then $\chi_{cd}(G) = k(G^*)$.*

Proof Sketch: By Observation 4, here it is enough to show that $\chi_{cd}(G) \leq k(G^*)$. Let $k(G^*) = t$, and let $K_1, K_2, \ldots, K_t$ be a clique cover of G^* having size t. Using the fact that G is $\mathcal{H}$-free, we then prove that for each $j \in \{1, 2, \ldots, t\}$, there exists a vertex $v_j \in V(G)$ such that $K_j \subseteq N_G(v_j)$. This further implies that the sets, $K_1, K_2, \ldots, K_t$, form the color classes for a cd-coloring of G, proving that $\chi_{cd}(G) \leq k(G^*)$.

The following corollary that provides *a sufficient condition for a graph G to be cd-perfect* is a consequence of Proposition 3 and Theorem 5.

Corollary 2. *Let G be an $\mathcal{H}$-free graph. If for any induced subgraph H of G, $k(H^*) = \alpha(H^*)$ then G is cd-perfect. Consequently, if G is $\mathcal{H}$-free and H^* is perfect for any induced subgraph H of G, then G is cd-perfect.*

It is known in the literature that if G is a co-bipartite graph (complement of a bipartite graph), then $\chi_{cd}(G) = \omega_s(G)$ [4]. Since co-bipartite graphs are $\mathcal{H}$-free and perfect, by Corollary 2, we now have a simple and shorter proof for the same.

Note: Consider a graph $H \in \mathcal{H}$. It is not difficult to see that $\chi_{cd}(H) = 3 = \omega_s(H)$, if $H \neq C_6$, and $\chi_{cd}(H) = 4 > 2 = \omega_s(H)$, if $H = C_6$. This is why the sufficient condition for cd-perfectness given in Corollary 2 is not a necessary condition. To obtain a necessary and sufficient condition, we consider the set $\mathcal{H}' = \{C_6^1, C_6^2, C_6^3\}$. Then, in the following theorem, we have a characterization for $\mathcal{H}'$-free graphs (a superclass of triangle-free graphs, $3K_1$-free graphs, etc.) to be cd-perfect.

Theorem 6. *Let G be an $\mathcal{H}'$-free graph. Then G is cd-perfect if and only if G is C_6-free and $k(H^*) = \alpha(H^*)$, for each induced subgraph H of G.*

In the following theorem, we propose some necessary conditions for a graph G to be cd-perfect. Observe that these conditions are *almost* consistent with the necessary conditions for a graph to be perfect. The proof of the theorem is majorly due to Theorem 5 and Proposition 3.

Theorem 7. *If a graph G is cd-perfect, then G is C_n-free for each $n \geq 4$ with $n \neq 4k$, and $\bar{C}_n$-free for each $n \geq 5$ with $n \neq 2k$, where k is a positive integer.*

4.1   A Unified Approach for Algorithmic Complexity on Some Special Graph Classes

Here, we see some implications of our generalized framework of relating $\chi_{cd}(G)$ and $\omega_s(G)$ to $k(G^*)$ and $\alpha(G^*)$, respectively, together with the notion of cd-perfectness. In particular, we use Theorem 5 and Proposition 3 as tools for deriving both positive and negative results concerning the algorithmic complexity of the problems CD-COLORING and SEPARATED-CLUSTER on some special classes of graphs. Note that the classes of graphs for which these results are applied to evaluate the time complexity of the problems CD-COLORING and SEPARATED-CLUSTER, may not be limited to those studied here.

Polynomial-Time Algorithms: In this section, we will see a unified approach for obtaining polynomial-time algorithms (mostly with an improvement) for some special classes of graphs, using the framework of cd-perfectness and the auxiliary graph G^*.

(1) Chordal Bipartite Graphs: The following theorem is proved in [17].

Theorem 8 [17]. *Let G be a P_6-free chordal bipartite graph. Then $\chi_{cd}(G) = \omega_s(G)$, and the problem* SEPARATED-CLUSTER *can be solved in $O(n^3)$ time for P_6-free chordal bipartite graphs.*

We improve and generalize the result above in Theorem 10. First, note the following theorem proved in [20].

Theorem 9 [20]. *The problem* TOTAL DOMINATION *can be solved in $O(n^2)$ time for chordal bipartite graphs.*

We then prove the following theorem.

Theorem 10. *Let $G = (A, B, E)$ be a chordal bipartite graph. Then G is cd-perfect. Consequently,* TOTAL DOMINATION, CD-COLORING, *and* SEPARATED-CLUSTER *are all equivalent problems for chordal bipartite graphs and can be solved in $O(n^2)$ time.*

Proof Sketch: Clearly, G is $\mathcal{H}$-free. We first prove that G^* is chordal, and therefore, perfect. Since any induced subgraph H of G is also chordal bipartite, it follows from Corollary 2 that G is cd-perfect. Further, as G is triangle-free, by Proposition 2, we have that $\chi_{cd}(G) = \gamma_t(G)$. Since $\chi_{cd}(G) = \omega_s(G)$, the latter statement of the theorem is now immediate from Theorem 9.

(2) Proper Interval Graphs and $3K_1$-Free Graphs: It is known that for the classes of proper interval graphs and $3K_1$-free graphs, CD-COLORING can be solved in $O(n^3)$ time [3] (as they are sub-classes of claw-free graphs). Again, using Theorem 5 as a tool, we show that CD-COLORING can be solved in $O(n^{2.5})$ time for the above classes of graphs. First, we observe the following:

Observation 11. *Let G be a proper interval graph or a $3K_1$-free graph. Then, G^* is triangle-free.*

Before evaluating the complexity of CD-COLORING for proper interval graphs and $3K_1$-free graphs, recall that for any graph G, we have $V(G^*) = V(G) = V(G^2)$, and $E(G^*) = E(G^2) \setminus E(G)$. Therefore, if G^2 can be computed in $O(n^2)$ time, then G^* can also be computed in $O(n^2)$ time (as the subtraction of the adjacency matrices of G^2 and G takes only $O(n^2)$ time). Note that the squares of proper interval graphs can be computed in $O(n^2)$ time [21]. We further note some additional observations.

Observation 12. *For a co-bipartite graph G, the auxiliary graph G^* can be computed in $O(n^2)$ time.*

Proof. Let G be a co-bipartite graph. For any pair of vertices $u, v \in V(G)$, we have $uv \in E(G^2)$ if and only if $u \in A$, $v \in B$, and either $N_B(u) \neq \emptyset$ or $N_A(v) \neq \emptyset$ or both. This implies that G^2 can be constructed in $O(n^2)$ time, and so does G^*.

In the following lemma, we observe an important property of $3K_1$-free graphs that are not co-bipartite, which is crucial in proving Theorem 13, but also interesting on its own.

Lemma 4. *Let G be a $3K_1$-free graph. If G is not co-bipartite, then G^2 is a clique.*

Main Idea: For any two non-adjacent vertices, say u and v in G, as G is $3K_1$-free but not co-bipartite, we prove that $d_G(u) + d_G(v) > n - 2$. This implies that there exists a vertex $w \in V(G)$ such that $uw, wv \in E(G)$, and therefore, $uv \in E(G^2)$.

Using the above lemma and Observation 12, we have that for a $3K_1$-free graph G, the auxiliary graph G^* can be computed in $O(n^2)$ time. We then have the following theorem due to Theorem 5 and some earlier observations.

Theorem 13. *The* CD-COLORING *problem can be solved in $O(n^{2.5})$ time for proper interval graphs and $3K_1$-free graphs.*

Remark 1. As noted before, the time complexity of CD-COLORING for the class of proper interval graphs and $3K_1$-free graphs provided in Theorem 13 is an improvement over the existing algorithms [3] for the problem in the same classes. Clearly, the class of $3K_1$-free graphs is not cd-perfect, as SEPARATED-CLUSTER is NP-Complete for $3K_1$-free graphs [13]. Further, we note that the parameters χ_{cd} and ω_s are not necessarily equal for proper interval graphs too (see an example in Fig. 6). Therefore, the class of proper interval graphs and, in general interval graphs are not cd-perfect.

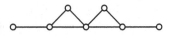

Fig. 6. An example of a proper interval graph G, with $\chi_{cd}(G) = 4 > 3 = \omega_s(G)$.

Hardness Implications: In contrast to the previous section, here we see an application of Proposition 3 and Theorem 5 in deriving hardness results for the problems CD-COLORING and SEPARATED-CLUSTER.

C_6-**Free Bipartite Graphs:** We prove the hardness of the problems CD-COLORING and SEPARATED-CLUSTER for C_6-free bipartite graphs by proposing a polynomial-time reduction respectively, from the problems CLIQUE COVER and INDEPENDENT SET for *diamond-free graphs*. Note that *both* CLIQUE COVER *and* INDEPENDENT SET *are known to be NP-hard for diamond-free graphs* [22,23], where a *diamond* is the graph obtained by deleting exactly one edge from K_4 (complete graph on 4 vertices). First, we note the following observation for diamond-free graphs [24].

Observation 14 [24]. *Any diamond-free graph with m edges can have at most m maximal cliques.*

Let G be a diamond-free graph and let $\mathcal{K} = \{K_1, K_2, \ldots, K_l\}$ denote the collection of all maximal cliques in G. By Observation 14, we have that $l \leq |E(G)|$. From G, we now construct a bipartite graph $B_G = (A, B, E)$ in polynomial-time as follows:

Construction 4. *Define $A = V(G) \cup \{u\}$ and $B = \mathcal{K} = \{K_1, K_2, \ldots, K_l\}$ (each vertex in the partite set B represents a maximal clique in G). For a pair of vertices, $a \in A \setminus \{u\}$ and $K_j \in B$ (where $j \in \{1, 2, \ldots, l\}$), we make the vertices a and K_j adjacent in B_G if and only if $a \in K_j \subseteq V(G)$. In addition, we make the vertex $u \in A$ adjacent to all the vertices in B. i.e. $E(B_G) = \{aK_j : a \in A \setminus \{u\}, K_j \in \mathcal{K} = B, \text{ and } a \in K_j \subseteq V(G)\} \cup \{uK_j : K_j \in \mathcal{K} = B\}$. Clearly, B_G is a bipartite graph.*

We first prove the following lemmas, which are crucial for proving Theorem 15.

Lemma 5. *Let G be a diamond-free graph and $B_G = (A, B, E)$ be the corresponding bipartite graph (as in Construction 4). Then, $B_G^2[A \setminus \{u\}] \cong G$ and $B_G^2[B]$ is a clique.*

Proof Sketch: The former part of the lemma is due to the fact that, for any $x, y \in V(G)$, we have $xy \in E(G) \iff x$ and y belong to a maximal clique K_j in G. Whereas the latter part $B_G^2[B]$ is a clique is trivial, since $B \subseteq N_{B_G}(u)$.

Lemma 6. *Let G be a diamond-free graph and $B_G = (A, B, E)$ be the corresponding bipartite graph. Then, B_G is a C_6-free bipartite graph.*

Proof Sketch: Suppose that $S = (a_1, K_1, a_2, K_2, a_3, K_3, a_1)$ is an induced C_6 in B_G. Then, using the facts that $\{a_1, a_2, a_3\}$ induces a triangle in $B_G^2[A]$ and K_1 is a maximal clique in G, we prove that there exists a vertex $a_4 \in V(G) \subseteq A$ such that $a_4 \in K_1$ in G and $a_3 a_4 \notin E(G)$. Then $\{a_1, a_2, a_3, a_4\}$ induces a diamond in G, a contradiction.

Theorem 15. *The problems* CD-COLORING *and* SEPARATED-CLUSTER *are NP-Compete for C_6-free bipartite graphs.*

Proof Sketch: Let G be a diamond-free graph, and let B_G be the corresponding bipartite graph. We prove that $\omega_s(B_G) = \alpha(B_G^*) = \alpha(G) + 1$. Since B_G is a C_6-free bipartite graph (by Lemma 6), the fact that INDEPENDENT SET is NP-hard for diamond-free graphs implies that SEPARATED-CLUSTER is NP-hard for C_6-free bipartite graphs. Also, we can infer that B_G is $\mathcal{H}$-free. Therefore, by Theorem 5, we have that $\chi_{cd}(B_G) = k(B_G^*)$. We then prove $\chi_{cd}(B_G) = k(B_G^*) = k(G) + 1$. Therefore, the fact that CLIQUE COVER is NP-hard for diamond-free graphs implies that CD-COLORING is NP-hard for C_6-free bipartite graphs.

References

1. Chudnovsky, M., Robertson, N., Seymour, P., Thomas, R.: The strong perfect graph theorem. Annals Math. **164**(1), 51–229 (2006)
2. Merouane, H.B., Haddad, M., Chellali, M., Kheddouci, H.: Dominated colorings of graphs. Graphs Comb. **31**(3), 713–727 (2015)
3. Shalu, M.A., Vijayakumar, S., Sandhya, T.P.: On the complexity of cd-coloring of graphs. Discret. Appl. Math. **280**, 171–185 (2020)
4. Shalu, M.A., Kirubakaran, V.K.: On cd-coloring of trees and co-bipartite graphs. In: Mudgal, A., Subramanian, C.R. (eds.) CALDAM 2021. LNCS, vol. 12601, pp. 209–221. Springer, Cham (2021). https://doi.org/10.1007/978-3-030-67899-9_16
5. Banik, A., Kasthurirangan, P.N., Raman, V.: Dominator coloring and CD coloring in almost cluster graphs. In: Morin, P., Suri, S. (eds.) WADS 2023. LNCS, vol. 14079, pp. 106–119. Springer, Cham (2023). https://doi.org/10.1007/978-3-031-38906-1_8
6. Krithika, R., Rai, A., Saurabh, S., Tale, P.: Parameterized and exact algorithms for class domination coloring. Discret. Appl. Math. **291**, 286–299 (2021)
7. Chen, Y.H.: The dominated coloring problem and its application. In: Murgante, B., et al. (eds.) ICCSA 2014. LNCS, vol. 8584, pp. 132–145. Springer, Cham (2014). https://doi.org/10.1007/978-3-319-09153-2_10
8. Klavžar, S., Tavakoli, M.: Dominated and dominator colorings over (edge) corona and hierarchical products. Appl. Math. Comput. **390**, 125647 (2021)
9. Hoppen, C., Mansan, G.: Total domination in regular graphs. Electron. Notes Theor. Comput. Sci. **346**, 523–533 (2019)
10. Henning, M.A., Yeo, A.: Total Domination in Graphs. Springer, New York (2013)
11. Zhu, J.: Approximation for minimum total dominating set. In: ICIS (2009)
12. Shamir, R., Sharan, R., Tsur, D.: Cluster graph modification problems. Discret. Appl. Math. **144**(1–2), 173–182 (2004)
13. Shalu, M.A., Vijayakumar, S., Sandhya, T.P.: A lower bound of the cd-chromatic number and its complexity. In: Gaur, D., Narayanaswamy, N.S. (eds.) Algorithms and Discrete Applied Mathematics. LNCS, vol. 10156, pp. 344–355. Springer, Cham (2017). https://doi.org/10.1007/978-3-319-53007-9_30
14. Chellali, M., Volkmann, L.: Relations between the lower domination parameters and the chromatic number of a graph. Discret. Math. **274**(1–3), 1–8 (2004)
15. Chellali, M., Maffray, F.: Dominator colorings in some classes of graphs. Graphs Comb. **28**(1), 97–107 (2012)
16. Mycielski, J.: Sur le coloriage des graphs. In: Colloquium Mathematicae (1955)
17. Shalu, M.A., Kirubakaran, V.K.: On cd-coloring of P_5,K_4-free chordal graphs. In: CALDAM (2022)
18. West, D.B.: Introduction to Graph Theory. Prentice Hall, Upper Saddle River (2000)
19. Cygan, M., et al.: Parameterized Algorithms. Springer, Cham (2015). https://doi.org/10.1007/978-3-319-21275-3
20. Damaschke, P., Müller, H., Kratsch, D.: Domination in convex and chordal bipartite graphs. Inf. Process. Lett. **36**(5), 231–236 (1990)
21. Paul, S., Pal, M., Pal, A.: A linear time algorithm to compute square of interval graphs and their colouring. AKCE Int. J. Graphs Comb. **13**(1), 54–64 (2016)
22. Král', D., Kratochvíl, J., Tuza, Z., Woeginger, G.J.: Complexity of coloring graphs without forbidden induced subgraphs. In: Brandstädt, A., Le, V.B. (eds.) Graph-Theoretic Concepts in Computer Science, pp. 254–262. Springer, Heidelberg (2001). https://doi.org/10.1007/3-540-45477-2_23

23. Poljak, S.: A note on stable sets and colorings of graphs. In: Commentationes Mathematicae Universitatis Carolinae, pp. 307–309 (1974)
24. Chiarelli, N., Martínez-Barona, B., Milanič, M., Monnot, J., Muršič, P.: Strong cliques in diamond-free graphs. Theoret. Comput. Sci. **858**, 49–63 (2021)

Generating Signed Permutations by Twisting Two-Sided Ribbons

Yuan Qiu and Aaron Williams[(✉)] [iD]

Williams College, Williamstown, MA 01267, USA
{yq1,aaron.williams}@williams.edu
https://csci.williams.edu/people/faculty/aaron-williams/

Abstract. We provide a simple approach to generating all $2^n \cdot n!$ signed permutations of $[n] = \{1, 2, \ldots, n\}$. Our solution generalizes the most famous ordering of permutations: plain changes (Steinhaus-Johnson-Trotter algorithm). In plain changes, the $n!$ permutations of $[n]$ are ordered so that successive permutations differ by swapping a pair of adjacent symbols, and the order is often visualized as a weaving pattern on n ropes. Here we model a signed permutation as n ribbons with two distinct sides, and each successive configuration is created by twisting (i.e., swapping and turning over) two neighboring ribbons or a single ribbon. By greedily prioritizing 2-twists of large symbols then 1-twists of large symbols, we create a signed version of plain change's memorable zig-zag pattern. We also provide a loopless implementation (i.e., worst-case $\mathcal{O}(1)$-time per object) by enhancing the well-known mixed-radix Gray code algorithm.

Keywords: plain changes · signed permutations · signed permutohedron · greedy Gray codes · combinatorial generation · loopless algorithms

1 Generating Permutations and Signed Permutations

The generation of permutations is a classic problem that dates back to the dawn of computer science (and several hundred years earlier). The goal is to create all $n!$ permutations of $[n] = \{1, 2, \ldots, n\}$ as efficiently as possible. A wide variety of approaches have been considered, some of which can be conceptualized using a specific physical model of the permutation. Let's consider three such examples.

Zaks' algorithm [38] can be conceptualized using a stack of n pancakes of varying sizes. Successive permutations are created by flipping some pancakes at the top of the stack, which is equivalent to a *prefix-reversal* in the permutation. For example, if 🥞 represents 1234, then flipping the top three pancakes gives 🥞 or $\overleftrightarrow{123}4 = 3214$. Table 1 shows the full order for $n = 4$. Zaks designed his 'new' order to have an efficient array-based implementation. Unknown to Zaks, Klügel had discovered this *pancake order* by 1796 [14]; see [1] for further details.

Corbett's algorithm [3] can be conceptualized using n marbles on a ramp. Successive permutations are created by moving a marble to the top of the ramp,

J. A. Soto and A. Wiese (Eds.): LATIN 2024, LNCS 14578, pp. 114–129, 2024.
https://doi.org/10.1007/978-3-031-55598-5_8

which is equivalent to a *prefix-rotation* in the permutation. For example, if represents 1234, then moving the fourth marble gives or $\overleftarrow{1234} = 4123$.

The algorithms by Zaks and Corbett are well-known, and have their own specific applications. For example, in interconnection networks [4], the algorithms give Hamilton cycles in the pancake network and rotator network, respectively.

Plain changes can be conceptualized using n parallel ropes. Successive permutations are obtained by crossing one rope over a neighboring rope as in weaving. This is equivalent to a *swap* (or *adjacent-transposition*) in the permutation. For example, if $\equiv$ represents 1234, then swapping the middle pair gives $\equiv$ or $1\overleftrightarrow{23}4 = 1324$. Plain changes dates to bell-ringers in the 1600 s [5]. Figure 3 shows the order for $n = 4$ and its zig-zag pattern. It is also known as the *Steinhaus-Johnson-Trotter algorithm* [16,31,34] due to rediscoveries circa 1960.

Many other notable approaches to permutation generation exist, with surveys by Sedgewick [30], Savage [27], and Mütze [22], and frameworks by Knuth [17] and Ganapathi and Chowdhury [8]. While some methods have specific advantages [15] or require less additional memory when implemented [19], there is little doubt that plain changes is <u>the</u> solution to permutation generation.

A *signed permutation* of $[n]$ is a permutation of $[n]$ in which every symbol is given a $\pm$ sign. We let S_n and $S_n^{\pm}$ be the sets of all permutations and signed permutations of $[n]$, respectively. Note that $|S_n| = n!$ and $|S_n^{\pm}| = 2^n \cdot n!$. For example, $231 \in S_3$ has eight different signings, including $+2-3-1 \in S_3^{\pm}$. For convenience, we also use bold or overlines for negatives, with **231** and $2\overline{31}$ denoting $+2-3-1$. Signed permutations arise in many contexts including genomics [7].

The efficient generation of signed permutations has been considered. Suzuki, Sawada, and Kaneko [33] treat signed permutations as stacks of n *burnt pancakes* and provide a signed version of Zaks' algorithm. Korsh, LaFollette, and Lipschutz [18] provide a Gray code that swaps two symbols (and preserves their signs) or changes the rightmost symbol's sign. Both approaches offer improvements over standard *lexicographic orders* (i.e., alphabetic orders) but neither is considered to be <u>the</u> solution for signed permutations. We define a *signed plain change order* to be any extension of plain changes to signed permutations.

Physical Model of Signed Permutations: Two-Sided Ribbons. A *two-sided ribbon* is glossy on one side and matte on the other[1], and we model a signed permutation using n two-sided ribbons in parallel. We modify the ribbons via twists. More specifically, a k-*twist* turns over k neighboring ribbons and reverses their order, as visualized in Fig. 1 for $k = 1, 2$. A twist performs a *complementing substring reversal*, or simply a *reversal* [11], on the signed permutation.

Our goal is to create a *twist Gray code* for signed permutations. This means that each successive entry of $S_n^{\pm}$ is created by applying a single twist. Equivalently, a sequence of $2^n n! - 1$ twists generates each entry of $S_n^{\pm}$ in turn. It should be obvious that 1-twists are insufficient for this task on their own, as they do not modify the underlying permutation. Similarly, 2-twists are insufficient on their

---

[1] Manufacturers refer to this type of ribbon as *single face* as only one side is polished.

(a) The 1-twist changes $1\,2\,3\,4$ into $1\,\overline{2}\,3\,4$.   (b) The 2-twist changes $1\,\overline{2}\,3\,\overline{4}$ into $1\,\overline{3}\,2\,\overline{4}$.

Fig. 1. Two-sided ribbons with distinct positive (i.e., glossy) and negative (i.e., matte) sides running in parallel. A k-twist reverses the order of k neighboring ribbons and turns each of them over, as shown for (a) $k = 1$ and (b) $k = 2$.

own, as they do not modify the number of positive symbols modulo two. However, we will show that 2-twists and 1-twists are sufficient when used together. Our solution is a signed plain change order that we name *twisted plain changes*.

Application: Train-Based Traveling Salesman Problems. Exhaustive generation is central to many applications, including testing and exact algorithms. Gray code algorithms can also improve the latter. For example, a traveling salesman problem on n cities can be solved by generating all $n!$ permutations of $[n]$, with each member of S_n providing a possible route through the cities (e.g., $p_1 p_2 \cdots p_n \in S_n$ represents the route $p_1 \to p_2 \to \cdots \to p_n$). Plain changes is advantageous because successive routes differ in at most three segments (e.g., swapping $p_i p_{i+1}$ to $p_{i+1} p_i$ replaces segment $p_i \to p_{i+1}$ with $p_i \to p_{i+2}$) [15]. Thus, the distance of each successive route can be *updated* in constant time.

Now consider a TSP-variant involving trains, where each of the n stations can be entered/exited in one of two orientations (e.g., the train may travel along the station's eastbound or westbound track). Note that the time taken to travel from one station to another depends on these orientations. As a result, there are $2^n \cdot n!$ possible routes and they correspond to the members of $S_n^{\pm}$. Our twist Gray code algorithm generates successive routes that differ in at most three segments.

1.1 Outline

Section 2 provides background on combinatorial generation. Section 3 defines our twist Gray code using a simple (but inefficient) greedy algorithm. Section 4 discusses ruler sequences and their applications. Section 5 uses a signed ruler sequence to generate our Gray code in worst-case $\mathcal{O}(1)$-time per signed permutation. A Python implementation of our final algorithm appears in the appendix. The proofs of Lemma 1–3 are left as exercises to the reader due to page limits.

2   Combinatorial Generation

As Ruskey explains in *Combinatorial Generation* [26], humans have been writing exhaustive lists of various kinds for thousands of years, and more recently, programming computers to do so. Here we review basic concepts and terminology, then we discuss two foundational results and modern reinterpretations of them.

2.1   Gray Codes and Loopless Algorithms

If successive objects in an order differ in a constant amount (by some metric), then it is a *Gray code*. If an algorithm generates each object in amortized or worst-case $\mathcal{O}(1)$-time, then it is *constant amortized time* (CAT) or *loopless* [6]. To understand these terms, note that a well-written generation algorithm shares one object with an application. It modifies the object and announces that the 'next' object can be *visited*, without using linear-time to create a new object. Loopless algorithms make constant-time modifications using a Gray code. For example, Zaks' order can be generated in CAT as its prefix-reversals have constant average length (see Ord-Smith's earlier ECONOPERM program [24]), but a loopless algorithm is not possible as a length n prefix-reversal takes $\Theta(n)$-time[2].

2.2   Binary Reflected Gray Code and Plain Changes

Plain change's stature in combinatorial generation is rivaled only by the *binary reflected Gray code*[3]. The BRGC orders n-bit binary strings by *bit-flips*, meaning successive strings differ in one bit. It is typically defined recursively as

$$\mathsf{brgc}(n) = 0 \cdot \mathsf{brgc}(n-1),\ 1 \cdot \mathsf{reflect}(\mathsf{brgc}(n-1)) \text{ with } \mathsf{brgc}(1) = 0,1 \qquad (1)$$

where reflect denotes *list reflection* (i.e., last string goes first). For example,

$$\mathsf{brgc}(2) = 0 \cdot \mathsf{brgc}(1),\ 1 \cdot \mathsf{reflect}(\mathsf{brgc}(1)) = 0 \cdot (0,1),\ 1 \cdot (1,0) = 0\overline{0}, \overline{0}1, 1\underline{1}, 10$$

where overlines and underlines have been added for flips from 0 to 1 and 1 to 0, respectively. The order for $n = 4$ is visualized in Fig. 2 using two-sided ribbons, where each bit-flip is a 1-twist of the corresponding ribbon.

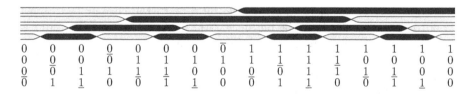

0	0	0	0	0	0	0	$\overline{0}$	1	1	1	1	1	1	1	1
0	0	0	$\overline{0}$	1	1	1	1	1	1	1	1	0	0	0	0
0	$\overline{0}$	1	1	1	1	0	0	0	$\overline{0}$	1	1	1	1	0	0
$\overline{0}$	1	1	0	$\overline{0}$	1	1	0	$\overline{0}$	1	1	0	$\overline{0}$	1	1	0

Fig. 2. Binary reflected Gray code using indistinct two-sided ribbons for $n = 4$.

Plain changes recursively zigs and zags n through permutations of $[n-1]$. In (2), zig and zag give length $n-1$ lists that repeatedly swap n to the left or right.

$$\mathsf{plain}(n) = \mathsf{zig}(p_1 \cdot n), \mathsf{zag}(n \cdot p_2), \ldots, \mathsf{zig}(p_{(n-1)!-1} \cdot n), \mathsf{zag}(n \cdot p_{(n-1)!}) \qquad (2)$$
$$\text{with } \mathsf{plain}(n-1) = p_1, p_2, \ldots, p_{(n-1)!}$$

---

[2] If the permutation is stored in a BLL instead of an array, then loopless is possible [35].
[3] The eponymous *Gray code* by Gray [10] also demonstrates Stigler's law [32]: [9,13].

Formula (2) assumes $(n-1)!$ is even, so we use base case $\mathsf{plain}(2) = \overleftarrow{12}, 21$. Here the arrow denotes a larger value swapping left past its smaller neighbor. Thus,

$$\mathsf{plain}(3) = \mathsf{zig}(12 \cdot 3), \mathsf{zag}(3 \cdot 21) = 1\overleftrightarrow{23}, \overleftarrow{132}, 3\overrightarrow{12}, \overrightarrow{321}, 2\overrightarrow{31}, 213,$$

Figure 3 visualizes $\mathsf{plain}(4)$ with distinct one-sided ribbons[4]. Note how 4 zigzags.

1	1	1	4	4	1	1	1	3	3	3	4	4	3	3	3	2	2	2	4	4	2	2	2
2	2	4	1	1	4	3	3	1	1	4	3	3	4	2	2	3	3	4	2	2	4	1	1
3	4	2	2	3	3	4	2	2	4	1	1	2	2	4	1	1	4	3	3	1	1	4	3
4	3	3	3	2	2	2	4	4	2	2	2	1	1	1	4	4	1	1	1	3	3	3	4

Fig. 3. Plain changes $\mathsf{plain}(n)$ using distinct one-sided ribbons for $n = 4$.

2.3   The Greedy Gray Code Algorithm

Historically, Gray codes have been created using recursion. In contrast, the *greedy Gray code algorithm* [36] attempts to create a Gray code one object at time. A list $\mathsf{greedy}(\mathbf{s}, \langle o_1, o_2, \ldots, o_k \rangle)$ is initialized with a *start object* $\mathbf{s}$, then it is repeatedly extended as follows: If $\mathbf{t}$ is the last object in the list, then add $o_i(\mathbf{t})$ to the end of the list, where i is the minimum index such that $o_i(\mathbf{t})$ is valid and not in the list. This continues until none of the *operations* $o_1, o_2, \ldots, o_k$ produce a new object.

The binary reflected Gray code is a *greedy Gray code*: start at $\mathbf{s} = 0^n$ and flip the rightmost possible bit [36]. That is, $\mathsf{brgc}(n) = \mathsf{greedy}(0^n, \langle f_1, f_2, \ldots, f_n \rangle)$ where f_i flips b_i in $b_n b_{n-1} \cdots b_1 \in B_n$. For example, the order for $n = 4$ begins

$$\mathsf{brgc}(4) = 000\overline{0}, 00\overline{0}1, 001\underline{1}, 0010, \ldots. \tag{3}$$

To continue (3) we consider applying the bit-flips to the current last object $\mathbf{t} = 0010$. We can't flip its right bit since $f_1(\mathbf{t}) = f_1(0010) = 001\overline{0} = 0011$ is already in the list. Similarly, $f_2(\mathbf{t}) = f_2(0010) = 00\underline{1}0 = 0000$ is also in the list. But $f_3(\mathbf{t}) = f_3(0010) = 0\overline{0}10 = 0110$ is not in the list, so it is the next string.

Plain changes is also greedy: start at $\mathbf{s} = 12 \cdots n$ and swap the largest possible value left or right [36]. That is, $\mathsf{plain}(n) = \mathsf{greedy}(12 \cdots n, \langle \overleftarrow{s_n}, \overrightarrow{s_n}, \ldots, \overleftarrow{s_2}, \overrightarrow{s_2} \rangle)^5$ where $\overleftarrow{s_v}$ and $\overrightarrow{s_v}$ swap value v to the left and right, respectively, when applied to any member of S_n. For example, the order for $n = 4$ begins

$$\mathsf{plain}(4) = 12\overleftarrow{34}, 1\overleftarrow{2}43, \overleftarrow{1}423, 41\overleftarrow{2}3, \overrightarrow{4}132, 14\overrightarrow{3}2, 134\overrightarrow{2}, 1324, \ldots. \tag{4}$$

We can't apply $\overleftarrow{s_4}$ to $\mathbf{t} = 1324$ since $\overleftarrow{s_4}(1324) = 13\overleftarrow{24} = 1342$ is already in the list. Nor can we apply the next highest-priority operation $\overrightarrow{s_4}$ as $\overrightarrow{s_4}(1324)$ is invalid. But $\overleftarrow{s_3}(\mathbf{t}) = \overleftarrow{1}324 = 3124$ is not in the list, so it is the next permutation. Plain change's greedy formula also holds when the swaps are replaced by *jumps* (i.e., values can only be swapped over smaller values) and with $\overleftarrow{s_1}$ and $\overrightarrow{s_1}$ omitted [12].

---

[4] Physically, a ribbon moves above or below its neighbor, but that is not relevant here.
[5] $\overleftarrow{s_1}$ and $\overrightarrow{s_1}$ are omitted as they equal other swaps. In fact, the swaps are all *jumps* [12].

Reflecting BRGC and Plain Changes. Interestingly, if either of the previous two greedy algorithms is started from their final object, then the entire order is reflected. For example, if $\mathbf{s} = 1000$ is chosen in Fig. 2, or $\mathbf{s} = 2134$ is chosen in Fig. 3, then the greedy algorithms generate the objects from right-to-left. This is in part explained by their palindromic change sequences (see Sect. 4).

Lemma 1. $\mathsf{greedy}(10^{n-1}, \langle f_1, f_2, \ldots, f_n \rangle) = \mathsf{reflect}(\mathsf{brgc}(n))$.

Lemma 2. $\mathsf{greedy}(2134 \cdots n, \langle \overleftarrow{s_n}, \overrightarrow{s_n}, \overleftarrow{s_{n-1}}, \overrightarrow{s_{n-1}}, \ldots, \overleftarrow{s_2}, \overrightarrow{s_2}, \rangle)$
$= \mathsf{reflect}(\mathsf{plain}(n))$.

3   A Signed Plain Change Order: Twisted Plain Changes

Now we present a greedy solution to generating a signed plain change order.

Definition 1. Twisted plain changes $\mathsf{twisted}(n)$ *is the signed permutation order visited by Algorithm 1. It starts with* $\mathbf{s} = +1 + 2 \cdots +n \in S_n^{\pm}$ *and prioritizes 2-twists of the largest possible value then 1-twists of the largest possible value. That is,* $\mathsf{greedy}(\mathbf{s}, \langle \overleftarrow{t_n}, \overrightarrow{t_n}, \overleftarrow{t_{n-1}}, \overrightarrow{t_{n-1}}, \ldots, \overleftarrow{t_2}, \overrightarrow{t_2}, t_n, \ldots, t_2, t_1 \rangle)$, *where* $\overleftarrow{t_v}$ *and* $\overrightarrow{t_v}$ *2-twist value v left or right, and* t_v *1-twists value v (i.e., v's sign is flipped).*

Algorithm 1. Greedy algorithm for generating twisted plain changes $\mathsf{twisted}(n)$.

```
 1: procedure Twisted(n)           ▷ Signed permutations are visited in twisted(n) order
 2:     T ← t⃖ₙ, t⃗ₙ, t⃖ₙ₋₁, t⃗ₙ₋₁,..., t⃖₂, t⃗₂, tₙ,...,t₂,t₁  ▷ List 2-twists then 1-twists
 3:     π ← +1 +2 ⋯ +n              ▷ Starting signed permutation s = π ∈ Sₙ±
 4:     visit(π)                    ▷ Visit π for the first and only time
 5:     S = {π}                     ▷ Add π to the visited set
 6:     i ← 1            ▷ 1-based index into T; T[1] = t⃖ₙ will 2-twist n left
 7:     while i ≤ 3n − 2 do         ▷ Index i iterates through the 3n − 2 twists in T
 8:         π′ ← T[i](π)            ▷ Apply the iᵗʰ highest priority twist to create π′
 9:         if π′ ∉ S then          ▷ Check if π′ is a new signed permutation
10:             π ← π′              ▷ Update the current signed permutation π
11:             visit(π)            ▷ Visit π for the first and only time
12:             S = S ∪ {π}         ▷ Add π to the visited set
13:             i ← 1               ▷ Reset the 1-based index into T
14:         else
15:             i ← i + 1           ▷ If π′ ∈ S, then consider the next twist
```

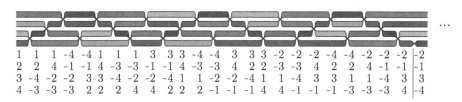

1	1	1	-4	-4	1	1	1	3	3	3	-4	-4	3	3	3	-2	-2	-2	-4	-4	-2	-2	-2	-2
2	2	4	-1	-1	4	-3	-3	-1	-1	4	-3	-3	4	2	2	-3	-3	4	2	2	4	-1	-1	-1
3	-4	-2	-2	3	3	-4	-2	-2	-4	1	1	-2	-2	-4	1	1	-4	3	3	1	1	-4	3	3
4	-3	-3	-3	2	2	2	4	4	2	2	2	-1	-1	-1	4	4	-1	-1	-1	-3	-3	-3	4	-4

Fig. 4. Twisted plain changes $\mathsf{twisted}(n)$ for $n = 4$ up to its 25th entry.

For example, our twist Gray code for $n = 4$ begins as follows

$$\text{twisted}(4) = 12\overleftarrow{34}, 1\overleftarrow{243}, \overleftarrow{1423}, 41\overleftarrow{23}, \overrightarrow{4132}, 14\overrightarrow{32}, 1324, \ldots. \quad (5)$$

with bold for negatives and arrows for twists. To continue the order note that $\overleftarrow{t_4}(t) = \overleftarrow{t_4}(1324) = 13\overleftarrow{24} = 1342$ is already in the list and $\overrightarrow{t_4}(t)$ is invalid. But $\overleftarrow{t_3}(t) = \overleftarrow{t_3}(1324) = \overleftarrow{1324} = 3124$ is new, so it is the next signed permutation.

Figure 4 shows the start of $\text{twisted}(n)$ for $n = 4$. Note that the first 24 entries are obtained by 2-twists. The result is a familiar zig-zag pattern, but with every ribbon turning over during each pass. The 25th entry is obtained by a 1-twist.

3.1   2-Twisted Permutohedron and Signed Permutohedra

At this point it is helpful to compare the start of plain changes and twisted plain changes. The *permutohedron of order n* is a graph whose vertices are permutations S_n and whose edges join two permutations that differ by a swap. Plain changes traces a Hamilton path in this graph, as illustrated in Fig. 5a.

Now consider signing each vertex $p_1 p_2 \cdots p_n$ in the permutohedron as follows:

$$p_j = i \text{ is positive if and only if } i \equiv j \mod 2. \quad (6)$$

In particular, the permutation $12 \cdots n$ is signed as $+1+2 \cdots +n$ due to the fact that odd values are in odd positions, and even values are in even positions. One way of interpreting (6) is that swapping a symbol changes its sign. Thus, after this signing, the edges in the resulting graph model 2-twists instead of swaps. For this reason, we refer to the graph as a 2-*twisted permutohedron of order n*.

Since twisted plain changes prioritizes 2-twists before 1-twists, the reader should be able to conclude that $\text{twisted}(n)$ starts by creating a Hamilton path in the 2-twisted permutohedron. This is illustrated in Fig. 5b.

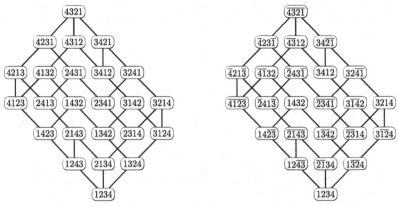

(a) The Hamilton path from $1234 \in S_n$ to $2134 \in S_4$ follows $\text{plain}(4)$.

(b) The Hamilton path from $1234 \in S_n^{\pm}$ to $\overline{2}134 \in S_n^{\pm}$ follows $\text{twisted}(4)$.

Fig. 5. The (a) permutohedron and (b) 2-twisted permutohedron for $n = 4$.

In general, there are 2^n *signed permutohedron of order n*. Each signed permutohedron contains $n!$ vertices, including a single signing of the vertex $12\cdots n$, and edges for every possible 2-flip. In particular, the 2-twisted permutohedron is the signed permutohedron with vertex $+1+2\cdots+n$ (i.e., $12\cdots n$ is fully positive).

3.2 Global Structure

Our greedy approach can be verified to work for small n. To prove that it works for all n we need to deduce the global structure of the order that is created. We'll see that the order navigates through successive signed permutohedron.

Theorem 1. *Algorithm 1 visits a twist Gray code of signed permutations. That is,* $\mathsf{twisted}(n) = \mathsf{greedy}(\mathsf{s}, \langle \overleftarrow{\mathsf{t}_n}, \overrightarrow{\mathsf{t}_n}, \overleftarrow{\mathsf{t}_{n-1}}, \overrightarrow{\mathsf{t}_{n-1}}, \ldots, \overleftarrow{\mathsf{t}_2}, \overrightarrow{\mathsf{t}_2}, \mathsf{t}_n, \ldots, \mathsf{t}_2, \mathsf{t}_1 \rangle)$ *orders* $S_n^{\pm}$.

Proof. Since 2-twists are prioritized before 1-twists, the algorithm proceeds in the same manner as plain changes, except for the signs of the visited objects. As a result, it generates sequences of $n!$ signed permutations using 2-twists until a single 1-twist is required. One caveat is that the first signed permutation in a sequence alternates between having the underlying permutation of $1234\cdots n$ or $2134\cdots n$. This is due to the fact that plain changes starts at $1234\cdots n$ and ends at $2134\cdots n$ and swaps 12 to 21 one time. As a result, 12 will be inverted while traversing every second sequence of length $n!$, and these traversals will be done in reflected plain changes order by Lemma 2. More specifically, the order generated by the algorithm appears in Fig. 6, with an example in Fig. 7. □

$$123\cdots n{-}2\,n{-}1\,n \rightarrow 123\cdots n{-}2\,\overline{n}\,\overline{n{-}1} \rightarrow \quad \cdots \quad \rightarrow \overline{2}\,\overline{1}\,3\cdots n{-}2\,n{-}1\,n$$
$$\downarrow$$
$$123\cdots n{-}2\,n{-}1\,\overline{n} \leftarrow \quad \cdots \quad \leftarrow \overline{2}\,\overline{1}\,3\cdots n{-}2\,n\,\overline{n{-}1} \leftarrow \overline{2}\,\overline{1}\,3\cdots n{-}2\,n{-}1\,\overline{n}$$
$$\downarrow$$
$$123\cdots n{-}2\,\overline{n{-}1}\,\overline{n} \rightarrow 123\cdots n{-}2\,n\,n{-}1 \rightarrow \quad \cdots \quad \rightarrow \overline{2}\,\overline{1}\,3\cdots n{-}2\,\overline{n{-}1}\,\overline{n}$$
$$\downarrow$$
$$123\cdots n{-}2\,\overline{n{-}1}\,n \leftarrow \quad \cdots \quad \leftarrow \overline{2}\,\overline{1}\,3\cdots n{-}2\,\overline{n}\,n{-}1 \leftarrow \overline{2}\,\overline{1}\,3\cdots n{-}2\,\overline{n{-}1}\,n$$
$$\downarrow$$
$$123\cdots \overline{n{-}2}\,\overline{n{-}1}\,n \rightarrow 123\cdots \overline{n{-}2}\,\overline{n}\,n{-}1 \rightarrow \quad \cdots \quad \rightarrow \overline{2}\,\overline{1}\,3\cdots \overline{n{-}2}\,\overline{n{-}1}\,n$$
$$\downarrow$$
$$\vdots$$
$$\downarrow$$
$$\overline{1}\,23\cdots n{-}2\,n{-}1\,n \leftarrow \quad \cdots \quad \leftarrow 2\,\overline{1}\,3\cdots n{-}2\,\overline{n}\,\overline{n{-}1} \leftarrow 2\,\overline{1}\,3\cdots n{-}2\,n{-}1\,n$$

Fig. 6. The global structure of twisted plain changes. Each row greedily applies 2-twists to the largest possible symbol, thus following plain changes. At the end of a row, no 2-twist can be applied, and the down arrows greedily 1-twist the largest possible symbol. The rows alternate left-to-right and right-to-left (i.e., in boustrophedon order) by Lemma 2 The leftmost column contains $12\cdots n$ signed according to successive strings in the binary reflected Gray code. The overall order is cyclic as a 1-twist on value 1 transforms the last entry into the first.

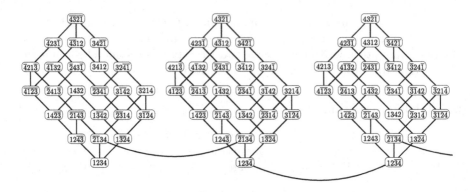

Fig. 7. Our twisted(4) order begins by traversing the above signed permutohedron, starting from the 2-twisted permutohedron on the left. Straight lines are the edges of a signed permutohedron (i.e., every possible 2-twist). Curved edges are 1-twists between the vertices shown, and they connect two signed permutohedron. Highlighted edges are used by the greedy algorithm: green subpaths start at a signed $1234\cdots n$ vertex and proceed in plain changes; red subpaths start at a signed $2134\cdots n$ vertex and proceed in reflected plain changes.

Theorem 1's proof can be used toward a CAT implementation of twisted(n). We'll instead develop a loopless implementation of twisted(n) in Sects. 4–5.

4   Ruler Sequences

Here we consider integer sequences called *ruler sequences*. The sequences are named after the tick marks on rulers and tape measures whose heights follow the *decimal ruler sequence* ruler($10, 10, \ldots, 10$). They are central to Algorithm 2 in Sect. 5, and relate twisted(n) to other Gray codes and lexicographic orders.

4.1   Ruler Sequences, Mixed-Radix Words, and Lexicographic Orders

The *ruler sequence* with bases $b_n, b_{n-1}, \ldots, b_1$ can be inductively defined as follows, where commas join sequences, and exponentiation denotes repetition.

$$\mathsf{ruler}(b_1) = 1^{b_1-1} = 1, 1, \ldots, 1 \quad (\text{i.e., } b_1 - 1 \text{ copies}) \tag{7}$$

$$\mathsf{ruler}(b_n, b_{n-1}, \ldots, b_1) = (s, n)^{b_n - 1}, s \text{ where } s = \mathsf{ruler}(b_{n-1}, b_{n-2}, \ldots, b_1) \tag{8}$$

Hence, ruler$(5, 3) = 1,1,2,1,1,2,1,1,2,1,1,2,1,1 = (s, 2)^4, s$ since $s = \mathsf{ruler}(3) = 1, 1$. The length of the ruler sequence is $|\mathsf{ruler}(b_n, b_{n-1}, \ldots, b_1)| = (\prod_{i=1}^{n} b_i) - 1$.

Bases can also be used to define the set of *mixed-radix words* $W_{b_n, b_{n-1}, \ldots, b_1}$, where $w_n \cdots w_2 w_1$ is in the set if its digits satisfy $0 \leq w_i < b_i$ for $1 \leq i \leq n$. The number of these words is $|W_{b_n, b_{n-1}, \ldots, b_1}| = \prod_{i=1}^{n} b_i = |\mathsf{ruler}(b_n, b_{n-1}, \ldots, b_1)| + 1$.

When mixed-radix words are written in lexicographic order, the ruler sequence is its *change sequence*. Each ruler entry is the number of digits that "roll over" to create the next word. In particular, the *binary ruler sequence* ruler$(2, 2, \ldots, 2)$ (OEIS A001511 [23]) gives the suffix lengths of the form $011 \cdots 1$ that change to $100 \cdots 0$ when counting in binary. This is shown below for $n = 3$.

$$\text{lex}(B_3) = 00\overline{0}, 00\overline{1}, 01\overline{0}, 0\overline{11}, 10\overline{0}, 1\overline{01}, 11\overline{0}, 111 \text{ since } \text{ruler}(2, 2, 2) = 1, 2, 1, 3, 1, 2, 1$$

The *upstairs ruler sequence* ruler$(1, 2, \ldots, n)$ (OEIS A235748) arises when listing *upstairs words* $W_{1,2,\ldots,n}$, and the *downstairs ruler sequence* ruler$(n, n-1, \ldots, 1)$ (OEIS A001511) arise when listing *downstairs words* $W_{n,n-1,\ldots,1}$. The start of these *factorial patterns* are below for $n = 4$ with full signed versions in Table 1.

$$\text{lex}(W_{1,2,3,4}) = 111\overline{1}, 111\overline{2}, 11\overline{13}, 112\overline{1}, 112\overline{2}, 1\overline{123}, 1211, \ldots \text{ as ruler}(1,2,3,4) = 1,1,2,1,1,3,\ldots$$
$$\text{lex}(W_{4,3,2,1}) = 11\overline{11}, 1\overline{121}, 12\overline{11}, 1\overline{221}, 13\overline{11}, 1\overline{321}, 2111, \ldots \text{ as ruler}(4,3,2,1) = 2,3,2,3,2,4,\ldots$$

Note that the *unary bases* $b_n = 1$ (which never changes) and $b_1 = 1$ (which always rolls over to itself) are often omitted from these patterns.

Ruler sequences provide change sequences for various Gray codes, including some from Sect. 1. The downstairs sequence gives the flip lengths in Zaks' order as seen in Table 1. Corbett's order uses the upstairs sequence but subtly [36]. Other change sequences are more fully understood as signed ruler sequences.

4.2 Signed Ruler Sequences and (Reflected) Gray Codes

We define the *signed ruler sequence* ruler$\pm$ as ruler with some entries negated. The overlines complement the sign of each entry, and the ^{R} reverses a sequence.

$$\text{ruler} \pm (b_1) = 1^{b_1 - 1} = 1, 1, \ldots, 1 \quad (\text{i.e., } b_1 - 1 \text{ copies}) \tag{9}$$

$$\text{ruler} \pm (b_n, b_{n-1}, \ldots, b_1) = \begin{cases} (s, n, \overline{s}^R, n)^{b_n/2}, s & \text{if } b_n \text{ is odd} \\ (s, n, \overline{s}^R, n)^{(b_n-1)/2}, s, n, \overline{s} & \text{if } b_n \text{ is even} \end{cases} \tag{10}$$

where $s = \text{ruler} \pm (b_{n-1}, b_{n-2}, \ldots, b_1)$. Note that the subsequence s is repeated b_n times in (10) just as in (8), but every second subsequence is complemented[6]. For example, ruler$\pm (3) = 1, 1$ so ruler$\pm (4, 3) = 1, 1, 2, -1, -1, 2, 1, 1, 2, -1, -1$. The specific sequences (and associated orders) discussed below are shown in Table 1.

Signed ruler sequences govern *reflected mixed-radix Gray codes*, which generalize (1) to non-binary bases $\mathbf{b} = b_n, b_{n-1}, \ldots, b_1$ by reflecting every 2nd sublist,

$$\text{mix}(\mathbf{b}) = \begin{cases} 0, 1, \ldots, b_1 - 1 & \text{if } n = 1 \\ 0 \cdot \text{mix}(\mathbf{b}'), 1 \cdot \text{reflect}(\mathbf{b}'), \ldots, (b_n - 1) \cdot \text{mix}(\mathbf{b}') & \text{odd } n > 1 \\ 0 \cdot \text{mix}(\mathbf{b}'), 1 \cdot \text{reflect}(\mathbf{b}'), \ldots, (b_n - 1) \cdot \text{reflect}(\text{mix}(\mathbf{b}')) & \text{even } n > 1 \end{cases}$$

where $\mathbf{b}' = b_{n-1}, b_{n-2}, \ldots, b_1$. The entries of ruler$\pm (\mathbf{b})$ specify how to change $w_n w_{n-1} \cdots w_1 \in W_{\mathbf{b}}$ into the next word: increment w_j for $+j$; decrement w_j for $-j$. The orders are also greedy: increment or decrement the rightmost digit.

---

[6] Unsigned ruler sequences are palindromes, so ^{R}'s can be added to (8) to mirror (10).

Table 1. Ruler sequences provide the change sequences of reflected Gray codes of mixed-radix words, and (greedy) Gray codes of various other objects. The left columns show that the unsigned downstairs ruler sequence ruler$(n, n-1, \ldots, 1)$ is the change sequence for the up-words $W_{4,3,2,1}$, and the prefix-reversal lengths (i.e., flip lengths) in Zaks' Gray code. The change sequences of the binary reflected Gray code and plain changes are usually given as unsigned ruler sequences. However, signed versions provide more information. The middle-left columns show that the signed binary ruler sequence ruler$\pm (2,2,\ldots,2)$ is the change sequence for brgc(n), with the sign providing the direction of the flip: $+j$ for $b_j = \overline{0} = 1$ and $-j$ for $b_j = \underline{1} = 0$. Similarly, the middle-right columns show that the signed upstairs ruler sequence ruler$\pm (1,2,\ldots,n)$ is the change sequence for plain(n), with the sign providing the direction of the swap: $+j$ for swapping x left and $-j$ for swapping x right where $x = n-j+1$. Our twisted plain change Gray code twisted(n) uses a signed factorial ruler sequence ruler$\pm(n, n-1, \ldots, 2, 1, 2, 2, \ldots, 2)$ (with the unary 1 omitted). Sequence entries from the factorial and binary portions give 2-twists and 1-twists, respectively. In particular, the last row in the right columns is the 1-twist in Fig. 3.

down words 4321	ruler	Zaks $p_4p_3p_2p_1$	ruler± 2222	BRGC $b_4b_3b_2b_1$	up words 123	ruler± 1234	plain changes	up± words 22221234	ruler± 22221234	twisted plain
0000	2	$\overleftrightarrow{12}34$	+1	0000	000	+1	$12\overleftarrow{34}$	0000000	+1	$12\overleftarrow{34}$
0010	3	$\overleftrightarrow{213}4$	+2	0001	001	+1	$1\overleftarrow{24}3$	0000001	+1	$1\overleftarrow{24}3$
0110	2	$\overleftrightarrow{31}24$	−1	0011	002	+1	$\overleftarrow{14}23$	0000002	+1	$\overleftarrow{14}23$
0100	3	$\overleftrightarrow{132}4$	+3	0010	003	+2	$41\overleftarrow{23}$	0000003	+2	$41\overleftarrow{23}$
0200	2	$\overleftrightarrow{23}14$	+1	0110	013	−1	$\overrightarrow{41}32$	0000013	−1	$\overrightarrow{41}32$
0210	4	$\overleftrightarrow{3214}$	−2	0111	012	−1	$1\overrightarrow{43}2$	0000012	−1	$1\overrightarrow{43}2$
1210	2	$\overleftrightarrow{41}23$	−1	0101	011	−1	$13\overrightarrow{42}$	0000011	−1	$13\overrightarrow{42}$
1200	3	$\overleftrightarrow{142}3$	+4	0100	010	+2	$\overleftarrow{13}24$	0000010	+2	$\overleftarrow{13}24$
1100	2	$\overleftrightarrow{24}13$	+1	1100	020	+1	$31\overleftarrow{24}$	0000020	+1	$31\overleftarrow{24}$
1110	3	$\overleftrightarrow{421}3$	+2	1101	021	+1	$3\overleftarrow{14}2$	0000021	+1	$3\overleftarrow{14}2$
1010	2	$\overleftrightarrow{12}43$	−1	1111	022	+1	$\overleftarrow{34}12$	0000022	+1	$\overleftarrow{34}12$
1000	4	$\overleftrightarrow{2143}$	−3	1110	023	+3	$43\overleftarrow{12}$	0000023	+3	$43\overleftarrow{12}$
2000	2	$\overleftrightarrow{34}12$	+1	1010	123	−1	$\overrightarrow{43}21$	0000123	−1	$\overrightarrow{43}21$
2010	3	$\overleftrightarrow{431}2$	−2	1011	122	−1	$3\overrightarrow{42}1$	0000122	−1	$3\overrightarrow{42}1$
2110	2	$\overleftrightarrow{13}42$	−1	1001	121	−1	$32\overrightarrow{41}$	0000121	−1	$32\overrightarrow{41}$
2100	3	$\overleftrightarrow{314}2$			120	−2	$\overrightarrow{32}14$	0000120	−2	$\overrightarrow{32}14$
2200	2	$\overleftrightarrow{41}32$			110	+1	$23\overleftarrow{14}$	0000110	+1	$23\overleftarrow{14}$
2210	4	$\overleftrightarrow{1432}$			111	+1	$2\overleftarrow{34}1$	0000111	+1	$2\overleftarrow{34}1$
3210	2	$\overleftrightarrow{23}41$			112	+1	$\overleftarrow{24}31$	0000112	+1	$\overleftarrow{24}31$
3200	3	$\overleftrightarrow{324}1$			113	−2	$42\overrightarrow{31}$	0000113	−2	$42\overrightarrow{31}$
3100	2	$\overleftrightarrow{42}31$			103	−1	$\overrightarrow{42}13$	0000103	−1	$\overrightarrow{42}13$
3110	3	$\overleftrightarrow{243}1$			102	−1	$2\overrightarrow{41}3$	0000102	−1	$2\overrightarrow{41}3$
3010	2	$\overleftrightarrow{34}21$			101	−1	$21\overrightarrow{43}$	0000101	−1	$21\overrightarrow{43}$
3000		4321			100		2134	0000100	+5	2134
								0001100	⋯	2134

The binary reflected Gray code $\mathsf{brgc}(n)$ is the special case where the *signed binary sequence* $\mathsf{ruler} \pm (2, 2, \ldots, 2)$ (OEIS A164677) gives bit increments and decrements. More interestingly, $\mathsf{plain}(n)$ follows the *signed upstairs sequence* $\mathsf{ruler} \pm (1, 2, \ldots, n)$: $+j$ swaps value $n-j+1$ left; $-j$ swaps value $n-j+1$ right[7].

A *signed basis* $\mathbf{b}$ contains $1, 2, \ldots, n$ plus n copies of 2. Note that $|\mathsf{ruler}\pm(\mathbf{b})| = 2^n n! - 1 = |S_n^{\pm}| - 1$. The *twisted basis* concatenates the signed binary and signed upstairs bases to give the *twisted ruler sequence* $\mathsf{ruler} \pm (2, 2, \ldots, 2, 1, 2, \ldots, n)$.

Lemma 3. *A change sequence for* $\mathsf{twisted}(n)$ *is* $\mathsf{ruler} \pm (2, 2, \ldots, 2, 1, 2, \ldots, n)$: $+j$ *and* $-j$ *respectively 2-twist value* $n-j+1$ *to the left and right for* $1 \leq |j| \leq n$; $+j$ *and* $-j$ *respectively 1-twist (flip) value* $n-j+1$ *down and up for* $n < |j| \leq 2n$.

Now we can looplessly generate twisted plain changes $\mathsf{twisted}(n)$ by looplessly generating the twisted ruler sequence $\mathsf{ruler}\pm(2,2,\ldots,2,1,2,\ldots,n)$ and its changes.

5   Loopless Generation of Gray Codes via Ruler Sequences

The greedy algorithm for $\mathsf{twisted}(n)$ in Sect. 2.3 is simple but inefficient. It requires exponential space, as all previously created objects must be remembered. Fortunately, greedy Gray codes can often be generated without remembering previous objects [21,28,29]. The loopless *history-free* implementation that we provide here uses a signed ruler sequence to generate the changes. Loopless algorithms for non-greedy Gray codes also exist using ruler sequence changes [8,15].

Algorithm 2 has procedures for generating Gray codes whose changes follow a ruler sequence with any bases $\mathbf{b}$. The start object is $\mathbf{s}$ and the change functions are in $\mathbf{fns}$. The ruler sequence is generated one entry at a time, and the current object is updated and visited accordingly. More specifically, if j is the next entry, then $\mathbf{fns}[j]$ is applied to $\mathbf{s}$ to create the next object. The pseudocode is adapted from Knuth's loopless reflected mixed-radix Gray code Algorithm M [17].

Algorithm 2 can looplessly generate various Gray codes in this paper. As a simple example, Zaks' pancake order uses RulerGrayCode with $\mathbf{b} = 2, 3, \ldots, n$, $\mathbf{s} = 12 \cdots n$, and $\mathbf{fns} = \overleftrightarrow{r_1}, \overleftrightarrow{r_2}, \ldots, \overleftrightarrow{r_n}$ where $\overleftrightarrow{r_i}$ reverses the prefix of length i. The $\mathsf{brgc}(n)$ can be looplessly generated using RulerGrayCode or RulerGrayCode$\pm$. When generating $\mathsf{plain}(n)$ with RulerGrayCode$\pm$ we maintain the inverse of the current permutation in order to swap a specific value left or right in $\mathcal{O}(1)$-time. Maintaining the inverse is also required to looplessly generate our new order.

---

[7] Surprisingly, this sequence is not yet in the *Online Encyclopedia of Integer Sequences*, nor is the *signed downstairs sequence* $\mathsf{ruler}\pm(n,n-1,\ldots,2) = \mathsf{ruler}\pm(n,n-1,\ldots,1)-1 = 1, 2, -1, 2, 1, 3, -1, -2, 1, -2, -1, 3, 1, 2, -1, 2, 1, 3, -1, -2, 1, -2, -1, 4, \ldots$.

Algorithm 2. Generating Gray codes using ruler sequences with bases **b**. The **fns** modify object **s** and are indexed by the sequence. For example, if **b** = 3, 2 then RulerGrayCode±(**b**) visits ruler ± (2, 3) = 1, 1, 2, −1, −1 alongside a Gray code that starts **s** and applies **fns** with indices 1, 1, 2, −1, −1. The signed version also generates the reflected mixed-radix Gray code mix(**b**) in **a**, with the **d** values providing ±1 directions of change. So in the previous example the mixed-radix words $\overline{0}0, \overline{1}0, 2\overline{0}, 2\underline{1}, 1\underline{1}, 10$ are generated in **a**. Focus pointers are stored in **f**. The overall algorithm is loopless if each function runs in worst-case $\mathcal{O}(1)$-time. Note that the indexing is reversed with respect to Sect. 4 with $\mathbf{b} = b_1, b_2, \ldots, b_n$. Unary bases should be omitted: $b_i \geq 2$ is required for $0 \leq i < n$.

1: **procedure** RulerGrayCode(**b**, **s**, **fns**)	1: **procedure** RulerGrayCode±(**b**, **s**, **fns**)
2: $a_1\ a_2\ \cdots\ a_n \quad \leftarrow 0\ 0\ \cdots\ 0$	2: $a_1\ a_2\ \cdots\ a_n \quad \leftarrow 0\ 0\ \cdots\ 0$
3: $f_1\ f_2\ \cdots\ f_{n+1} \leftarrow 1\ 2\ \cdots\ n{+}1$	3: $f_1\ f_2\ \cdots\ f_{n+1} \leftarrow 1\ 2\ \cdots\ n{+}1$
4:	4: $d_1\ d_2\ \cdots\ d_n \quad \leftarrow 1\ 1\ \cdots\ 1$
5: visit(**s**)	5: visit(**s**)
6: **while** $f_1 \leq n$ **do**	6: **while** $f_1 \leq n$ **do**
7: $j \leftarrow f_1$	7: $j \leftarrow f_1$
8: $f_1 \leftarrow 1$	8: $f_1 \leftarrow 1$
9: $a_j \leftarrow a_j + 1$	9: $a_j \leftarrow a_j + d_j$
10: $\mathbf{s} \leftarrow \mathbf{fns}[j](\mathbf{s})$	10: $\mathbf{s} \leftarrow \mathbf{fns}[d_j \cdot j](\mathbf{s})$
11: visit($j, \mathbf{s}$)	11: visit($d_j \cdot j, \mathbf{s}$)
12: **if** $a_j = b_j - 1$ **then**	12: **if** $a_j \in \{0, b_j - 1\}$ **then**
13: $a_j \leftarrow 0$	13: $d_j \leftarrow -d_j$
14: $f_j \leftarrow f_{j+1}$	14: $f_j \leftarrow f_{j+1}$
15: $f_{j+1} \leftarrow j + 1$	15: $f_{j+1} \leftarrow j + 1$

Theorem 2. *Twisted plain changes* twisted(n) *and its change sequence are generated looplessly by* RulerGrayCode±(**b**, **s**, **fns**) *with twisted bases* **b**, *the positive identity permutation* $\mathbf{s} \in S_n^{\pm}$, *and the change functions* **fns** *given in Lemma 3.*

6   Final Remarks

Alternate Gray codes for signed permutations can be generated using other signed ruler sequences, and some of these generalize to colored permutations [25]. For additional new results involving greedy Gray codes see Merino and Mütze [20].

Open question: Does $S_n^{\pm}$ have a doubly-adjacent Gray code [2] using twists? We thank the reviewers for their helpful comments, proofreading, and debugging.

A   Python Implementation

A loopless implementation of our signed plain change order twisted(n) in Python 3. Entries in the twisted ruler sequence ruler $\pm (n, n-1, \ldots, 2, 1, 2, 2, \ldots, 2)$ select the 2-twist or 1-twist (i.e., flip) to apply[8]. Programs are available online [37].

```
Flip sign of value v in signed permutation p with unsigned inverse q
def flip(p, q, v):  # with 1-based indexing, ie p[0] and q[0] are ignored.
    p[q[v]] = -p[q[v]]
    return p, q


# 2-twists value v to the left / right using delta = -1 / delta = 1
def twist(p, q, v, delta):  # with 1-based indexing into both p and q.
    pos = q[abs(v)]   # Use inverse to get the position of value v.
    u = p[pos+delta]  # Get value to the left or right of value v.
    p[pos], p[pos+delta] = -p[pos+delta], -p[pos]  # Twist u and v.
    q[abs(v)], q[abs(u)] = pos+delta, pos  # Update unsigned inverse.
    return p, q  # Return signed permutation and its unsigned inverse.


# Generate each signed permutation in worst-case O(1)-time.
def twisted(n):
    m = 2*n-1  # The mixed-radix bases are n, n-1, ..., 2, 1, 2,..., 2
    bases = tuple(range(n,1,-1)) + (2,) * n  # but the 1 is omitted.
    word = [0] * m          # The mixed-radix word is initially 0^m.
    dirs = [1] * m          # Direction of change for digits in word.
    focus = list(range(m+1))  # Focus pointers select digits to change.
    flips =   [lambda p,q,v=v: flip(p,q,v)      for v in range(n,0,-1)]
    twistsL = [lambda p,q,v=v: twist(p,q,v,-1) for v in range(n,1,-1)]
    twistsR = [lambda p,q,v=v: twist(p,q,v, 1) for v in range(n,1,-1)]
    fns = [None] + twistsL + flips + flips[-1::-1] + twistsR[-1::-1]
    p = [None] + list(range(1,n+1))  # To use 1-based indexing we set
    q = [None] + list(range(1,n+1))  # and ignore p[0] = q[0] = None.
    yield p[1:]  # Pause the function and return signed permutation p.
    while focus[0] < m:  # Continue if the digit to change is in word.
        index = focus[0]   # The index of the digit to change in word.
        focus[0] = 0       # Reset the first focus pointer.
        word[index] += dirs[index]  # Adjust the digit using its direction.
        change = dirs[index] * (index+1)  # Note: change can be negative.
        if word[index] == 0 or word[index] == bases[index]-1:  # If the
            focus[index] = focus[index+1]  # mixed-radix word's digit is at
            focus[index+1] = index+1       # its min or max value, then update
            dirs[index] = -dirs[index]     # focus pointers, change direction.
        p, q = fns[change](p, q)  # Apply twist or flip encoded by change.
        yield p[1:]


# Demonstrating the use of our twisted function for n = 4.
for p in twisted(4): print(p)  # Print all 2^n n! signed permutations.
```

---

[8] Negative indices give right-to-left access in Python. So the ruler entry -1 selects the last function fns[-1] = twist(p,q,n,1) (i.e., 2-twist n right). Notes: v=v is for binding; slice notation [-1::-1] reverses a list; indices are reversed from Sect. 4.

References

1. Cameron, B., Sawada, J., Therese, W., Williams, A.: Hamiltonicity of k-sided pancake networks with fixed-spin: efficient generation, ranking, and optimality. Algorithmica **85**(3), 717–744 (2023)
2. Compton, R.C., Gill Williamson, S.: Doubly adjacent Gray codes for the symmetric group. Linear Multilinear Algebra **35**(3–4), 237–293 (1993)
3. Corbett, P.F.: Rotator graphs: an efficient topology for point-to-point multiprocessor networks. IEEE Trans. Parallel Distrib. Syst. **3**(5), 622–626 (1992)
4. Duato, J., Yalamanchili, S., Ni, L.: Interconnection Networks. Morgan Kaufmann, Burlington (2003)
5. Duckworth, R., Stedman, F.: Tintinnalogia: Or, The Art of Ringing. London (1668)
6. Ehrlich, G.: Loopless algorithms for generating permutations, combinations, and other combinatorial configurations. J. ACM **20**(3), 500–513 (1973)
7. Fertin, G., Labarre, A., Rusu, I., Vialette, S., Tannier, E.: Combinatorics of Genome Rearrangements. MIT Press, Cambridge (2009)
8. Ganapathi, P., Chowdhury, R.: A unified framework to discover permutation generation algorithms. Comput. J. **66**(3), 603–614 (2023)
9. Gardner, M.: Curious properties of the Gray code and how it can be used to solve puzzles. Sci. Am. **227**(2), 106 (1972)
10. Gray, F.: Pulse code communication. United States Patent Number 2632058 (1953)
11. Hannenhalli, S., Pevzner, P.A.: Transforming cabbage into turnip: polynomial algorithm for sorting signed permutations by reversals. In: Proceedings of the 27th Annual ACM Symposium on Theory of Computing (STOC 1995), pp. 178–189. ACM (1995)
12. Hartung, E., Hoang, H., Mütze, T., Williams, A.: Combinatorial generation via permutation languages. I. fundamentals. Trans. Am. Math. Soc. **375**(04), 2255–2291 (2022)
13. Heath, F.: Origins of the binary code. Sci. Am. **227**(2), 76–83 (1972)
14. Hindenburg, C.F.: Sammlung combinatorisch-analytischer Abhandlungen, vol. 1. ben Gerhard Fleischer dem Jungern (1796)
15. Holroyd, A.E., Ruskey, F., Williams, A.: Shorthand universal cycles for permutations. Algorithmica **64**, 215–245 (2012)
16. Johnson, S.M.: Generation of permutations by adjacent transposition. Math. Comput. **17**(83), 282–285 (1963)
17. Knuth, D.E.: Art of Computer Programming, Volume 4, Fascicle 4, The: Generating All Trees-History of Combinatorial Generation. Addison-Wesley, Boston (2013)
18. Korsh, J., LaFollette, P., Lipschutz, S.: A loopless implementation of a Gray code for signed permutations. Publications de l'Institut Mathematique **89**(103), 37–47 (2011)
19. Liptak, Z., Masillo, F., Navarro, G., Williams, A.: Constant time and space updates for the sigma-tau problem. In: Nardini, F.M., Pisanti, N., Venturini, R. (eds.) SPIRE 2023. LNCS, vol. 14240, pp. 323–330. Springer, Cham (2023). https://doi.org/10.1007/978-3-031-43980-3_26
20. Merino, A., Mutze, T.: Traversing combinatorial 0/1-polytopes via optimization. In: 2023 IEEE 64th Annual Symposium on Foundations of Computer Science (FOCS), pp. 1282–1291 (2023)
21. Merino, A., Mutze, T., Williams, A.: All your bases are belong to us: listing all bases of a matroid by greedy exchanges. In: 11th International Conference on Fun

with Algorithms (FUN 2022), vol. 226, p. 22. Schloss Dagstuhl-Leibniz-Zentrum für Informatik (2022)

22. Mütze, T.: Combinatorial Gray codes-an updated survey. arXiv preprint arXiv:2202.01280 (2022)
23. OEIS Foundation Inc.: The On-Line Encyclopedia of Integer Sequences (2023). http://oeis.org
24. Ord-Smith, R.: Generation of permutation sequences: part 1. Comput. J. **13**(2), 152–155 (1970)
25. Qiu, Y.F.: Greedy and speedy: new iterative gray code algorithms. Bachelor's thesis, Williams College (2024)
26. Ruskey, F.: Combinatorial generation. Preliminary working draft. University of Victoria, Victoria, BC, Canada 11, 20 (2003)
27. Savage, C.: A survey of combinatorial Gray codes. SIAM Rev. **39**(4), 605–629 (1997)
28. Sawada, J., Williams, A.: Greedy flipping of pancakes and burnt pancakes. Discret. Appl. Math. **210**, 61–74 (2016)
29. Sawada, J., Williams, A.: Successor rules for flipping pancakes and burnt pancakes. Theoret. Comput. Sci. **609**, 60–75 (2016)
30. Sedgewick, R.: Permutation generation methods. ACM Comput. Surv. (CSUR) **9**(2), 137–164 (1977)
31. Steinhaus, H.: One hundred problems in elementary mathematics. Courier Corporation (1979)
32. Stigler, S.M.: Stigler's law of eponymy. Trans. New York Acad. Sci. **39**(1 Series II), 147–157 (1980)
33. Suzuki, Y., Sawada, N., Kaneko, K.: Hamiltonian cycles and paths in burnt pancake graphs. In: Proceedings of the ISCA 18th International Conference on Parallel and Distributed Computing Systems, pp. 85–90 (2005)
34. Trotter, H.F.: Algorithm 115: perm. Commun. ACM **5**(8), 434–435 (1962)
35. Williams, A.: $O(1)$-time unsorting by prefix-reversals in a boustrophedon linked list. In: Boldi, P., Gargano, L. (eds.) FUN 2010. LNCS, vol. 6099, pp. 368–379. Springer, Heidelberg (2010). https://doi.org/10.1007/978-3-642-13122-6_35
36. Williams, A.: The greedy Gray code algorithm. In: Dehne, F., Solis-Oba, R., Sack, J.R. (eds.) WADS 2013. LNCS, vol. 8037, pp. 525–536. Springer, Heidelberg (2013). https://doi.org/10.1007/978-3-642-40104-6_46
37. Williams, A.: Signed-plain-changes (2024). https://gitlab.com/combinatronics/signed-plain-changes
38. Zaks, S.: A new algorithm for generation of permutations. BIT Numer. Math. **24**(2), 196–204 (1984)

A BWT-Based Algorithm for Random de Bruijn Sequence Construction

Zsuzsanna Lipták[1]([✉])[ID] and Luca Parmigiani[2][ID]

[1] Dipartimento di Informatica, University of Verona, Verona, Italy
zsuzsanna.liptak@univr.it
[2] Center for Biotechnology, Bielefeld University, Bielefeld, Germany

Abstract. A binary de Bruijn sequence (dB sequence) of order k is a circular binary string that contains each k-length word exactly once as a substring. Most existing algorithms construct a specific dB sequence, or members of a specific class of dB sequences, representing only a tiny fraction of the complete set. The only algorithms capable of generating *all* dB sequences are based on finding Euler cycles in de Bruijn graphs. Here, we present an algorithm for constructing *random* binary dB sequences which uses the extended Burrows-Wheeler Transform. Our method is simple to implement (less than 120 lines of C++ code) and can produce random dB sequences of any order. Even though it does not output dB sequences uniformly at random, it provably outputs each dB sequence with positive probability. The algorithm runs in linear space and near-linear time in the length of the dB sequence and needs less than one second on a laptop computer for orders up to 23, including outputting the sequence. It can be straightforwardly extended to any constant-size alphabet. To the best of our knowledge, this is the first practical algorithm for generating random dB sequences which is capable of producing *all* dB sequences. Apart from its immediate usefulness in contexts where it is desirable to use a dB sequence that cannot be guessed easily, we also demonstrate our algorithm's potential in theoretical studies, giving hitherto unknown estimates of the average discrepancy of binary dB sequences. The code is available (in C++ and python) at https://github.com/lucaparmigiani/rnd_dbseq.

Keywords: De Bruijn sequence · Burrows-Wheeler Transform · extended BWT · random generation · spanning tree · standard permutation · Lyndon words

1 Introduction

A *binary de Bruijn sequence of order k* (or *dB sequence*) is a circular binary string in which every k-length string occurs exactly once as a substring. For example, aabb, aaababbb, aaaabbababbbbaab are dB sequences of order $2, 3,$ and 4 respectively, and the following is a dB sequence of order 8:

© The Author(s), under exclusive license to Springer Nature Switzerland AG 2024
J. A. Soto and A. Wiese (Eds.): LATIN 2024, LNCS 14578, pp. 130–145, 2024.
https://doi.org/10.1007/978-3-031-55598-5_9

aaaaaaaabbbbbbbbaabbbaaaabaababaababbbaaababababbaabaaaabbbabaaaab
abbaaaabbaaaaabbabaabbbbaaabbbaabbaababababbbbaabaabbabbaabbababbabb
abbbabbaaabaaaaabaaabbbbabbbaabababbabaaabaabbbabbbbabaabaabaaababbb
bbabbababaabbaaabbabbbbbbababbbabababbaaabbaabbbbbaaaaaabab

It is well known that there exist $2^{2^{k-1}-k}$ distinct binary dB sequences of order k [7]. De Bruijn sequences are among the best studied families of binary strings, due to their multitudinous applications, in particular as pseudo-random number generators, for bias-free experiment setup [9, 26], cryptographic protocols [20, 29], and computational biology [2, 3, 24], to name just a few. Accordingly, much research effort has gone into constructing them, and many construction algorithms exist; however, most construct a particular dB sequence, such as the lex-least dB sequence, or some very restricted class of dB sequences, in particular, dB sequences that stem from linear feedback shift registers based on primitive polynomials (LFSRs). But only a tiny fraction of all dB sequences are LFSRs. To get an idea, in Table 1, we compare the number of LFSRs to the total number of dB sequences, for small values of k.

Table 1. Number of de Bruijn sequences generated by LFSRs vs. the total number, for some small values of k. Values in last row for $k \geq 7$ are approximate.

k	4	5	6	7	10	15	20
#LFSRs	2	6	6	18	60	1 800	24 000
#dbseqs	16	2048	67 108 864	$1.44 \cdot 10^{17}$	$1.3 \cdot 10^{151}$	$3.63 \cdot 10^{4927}$	$2.47 \cdot 10^{157820}$

In this paper, we present a simple and practical random dB sequence construction algorithm that is able to produce *all* dB sequences. It is based on an insight regarding the Burrows-Wheeler Transform (BWT [4]) of dB sequences, due to Higgins [16]. Higgins introduced a generalization of dB sequences, called *de Bruijn sets*, and proved that these are exactly the preimages of certain type of strings under the *extended BWT* (eBWT) of Mantaci et al. [21]. We connect this result with the *swap method* of [14] to manipulate the eBWTs of strings, and employ it to produce a dB sequence, starting from a random bitstring.

Our approach turns out to be similar to the so-called cycle-joining [18, 30], in a new guise, and applied in a more general manner. To the best of our knowledge, this is the first time that the connection between cycle-joining and the eBWT has been made. Moreover, the insights of [16] on the eBWT of dB sequences have not been previously used to *generate* dB sequences, even less to generate *random* dB sequences. Our algorithm is simple, easy to implement, and runs in near-linear time and linear space in the length of the dB sequence. Additionally, it can be straightforwardly adapted to any constant-size alphabet. This is the first practical algorithm able to generate *all* dB sequences. An implementation, in C++ (less than 120 lines) and python, for binary as well as general alphabets, can be found at https://github.com/lucaparmigiani/rnd_dbseq.

1.1   Related Work

There are many algorithms for constructing dB sequences in the literature, see [11] for a classic survey, or the introductions of e.g. [12,13,30] for more recent ones. Many of these constructions are described, and some implemented, at debruijnsequence.org [25]; most use only $\mathcal{O}(k)$ space, i.e. logarithmic in the output size.[1] None of these algorithms, except those based on generating Euler cycles in de Bruijn graphs, are capable of generating *all* dB sequences. This is important because it may explain why there appear to be no efficient algorithms for *random* dB sequence generation: adding randomization to one of these construction algorithms will not produce a random generator capable of producing *any* dB sequence.

One reason why much of the research has been restricted to algorithms that require $\mathcal{O}(k)$ space is that the length of a dB sequence of order k is $n = 2^k$, i.e. exponential in k. Thus, when dB sequences with large order are sought, one may not be able to afford to store the entire sequence. Further, many construction algorithms stand in the tradition of *feedback shift register sequences*, which use k binary memory cells and a feedback function $f : GF(2)^k \to GF(2)$, which, from the current state of the cells, determines the next bit. When f is a linear function based on a primitive polynomial, then the resulting sequence is a so-called *m-sequence*, a dB sequence without the substring 0^k; adding a 0 in the right place will produce a dB sequence. These special dB sequences are called LFSRs. They can be produced in $\mathcal{O}(k)$ space and $\mathcal{O}(k)$ time per bit, but have the drawback that after only $2k$ bits, the underlying function can be determined by an adversary. For more on shift registers, see the classic book by Golomb [15].

In many real-life situations, such as experimental setup or computational biology applications, orders up to $k = 20$ or so are amply sufficient. In these situations, a truly random dB sequence would be preferable, chosen from the vast set of all dB sequences. However, no such algorithm appears to be available, so users have to turn to naive implementations of finding Euler cycles [9,28].

The fact that it is only possible to generate such a small subset of all de Bruijn sequences also leads to a knowledge gap, since we simply do not know much about other dB sequences. This point is argued in [24], where dB sequences are used for the specific purpose of designing microarrays: "[...] only a small fraction of de Bruijn sequences correspond to sequences generated by an LFSR, and the utility of such non-LFSR-generated de Bruijn sequences remains largely unexplored. [...] there may be additional desirable properties of de Bruijn sequences that we have not yet considered, and for which LFSRs might not be optimal."

The only construction algorithms that can generate *all* de Bruijn sequences are those that construct an Euler cycle in the de Bruijn graph, such as Fleury's algorithm [10]. Fleury's algorithm uses a spanning tree of the de Bruijn graph, which is used to define the last edge to be used for each vertex. There is a one-to-one-correspondence between spanning in-trees and binary de Bruijn sequences,

---

[1] Note that in this tradition, an algorithm which runs in time and space $\mathcal{O}(n)$ is considered exponential, since it is exponential in k; however, if one wants to output or even store the sequence, then it is de facto optimal.

so choosing a spanning tree uniformly at random would result in a uniformly random dB sequence. Unfortunately, the fastest algorithms for random spanning tree construction run in superquadratic time in the number of vertices [5,8], which in our case is $2^{k-1} = n/2$, where n is the length of the dB sequence.

Our method differs from previous work using the so-called cycle-joining [18, 30] in at least two important ways: One, we (conceptually) start from *any* cycle cover of the de Bruijn graph, and not a specific one such as the pure cycle transition. Two, when joining cycles, we choose from all intersecting nodes, and do not restrict ourselves to specific ones (such as the 'representatives' of Jansen [18]).

2 The Burrows-Wheeler Transform of a dB Sequence

A *binary string* is a finite sequence of elements from the alphabet {a, b}. We index strings from 0, denote the ith character of t by t_i, and its length by $|t|$. String u is a *substring* of string t if $t = xuy$ for some strings x, y; u is called a *prefix* of t if x is empty, and a *suffix* of t if y is empty. The notation $u^i = u \cdots u$ denotes the i-fold concatenation of string u, also called a *power* when $i > 1$. A string t is called *primitive* if it is not a power, i.e. if $t = u^i$ implies $u = t$ and $i = 1$. Two strings u, v are *conjugates* (or *rotations*) if there exist words x, y such that $u = xy$ and $v = yx$. A string is called a *necklace* if it is lexicographically smaller or equal to all of its rotations, and a *Lyndon word* if it is strictly smaller than all of its rotations. Every necklace t can be written uniquely as $t = u^d$ for some $d \geq 1$ and some Lyndon word u. We use $\text{dist}_H(s, t)$ to denote the Hamming distance between two equal-length strings s and t.

A *circular string* is an equivalence class of its conjugates. Often, a circular string is simply viewed as a linear string which is read circularly. A *substring* of a circular string is a substring of any of the conjugates.

The *Burrows-Wheeler Transform (BWT)* [4] of a primitive string t is a permutation of the characters of t, defined as the concatenation of the last columns of a matrix, the so-called BW-matrix, which contains all of t's rotations in lexicographic order. Thus, if s is the BWT of t, then s_i is the last character of the i'th rotation of t in lexicographic order, see Fig. 1 (left and middle). It follows from the definition that the BWT is invariant w.r.t. conjugacy. In fact, $\text{bwt}(t) = \text{bwt}(t')$ if and only if t and t' are conjugates, i.e., the BWT is unique up to rotation; moreover, it is reversible in linear time in the length of the string [22].

Given a string s, its *standard permutation* [23] π_s is defined by: $\pi_s(i) < \pi_s(j)$ if $s_i < s_j$, or $s_i = s_j$ and $i < j$.[2] For example, the standard permutation of $s = \text{baabbbaba}$ is $\pi_s = \left(\begin{smallmatrix} 0 & 1 & 2 & 3 & 4 & 5 & 6 & 7 \\ 4 & 0 & 1 & 5 & 6 & 2 & 7 & 3 \end{smallmatrix}\right) = (0, 4, 6, 7, 3, 5, 2, 1)$, where we give both the two-line and the cycle notation of π_s. For a permutation π, we denote the number of its cycles by $c(\pi)$. It is a well-known fact that a string s is the BWT of some primitive string t if and only if π_s is cyclic, i.e. if $c(\pi_s) = 1$ [19]. In this case, $\text{baabbbaba} = \text{bwt}(\text{aaababbbb})$. Given π_s, such a t can be computed as follows: start with $t_{n-1} = s_0$, and list the characters of t back to front, using

---

[2] The standard permutation is also called *LF-mapping* if s is the BWT of some string.

$t_{n-j-1} = s_{\pi_s^j(0)}$. In general, data structures for fast rank-queries, such as wavelet trees, are needed to compute $\pi_s(j)$ efficiently [22].

Mantaci et al. [21] introduced the *extended BWT (eBWT)*, which generalizes the BWT to multisets. The eBWT of a multiset of primitive strings $\mathcal{M}$ consists of the concatenation of the last characters of the conjugates of the strings in $\mathcal{M}$, taken in ω-order. The ω-order $<_\omega$ is defined based on the infinite concatenation $t^\omega = t \cdot t \cdot t \cdots$, as follows: $t <_\omega s$ if either $t^\omega <_{\text{lex}} s^\omega$, or there is a string u such that $t = u^i$, $s = u^j$, and $i < j$ (in this case, $t^\omega = s^\omega$). See Fig. 1 (right) for an example. Again, ebwt($\mathcal{M}$) is a permutation of the characters of the strings in $\mathcal{M}$. Similarly to the BWT, the eBWT is unique up to rotation, and the number and length of the strings in $\mathcal{M}$ corresponds to the number and length of the cycles of the standard permutation of the string ebwt($\mathcal{M}$) [21]. For example, $\pi_{\text{abbababa}} = \left(\begin{smallmatrix} 0 & 1 & 2 & 3 & 4 & 5 & 6 & 7 \\ 0 & 4 & 5 & 1 & 6 & 2 & 7 & 3 \end{smallmatrix} \right) = (0)(1, 4, 6, 7, 3)(2, 5)$, and abbababa = ebwt({a, aabbb, ab}). Every string is the eBWT of some multiset $\mathcal{M}$. Note that if $\mathcal{M}$ is a singleton, then eBWT and BWT conincide: ebwt({t}) = bwt(t).

a dB seq. of order 3		a string of length 2^3		a dB set of order 3	
aaababbb		ababbaab		{a, ab, aabbb}	
rotations	bwt	rotations	bwt	rotations	ebwt
aaababbb	b	aabababb	b	a	a
aababbba	a	abababba	a	aabbb	b
ababbbaa	a	ababbaab	b	ab	b
abbbaaab	b	abbaabab	b	abbba	a
baaababb	b	baababab	b	baabb	b
babbbaaa	a	bababbaa	a	ba	a
bbaaabab	b	babbaaba	a	bbaab	b
bbbaaaba	a	bbaababa	a	bbbaa	a

Fig. 1. The BW-matrices of the order-3 dB sequence aaababbb (left), of a binary string of the same length (center), and the order-3 dB-set {a, ab, aabbb} (right).

Higgins [16] introduced *de Bruijn sets*, which are a generalization of dB sequences. We give the definition restricted to binary, and adjusting terminology:[3]

Definition 1. *A set $\mathcal{M}$ of primitive binary strings is called a de Bruijn set of order k if the total length of strings in $\mathcal{M}$ is 2^k and for every k-length string u there is a string $t \in \mathcal{M}$ s.t. u is the prefix of some power of some conjugate of t.*

It follows from the definition that no string can occur more than once, hence this is always a set. An example is given in Fig. 1 (right). The reader is invited to check that every 3-mer is the prefix of some power of some row. Singleton sets of one dB sequence are special cases of de Bruijn sets. We note that the BWT of the dB sequence (left), and the eBWT of the de Bruijn set (right), have

---

[3] In particular, Higgins uses the term *necklace* in a non-standard meaning.

a special form, not shared by the bitstring (center): they are concatenations of ab's and ba's. Indeed, Higgins proved the following beautiful result (here given in the restricted version for binary alphabets):

Theorem 1 (Higgins 2012 [16]). *A multiset $\mathcal{M}$ is a binary de Bruijn set of order k if and only if $ebwt(\mathcal{M}) \in \{ab, ba\}^{2^{k-1}}$.*

Let $m = 2^{k-1}$, and let $F_k = \{ab, ba\}^m$. We will encode elements of F_k using a bitstring of length m: for $v \in F_k$, define enc(v) by setting the ith bit to 0 if $v[2i, 2i + 1] = $ ab, and to 1 otherwise (i.e., if $v[2i, 2i + 1] = $ ba). For example, enc(abbababa) $= $ 0111. For a bitstring $b \in \{0, 1\}^m$, dec(b) is the inverse map, e.g. dec(1101) $= $ babaabba. Let $S_k = \{s \mid s$ is the BWT of a dB sequence of order $k\}$, thus, $S_k \subseteq F_k$. Since de Bruijn sequences are primitive, we get:

Corollary 1. *Let $v \in F_k$. Then $v \in S_k$ if and only if π_v is cyclic.*

Our strategy for constructing a random dB sequence will be to produce an element from $v \in F_k$ uniformly at random, turn it into an element from $s \in S_k$, invert $s = bwt(t)$, and output the dB sequence t. From now on we will use $n = 2^k$, the length of the dB sequence, and $m = 2^{k-1}$, the length of its encoding.

3   Swapping Characters in the eBWT

The following is a generalization of a technique from [14], where it was used only for swapping a specific character (the dollar) with its right neighbor.

Lemma 1 (Swap Lemma). *Let $v \in \{a, b\}^n$, $v_i \neq v_{i+1}$, v' the string that results from swapping v_i with v_{i+1}. Then $\pi_{v'} = (\pi_v(i), \pi_v(i + 1)) \circ \pi_v$. Moreover, if i and $i + 1$ are in distinct cycles in the cycle decomposition of π, then $c(\pi_{v'}) = c(\pi_v) - 1$, otherwise $c(\pi_{v'}) = c(\pi_v) + 1$.*

Proof. The first statement follows from the definition of the standard permutation, which is equivalent to a stable sort of the characters of the string: since two distinct and neighboring characters are exchanged, the only positions affected are i and $i + 1$. It is a well-known property of permutations that a transposition (j, j') either splits a cycle (if j and j' are in the same cycle), or merges two cycles (if they are in different cycles), for an explicit proof see [14].

Example 1. Let $v = $ babbbaaa and $v' = $ bbabbaaa. Then $\pi_v = \left(\begin{smallmatrix} 0 & 1 & 2 & 3 & 4 & 5 & 6 & 7 \\ 4 & 0 & 5 & 6 & 7 & 1 & 2 & 3 \end{smallmatrix}\right) = (0, 4, 7, 3, 6, 2, 5, 1)$, and $\pi_{v'} = \left(\begin{smallmatrix} 0 & 1 & 2 & 3 & 4 & 5 & 6 & 7 \\ 4 & 5 & 0 & 6 & 7 & 1 & 2 & 3 \end{smallmatrix}\right) = (0, 4, 7, 3, 6, 2)(1, 5) = (0, 5) \circ \pi_v$.

The next lemma shows that for $v \in F_k$, the standard permutation is easy to compute. Let us refer to the interval $[2i, 2i + 1]$ as the ith *block* of v.

Lemma 2. *Let $v \in F_k$ and $b = enc(v)$. If $b_i = 0$, then $\pi_v(2i) = i$ and $\pi_v(2i + 1) = m+i$; otherwise, $\pi_v(2i) = m+i$ and $\pi_v(2i+1) = i$. In particular, $\pi_v(\{2i, 2i+1\}) = \{i, m + i\}$.*

Proof. Since $v \in F_k$, we know that each substring $v_{2i}v_{2i+1}$ has the form ab or ba. Thus, the ith block contains the ith a, which is mapped to position i by π_v, and the ith b, which is mapped to position $m+i$. Whether $v_{2i} = $ a or $v_{2i+1} = $ a is encoded in b_i.

We call the ith block *unhappy (w.r.t. v)* if $2i$ and $2i+1$ belong to distinct cycles in π_v. We now show that, in order to arrive at a standard permutation with only one cycle, it is sufficient to swap positions within blocks.

Lemma 3. *Let $v, v' \in F_k$. Then*

1. *If $v \notin S_k$, then there exists an unhappy block w.r.t. v.*
2. *If $dist_H(enc(v), enc(v')) = 1$, then $c(\pi_{v'}) = c(\pi_v) - 1$ if and only if i is unhappy w.r.t. v, where i is the unique position s.t. $enc(v)_i \neq enc(v')_i$.*

Proof. 1. Assume otherwise, and choose j minimal such that j is not in the same cycle as 0. Note that this implies that j is even, since there are no unhappy blocks. Let $i = j/2$, then j is in the ith block, which, by Lemma 2, is mapped to $\{i, m+i\}$, hence $j, j+1, i, m+i$ are all in the same cycle. But since $i < j$, i is in the same cycle as 0, a contradiction. 2. follows from Lemma 1.

Proposition 1. *Let $v \in F_k, b = enc(v)$, and let $d = \min\{dist_H(b, b') \mid b' \in enc(S_k)\}$. Then $d = c(\pi_v) - 1$.*

Proof. Note that $c(\pi_v) = 1$ iff $v \in S_k$ iff $d = 0$. Let therefore $c(\pi_v) > 1$. Note that every flip of a bit in $enc(v)$ corresponds to a swap of neighboring distinct characters, and thus changes the number of cycles by 1 (Lemma 1), decreasing it if and only if the corresponding block is unhappy (Lemma 3). By Lemma 3, part 1., there is always an unhappy block as long as $c(\pi_v) > 1$.

Let $v \in F_k, c(\pi_v) > 1$. Define the *cycle graph* $\Gamma_v = (V, E)$ as an undirected multigraph whose vertex set equals the set of cycles of π_v, and there is an edge for every unhappy block i, connecting the cycles $C_j, C_{j'}$, where $2i$ is in C_j and $2i+1$ in $C_{j'}$. Let $B(v) = \{s \in S_k \mid dist_H(enc(v), enc(s)) = c(\pi_v) - 1\}$.

The following is a new formulation of a theorem due to Aardenne-Ehrenfest and de Bruijn (1951) [1], there given in terms of cycle joining (CJM), see the Introduction. We give a new and simple proof in terms of the eBWT.

Theorem 2 (Aardenne-Ehrenfeucht and de Bruijn, 1951). *There is a one-to-one correspondence between spanning trees of Γ_v and $s \in B(v)$.*

Proof. Swapping the two characters in an unhappy block merges the two corresponding cycles. Therefore, the edges of a spanning tree of Γ_v represent a set of blocks which result in merging all cycles of π_v, and thus will yield an element $s \in S_k$. Since the number of edges of any spanning tree is $c(\pi_v) - 1$, $s \in B(v)$. Conversely, two distinct elements of $B(v)$ necessarily differ in their BWT, and thus, in the blocks which need to be flipped, yielding distinct spanning trees.

4   Algorithm

Conceptually, our algorithm works as follows.

1. Choose a random bitstring b of length 2^{k-1}.
2. Compute the standard permutation π_v of $v = \text{dec}(b)$.
3. Construct the cycle graph Γ_v.
4. Choose a random spanning tree T of Γ_v.
5. Flip the bits of b corresponding to the edges of T, resulting in b'.
6. Invert $s = \text{dec}(b')$, resulting in dB sequence t.
7. Output t^{rev}.

We will see that several steps can be skipped, however. A pseudocode is given in Algorithm 1, and an example in Fig. 2.

First and Last Bit Must Be 1: It is easy to see that if a string starts with a or ends with b, then its standard permutation cannot be cyclic, since either of these creates a cycle of length 1. Therefore, every $s \in S_k$ has first and last bit 1.

Computing the Standard Permutation: We compute directly the cycle decomposition of π_v, using Lemma 2 and passing conceptually through v. In Step 2, we fill in an array $cycle$ of length n, which stores in $cycle[j]$ the number of the cycle in which j is contained (see Algorithm 2). In Step 6, we use Lemma 2 to invert s, listing the characters of s in the order given by the unique cycle of π_s: $0, \pi_s(0), \pi_s^2(0), \pi_s^3(0), \ldots$, and thus constructing t back-to-front. Since for every dB sequence, its reverse is also a dB sequence, we can output the reverse of the dB sequence whose BWT equals s directly to the standard output.

Computing the Cycle Graph: We store an array $edges$, containing position $2i$ whenever i is an unhappy block. The two ends of the edge are not stored and can be identified on the fly using the $cycle$ array.

Choosing a Random Spanning Tree: As discussed in the Introduction, choosing a spanning tree of an undirected multigraph uniformly at random is complex and slow (even the fastest algorithms have running time $\omega(c^2)$, where $c = c(\pi_v)$ is the number of vertices of Γ_v). Instead we construct a spanning tree by randomly choosing and removing edges without replacement from the set of edges. This can be done by moving the chosen edge to the end of the $edges$ array. We maintain a Union-Find data structure to keep track of the connected components of the graph, updating it whenever an edge is chosen from different connected components.

Example 2. Figure 2 shows an example of possible de Bruijn sequences constructed from a random bitstring b. In the first case the following edges are chosen: $10, 4, 28, 22$ (terminates after 4 edges). The first two choices are accepted, 28 is not accepted, because C_1 and C_2 are already in the same connected component at that point, 22 is accepted. So the following three bits will be flipped in b: $2, 5, 11$. Therefore, the bitstring $b' = 1001100111001101$ is the encoding of some $s \in S_k$, and the algorithm returns: aaaaabaabbaababaaabbbbbababbabbb. In the second case the choice is the following: $14, 4, 16, 8, 24, 28, 26$ (terminates after 7

Algorithm 1: Random binary de Bruijn sequence

```
 1 function Random_deBruijn():
 2     bwt_enc ← random_bitstring()              ▷ of length m, start and
                                                    end with 1
 3     (cycle, num_cycles) ← bwt2cycle(bwt_enc)
 4     if num_cycles = 1 then
 5         inv_bwt(bwt_enc)
 6         return

 7     edges ← [ ]
 8     for i ← 0 to n − 1 by 2 do
 9         if cycle[i] ≠ cycle[i + 1] then
10             edges.push(i)

11     SpanTree ← UnionFind(num_cycles)
12     while num_cycles > 1 do
13         j ← random_integer(num_edges)
14         e ← edges[j]
15         if SpanTree.union(cycle[e], cycle[e + 1]) then
16             bwt_enc[e/2] ← ! bwt_enc[e/2]
17             num_cycles ← num_cycles − 1
18         num_edges ← num_edges − 1
19         swap(edges[j], edges[num_edges])
20     inv_bwt(bwt_enc)
```

---

edges), then we accept 14 and 4, reject 16, 8, 24, 28, and accept 26. Therefore, bits 2, 7, and 13 are flipped, resulting in: 1001110011011001, and the output is: aaaaabaabbaababbbbbabbababaaabbb.

Analysis. The data structures used by the algorithm are: the bitstring of length $m = 2^{k-1}$, the *cycle*-array of length $n = 2^k$, the *edges*-array of length at most m, and the Union-Find data structure for the vertices of Γ_v, i.e. of size $c(\pi_v)$. Therefore, the total space required is $\mathcal{O}(n)$.

Regarding the running time, (1) constructing the *cycle*-array takes $\mathcal{O}(n)$ time, since each position j is visited at most four times: once to check if it is filled (Line 6 in Algorithm 2), once to fill it (Line 7), once as the next to be checked (Line 10), and a fourth time exclusively for the smallest j in each cycle when the cycle closes. Note in particular that $\pi(j)$ is computed in constant time (Lemma 2). (2) Filling in the *edges*-array takes $\mathcal{O}(m)$ time, since we check for each block whether it is unhappy in one scan over the *cycle*-array. (3) Constructing the spanning tree takes $\mathcal{O}(m\alpha(m))$ time, since we choose each edge at most once, and need to query and possibly update the Union-Find data structure. Here, α is the inverse Ackermann function [27]. Finally, inverting the BWT takes $\mathcal{O}(n)$ time, again because $\pi(j)$ can be computed in constant time. Altogether, we have $\mathcal{O}(n\alpha(n))$ time, as $n = 2m$.

Algorithm 2: eBWT to *cycle*

```
1  function bwt2cycle(bwt_enc):
2      cycle ← integer[n]                    ▷ initialized with zeros
3      i ← 0
4      num_cycles ← 1
5      while i < n do
6          while cycle[i] = 0 do            ▷ cycle[i] not yet assigned
7              cycle[i] ← num_cycles
8              i ← π(i)
9          i ← i + 1
10         if i < n and cycle[i] = 0 then
11             num_cycles ← num_cycles + 1
12     return (cycle, num_cycles)
```

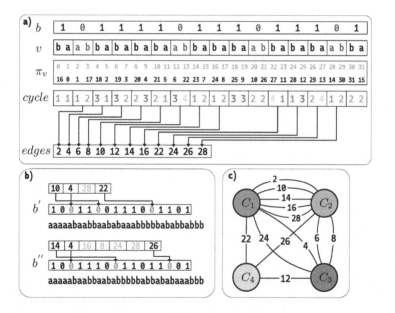

Fig. 2. a) Random bitstring b of length 2^{k-1}, with its corresponding eBWT $v = \text{dec}(b)$, its permutation π_v and the arrays *cycle* and *edges*. **b)** Example of possible de Bruijn sequence constructed from b. **c)** the cycle graph Γ_v.

Output Distribution. The algorithm does not output dB sequences according to the uniform distribution, and there are two reasons for this. First, let us assume that we choose the spanning tree of the cycle uniformly at random. Let $b = \text{enc}(v)$ be the random bitstring chosen, then the conditional probability that a particular dB sequence t is output equals $p(t \mid v) = 1/|B(v)|$ if $\text{bwt}(t) \in B(v)$, and $p(t \mid v) = 0$ otherwise. This leads us to define, for a dB sequence t, the *prestige of* t, given by $pres(t) = \frac{1}{|F_k|} \sum_{v \in F_k} p(t \mid v)$. Clearly, *pres* is a probability distribution, and *if* we chose the spanning tree uniformly at random, then it would equal the output distribution asymptotically. The second reason is that the spanning tree is not chosen uniformly at random. In the next section, we give comparisons of prestige, the output distribution, and the uniform distributions.

5 Experimental Results

Due to lack of space, we give here a short summary of our experimental results. Full details will be given in the full version of the paper.

Running Times. In Table 2, we report running times (real time) for $\sigma = 2$ and k up to 30. All code was compiled with g++ with the flag -O2. It ran on a portable computer equipped with 12 Intel Core i7-8750H (2.20 GHz) and 16 GB of RAM.

Table 2. Average running times in seconds, for $\sigma = 2$, taken over 100 randomly generated dB sequences, without (w/o) and with (w) the time for outputting.

k	17	18	19	20	21	22	23	24	25	26	27	28	29	30
w/o (s)	0.003	0.01	0.02	0.04	0.10	0.29	0.87	2.63	6.07	12.42	27.49	57.19	125.38	247.10
w (s)	0.01	0.02	0.03	0.07	0.16	0.39	0.96	3.11	7.31	15.44	32.32	67.20	144.72	293.49

For $\sigma = 2$, we compared our approach with an implementation of Fleury's algorithm provided at [25], which we modified by adding randomization. We refer to this implementation as 'random-Fleury'. Even though that algorithm cannot generate all possible dB sequences, given its reliance on DFS for spanning tree generation, it serves as the closest available method for comparison with our approach. We show running times and memory peaks consumption in Fig. 3.

Our algorithm is approximately 10–12 times faster than random-Fleury for k values between 17 and 23, and five times faster for $k = 29$. Additionally, it utilizes only half the memory. Memory consumption was evaluated using the Valgrind Massif tool. The data point of random-Fleury for $k = 30$ is missing in both graphs due to exceeding the available RAM (16 GB).

Output Distribution. For $\sigma = 2$, we computed the prestige for $k = 4, 5$ and estimated experimentally the output distribution for $k = 4, 5, 6$ (*empirical distribution* P_e), comparing both to the uniform distribution $P_u \equiv 1/|S_k|$. Estimates of P_e are based on 10^8 trials each, except for $k = 6$, where they are based on 10^{10} trials. We note that we computed prestige explicitly, computing $p(t \mid v)$ for all v and t, which is prohibitive for $k > 5$.

There are 16 binary dB sequences of order $k = 4$, of which 8 have $pres(t) = 0.0724$ and 8 have $pres(t) = 0.0526$. For $k = 5$, there are 132 different prestige values. In Table 3 we report the maximum and minimum prestige and empirical probabilities, for $k = 4, 5, 6$, and their normalized variants w.r.t. P_u (using $f_{\text{norm.}}(t) = (f(t) - \frac{1}{|S_k|})|S_k|$, where f is *pres* or P_e).

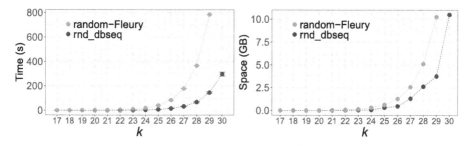

Fig. 3. Left: real-time performance comparison between our algorithm (rnd_dbseq) and a modified C implementation of Fleury's algorithm from [25] (random-Fleury) across different values of k. For rnd_dbseq, each data point represents the average over 100 randomly generated binary dB sequences, while for random-Fleury, each data point is based on 10 binary dB sequences. Timings include outputting the sequences. Bars about the points indicate standard deviation. Right: peak memory consumption of rnd_dbseq and random-Fleury.

Table 3. Min and max values P_e, *pres* for $k = 4, 5, 6$, and normalized w.r.t. P_u.

k	$P_e(s)$		$pres(s)$		$P_u(s)$	Norm. P_e		Norm. *pres*	
	min	max	min	max		min	max	min	max
4	0.0510	0.0740	0.0526	0.0724	0.0625	−0.184	0.184	−0.158	0.158
5	1.82e–4	6.68e–4	2.01e–4	6.48e–4	4.88e–4	−0.628	0.367	−0.589	0.328
6	1.70e–9	4.74e–8			1.59e–8	−0.886	2.180		

6 A Case Study: Estimating the Discrepancy of a Random de Bruijn Sequence

The *discrepancy* of a binary string w is defined as the maximum absolute difference between the number of a's and b's, taken over all substrings u of w. In certain contexts where dB sequences are applied, high discrepancy can lead to undesirable effects, such as interference in spread spectrum communications or unwanted biochemical properties in certain biological settings [6]. In these cases, one is interested in generating dB sequences with lowest possible discrepancy. In [6], the discrepancy of the lex-least dB sequence was studied. This was extended to a number of dB sequence constructions in [12], with the average discrepancy taken over 10^4 random bitstrings and the average discrepancy of LFSR-based dB sequences used as baselines for comparison.

With our method, it is now possible to estimate the *expected discrepancy of a random de Bruijn sequence*, which is clearly a more adequate baseline. We note that the discrepancy for a dB sequence, being a circular string, should be measured over all *cyclic* substrings. It was shown in [12] that for dB sequences, the linear and the cyclic discrepancies coincide, and a linear-time algorithm for computation of the (linear) discrepancy was given.

Table 4. Discrepancy of random dB sequences vs. LFSRs and random bitstrings. Values taken over 10^5 trials, except for columns 4 (10^4 trials) and 5 (all LFSRs).

k	random de Bruijn seq			r. de Bruijn seq	LFSRs [12]	r. bitstring
	min	*max*	*mean*	mean over 10^4	mean	mean
10	17	82	39	39	41	50
11	24	125	56	56	58	71
12	36	177	79	79	84	101
13	47	255	113	112	118	143
14	66	360	159	160	167	202
15	96	519	226	225	236	287
16	136	740	319	320	335	406
17	197	1058	453	452	473	575
18	283	1485	641	641	669	813
19	384	2299	904	905	947	1148
20	515	2830	1282	1282	1341	1621
21	810	4210	1811	1819	1986	2287
22	1140	5676	2566	2565	2681	3233
23	1676	8143	3627	3638	3793	4491
24	2173	11162	5134	5134	5362	6354
25	3248	17589	7264	7245	7586	9134
26	4476	24036	10272	10258		12730

We report our results in Table 4: for $k \leq 26$, we list the minimum, maximum, and average discrepancy over 10^5 dB sequences of order k, as well as the average discrepancy of 10^5 bitstrings of the same length. We also ran the same experiment over 10^4 trials (column 4) and found that the values were very close, indicating that already a sample size of 10^4 could be sufficient. It can be seen that the average discrepancy of a dB sequence appears to be lower than that of a bitstring or an LFSR-based dB sequence. While the first result is not surprising, the second was hitherto unknown.

We conclude that for studying properties of dB sequences such as the discrepancy, it might not be realistic to use random bitstrings as a base for comparison, nor LFSRs. Even though our trial number of 10^4 resp. 10^5, for most k is many orders of magnitude below the total number of dB sequences, this is still a better estimate of the mean discrepancy than the tests reported in [12], based on bitstrings and LFSRs only. Consider also that for these values of k, it is very likely that our algorithm never produced the same dB sequence twice.

We further note that our experiments did not produce any counterexamples for the conjectured minimum and maximum discrepancy of dB sequences, that of the Huang construction [17] (minimum) and $\binom{k-1}{\lfloor k/2 \rfloor} + \lfloor k/2 \rfloor$ (maximum), both of which are listed as open problems in [12].

7   Conclusion

We presented a remarkably simple algorithm for constructing a random de Bruijn sequence, which, surprisingly, appears to be the first practical algorithm that can output all dB sequences. On top of the obvious practical utility of such an algorithm, we also demonstrated its potential in theoretical research by giving the first estimate on the mean discrepancy of de Bruijn sequences. Our algorithm is based on the fact that dB sequences have very regular BWTs and uses what we termed the swapping technique for manipulating the BWTs of strings.

The algorithm can be adapted to any constant-size alphabet. A straightforward adaptation has worst-case running time $\mathcal{O}(\sigma \cdot n)$, where σ is the alphabet size, since the number of edges of the cycle graph can become $\binom{\sigma}{2}\sigma^{k-1} = \Theta(\sigma n)$.

Our approach does not lead to construction of dB sequences uniformly at random; even if we chose a spanning tree of the cycle graph uniformly at random, we would still output dB sequences according to their prestige, which is not uniform. If we were able to efficiently compute the prestige of a dB sequence of any order k, this might allow us to get a uniform sampler via rejection sampling (albeit a Las Vegas one). We leave it as an open problem to compute the distribution of the prestige function for general k.

Acknowledgements. ZsL would like to thank Joe Sawada for awakening her interest in de Bruijn sequences. We thank the anonymous reviewers for some insightful suggestions, and the participants of the Monday Meetings of the Algorithms Group of Verona University for useful discussions. This work has been supported in part by the European Union's Horizon 2020 research and innovation programme under the Marie Skłodowska-Curie grant agreement No 956229 and by the MUR PRIN Project 'PINC, Pangenome INformatiCs: from Theory to Applications' (Grant No. 2022YRB97K).

References

1. Aardenne-Ehrenfest, T.v., Bruijn, N.G.d.: Circuits and trees in oriented linear graphs. Simon Stevin, Wisen Natuurkundig Tijdschrift **28**, 203–217 (1951)
2. Aguirre, G.K., Mattar, M.G., Magis-Weinberg, L.: De Bruijn cycles for neural decoding. Neuroimage **56**(3), 1293–1300 (2011)
3. Ben-Dor, A., Karp, R., Schwikowski, B., Yakhini, Z.: Universal DNA tag systems: a combinatorial design scheme. J. Comp. Biol. **7**(3/4), 503–519 (2000)
4. Burrows, M., Wheeler, D.: A block sorting lossless data compression algorithm. Technical Report 124, Digital Equipment Corporation (1994)
5. Colbourn, C.J., Myrvold, W.J., Neufeld, E.: Two algorithms for unranking arborescences. J. Algorithms **20**(2), 268–281 (1996)

6. Cooper, J.N., Heitsch, C.E.: The discrepancy of the lex-least de Bruijn sequence. Discret. Math. **310**(6–7), 1152–1159 (2010)
7. de Bruijn, N.G.: A combinatorial problem. Proc. Sect. Sci. **49**(7), 758–764 (1946)
8. Durfee, D., Kyng, R., Peebles, J., Rao, A.B., Sachdeva, S.: Sampling random spanning trees faster than matrix multiplication. In: Hatami, H., McKenzie, P., King, V. (eds.) Proceedings of the 49th Annual ACM SIGACT Symposium on Theory of Computing, STOC 2017, pp. 730–742. ACM (2017)
9. Emerson, P.L., Tobias, R.D.: Computer program for quasi-random stimulus sequences with equal transition frequencies. Behav. Res. Methods Instrum. Comput. **27**(1), 88–98 (1995)
10. Fleury, P.-H.: Deux problèmes de géométrie de situation. J. Mathématiq. élément. **2**, 257–261 (1883)
11. Fredricksen, H.: A survey of full length nonlinear shift register cycle algorithms. SIAM Rev. **24**(2), 195–221 (1982)
12. Gabric, D., Sawada, J.: Investigating the discrepancy property of de Bruijn sequences. Discret. Math. **345**(4), 112780 (2022)
13. Gabric, D., Sawada, J., Williams, A., Wong, D.: A framework for constructing de Bruijn sequences via simple successor rules. Discret. Math. **341**(11), 2977–2987 (2018)
14. Giuliani, S., Lipták, Zs., Masillo, F., Rizzi, R.: When a dollar makes a BWT. Theor. Comput. Sci. **857**, 123–146 (2021)
15. Golomb, S.: Shift Register Sequences, 3rd edn. World Scientific (2016)
16. Higgins, P.M.: Burrows-Wheeler transformations and de Bruijn words. Theor. Comput. Sci. **457**, 128–136 (2012)
17. Huang, Y.: A new algorithm for the generation of binary de Bruijn sequences. J. Algorithm. **11**(1), 44–51 (1990)
18. Jansen, C.J., Boekee, D.E.: An efficient algorithm for the generation of DeBruijn cycles. IEEE Trans. Inf. Theory **37**(5), 1475–1478 (1991)
19. Lothaire, M.: Algebraic Combinatorics on Words. Cambridge University Press (2002)
20. Mandal, K., Gong, G.: Cryptographically strong de Bruijn sequences with large periods. In: Knudsen, L.R., Wu, H. (eds.) Selected Areas in Cryptography: 19th International Conference, SAC 2012, pp. 104–118. Springer, Heidelberg (2013). https://doi.org/10.1007/978-3-642-35999-6_8
21. Mantaci, S., Restivo, A., Rosone, G., Sciortino, M.: An extension of the Burrows-Wheeler Transform. Theor. Comput. Sci. **387**(3), 298–312 (2007)
22. Navarro, G.: Compact Data Structures: A Practical Approach. Cambridge University Press (2016)
23. Perrin, D., Restivo, A.: Words. In: Bóna, M. (ed.) Enumerative Combinatorics, chapter 8, pp. 485–540. CRC Press (2015)
24. Philippakis, A., Qureshi, A.M., Berger, M.F., Bulyk, M.L.: Design of compact, universal DNA microarrays for protein binding microarray experiments. J. Comp. Biol. **15**(7), 655–665 (2008)
25. Sawada, J.: De Bruijn sequence and universal cycle constructions. https://debruijnsequence.org
26. Sohn, H.-S., Bricker, D.L., Simon, J.R., Hsieh, Y.-C.: Optimal sequences of trials for balancing practice and repetition effects. Behav. Res. Methods Instrum. Comput. **29**(4), 574–581 (1997)
27. Tarjan, R.E., van Leeuwen, J.: Worst-case analysis of set union algorithms. J. ACM **31**(2), 245–281 (1984)

28. Turan, M.S.: Evolutionary construction of de Bruijn sequences. In: Proceedings of ACM-AISec, pp. 81–86 (2011)
29. Yang, B., Mandal, K., Aagaard, M.D., Gong, G.: Efficient composited de Bruijn sequence generators. IEEE Trans. Computers **66**(8), 1354–1368 (2017)
30. Zhu, Y., Chang, Z., Ezerman, M.F., Wang, Q.: An efficiently generated family of binary de Bruijn sequences. Discret. Math. **344**(6), 112368 (2021)

Space-Efficient Conversions from SLPs

Travis Gagie[1,5], Adrián Goga[2(✉)], Artur Jeż[3], and Gonzalo Navarro[4,5]

[1] Faculty of Computer Science, Dalhousie University, Halifax, Canada
[2] Department of Computer Science, Comenius University in Bratislava, Bratislava, Slovakia
adriangoga@gmail.com
[3] Institute of Computer Science, University of Wrocław, Wroclaw, Poland
[4] Department of Computer Science, University of Chile, Santiago, Chile
[5] CeBiB—Center for Biotechnology and Bioengineering, Santiago, Chile

Abstract. We give algorithms that, given a straight-line program (SLP) with g rules that generates (only) a text $T[1..n]$, build within $O(g)$ space the Lempel-Ziv (LZ) parse of T (of z phrases) in time $O(n \log^2 n)$ or in time $O(gz \log^2(n/z))$. We also show how to build a locally consistent grammar (LCG) of optimal size $g_{lc} = O(\delta \log \frac{n}{\delta})$ from the SLP within $O(g + g_{lc})$ space and in $O(n \log g)$ time, where δ is the substring complexity measure of T. Finally, we show how to build the LZ parse of T from such an LCG within $O(g_{lc})$ space and in time $O(z \log^2 n \log^2(n/z))$. All our results hold with high probability.

1 Introduction

With the rise of enormous and highly repetitive text collections [32], it is becoming practical, and even necessary, to maintain the collections compressed all the time. This requires being able to perform all the needed computations, like text searching and mining, directly on the compressed data, without ever decompressing it.

As an example, consider the modest (for today's standards) genomic repository *1000 Genomes* [12] containing the genomes of 2,500 individuals. At the typical rate of about 3 billion bases each, the collection would occupy about 7

Research supported by the European Union's Horizon 2020 research and innovation program under Marie Skłodowska-Curie grant agreement No 956229 (ALPACA) and by grants 1/0463/20 and 1/0538/22 from the Scientific Grant Agency of the Ministry of Education, Science, Research, and Sport of the Slovak Republic and Slovak Academy of Sciences (VEGA) and grant APVV-22-0143 from the Slovak Research and Development Agency. T.G. and G.N. funded in part by Basal Funds FB0001, ANID, Chile. T.G. funded in part by NSERC RGPIN-07185-2020. G.N. funded in part by Fondecyt Grant 1-230755, ANID, Chile.

J. A. Soto and A. Wiese (Eds.): LATIN 2024, LNCS 14578, pp. 146–161, 2024.
https://doi.org/10.1007/978-3-031-55598-5_10

terabytes. Recent projects like the *Million Genome Initiative*[1] would then require petabytes. The 1000 Genomes project stores and distributes its data already in a compressed form[2] to exploit the fact that, compared to a reference genome, each individual genome has only one difference every roughly 500 bases, on average. Certainly one would like to manipulate even such a modest collection always in a compressed form, using gigabytes instead of terabytes of memory!

Some compression formats are more useful for some tasks than others, however. For example, Lempel-Ziv compression [29] tends to achieve the best compression ratios, which makes it more useful for storage and transmission. Grammar compression [26] yields slightly larger files, but in exchange it can produce T in streaming form, and provide direct access to any text snippet [7], as well as indexed searches [11]. Locally consistent grammars provide faster searches, and support more complex queries, while still being bounded by well-known repetitiveness measures [10,16,23,27,28,33]. The run-length-encoded Burrows-Wheeler Transform of T requires even more space [22], but in exchange it enables full suffix tree functionality [13].

It is of interest, then, to *convert* from one format to another. Doing this conversion by decompressing the current format and then compressing to the new one is impractical, as it is bound to use $\Omega(n)$ space, which in practice implies running $\Theta(n)$-time algorithms on secondary storage. Thus the interest in algorithms whose running time and space usage can be bounded in terms of input and output size. We say that a conversion between different compression formats is a *fully compressed conversion* when it uses space and time polynomial in the size of the (compressed) input, the size of the (compressed) output, and $\log n$; it is a *compressed conversion* when the bound applies only to space (and so the running time may polynomially depend on n). There is a long line of research on compressed conversions, we recall it below. For brevity we omit a large body of work on producing compressed representations from the original string S, aiming to use little space on top of S itself, and the work on compression formats that are too weak for repetitive data, like LZ78 or run-length compression of the text.

Let z, g, g_{lc} and r be the asymptotic (i.e., up to constant factors) sizes of the Lempel-Ziv (LZ) parse of a string $T[1..n]$, a straight-line program (SLP) or context-free grammar that expands to T, a locally consistent grammar (LCG) that expands to T, and the run-length encoded Burrows-Wheeler Transform (RLBWT) for T, respectively. On highly repetitive texts, all the given measures can be exponentially smaller than n, hence the relevance of such conversions. We refer to some SLP because finding the smallest SLP generating a given string is NP-complete [8]. It holds that $z \leq g \leq g_{lc} \leq r$ in practice. The first such conversion was implicitly given by Mehlhorn, Sundar and Uhrig [30], who proposed a data structure for a dynamic collection of strings allowing adding concatenations and substrings of strings in the collection in polylogarithmic time. The data structure implicitly used (a variant of) a LCG and so it allowed compressed conversions from SLP and LZ

---

[1] https://digital-strategy.ec.europa.eu/en/policies/1-million-genomes.

[2] In VCF, https://github.com/samtools/hts-specs/blob/master/VCFv4.3.pdf.

to LCG, in time $O(g \log n(\log g \log^* n + \log n))$ and space $O(g \log n \log^* n)$ and time $O(z \log n(\log g \log^* n + \log n))$ and space $O(z \log n \log^* n)$, respectively. They also proposed a randomised variant of the data structure, with which the conversion had expected time and space $O(g \log^2 n)$, $O(g \log n)$ and $O(z \log^2 n)$, $O(z \log n)$, respectively. Their data structure was improved by Alstrup, Brodal and Rauhe [1], who mainly added new functionalities and improved the conversion times to $O(g \log n \log^* n)$ and $O(z \log n \log^* n)$ w.h.p. (the space usage remained the same). Rytter [38] studied the problem of constructing the smallest SLP for a given string and showed how to build an SLP of size $g = O(z \log(n/z))$ within $O(g)$ space and time from the LZ parse of T in the non-overlapping case (i.e., when phrases cannot overlap their sources), and Gawrychowski [15, Lemma 8] extended this result to the general LZ parse. Nishimoto et al. [36] gave an algorithm constructing the LZ parse from the LCG of Mehlhorn et al. [30], with running time $O(z \log g_{lc} \log^3 n(\log^* n)^2)$ and linear-space. It can also be used to convert an SLP to the LZ parse in time $O(n(\log \log n)^2 + z \log^4 n(\log^* n)^2)$ or $O(n(\sqrt{\log z + \log \log n} + z \log^4 n(\log^* n)^2))$ and $O(z \log n \log^* n)$ space. Tomohiro I [18] proposed a conversion algorithm from an SLP to (a variant of) a LCG using $O(g \log(n/g))$ time and $O(g + z \log(n/z))$ space; one can also transform an LZ77 to SLP using with $\log(n/z)$ blowup and then apply the reduction to LCG, using $O(z \log^2(n/z))$ time and $O(z \log(n/z))$ space. Kempa and Kociumaka [23] built on the produced LCG, showing how to convert a LCG or a SLP to the LZ parse in time $O(g_{lc} \log^4 n)$ or $O(g \log^4 n)$, respectively. They also gave a fully compressed conversion from a SLP or the LZ parse to a LCG (of optimal size $O(\delta \log n/\delta)$) in time $O(\delta \log^7 n)$ (δ is another compression measure with $\delta \leq z \leq \delta \log n$ [28]). Policriti and Prezza [37] showed how to convert from the RLBWT to the LZ parse in $O(r + z)$ space and $O(n \log r)$ time, and back in the same space and $O(n \log(rz))$ time. The earlier mentioned paper of Kempa and Kociumaka [22] also converts from the LZ parse to the RLBWT in $O(z \log^8 n)$ expected time. Arimira et al. [2] recently showed how to convert from the compressed directed acyclic word graph (CDAWG) of size e to either RLBWT or LZ, both in $O(e)$ time and space, though e is the weakest among the commonly accepted repetitiveness measures [32].

Note that our contribution deals only with LZ, SLP and LCG; we recalled results for other compression formats (RLBWT, CDAWG) for comparison and to present the state of the art in the area.

In this paper we contribute to the state of the art with compressed and fully-compressed conversions between various formats, all of which then use space *linear in the input plus the output*, and work correctly with high probability:

1. A compressed conversion from any SLP to the LZ parse in $O(n \log^2 n)$ time.
2. A fully-compressed conversion from any SLP to the LZ parse in $O(gz \log^2(n/z))$ time.
3. A compressed conversion from any SLP to a certain (particularly small) LCG [10] in $O(n \log g_{lc})$ time.
4. A fully-compressed conversion from LCGs of some particular kind [10,27] to the LZ parse in $O(z \log^2(n/z) \log^2 n)$ time.

The third conversion builds a particular LCG whose size is the optimal $O(\delta \log \frac{n}{\delta})$ [27], other similar LCGs [10] can be produced analogously; note that there is a fully-compressed conversion from SLP to LCG [18]; it is for a different LCG, though, and it is not clear, whether it generalizes to other LCGs within given bounds. Also, while the running time of our fourth conversion is larger than Nishimoto et al. [36], we work with a particular LCG, which can be up to $\log n$ times smaller than the LCG of Mehlhorn et al. [30] (we use LCG with a bound of $O(\delta \log \frac{n}{\delta})$, while the latter is only known to be $O(z \log n \log^* n)$ and the only bound on z in terms of δ is $O(\delta \log \frac{n}{\delta})$ [28]). Our contributions together with previously known conversions are depicted in Table 1.

Table 1. The running times of compressed and fully compressed conversions between LZ, SLP, and LCG, with our contributions in bold. The bounds of Nishimoto et al. [36] are slightly simplified.

Fully compressed conversions			
From \ To	LZ	SLP	LCG
LZ		$O(z \log(n/z))$ [15,38]	$O(z \log^2 n)$ expected [30] $O(z \log n \log^* n)$ w.h.p. [1] $O(z \log^2(n/z))$ [18]
SLP	$O(g \log^4 n)$ [23] $\mathbf{O(gz \log 2(n/z))}$ **w.h.p.**		$O(g \log(n/g))$ [18]
LCG	$O(g_{lc} \log^4 n)$ [23] $O(z \log g_{lc} \log^3 n (\log^* n)^2)$ [36] $\mathbf{O(z \log 2(n/z) \log 2n)}$ **w.h.p.**		

Compressed conversions		
From \ To	LZ	LCG
SLP	$O(n(\log \log n)^2 + z \log^4 n (\log^* n)^2)$ [36] $O(n\sqrt{\log z + \log \log n} + z \log^4 n (\log^* n)^2)$ [36] $\mathbf{O(n \log 2n)}$ **w.h.p.**	$\mathbf{O(n \log g_{lc})}$ **w.h.p.**
LCG	$\mathbf{O(n \log 2n)}$ **w.h.p.**	

2   Preliminaries

A *string* $T[1..n]$ is a sequence of symbols $T[1] T[2] \ldots T[n]$ over an ordered alphabet Σ. For every $1 \le i, j \le n$, $T[1..i] = T[..i]$ is a *prefix* of T, $T[j..n] = T[j..]$ is a *suffix* of T, and $T[i..j]$ is a *substring* of T, which is the empty string ε if $i > j$. The *length* of $T[1..n]$ is $|T| = n$; the length of ε is $|\varepsilon| = 0$. The *concatenation* of two strings $S \cdot S'$ is defined as $S[1] S[2] \ldots S[|S|] S'[1] S'[2] \ldots S'[|S'|]$. The *lexicographic* order between strings $S \ne S'$ is defined as that between $S[1]$ and $S'[1]$ if these are different, or as the lexicographic order between $S[2..]$ and $S'[2..]$ otherwise; the empty string ε is smaller than every other string. The *co-lexicographic* order is defined as the lexicographic order between the reversed strings.

The *Karp-Rabin fingerprint* or the *Karp-Rabin hash* of a string $S[1..n]$ is a value $\phi(S) = \sum_{i=1}^{n} \left(S[i] x^i \right) \bmod p$, for a prime p and $x < p$ [21]. The crucial property of this hash is that that if $X \neq Y$, then $\phi(X) \neq \phi(Y)$ with high probability. Another well-known and useful property is that for strings S, S', S' for which $S = S' \cdot S''$ holds, we can compute the hash of any of the strings knowing the hashes of the other two, in $O(1)$ time (see, e.g., [35]).

A *straight-line program (SLP)* of a text T is a context-free grammar in Chomsky normal form (so, in particular, each rule is at most binary) generating only T, which contains exactly one rule for each nonterminal and the rules can be linearly ordered, such that for any rule $X \to YZ$ it holds that the rules for both Y and Z precede the rule for X in the ordering.

The height of the SLP is the height of the derivation tree, i.e. the height of a letter is 0 and the height of a nonterminal X with a (unique) rule $X \to YZ$ is 1 plus maximum of height of Y and Z. The size of the SLP is the number of its rules. We define the *expansion* of a nonterminal X as the string it produces: $\exp(a) = a$ if a is a terminal symbol, and $\exp(X) = \exp(Y) \cdot \exp(Z)$ if $X \to YZ$.

We say that a grammar is a *locally consistent grammar (LCG)* if it is constructed by iteratively applying rounds of a particular locally consistent parsing, which guarantees that matching fragments $S[i..j] = S[i'..j']$ are parsed the same way, apart from the $O(1)$ blocks from either end. This key property is lifted to such grammars, for which matching fragments are spanned by almost identical subtrees of the parse tree, differing in at most $O(1)$ flanking nonterminals at each level [10, 16]. Such a parsing is defined in Sect. 5.

The Lempel-Ziv (LZ) parse of a string T [29] is a sequence $F_1, F_2, \ldots, F_z$ of phrases, such that $F_1 \cdot F_2 \cdots F_z = T[1..n]$ and F_i is either a single letter, when this letter is not present in $F_1 \cdot F_2 \cdots F_{i-1}$, or else F_i is the maximal string that occurs twice in $F_1 \cdot F_2 \cdots F_i$, that is, it has an occurrence starting within $F_1 \cdot F_2 \cdots F_{i-1}$; in non-overlapping LZ we additionally require that F_i occurs within $F_1 \cdot F_2 \cdots F_{i-1}$. It is known that $z \leq g = O(z \log(n/z))$, where g is the size of the smallest grammar generating T [8, 15, 38].

We assume the standard word-RAM model of computation with word length $\Theta(\log n)$, in which basic operations over a single word take constant time. Some of our results hold with high probability (w.h.p.), meaning with probability over $1 - n^{-c}$ for any desired constant c. We can make the constant arbitrarily large at the cost of increasing the constant multiplying the running time.

3   Building the LZ Parse from an SLP in $\tilde{O}(n)$ Time

Our first result computes the LZ parse of a text $T[1..n]$ given an arbitrary SLP of size g that represents T, in time $O(n \log^2 n)$ and space $O(g)$; note that the classic LZ constructions use suffix trees or arrays and use $\Omega(n)$ space. We first describe a couple of tools we need to build on the SLP before doing the conversion.

Lemma 1. *Given an SLP of size g for $T[1..n]$ we can construct in $O(g)$ time and space a new SLP G, and augment it with a data structure such that:*

- *G has height $O(\log n)$.*
- *Any $T[i]$ can be accessed in $O(\log n)$ time.*

- *The Karp-Rabin fingerprint of any $T[i..j]$ can be computed in $O(\log n)$ time.*
- *The longest common prefix of any $T[i..j]$ and $T[i'..j']$ can be computed (w.h.p.) in $O(\log^2 n)$ time.*
- *Any $T[i..j]$ and $T[i'..j']$ can be compared lexicographically and co-lexicographically (w.h.p.) in $O(\log^2 n)$ time.*

Proof. Assume that we are given an SLP with g rules for a text $T[1..n]$. Ganardi et al. [14] showed that in $O(g)$ time and space we can turn it into an SLP G of size $O(g)$ and height $O(\log n)$ and augment G with $O(g)$-space structures that, in $O(\log n)$ time, finds any character $T[i]$ and returns the Karp-Rabin hash of any substring $T[i..j]$ (see Ganardi et al. [14], which refers to a simple data structure from Bille et al. [6]). We work with such augmented G from now on. Given two substrings of T, we can then compute their longest common prefix in $O(\log^2 n)$ time—w.h.p. of obtaining the correct answer—by exponentially searching for its length ℓ [6, Thm. 3]; by checking their characters at offset $\ell + 1$ we can also compare the substrings of T lexicographically within the same time complexity. We can similarly compute the longest common suffix of two substrings and thus compare them co-lexicographically (by comparing the preceding characters). □

We will also use a variant of a z-fast trie.

Lemma 2. *Let S be a lexicographically sorted multiset of m strings of total length n. Then one can build, in $O(n)$ time w.h.p., a data structure of size $O(m)$ that, given a string P, finds in $O(f_h \log |P|)$ time the lexicographic range of the strings in S prefixed by P, where f_h is the time to compute a Karp-Rabin fingerprint of a substring of P. If this range is nonempty, the answer is correct w.h.p.; if this range is empty, there are no guarantees on correctness of the answer, i.e. the answer could be incorrect.*

Proof. The structure is the z-fast trie of Belazzougui et al. [3, Theorem 5], and the query is the *fat binary search*. A simpler construction was given by Kempa and Kosolobov [24], and it was then fixed, and its construction analyzed, by Navarro and Prezza [35, Sec. 4.3]. □

We will resort to a classic grammar-based indexing method [11], for which we need a few definitions and properties.

Definition 1. *The grammar tree of an SLP G is formed by pruning the parse tree, converting to leaves, for every nonterminal X, all the nodes labeled X but the leftmost one. An occurrence of a string P in T is primary if it spans more than one leaf in the grammar tree; otherwise it is contained in the expansion of a leaf and is secondary. If a primary occurrence of P occurs in $\exp(X)$, with rule $X \to YZ$, starting within $\exp(Y)$ and ending within $\exp(Z)$, then the position $P[j]$ aligning to the last position of $\exp(Y)$ is the splitting point of the occurrence.*

A small exception to this definition is that, if $|P| = 1$, we say that its primary occurrences are those where it appears at the end of $\exp(X)$ in any leaf X of the grammar tree. We now give a couple of results on primary occurrences.

Lemma 3 ([11]). *A pattern occurring in T has at least one primary occurrence.*

Observation 4. *If X is the lowest nonterminal containing a primary occurrence of P with splitting point j, then, by the way we form the grammar tree, this is the leftmost occurrence of P under X with splitting position j.*

The index sorts all rules $X \to YZ$ twice: once by the lexicographical order of $\exp(Z)$, while collecting those expansions in a multiset $\mathcal{Z}$, and once by the co-lexicographical order of $\exp(Y)$, while collecting the reversed expansions in a multiset $\mathcal{Y}$. It builds separate z-fast tries (Lemma 2) on $\mathcal{Y}$ and $\mathcal{Z}$, and creates a discrete $g \times g$ grid $\mathcal{G}$, where the cell (x, y) stores the position p iff the xth rule $X \to YZ$ in the first order is the yth rule in the second order, and $T[p]$ is aligned to the last symbol of $\exp(Y)$ within the occurrence of X as an internal node in the grammar tree. The grid supports orthogonal range queries. The key idea of the index is that, given a search pattern P, for every $1 \le j \le |P|$, the lexicographic range $[y_1, y_2]$ of $P[j+1..]$ in $\mathcal{Z}$ and the lexicographic range $[x_1, x_2]$ of the reverse of $P[..j]$ in $\mathcal{Y}$, satisfy that there is a point in the range $[x_1, x_2] \times [y_1, y_2]$ of $\mathcal{G}$ per primary occurrence of P in T with splitting point $P[j]$. The structure $\mathcal{G}$ can determine if the area is empty, or else return a point in it, in time $O(\log g)$. We now build our first tool towards our goal.

Lemma 5. *Given an SLP of size g generating string $T[1..n]$ we can, in space $O(g)$ and time $O(n + g \log g)$ construct w.h.p. a data structure that, given $1 \le i \le j \le k \le n$, in $O(\log n \log(k - i) + \log^{1+\varepsilon} g)$ time finds w.h.p. the leftmost occurrence of $T[i..k]$ in T that is a primary occurrence with splitting point $T[j]$.*

Proof. We build the components $\mathcal{Y}$, $\mathcal{Z}$, and $\mathcal{G}$ of the described index, following the approach in (see [35, Sec. 4.3–4.4]), all time and space complexities are given there. The data structure is correct w.h.p.. We sort w.h.p. the sets $\mathcal{Y}$ and $\mathcal{Z}$ in $O(g)$ space and $O(n)$ time ([17]), we build the z-fast tries in $O(g)$ space and time $O(n)$ (Lemma 2), and we build the grid data structure in $O(g)$ space and $O(g\sqrt{\log g})$ time ([35, Sec. 4.4], [4]). We note that, by using Lemma 1, we can also do the sorting correctly w.h.p. in $O(g)$ space and $O(g \log g \cdot \log^2 n)$ time.

We use those structures to search for $P = T[i..k]$ with splitting point $T[j]$, that is, we search the z-fast trie of $\mathcal{Z}$ for $T[j+1..k]$ and the z-fast trie of $\mathcal{Y}$ for $T[i..j]$ reversed, in time $O(f_h \log |P|)$; recall Lemma 2. Since the substrings of $T[i..j]$ are also substrings of T, we can compute the Karp-Rabin hash of any substring of $T[i..j]$ in time $f_h = O(\log n)$ by Lemma 1, so this first part of the search takes time $O(\log n \log(k-i))$. Recall from Lemma 2 that this search yields correct results w.h.p., unless the ranges sought are empty, in which case there are no guarantees on correctness.

We now use $\mathcal{G}$ to determine if there are points in the corresponding area. If there are none, then w.h.p. $T[i..k]$ does not occur in T with splitting point $T[j]$. If there are some, then we obtain the value p associated with any point in the range, and compare the Karp-Rabin hash of $T[p - (j - i)..p + (k - j)]$ with that of $T[i..k]$. If they differ, then $T[i..k]$ has no occurrences with splitting point $T[j]$; otherwise w.h.p. the z-fast tries gave the correct range and there are occurrences. This check takes $O(\log n)$ time.

Once we know that (w.h.p.) there are occurrences with splitting point $T[j]$, we want the leftmost one. Each point within the grid range may correspond to a different rule $X \to YZ$ that splits $T[i..k]$ at $T[j]$; therefore, by Observation 4, we want the minimum of the p values stored for the points within the range. This kind of two-dimensional range minimum query can be solved in time $O(\log^{1+\varepsilon} g)$ and $O(g)$ space, for any constant $\varepsilon > 0$, with an enhancement of $\mathcal{G}$ that uses $O(g)$ space and can be built in time $O(g \log g)$ [9, 31]. This completes the query. $\qquad\square$

Finally, we will need the following observation on the monotonicity of occurrences in T, even when we stick to some splitting point.

Observation 6. *If P has a primary occurrence in T with splitting point $P[j]$, then any prefix $P' = P[..k]$, for any $j < k < |P|$, also has a primary occurrence with splitting point $P'[j]$.*

Proof. Let the primary occurrence of P appear in $\exp(X)$ and the occurrence start in $\exp(Y)$ and end in $\exp(Z)$, with $P[j]$ aligned to the last position of $\exp(Y)$. Then $P' = P[..k]$ satisfies the same conditions: a primary occurrence of P' with splitting point $P'[j]$ starts at the same text position. $\qquad\square$

We are now ready to give the final result.

Theorem 1. *Given an SLP with g rules for a text $T[1..n]$, w.h.p. we can build the LZ parse of T in $O(n \log^2 n)$ time and within $O(g)$ space.*

Proof. We first build the data structures of Lemma 5 in $O(n + g \log g)$ time and $O(g)$ space, correctly w.h.p.. We then carry out the LZ parse by sliding three pointers left-to-right across T, $i \le j \le k$, as follows: suppose that the parse for $T[1..i-1]$ is already constructed, so a new phrase must start at i. We first check whether $T[i]$ appeared already in $T[1..i-1]$,[3] if not then we create a one-letter phrase and proceed to $i + 1$.

If $T[i]$ has occurred earlier, we start the main process of building the next phrase. The invariant is that we have found $T[i..k]$ starting before i in T with splitting point $T[j]$, and there is no primary occurrence of $T[i..k]$ (nor of $T[i..k']$ for any $k' > k$, by Observation 6) with a splitting point in $T[i..j-1]$. To establish the invariant, we initialize j to i and try k from i onwards, using Lemma 5 and advancing k as long as the leftmost occurrence of $T[i..k]$ with splitting point $T[i]$ starts to the left of i.

Note that we will succeed the first time, for $k = i$. We continue until we reach $k = n$ (and output $T[i..n]$ as the last phrase of the LZ parse) or we cannot find $T[i..k+1]$ starting before i with splitting point $T[i]$. We then try successive

---

[3] This is easily done in $O(1)$ time and $|\Sigma| \in O(g)$ space by just storing an array with the leftmost occurrence of every distinct symbol in T. This array is built in $O(g)$ time from the leaves of the grammar tree.

values of j, from $i+1$ onwards, using Lemma 5 to find $T[i..k+1]$ starting before i with splitting point $T[j]$. If we finally succeed for some $j \leq k$, we reestablish the invariant by increasing k and return to the first loop, which again increases k with fixed j, and so on.

When j reaches $k+1$, it follows that $T[i..k]$ occurs before i and $T[i..k+1]$ does not, with any possible splitting point. The next phrase is then $T[i..k]$, which we output, reset $i = k+1$, and resume the parsing.

Since j and k never decrease in the process, we use queries from Lemma 5 $O(n)$ times for a total time of $O(n + g \log g + n(\log^2 n + \log^{1+\varepsilon} g)) = O(n \log^2 n)$ to build the LZ parse. $\qquad\square$

4   Building the LZ Parse from an SLP in $\tilde{O}(gz)$ Time

If T is highly compressible, the running time $O(n \log^2 n)$ in Theorem 1 could be exponential in the size $O(g)$ of the input. We can build the parse in $O(gz \log^2 \frac{n}{z}) \subset \text{poly}(g)$ time by using, instead of the machinery of the preceding section, Jeż's [19] algorithm for fully-compressed pattern matching. We will only balance the SLP if needed [14] so that its height is $O(\log n)$. We start by reminding some tools.

Lemma 7 ([38]). *Given an SLP of height h for T, we can in $O(h)$ time and space produce an SLP of size $O(h)$ for any desired substring $T[i..j]$ (without modifying the SLP of T).*

Note that the SLP constructed in the Lemma above may use some of the nonterminals of the original SLP for T, i.e. its size is in principal $g + O(h)$.

Lemma 8 ([19]). *If T and P have SLPs of size g and g', then we can find the leftmost occurrence of P in T in time $O((g+g') \log |P|)$, within $O(g+g')$ space.*

Note that [19] does not state the space complexity, however, the analysis [19, Sec. 6] bounds intermediate SLPs to be of size $O(g + g')$ ([19, Lem. 6.5] and the running time of the subprocedures (and so their space usage) to be linear; hence the linear space consumption follows.

Assume again we have already parsed $T[1..i-1]$, and aim to find the next phrase, $T[i..k]$. We will exponentially search for k using $O(\log(k-i))$ steps. Each step implies determining whether some $T[i..j]$ occurs in T starting to the left of i (so that k is the maximum such j). To do this we exploit the fact that our SLP is of height $h = O(\log n)$ and use Lemma 7 to extract an SLP for $T[i..j]$, of size $g' \leq g + O(h) = g + O(\log n)$, in $O(h) = O(\log n)$ time[4]. We then search for the SLP of $T[i..j]$ in the SLP of size g of T using Lemma 8, in time $O((g + g') \log(j - i)) \subseteq O(g \log(k - i))$ (because $g' \subseteq O(g)$, as g is always $\Omega(\log n)$). By comparing the leftmost occurrence position with i we drive the exponential search, finding k in time $O(g \log^2(k - i))$ and space $O(g)$.

---

[4] Rytter [38] rebalances the grammar he extracts, but we do not need to do this.

Repeating this for each LZ phrase we get $\sum_{i=1}^{z} g \log^2 n_i$, where $n_1, n_2, \ldots, n_z$ denote the consecutive phrase lengths. By Jensen's inequality (since $\log^2(\cdot)$ is concave), the sum is maximized when all $n_i = n/z$.

Theorem 2. *Given an SLP with g rules for a text $T[1..n]$ whose LZ parse has z phrases, we can build that parse in $O(gz \log^2(n/z))$ time and $O(g)$ space.*

5   Building an LCG from an SLP in $\tilde{O}(n)$ Time

Locally consistent grammars (LCGs) are actually run-length context-free grammars, that is, they allow rules $X \rightarrow Y_1 \cdots Y_t$ (of size t) and run-length rules of the form $X \rightarrow Y^t$, equivalent to $X \rightarrow Y \cdots Y$ (t copies of Y), of size 2. A particular kind of LCG can be obtained from T with the following procedure [27]. First, define $\ell_k = (4/3)^{\lceil k/2 \rceil - 1}$ and call $S_0 = T$. Then, for increasing levels $k > 0$, create S_k from S_{k-1} as follows:

1. If k is odd, find the maximal runs of (say, $t > 1$ copies of) equal symbols Y in S_{k-1} such that $|\exp(Y)| \leq \ell_k$, create a new grammar rule $X \rightarrow Y^t$, and replace the run by X. The other symbols are copied onto S_k as is.
2. If k is even, generate a function π_k that randomly reorders the symbols of S_{k-1} and define local minima as the positions $1 < i < |S_{k-1}|$ such that $\pi_k(S_{k-1}[i-1]) > \pi_k(S_{k-1}[i]) < \pi_k(S_{k-1}[i+1])$. Place a block boundary after each local minimum, and before and after the symbols Y with $|\exp(Y)| > \ell_k$. Create new rules for the resulting blocks of length more than 1 and replace them in S_k by their corresponding nonterminals. Leave other symbols as is.

Our plan is to extract T left to right from its SLP, in $O(n)$ time, and carry out the described process in streaming form. The only obstacle to perform the process at level k in a single left-to-right pass is the creation of the functions π_k without knowing in advance the alphabet of S_{k-1}. We can handle this by maintaining two balanced trees. The first, T_{id}, is sorted by the actual symbol identifiers, and stores for each symbol a pointer to its node in the second tree, T_{pos}. The tree T_{pos} is sorted by the current π_k values (which evolve as new symbols arise), that is, the π_k value of a symbol is its inorder position in T_{pos}. We can know the current value of a symbol in π_k by going up from its node in T_{pos} to the root, adding up one plus the number of nodes in the left subtree of the nodes we reach from their right child (so T_{pos} stores subtree sizes to enable this computation). Two symbols are then compared in logarithmic time by computing their π_k values using T_{pos}.

When the next symbol is not found in T_{id}, it is inserted in both trees. Its rank r in T_{pos} is chosen at random in $[1, |T_{id}| + 1]$. We use the subtree sizes to find the insertion point in T_{pos}, starting from the root: let t_l be the size of the left child of a node. If $r \leq t_l + 1$ we continue by the left child, otherwise we subtract $t_l + 1$ from r and continue by the right child. The balanced tree rotations maintain the ranks of the nodes, so the tree can be rebalanced after the insertion adds a leaf.

Our space budget does not allow us maintaining the successive strings S_k. Rather, we generate $S_0 = T$ left to right in linear time using the given SLP and have one iterator per level k (the number of levels until having a single nonterminal is logarithmic [27, Remark 3.16]). Each time the process at some level $k - 1$ produces a new symbol, it passes that new symbol on to the next level, k. When the last symbol of T is consumed, all the levels in turn close their processes, bottom-up; the LCG comprises the rules produced along all levels.

The total space used is proportional to the number of distinct symbols across all the levels of the grammar. This can be larger than the grammar size because symbols X with $|\exp(X)| > \ell_k$ are not replaced in level k, so they exist in the next levels as well. To avoid this, we perform a twist that ensures that every distinct grammar symbol is stored only in $O(1)$ levels. The twist is not to store in the trees the symbols that cannot form groups in this level, that is, those X for which $|\exp(X)| > \ell_k$. Since then the symbols stored in the tree for even levels k are forced to form blocks (no two consecutive minima can exist), they will no longer exist in level $k + 1$. Note that the sizes of the trees used for the symbols at level k are then proportional to the number of nonterminals of that level in the produced grammar.

There is a deterministic bound $O(\delta \log \frac{n}{\delta})$ on the total number of nontermi-nals in the generated grammar [27, Corollary 3.12], and thus on the total sizes of the balanced trees. Here, δ is the compressibility measure based on substring complexity, and size $O(\delta \log \frac{n}{\delta})$ is optimal for every n and δ [27,28]. The size g_{lc} of the produced LCG could be higher, as for some choices of letter permutations on various levels some right-hand of the productions can be of not-constant length, however, but it is still $O(\delta \log \frac{n}{\delta})$ in expectation and with high probability [27, Theorem 3.13]. Because the sum of the lengths of the strings S_k is $O(n)$ [27, Corollary 3.15], we produce the LCG in time $O(n \log g_{lc})$; the $\log g_{lc}$ comes from the cost of balanced tree operations.

Theorem 3. *Given an SLP with g rules for a text $T[1..n]$, we can build w.h.p. an LCG of size $g_{lc} = O(\delta \log \frac{n}{\delta})$ for T in $O(n \log g_{lc})$ time and $O(g + g_{lc})$ space.*

If we know δ, we can abort the construction as soon as its total size exceeds $c \cdot \delta \log \frac{n}{\delta}$ for some suitable constant c, and restart the process afresh. After $O(1)$ attempts in expectation, we will obtain a locally consistent grammar of size $O(\delta \log \frac{n}{\delta})$ [27, Corollary 3.15]. The grammar we produce, in $O(n \log g_{lc})$ expected time, is then of guaranteed size $g_{lc} = O(\delta \log \frac{n}{\delta})$. Note that we need a structure mapping blocks and runs to new symbols: using a simple trie for the rules uses $O(g_{lc})$ space and can be constructed in $O(g_{lc} \log g_{lc})$ time.

6   Building the LZ Parse from an LCG in $\tilde{O}(z)$ Time

One of the many advantages of LCGs compared to general SLPs is that, related to Definition 1, they may allow trying out only $O(\log |P|)$ splitting positions of P in order to discover all their primary occurrences, as opposed to $m - 1$ if using a generic SLP. This is the case of the LCG of size $O(\delta \log \frac{n}{\delta})$ of the previous section [27], which specializes [10], in the sense that any grammar produced with the first method [27] can be produced by the second [10], and therefore every property we prove for the second method holds for the first as well. The first method introduces a restriction to produce grammars of size $O(\delta \log \frac{n}{\delta})$, whereas the second kind has a weaker space bound of $O(\gamma \log \frac{n}{\gamma})$, where $\gamma \geq \delta$ is the size of the smallest string attractor of T [25] (concretely, the parsing is as in Sect. 5 but does not enforce the condition $\exp(X) \leq \ell_k$). We now show how the bound on the splitting positions number enables us to find the LZ parse of those LCGs in time $O(z \log^4 n)$. We will then stick to the more general LCG [10]; the results hold for the other too [27], as explained.

Our technique combines results used for Theorems 1 and 2: we will use exponential search, as in Sect. 4, to find the next phrase $T[i..k]$, and will use the data structures of Sect. 3 to search for its leftmost occurrence in T; the fact that we will need to check just a logarithmic number of splitting points will yield the bound. We start with an analogue of Lemma 1 for our LCG; we get better bounds in this case.

Lemma 9. *Given the LCG [10] of size g_{lc} of $T[1..n]$, we can build in $O(g_{lc} \log g_{lc})$ time and $O(g_{lc})$ space a data structure supporting the same operations listed in Lemma 1, all in $O(\log n)$ time.*

Proof. Since the LCG is already balanced, accessing $T[i]$ in $O(\log n)$ time is immediate. The Karp-Rabin fingerprints can be computed with the structure of Christiansen et al. [10, Thm. A.3], which can be built in $O(g_{lc})$ space and time.

To compute longest common prefixes (LCPs) we use a similar approach as Kempa and Kociumaka [23, Thm. III.3] or earlier Alstrup, Brodal and Rauhe [1]. To deal with rules of non-constant size, we build a data structure for answering the LCE queries on the (right-hand sides of) non-runs *rules* of the LCG. This is a standard construction (using suffix arrays and LCA queries [5]) and can be done in $O(g_{lc} \log g_{lc})$ time and $O(g_{lc})$ space, or even in $O(g_{lc})$ time, when the letters can be identified with numbers that are polynomial in g_{lc} [20]. □

Consider the cost to build the data structures of Sect. 3. Using Lemma 9, we sort the multisets $\mathcal{Y}$ and $\mathcal{Z}$ in time $O(g_{lc} \log g_{lc} \cdot \log n)$. This time dominates the construction time of the z-fast tries for $\mathcal{Y}$ and $\mathcal{Z}$, the grid structure $\mathcal{G}$, and the two-dimensional range minimum query mentioned in Lemma 5. Further, because $g_{lc} \leq \gamma \log \frac{n}{\gamma}$ [10] and $\gamma \leq z$ [25], this time is in $O(z \log^2 n \log(n/z))$.

After building those components, we start parsing the text using the exponential search of Sect. 4. To test whether the candidate phrase $T[i..j]$ occurs starting to the left of i, we use the LCG search algorithm for $T[i..j]$ provided by the LCG. Christiansen et al. [10] observed that we need to check only $O(\log(j-i))$ splitting points to find every primary occurrence of $T[i..j]$. They find the splitting points through a linear-time parse of $T[i..j]$, but we can do better by reusing the locally consistent parsing used to build the LCG. While we do not store the strings S_k of Sect. 5, we can recover the pieces that cover $T[i..j]$ by traversing the (virtual) grammar tree from the root towards that substring of T.

Lemma 10 ([10]). *Let $M_0(i,j) = \{i, j-1\}$. For any $k > 0$, let $M_k(i,j)$ contain the first and last positions ending a block of S_k that are within $T[i..j-1]$ but* do not belong to $M_{k'}(i,j)$ for any $k' < k$. Then, $M(i,j) = \cup_k M_k(i,j)$ is of size $O(\log(j-i))$ and the splitting point of every primary occurrence of $T[i..j]$ in T belongs to $M(i,j)$.

Proof. Our definition of $M(i,j)$ includes the positions in Definitions 4.7 and 4.8 of Christiansen et al. [10] (they use B_r and $\hat{B}_r$ instead of our even and odd levels S_k). The property we state corresponds to their Lemma 6.4 [10]. □

To compute $M(i,j)$, then, we descend from the root of the (virtual) parse tree of the LCG towards the lowest nonterminal X that fully contains $T[i..j]$, and continue from X towards the leaf L that contains $T[i]$. We then start adding to $M(i,j)$ the endpoint of L (which is i), and climb up to its parent P. If P ends in the same position of L, we shift P to its next sibling. We now set $L = P$, add the last position of L to $M(i,j)$, climb up to its parent P, and so on until the last position of L exceeds $j-1$ (which may occur when reaching X or earlier). We proceed analogously with the path from X to the leaf that contains $T[j-1]$.

We visit $O(\log n)$ nodes in this process, but since the LCG may not be binary, we may need $O(\log n)$ time to find the proper children of a node. The total time is then $O(\log^2 n)$.

Once the set $M(i,j)$ of splitting points is found, we search for each of them as in Lemma 5, each in time $O(\log(j-i)\log n + \log^{1+\varepsilon} g_{lc})$. Therefore, the total time to check a candidate $T[i..j]$ is $O\left(\log^2 n + \log(j-i)\left(\log(j-i)\log n + \log^{1+\varepsilon} g_{lc}\right)\right)$. In turn, the exponential search that finds the next phrase $T[i..k]$ carries out $O(\log(k-i))$ such checks, with $j-i \leq 2(k-i)$, thus the total time to find the next phrase is $O(\log(k-i)\log^2 n + \log^3(k-i)\log n + \log^2(k-i)\log^{1+\varepsilon} g_{lc})$. Using Jensen's inequality again and simplifying, this yields the running time of $O(z\log^2(n/z)\log^2 n)$.

Theorem 4. *Given the LCG of Christiansen et al. [10] of size g_{lc} of $T[1..n]$, we can build w.h.p. the LZ parse of T in $O(z\log^2(n/z)\log^2 n)$ time and $O(g_{lc})$ extra space. The result also holds verbatim for the LCG of Kociumaka et al. [27].*

7   Conclusions

We have contributed to the problem of *compressed conversions*, that is, using asymptotically optimal space, between various compression formats for repetitive data. Such a space means linear in the input plus output size, which outrules the possibility of decompressing the data. This is crucial to face the sharp rise the size of data in sequence form has experienced in the last decades, which requires manipulating the data always in compressed form. To the best of our knowledge, we are the first to propose methods to build the Lempel-Ziv parse of a text directly from its straight-line program representation. Our methods work in time $O(n\log^2 n)$ and $O(gz\log^2 n)$. The second is polynomial on the size of the compressed data and we thus call it a *fully-compressed conversion*; such methods can be considerably faster when the data is highly compressible. We also gave methods to convert from straight-line programs to locally consistent grammars, which enable faster and more complex queries, in $O(n \log n)$ time. As a showcase for their improved search capabilities, we show how to produce the Lempel-Ziv parse from those grammars in time $O(z \log^4 n)$, another fully-compressed conversion. All of our conversions work with high probability.

Obvious open problems are obtaining better running times without using more space. Furthermore, we think that approaches similar to those described in this article can be applied to effectively compute other parses, such as the lexicographic parse [34]. We plan to address these in the extended version.

Acknowledgements. We thank the anonymous reviewers, whose comments helped to improve the presentation, correct several errors and give a much better and more detailed exhibition of the state of the art in the introduction.

References

1. Alstrup, S., Brodal, G.S., Rauhe, T.: Pattern matching in dynamic texts. In: SODA, pp. 819–828 (2000)
2. Arimura, H., Inenaga, S., Kobayashi, Y., Nakashima, Y., Sue, M.: Optimally computing compressed indexing arrays based on the compact directed acyclic word graph. In: International Symposium on String Processing and Information Retrieval, pp. 28–34 (2023)
3. Belazzougui, D., Boldi, P., Pagh, R., Vigna, S.: Fast prefix search in little space, with applications. In: 18th European Symposium on Algorithms (ESA), Part I, pp. 427–438 (2010)
4. Belazzougui, D., Puglisi, S.J.: Range predecessor and Lempel-Ziv parsing. In: 27th Symposium on Discrete Algorithms (SODA), pp. 2053–2071 (2016)
5. Bender, M.A., Farach-Colton, M.: The LCA problem revisited. In: LATIN, pp. 88–94 (2000)
6. Bille, P., Gørtz, I.L., Cording, P.H., Sach, B., Vildhøj, H.W., Vind, S.: Fingerprints in compressed strings. J. Comput. Syst. Sci. **86**, 171–180 (2017)
7. Bille, P., Landau, G.M., Raman, R., Sadakane, K., Rao, S.S., Weimann, O.: Random access to grammar-compressed strings and trees. SIAM J. Comput. **44**(3), 513–539 (2015)
8. Charikar, M., et al.: The smallest grammar problem. IEEE Trans. Inf. Theory **51**(7), 2554–2576 (2005)
9. Chazelle, B.: A functional approach to data structures and its use in multidimensional searching. SIAM J. Comput. **17**(3), 427–462 (1988)
10. Christiansen, A.R., Ettienne, M.B., Kociumaka, T., Navarro, G., Prezza, N.: Optimal-time dictionary-compressed indexes. ACM Trans. Algorithms **17**(1), article 8 (2020)
11. Claude, F., Navarro, G., Pacheco, A.: Grammar-compressed indexes with logarithmic search time. J. Comput. Syst. Sci. **118**, 53–74 (2021)
12. Durbin, R.M., Auton, A., Brooks, L.D.: A global reference for human genetic variation. Nature **526**(7571), 68–74 (2015)

13. Gagie, T., Navarro, G., Prezza, N.: Fully-functional suffix trees and optimal text searching in BWT-runs bounded space. J. ACM **67**(1), article 2 (2020)
14. Ganardi, M., Jeż, A., Lohrey, M.: Balancing straight-line programs. J. ACM **68**(4), 27:1–27:40 (2021)
15. Gawrychowski, P.: Pattern matching in Lempel-Ziv compressed strings: fast, simple, and deterministic. In: European Symposium on Algorithms, pp. 421–432 (2011)
16. Gawrychowski, P., Karczmarz, A., Kociumaka, T., Łącki, J., Sankowski, P.: Optimal dynamic strings. In: 29th ACM-SIAM Symposium on Discrete Algorithms, pp. 1509–1528 (2018)
17. Gawrychowski, P., Kociumaka, T.: Sparse suffix tree construction in optimal time and space. In: 28th ACM-SIAM Symposium on Discrete Algorithms, pp. 425–439. SIAM (2017)
18. Tomohiro, I.: Longest common extensions with recompression. In: 28th Symposium on Combinatorial Pattern Matching (CPM), pp. 18:1–18:15 (2017)
19. Jeż, A.: Faster fully compressed pattern matching by recompression. ACM Trans. Algorithms (TALG) **11**(3), 1–43 (2015)
20. Kärkkäinen, J., Sanders, P., Burkhardt, S.: Linear work suffix array construction. J. ACM **53**(6), 918–936 (2006)
21. Karp, R.M., Rabin, M.O.: Efficient randomized pattern-matching algorithms. IBM J. Res. Dev. **31**(2), 249–260 (1987)
22. Kempa, D., Kociumaka, T.: Resolution of the Burrows-Wheeler Transform conjecture. In: 61st IEEE Symposium on Foundations of Computer Science (FOCS), pp. 1002–1013 (2020)
23. Kempa, D., Kociumaka, T.: Collapsing the hierarchy of compressed data structures: Suffix arrays in optimal compressed space. In: 63rd IEEE Symposium on Foundations of Computer Science (FOCS), pp. 1877–1886 (2023)
24. Kempa, D., Kosolobov, D.: LZ-End parsing in compressed space. In: 27th Data Compression Conference (DCC), pp. 350–359 (2017)
25. Kempa, D., Prezza, N.: At the roots of dictionary compression: string attractors. In: 50th ACM SIGACT Symposium on Theory of Computing, pp. 827–840 (2018)
26. Kieffer, J.C., Yang, E.-H.: Grammar-based codes: a new class of universal lossless source codes. IEEE Trans. Inf. Theory **46**(3), 737–754 (2000)
27. Kociumaka, T., Navarro, G., Olivares, F.: Near-optimal search time in δ-optimal space. Algorithmica (2023). accepted, available online
28. Kociumaka, T., Navarro, G., Prezza, N.: Toward a definitive compressibility measure for repetitive sequences. IEEE Trans. Inf. Theory **69**(4), 2074–2092 (2023)
29. Lempel, A., Ziv, J.: On the complexity of finite sequences. IEEE Trans. Inf. Theory **22**(1), 75–81 (1976)
30. Mehlhorn, K., Sundar, R., Uhrig, C.: Maintaining dynamic sequences under equality tests in polylogarithmic time. Algorithmica **17**(2), 183–198 (1997)
31. Navarro, G.: Wavelet trees for all. J. Dis. Algorithms **25**, 2–20 (2014)
32. Navarro, G.: Indexing highly repetitive string collections, part I: repetitiveness measures. ACM Comput. Surv. **54**(2), article 29 (2021)
33. Navarro, G.: Computing MEMs on repetitive text collections. In: 34th Symposium on Combinatorial Pattern Matching (CPM), page article 22 (2023)
34. Navarro, G., Ochoa, C., Prezza, N.: On the approximation ratio of ordered parsings. IEEE Trans. Inf. Theory **67**(2), 1008–1026 (2020)
35. Navarro, G., Prezza, N.: Universal compressed text indexing. Theoret. Comput. Sci. **762**, 41–50 (2019)

36. Nishimoto, T., Tomohiro, I., Inenaga, S., Bannai, H., Takeda, M.: Dynamic index and LZ factorization in compressed space. Discret. Appl. Math. **274**, 116–129 (2020)
37. Policriti, A., Prezza, N.: From LZ77 to the run-length encoded burrows-wheeler transform, and back. In: 28th Symposium on Combinatorial Pattern Matching (CPM). LIPIcs, vol. 78, pp. 17:1–17:10 (2017)
38. Rytter, W.: Application of Lempel-Ziv factorization to the approximation of grammar-based compression. Theoret. Comput. Sci. **302**(1–3), 211–222 (2003)

Sparse Suffix and LCP Array: Simple, Direct, Small, and Fast

Lorraine A. K. Ayad[1], Grigorios Loukides[2], Solon P. Pissis[3,4(✉)], and Hilde Verbeek[3]

[1] Brunel University London, London, UK
lorraine.ayad@brunel.ac.uk
[2] King's College London, London, UK
grigorios.loukides@kcl.ac.uk
[3] CWI, Amsterdam, The Netherlands
{solon.pissis,hilde.verbeek}@cwi.nl
[4] Vrije Universiteit, Amsterdam, The Netherlands

Abstract. Sparse suffix sorting is the problem of sorting $b = o(n)$ suffixes of a string of length n. Efficient sparse suffix sorting algorithms have existed for more than a decade. Despite the multitude of works and their justified claims for applications in text indexing, the existing algorithms have not been employed by practitioners. Arguably this is because there are no simple, direct, *and* efficient algorithms for sparse suffix array construction. We provide two new algorithms for constructing the sparse suffix and LCP arrays that are simultaneously simple, direct, small, and fast. In particular, our algorithms are: *simple* in the sense that they can be implemented using only basic data structures; *direct* in the sense that the output arrays are not a byproduct of constructing the sparse suffix tree or an LCE data structure; *fast* in the sense that they run in $\mathcal{O}(n \log b)$ time, in the worst case, or in $\mathcal{O}(n)$ time, when the total number of suffixes with an LCP value greater than $2^{\lfloor \log \frac{n}{b} \rfloor + 1} - 1$ is in $\mathcal{O}(b/ \log b)$, matching the time of optimal yet much more complicated algorithms [Gawrychowski and Kociumaka, SODA 2017; Birenzwige et al., SODA 2020]; and *small* in the sense that they can be implemented using *only* $8b + o(b)$ machine words. We also show that our second algorithm can be trivially amended to work in $\mathcal{O}(n)$ time for any uniformly random string. Our algorithms are non-trivial space-efficient adaptations of the Monte Carlo algorithm by I et al. for constructing the sparse suffix tree in $\mathcal{O}(n \log b)$ time [STACS 2014].

Keywords: suffix array · LCP array · suffix sorting · sparse suffix sorting

SPP and HV are supported by the PANGAIA project (GA 872539). SPP is supported by the ALPACA project (GA 956229). HV is supported by a Constance van Eeden Fellowship.

J. A. Soto and A. Wiese (Eds.): LATIN 2024, LNCS 14578, pp. 162–177, 2024.
https://doi.org/10.1007/978-3-031-55598-5_11

1   Introduction

Let $T = T[1 .. n]$ be a string of length n over an ordered alphabet Σ. Further let $\mathcal{B} \subseteq [1, n]$ be a set of $b > 1$ positions in T. *Sparse suffix sorting* is the problem of sorting the set of suffixes $T_\mathcal{B} = \{T[i .. n] : i \in \mathcal{B}\}$ lexicographically [18]. This is achieved by constructing the sparse suffix array. The *sparse suffix array* $\mathsf{SSA} = \mathsf{SSA}[1 .. b]$ is the array containing the positions in $\mathcal{B}$ in the lexicographical order of the suffixes in $T_\mathcal{B}$. The associated *sparse longest common prefix array* $\mathsf{SLCP} = \mathsf{SLCP}[1 .. b]$ stores the length $\mathsf{SLCP}[i]$ of the longest common prefix of $T[\mathsf{SSA}[i-1] .. n]$ and $T[\mathsf{SSA}[i] .. n]$ when $i \in [2, n]$ or 0 when $i = 1$. The SSA and SLCP array can be used to construct the sparse suffix tree in linear time using the algorithm by Kasai et al. [20]. The *sparse suffix tree* is the compacted trie of the set $T_\mathcal{B}$. Vice-versa, the SSA and SLCP array can be obtained in linear time via a pre-order traversal of the sparse suffix tree.

Sparse suffix sorting was introduced as a fundamental step in the construction of compressed or sparse text indexes [18]. Modern compressed text indexes [10,24], practical indexes for long patterns [2,15,22,23], and sublinear-space string algorithms [3,5] rely on sparse suffix sorting: they first sample a sublinear number of "important" suffixes, which they next sort to construct their final solution. Efficient sparse suffix sorting algorithms have existed for more than a decade. The following algorithms construct SSA explicitly, or implicitly by first constructing the sparse suffix tree. Since the size of SSA (and the size of sparse suffix tree) is $\Theta(b)$, the goal of these algorithms is to use $\mathcal{O}(b)$ words of space assuming read-only random access to T. Kärkkäinen et al. presented a deterministic $\mathcal{O}(n^2/s)$-time and $\mathcal{O}(s)$-space algorithm, for any $s \in [b, n]$ [17, Section 8]. Bille et al. presented a Monte Carlo $\mathcal{O}(n \log^2 b)$-time and $\mathcal{O}(b)$-space algorithm [6], as well as a Las Vegas $\mathcal{O}(n \log^2 n + b^2 \log b)$-time and $\mathcal{O}(b)$-space algorithm. I et al. presented a Monte Carlo $\mathcal{O}(n + (bn/s) \log s)$-time and $\mathcal{O}(s)$-space algorithm, for any $s \in [b, n]$ [16] and a Las Vegas $\mathcal{O}(n \log b)$-time and $\mathcal{O}(b)$-space algorithm. Gawrychowski and Kociumaka [14] presented a Monte Carlo $\mathcal{O}(n)$-time and $\mathcal{O}(b)$-space algorithm and a Las Vegas $\mathcal{O}(n\sqrt{\log b})$-time and $\mathcal{O}(b)$-space algorithm. Birenzwige et al. [7] presented a Las Vegas algorithm running in $\mathcal{O}(n)$ time using $\mathcal{O}(b)$ space. Besides this they also presented a deterministic $\mathcal{O}(n \log \frac{n}{b})$-time and $\mathcal{O}(b)$-space algorithm, for any $b = \Omega(\log n)$.

The following algorithms also construct SSA, but they work in the *restore model* [9]: an algorithm is allowed to overwrite parts of the input, as long as it can restore it to its original form at termination. Fischer et al. [12] presented a deterministic $\mathcal{O}(c\sqrt{\log n} + b \log b \log n \log^* n)$-time and $\mathcal{O}(b)$-space algorithm, where c is the number of letters that must be compared for distinguishing the suffixes in $T_\mathcal{B}$. In some cases, this runs in sublinear extra time; extra refers to the linear cost of loading T in memory. Prezza [26] presented a Monte Carlo $\mathcal{O}(n + b \log^2 n)$-time algorithm using $\mathcal{O}(1)$ words of space.

Motivation.   Despite the multitude of works on sparse suffix sorting and their justified claims for applications in text indexing, the existing algorithms have not been employed by practitioners. Arguably this is because there are no simple, direct, *and* efficient algorithms for SSA construction. The $\mathcal{O}(n)$-time algorithms of Gawrychowski and Kociumaka [14] and of Birenzwige et al. [7] are far from simple and do not seem to be practically promising either. The former (Monte Carlo) algorithm relies heavily on the construction of compacted tries, which induce high constants in space usage, and on a recursive application of difference cover to construct a Longest Common Extension (LCE) data structure. The latter (Las Vegas) algorithm relies on an intricate partitioning scheme (sampling) to construct SSA and on an LCE data structure to compute the SLCP array. The Monte Carlo $\mathcal{O}(n \log b)$-time algorithm of I et al. [16] is simple but it also relies heavily on compacted tries, which makes it less likely to be employed by practitioners for SSA construction. The Monte Carlo $\mathcal{O}(n + b \log^2 n)$-time algorithm of Prezza [26] makes heavy usage of an LCE data structure as well: constructing the SSA and SLCP array is a byproduct of an in-place LCE data structure. The latter algorithm is, to the best of our knowledge, the only algorithm which has been implemented (at least in a simplified form). Due to the interest in sparse suffix sorting and the above characteristics of the existing algorithms, we were motivated to revisit this problem to develop efficient, yet simple and direct, algorithms for SSA construction. Such algorithms may serve as baselines for practitioners to engineer the SSA and SLCP array construction.

Our Model and Results. We assume the standard word RAM model with word size $\Theta(\log n)$; basic arithmetic and bit-wise operations on $\mathcal{O}(\log n)$-bit integers take $\mathcal{O}(1)$ time. We assume that we have a read-only random access string T of length n over an integer alphabet $\Sigma = \{1, \ldots, n^{\mathcal{O}(1)}\}$, a read-only integer array A of size b storing the b elements of $\mathcal{B}$, and two write-only integer arrays SSA and SLCP, each of size b. We thus count the amount of *extra space in machine words* used to construct the SSA and SLCP array. We present two algorithms:

1. Our first algorithm, MAIN-ALGO, constructs SSA and SLCP *directly*; i.e., without first explicitly constructing the sparse suffix tree or an LCE data structure (see Sect. 3). Its time complexity is $\mathcal{O}(n + (bn/s) \log s)$ and its space complexity is $s + 7b + o(b)$ machine words, for any chosen $s \in [b, n]$. It is a Monte Carlo algorithm that returns the correct output *with high probability*; i.e., with probability at least $1 - n^{-c}$, for any constant $c \geq 1$ chosen at construction time. MAIN-ALGO is *simple* in the sense that it can be implemented using only *basic data structures* (e.g., dictionaries and arrays) readily available in widely-used programming languages (e.g., C++, Java, or Python). MAIN-ALGO is a non-trivial space-efficient simulation of the algorithm by I et al. for sparse suffix tree construction [16]. A disadvantage of these two algorithms is that they attain the $\Theta(n \log b)$ time bound for $s = b$ *in any case*. To address this, we develop PARAMETERIZED-ALGO, a parameterized algorithm which is input-sensitive.

2. Our second algorithm, PARAMETERIZED-ALGO, also constructs SSA and SLCP directly (see Sect. 4). Its time complexity is $\mathcal{O}(n + (b'n/b) \log b)$ and its space complexity is $8b + 4b' + o(b)$ machine words, where b' is the total number of suffixes $\mathsf{SSA}[i] \in \mathcal{B}$ with $\mathsf{SLCP}[i] \geq \ell$ or $\mathsf{SLCP}[i + 1] \geq \ell$, where

$\ell = 2^{\lfloor \log \frac{n}{b} \rfloor + 1} - 1$. When $b' = \mathcal{O}(b/\log b)$, PARAMETERIZED-ALGO runs in $\mathcal{O}(n)$ time, thus matching the time of the optimal yet much more complicated algorithms in [7,14], using *only* $8b + o(b)$ machine words. It is a Monte Carlo algorithm that returns the correct output with high probability. It is *remarkably simple* as it consists of two calls of MAIN-ALGO and a linear-time step that merges the partial results (however, the proof of correctness requires some work). The running time of PARAMETERIZED-ALGO is good in the following sense: if the instance is reasonably sparse, then ℓ is large and likely $b' = \mathcal{O}(b/\log b)$, thus it runs in $\mathcal{O}(n)$ time. In any case, it runs in $\mathcal{O}(n \log b)$ time. For instance, for the full human genome (v. GRCh38) as T, where $n \approx 3 \cdot 10^9$, and for $b = \lfloor \sqrt{n} \rfloor = 56137$ suffixes selected uniformly at random, $b' = 2525 < \lfloor b/\log b \rfloor = 3558$. We also analyze the time complexity of PARAMETERIZED-ALGO on random strings and show that it works in $\mathcal{O}(n)$ time (after a trivial amendment), for any string chosen uniformly at random from Σ^n and any set $T_{\mathcal{B}}$ of b suffixes of T, with high probability.

2 Preliminaries

We consider strings over an integer alphabet $\Sigma = \{1, \ldots, n^{\mathcal{O}(1)}\}$. The elements of Σ are called *letters*. A *string* $T = T[1 .. n]$ is a sequence of letters from Σ; we denote by $|T| = n$ the *length* of T. The fragment $T[i .. j]$ of T is an *occurrence* of the underlying *substring* $P = T[i] \ldots T[j]$ occurring at *position* i in T. A *prefix* of T is a substring of T of the form $T[1 .. j]$ and a *suffix* of T is a substring of T of the form $T[i .. n]$.

Karp-Rabin Fingerprints. Let T be a string of length n over an integer alphabet. Let p be a prime and choose $r \in [0, p-1]$ uniformly at random. The Karp-Rabin (KR) fingerprint [19] of $T[i .. j]$ is: $\phi_T(i, j) = (\sum_{k=i}^{j} T[k] r^{j-k} \mod p, r^{j-i+1} \mod p)$. Clearly, if $T[i .. i + \ell] = T[j .. j + \ell]$ then $\phi_T(i, i + \ell) = \phi_T(j, j + \ell)$. On the other hand, if $T[i .. i + \ell] \neq T[j .. j + \ell]$ then $\phi_T(i, i + \ell) \neq \phi_T(j, j + \ell)$ with probability at least $1 - \ell/p$ [11]. Since we are comparing only substrings of equal length, the number of different possible substring comparisons is less than n^3. Thus, for any constant $c \geq 1$, we can set p to be a prime larger than $\max(|\Sigma|, n^{c+3})$ to make the KR fingerprint function perfect with probability at least $1 - n^{-c}$. Any KR fingerprint or p fit in one machine word of size $\Theta(\log n)$.

Lemma 1 ([16]). *Any string $T \in \Sigma^n$ can be preprocessed in $\mathcal{O}(n)$ time using $s + \mathcal{O}(1)$ machine words, for any $s \in [1, n]$, so that the KR fingerprint of any length-k fragment of T is computed in $\mathcal{O}(\min\{k, n/s\})$ time.*[1]

I et al. [16] employ the *distribute-and-collect* technique [25] to group b suffixes, according to a fixed-length common prefix by using their KR fingerprints, in $\mathcal{O}(b \log_s n)$ time. We instead use hashing to achieve the same result in $\mathcal{O}(b)$ time with high probability. This gives improved running times in some special regimes (see Theorem 2 and Theorem 3).

---

[1] I et al. [16] claim $\mathcal{O}(s)$ space but from their construction it is evident that in fact $s + \mathcal{O}(1)$ machine words are used.

3   Main Algorithm

Overview. A summary of our main algorithm (MAIN-ALGO) follows with references to the pseudocode given in Algorithms 1 and 2.[2] It takes as input a string T from Σ^n and an array A of b elements, indicating the starting positions of the suffixes to be sorted. It also takes an integer j_{start}, which defines the number of iterations. In this section, j_{start} is set to $\lfloor \log n \rfloor$ (the default value), but a different value is used for the parameterized algorithm presented in Sect. 4.

Algorithm 1. MAIN-ALGO

Input: string $T \in \Sigma^n$, integer b, array A of b integers, and integer j_{start} (default $\lfloor \log n \rfloor$)
Output: SSA and SLCP

```
1:  m ← b + 1
2:  Lm ← (1, ..., b)
3:  B ← {(m, 0, Lm)}
4:  A[m] ← A[1]
5:  for j = jstart, ..., 0 do
6:      B' ← ∅
7:      for (i, k, Li) ∈ B do
8:          Hi ← empty hash table
9:          s ← |Li|
10:         for l ∈ Li do
11:             h ← φT(A[l] + k, A[l] + k + 2^j − 1)
12:             Hi[h].append(l)
13:             Li.erase(l)
14:         for h ∈ Hi do
15:             f ← Hi[h]
16:             if |f| = s then
17:                 B ← B \ {(i, k, Li)} ∪ {(i, k + 2^j, f)}
18:             else if |f| ≥ 2 then
19:                 m ← m + 1
20:                 Li.append(m)
21:                 B' ← B' ∪ {(m, k + 2^j, f)}
22:                 A[m] ← A[f[1]]
23:             else if |f| = 1 then
24:                 Li.append(f)
25:     B ← B ∪ B'
26: for (i, k, Li) ∈ B do
27:     Li.sort(l ↦ T[A[l] + k])
28: return OUTPUT-ARRAYS(B, b, A)
```

During the first phase (Algorithm 1, Lines 1–25), the suffixes are distributed into groups such that all suffixes belonging to a particular group share a common prefix. At the end of this process, we are left with a hierarchy of groups that describes the exact longest common prefixes between suffixes. The members of each group are then sorted lexicographically, which is made possible by knowing their longest common prefixes (Algorithm 1, Lines 26–27), such that a traversal of the hierarchy will yield the suffixes in lexicographic order. This is the second phase (Algorithm 1, Line 28 and Algorithm 2): a simple depth-first search is used to construct the sparse suffix array and accompanying sparse LCP array from the hierarchy.

3.1   Computing and Sorting the LCP Groups

During the first phase, the suffixes of A are organized into several LCP groups stored in set B. Each group in B is represented by a triple $(i, k, \{v_1, \ldots, v_{n_i}\})$,

---

[2] We stress that the pseudocode is complete in the sense that it only assumes the implementation of Lemma 1 (Line 11).

where i is the index (id) of the group, k is its associated LCP value and v_1 through v_{n_i} are its members, which are either suffixes or other groups. To distinguish between suffixes and groups, the indices 1 through b are reserved for the suffixes in A and the group numbering starts at $b + 1$. At every point of the algorithm, it holds that in a group $(i, k, \{v_1, \ldots, v_{n_i}\})$, all suffixes and groups (with their respective suffixes) in $\{v_1, \ldots, v_{n_i}\}$ share a prefix of length *at least* k. At the start (base case), there exists just one group $(b + 1, 0, \{1, \ldots, b\})$ (Line 3) containing all suffixes as members.

The LCP groups are then "refined" (the refinement process will be explained shortly) over the course of $\lfloor \log n \rfloor$ iterations, such that in the end each group describes the exact longest common prefix of its members rather than just a lower bound. Specifically, by the end of iteration j (where j descends from $\lfloor \log n \rfloor$ down to zero), in a group with LCP value k, two suffixes will have an actual longest common prefix of at least k and at most $k + 2^j - 1$ letters. This gap is closed once j has reached zero, at which point the refinement process is completed. The algorithm allows specifying a different starting value for j than $\lfloor \log n \rfloor$, through the parameter j_{start}. This is used in the parameterized algorithm described in Sect. 4.

The refinement process works as follows in iteration j. We refine every existing group; let one such group be $(i, k, \{v_1, \ldots, v_{n_i}\})$. We create a hash table (Line 8) and for every group member, with index v_i, we take the KR fingerprint as per Lemma 1 of $T[A[v_i] + k .. A[v_i] + k + 2^j - 1]$ (Line 11).[3] If v_i denotes a group, we do the same thing using any given suffix belonging to that group; this is easily achieved by appending "witness" suffixes to A for every created group as seen in Lines 4 and 22. (Any suffix can be a witness but we choose the one with the smallest index.) All the members are grouped in the hash table based on their KR fingerprints: if two suffixes have the same KR fingerprint, they will end up in the same entry of the hash table and with high probability have the same prefix of length $k + 2^j$. To save space, entries are removed from the group as they are added to the hash table (Line 13). All entries of the hash table are then inspected (Line 14). We distinguish three cases. In case 1 (Lines 16–17), if all suffixes in a group end up having the same KR fingerprint, we update the LCP value of the old group to $k + 2^j$ rather than creating a new group. In case 2 (Lines 18–22), if two or more suffixes have the same KR fingerprint, a new group is made with LCP value $k + 2^j$, containing these suffixes, and added to B. After removing the suffixes from their original group, we replace them with the index of the newly created group (Line 20). In case 3 (Lines 23–24), if a suffix is not grouped with any other suffix, we append it back to its original group.

Once the iteration with $j = 0$ ends, all LCP groups describe the exact longest common prefix of their members.[4] We now sort the members of every group lexicographically (Lines 26–27). Sorting can be done using merge sort or radix sort, because these algorithms can be performed in place using $\mathcal{O}(1)$ additional

---

[3] We assume that $A[v_i] + k + 2^j - 1 \leq n$; otherwise, the suffix ends at position n.

[4] This is generally not true when j_{start} was set to a value less than $\lfloor \log n \rfloor$; in this case, the LCP values are only correct if they are at most $2^{j_{\text{start}}+1} - 1$; see Sect. 4.

memory. Moreover, since we now know the exact LCP value for each group, two members in the same group can easily be compared in constant time: if they have a longest common prefix of length k, then the first position in which they differ is $k+1$, meaning they can be compared by only comparing their $k+1$-th letters. After this, set B contains the complete and sorted LCP groups, which are passed on to the second step of the algorithm.

3.2 Constructing the SSA and SLCP Array

The second phase (Algorithm 2) of the main algorithm involves traversing the groups created in the previous phase in order to construct the SSA and SLCP array. At this point, the members of each group are sorted lexicographically, which means that the SSA can be obtained by a simple pre-order walk along the hierarchy of the groups. For any two members, their exact longest common prefix is stored by their lowest common ancestor; that is, the group with the greatest LCP value that both suffixes fall under.

This part of the algorithm is thus a simple depth-first search of the underlying hierarchy that records all encountered suffixes in SSA in the order they appear. For every group that is visited, the LCP value of its direct "parent" is stored with it (Lines 4 and 17). Throughout, a value ℓ is tracked that takes the value of the lowest LCP value that has been seen since the last suffix was encountered (Lines 8–9); every time a suffix is appended to SSA, ℓ is appended to the SLCP array (Lines 11–12). This completes the construction.

Algorithm 2. OUTPUT-ARRAYS

Input: Set B of tuples (i, k, L_i) in ascending order by i, integer b, and array A
Output: SSA and SLCP

```
1:  SSA ← empty array
2:  SLCP ← empty array
3:  S ← empty stack
4:  S.push((b + 1, 0))
5:  ℓ ← 0
6:  while S is not empty do
7:      (i, ℓ') ← S.pop()
8:      if ℓ' < ℓ then
9:          ℓ ← ℓ'
10:     if i ≤ b then
11:         SSA.append(A[i])
12:         SLCP.append(ℓ)
13:         ℓ ← ∞
14:     else
15:         (i, k, L_i) ← B[i − b]
16:         for i' ∈ L_i in reverse order do
17:             S.push((i', k))
18: return SSA and SLCP
```

3.3 Analysis

We prove the following result (Theorem 1) by analyzing the time (Lemma 2) and space (Lemma 3) complexity of MAIN-ALGO. (The correctness of the algorithm follows directly from [16].)

Theorem 1. *For any string $T \in \Sigma^n$, any set T_B of b suffixes of T, and any $s \in [b, n]$, MAIN-ALGO with j_{start} set to $\lfloor \log n \rfloor$ computes the SSA and SLCP of T_B in $\mathcal{O}(n + (bn/s) \log s)$ time using $s + 7b + o(b)$ machine words. The output is correct with high probability.*

Lemma 2. *MAIN-ALGO with j_{start} set to $\lfloor \log n \rfloor$ runs in $\mathcal{O}(n + (bn/s) \log s)$ time.*

Proof. The first phase of the algorithm consists of $\mathcal{O}(\log n)$ iterations and a sorting step. During every iteration, each existing group is considered and every member within the group is hashed. After being hashed, it is either re-added to the same group or put into a new group. The total number of groups is at most $b - 1$, as the group structure represents a conceptual tree (hierarchy) with b leaves in which all internal nodes have at least two children. The number of members in each group is at most b. However, by amortization, it can be seen that every member (that is, every suffix and every group other than the "root") is processed precisely once during every iteration. Thus, we have $2b - 2 = \mathcal{O}(b)$ members in total.

For every group member, a KR fingerprint is computed. After the one-time pre-processing of T in $\mathcal{O}(n)$ time, the KR fingerprint of a length-k substring can be computed in $\mathcal{O}(\min\{k, n/s\})$ time (Lemma 1). In the first $\log s$ iterations, the cost is $\mathcal{O}(n/s)$, so the total cost of these iterations is $\mathcal{O}((bn/s)\log s)$. After $\log s$ iterations, the length k of the substring whose KR fingerprint is computed is $k < n/2^{\log s} = n/s$ and so the total cost of all remaining iterations is $bn/s + bn/(2s) + bn/(4s) + \cdots + b = \mathcal{O}(bn/s)$. Thus the total cost of computing all KR fingerprints is $\mathcal{O}(n + (bn/s)\log s)$.

Every group member has its KR fingerprint taken and added to a hash table supporting constant worst-case operations with high probability [1,4]. Afterwards, all members from the hash table are re-added to the groups; for every KR fingerprint collision a new group is created with its respective members, and all other members are re-added to the original group. The number of newly created groups is at most half the number of members in the original group, as every new group has to contain at least two members. So other than the fingerprinting, all operations for a single member are performed in constant time with high probability,[5] meaning that the total time for every iteration is $\mathcal{O}(b \log n) = \mathcal{O}((bn/s)\log s)$.

In the sorting step, we have two cases: (a) $b < n/\log n$ and (b) $b \geq n/\log n$. The members in each group are sorted using in-place merge sort [21,27] (Case (a)) or in-place radix sort [13] (Case (b)) in $\mathcal{O}(n)$ time.

Case (a): $b < n/\log n$. Sorting k members with merge sort takes $\mathcal{O}(k \log k)$ time. Recall that there are at most $b-1$ groups and that the total number of members over all groups is at most $2b - 2 = \mathcal{O}(b)$. If the number of members to be sorted in group i is k_i, then $k_1 + \cdots + k_{b-1} = \mathcal{O}(b)$ so the time needed to sort all groups is $\mathcal{O}(k_1 \log k_1 + \cdots + k_{b-1} \log k_{b-1}) = \mathcal{O}((k_1 + \cdots + k_{b-1}) \log b) = \mathcal{O}(b \log b) = \mathcal{O}(n)$.

Case (b): $b \geq n/\log n$. We employ the algorithm by Franceschini et al. [13], which, given an array A of k $\mathcal{O}(\log k)$-bit integers, sorts A in place in $\mathcal{O}(k)$ time. Sorting the $2b - 2$ members takes $\mathcal{O}(b) = \mathcal{O}(n)$ time, because every member can be encoded by its group id, which is a $\mathcal{O}(\log b)$-bit integer, and a letter, which is also a $\log \sigma = \mathcal{O}(\log n) = \mathcal{O}(\log b)$-bit integer, where $\sigma = |\Sigma|$.

---

[5] If this is not the case, we output incorrect arrays deliberately to ensure that our algorithm is Monte Carlo.

The second phase of the algorithm is a simple stack-based DFS. Each of the $\mathcal{O}(b)$ members is pushed to and popped from the stack precisely once. The further operations applied to each member all take $\mathcal{O}(1)$ time, so this step takes $\mathcal{O}(b)$ time.

Adding all this together gives $\mathcal{O}(n) + \mathcal{O}(bn/s) \cdot \mathcal{O}(\log s) + \mathcal{O}(n) + \mathcal{O}(b) = \mathcal{O}(n + (bn/s)\log s)$ time.   □

We remark that, like the algorithm by I et al. [16], MAIN-ALGO can be amended to work in $\mathcal{O}(n)$ time, when $s = b\log b$.

Lemma 3. MAIN-ALGO *can be implemented using* $s + 7b + o(b)$ *machine words, excluding the read-only string* T, *the array* A *representing the set of* b *suffixes, and the write-only output arrays* SSA *and* SLCP.

Proof. We analyze the peak space used by the algorithm neglecting the use of $\mathcal{O}(1)$ machine words:

- **KR fingerprints**: Pre-processing T to compute KR fingerprints takes s machine words by Lemma 1.
- **Array A**: Array A starts with b integers as input, but at most $b - 1$ more integers are appended to it during the algorithm to store witness suffixes for new groups, so it stores at most $b - 1$ extra integers. (Even if A is read-only we can simulate the append operation by using an extra array.)
- **Set B**: We implement set B using three integer arrays: K of size $b - 1$; C of size $b - 1$; and L of size $2b - 2$. $K[i]$ stores the LCP value for the group with id $i + b$; and $L[C[i - 1] + 1], \ldots, L[C[i]]$ are the group's member id's.[6] There are at most $b - 1$ groups; for every group one integer is stored in K and C as well as the group member id's in L. The total number of group members is at most $2b - 2$, since all groups except the "root" group are a member. Thus B can be implemented using $4b - 4$ integers.
- **Hash table H_i**: While processing group $i + b$, every group member is (removed from the group and) added to a hash table H_i as satellite value of the corresponding KR fingerprint key. We use a space-efficient hash table storing $c_i = C[i] - C[i - 1]$ integers (KR fingerprints) as keys: By using [1,4], we implement H_i using $(1 + \epsilon)c_i$ machine words, for any $\epsilon = \Omega(\log\log c_i / \log c_i)$. We need at most b integers to maintain the size of the satellite values per KR fingerprint because every group can have at most b members. By choosing $\epsilon = \log\log c_i / \log c_i$ we need at most $2b + o(b)$ machine words in total.

We can delete the fingerprint data structure and the hash table before moving to the sorting step. Sorting does not use any additional space because merge sort and radix sort can be implemented in-place [13,21,27], thus using only $\mathcal{O}(1)$ additional machine words. The first phase of the algorithm uses at most $s + 7b + o(b)$ machine words but at the end of it we have $5b + \mathcal{O}(1)$ machine words stored: array A and set B.

---

[6] If $i = 1$ then the group member id's are $L[1], \ldots, L[C[i]]$.

We now analyze the space used in the second phase (Algorithm 2); in particular, the space taken by the search stack. The stack stores at most every group and every suffix. However, the stack never simultaneously stores a member and one of its ancestors, meaning the maximum size of the stack at any point is at most the maximum width of the sparse suffix tree, which is b. Every element in the stack consists of two integers, so the stack takes up at most $2b$ machine words. No other machine words need to be stored as the maximum stack size b is known in advance and so the stack is implemented using an array.

Adding this together gives at most $s + 7b + o(b)$ machine words in total. $\square$

4   A Simple Parameterized Algorithm

Motivation. Let us start by motivating the parameterization. In real-world datasets, the b suffixes in $T_\mathcal{B}$ will generally not share very long prefixes. Even when they do, it is highly unlikely that all of them have this property. While MAIN-ALGO is theoretically efficient, it would waste a lot of time with such datasets by considering large overlaps between suffixes when in reality the longest common prefixes are much shorter or when only very few suffixes share very long prefixes. Below, we show a simple method to take advantage of this, by only considering short common prefixes in the beginning and then extending them *only* for the suffixes that happen to share longer prefixes. By considering an extra parameter b' indicating the number of suffixes that share longest common prefixes longer than a certain threshold, we arrive at a time and space complexity that appears favorable for such real-world datasets.

Main Idea. We design an algorithm for constructing SSA and SLCP which is parameterized by the total number $b' \leq b$ of suffixes which have an LCP value of at least $\ell = 2^{\lfloor \log \frac{n}{b} \rfloor + 1} - 1$ with some other suffix. We show that partitioning the b suffixes into two classes (one with suffixes with LCP value strictly less than ℓ; and another with suffixes with LCP value greater than or equal to ℓ) can be done in $\mathcal{O}(n)$ time. In particular, we show that it suffices to invoke Theorem 1 twice: once (with a small change) for the b suffixes; and once (as is) for the b' suffixes; and then merge the partial results in $\mathcal{O}(b)$ time to obtain the final SSA and SLCP array.

Description and Pseudocode. The pseudocode is given as PARAMETERIZED-ALGO (Algorithm 3); it is complete in the sense that it only assumes the implementation of MAIN-ALGO. A line-by-line explanation of the algorithm follows.

PARAMETERIZED-ALGO invokes the original algorithm MAIN-ALGO twice with different arguments. In Line 1, it calls MAIN-ALGO with the full array A as argument (and $s = b$). We set the parameter j_{start} that indicates the starting value of j (Line 5 of Algorithm 1) to $\lfloor \log \frac{n}{b} \rfloor$, meaning that j starts at a lower value than the value $\lfloor \log n \rfloor$ used in the MAIN-ALGO and so it will take less time to complete. The result of this is that SSA will only be sorted up to $\ell = 2^{\lfloor \log \frac{n}{b} \rfloor + 1} - 1$ positions. This means that for every consecutive pair of suffixes in SSA, if their LCP value is less than ℓ, they will already be sorted correctly, whereas the other suffixes, with associated LCP values of ℓ, will need to be further sorted in the second phase (Lines 8 to 13) of PARAMETERIZED-ALGO.

Algorithm 3. PARAMETERIZED-ALGO

Input: string $T \in \Sigma^n$, integer b, and array A of b integers

Output: SSA and SLCP

1: SSA, SLCP $\leftarrow$ MAIN-ALGO$(T, A, b, j_{\text{start}} = \lfloor \log \frac{n}{b} \rfloor)$
2: $\ell \leftarrow 2^{\lfloor \log \frac{n}{b} \rfloor + 1} - 1$
3: $P, A' \leftarrow$ empty arrays
4: **for** $i = 1, \ldots, b$ **do**
5:    **if** SLCP$[i] = \ell \vee (i < b \wedge$ SLCP$[i + 1] = \ell)$ **then**
6:        P.append(i)
7:        A'.append(SSA$[i]$)
8: **if** $|A'| > 0$ **then**
9:    SSA$'$, SLCP$' \leftarrow$ MAIN-ALGO$(T, A', |A'|)$
10:    **for** $i = 1, \ldots, |A'|$ **do**
11:        SSA$[P[i]] \leftarrow$ SSA$'[i]$
12:        **if** SLCP$[P[i]] = \ell$ **then**
13:            SLCP$[P[i]] \leftarrow$ SLCP$'[i]$
14: **return** SSA and SLCP

What remains is to identify the suffixes that need to be further sorted, sort these suffixes separately from the others, and re-insert them into the output arrays along with the corrected LCP values. We use two arrays A' and P for this purpose: A' contains the suffixes; and P tracks the positions in SSA that these suffixes are taken from, to ensure that they will later be re-inserted at the correct positions. In Line 5, we ensure that the right suffixes are tracked in these arrays, namely those that have an LCP value of ℓ with their predecessor or successor suffix. If any such suffixes are found, we invoke MAIN-ALGO again (Line 9), but with just these suffixes (those in array A') as input, and with the default value of $j_{\text{start}} = \lfloor \log n \rfloor$. This means that the suffixes of A' will now be fully sorted rather than being sorted up to ℓ positions. Then, in Lines 10 and 11, we insert these re-sorted suffixes at the same positions that they were taken from before, but in the corrected order. In Lines 12 and 13, we also copy the associated LCP values, but only at the positions in-between two re-sorted suffixes, as all other LCP values were already correct.

We next state and prove Theorem 2.

Theorem 2. *For any string $T \in \Sigma^n$ and any set T_B of b suffixes of T, PARAMETERIZED-ALGO computes the SSA and SLCP of T_B in $\mathcal{O}(n + (b'n/b) \log b)$ time using $8b + 4b' + o(b)$ machine words, where b' is the total number of i such that SSA$[i] \in \mathcal{B}$ and SLCP$[i] \geq \ell$ or SLCP$[i+1] \geq \ell$, with $\ell = 2^{\lfloor \log \frac{n}{b} \rfloor + 1} - 1$. The output is correct with high probability. When $b' = \mathcal{O}(b/ \log b)$, PARAMETERIZED-ALGO runs in $\mathcal{O}(n)$ time using $8b + o(b)$ machine words.*

Time Complexity. The first phase of the algorithm (Line 1) runs in $\mathcal{O}(\log \frac{n}{b})$ iterations. The longest prefixes whose KR fingerprints are computed have length $\mathcal{O}(\frac{n}{b})$, and there are $\mathcal{O}(b)$ KR fingerprints computed in each iteration. This means that computing the KR fingerprints during the first phase takes $\mathcal{O}(b) \cdot (\mathcal{O}(\frac{n}{b}) + \mathcal{O}(\frac{n}{2b}) + \mathcal{O}(\frac{n}{4b}) + \ldots) = \mathcal{O}(n)$ time. Hashing the fingerprints takes $\mathcal{O}(b \log \frac{n}{b}) = \mathcal{O}(n)$ worst-case time in total with high probability. (Grouping the fingerprints via distribute-and-collect, like the algorithm by I et al. [16], would incur a multiplicative factor of $\log_s n$.) Sorting takes $\mathcal{O}(n)$ time (see Lemma 2). Therefore the entire first phase runs in $\mathcal{O}(n)$ time. The second phase (Lines 8 to 13) computes KR fingerprints of longer prefixes as well and otherwise runs the same as MAIN-ALGO, with the exception that only b' suffixes are now sorted. By Lemma 2, for $s = b$, this takes $\mathcal{O}(n + (b'n/b) \log b)$ time. All other operations run in single loops over arrays of size b or b' with constant-time operations, and thus take $\mathcal{O}(b)$ time. Adding everything together gives $\mathcal{O}(n + (b'n/b) \log b)$ time. When $b' = \mathcal{O}(b/\log b)$, the running time becomes $\mathcal{O}(n)$.

Space Complexity. The first phase of the algorithm uses $s + 7b + o(b)$ machine words (Lemma 3). The additional arrays P, A', SSA$'$ and SLCP$'$ use $4b'$ machine words in total. The second invocation of MAIN-ALGO uses $s + 7b' + o(b')$ machine words (Lemma 3). By setting $s = b$, the algorithm uses $8b + 4b' + o(b)$ machine words in total. If $b' = \mathcal{O}(b/\log b)$, the algorithm uses $8b + o(b)$ machine words.

Correctness. We prove the correctness of SSA by Lemma 6 and that of SLCP by Lemma 7. To prove these lemmas, we first show the auxiliary Lemmas 4 and 5.

Lemma 4. *Let* SSA_1 *be the instance of* SSA *after the first invocation of* MAIN-ALGO *(Line 1). The strings* $T[SSA_1[i] \mathinner{..} n]$, $i \in [1, b]$, *are sorted up to their prefix of length* $\ell = 2^{\lfloor \log \frac{n}{b} \rfloor + 1} - 1$.

Proof. In MAIN-ALGO, all LCP values can be increased by powers of two in each iteration. With the starting value $j_{\text{start}} = \lfloor \log \frac{n}{b} \rfloor$, this adds up to a maximum LCP value of ℓ in any group. At any point during MAIN-ALGO, two suffixes that are in the same group with LCP value k share a longest common prefix of length at least k. Thus, this invocation of MAIN-ALGO will compute the LCP values between suffixes correctly if they are at most ℓ, and all other LCP values will be ℓ. The sorting step takes into account only the letter which appears after the computed (longest) common prefix, so if the LCP between any two suffixes is less than ℓ the suffixes are sorted correctly. □

Lemma 5. *Let* SSA_1 *be the instance of* SSA *after the first invocation of* MAIN-ALGO *(Line 1), and let* SSA_2 *be the instance of* SSA *returned at the end of* PARAMETERIZED-ALGO *(Line 14). For every* $i \in [1, b]$, *either* $SSA_1[i] = SSA_2[i]$ *and* $SSA_1[i]$ *and* $SSA_2[i]$ *have a longest common prefix of length* $n - SSA_1[i] + 1$, *or* $SSA_1[i] \neq SSA_2[i]$ *and* $SSA_1[i]$ *and* $SSA_2[i]$ *have a longest common prefix of length at least* ℓ.

Proof. If $SSA_1[i] = SSA_2[i]$, this is trivial, so we only concern ourselves with the case $SSA_1[i] \neq SSA_2[i]$. In this case, the value was overwritten in Line 11, meaning that the suffix $SSA_1[i]$ was stored in A' in Line 7 to be re-sorted in the second invocation of MAIN-ALGO. The same must hold for $SSA_2[i]$.

Consider the array A' as it is built in Lines 4–7. By Lemma 4, the suffixes of SSA_1 are sorted up to their length-ℓ prefix; since the entries of A' appear in the same order as they appear in SSA_1, this must also be the case for A'. Because the suffixes of A' are already sorted correctly up to their length-ℓ prefix, it must be that for every position $j \in [1, b']$, $A'[j]$ and $\mathsf{SSA}'[j]$ have the same length-ℓ prefix. Now note that if $\mathsf{SSA}_1[i]$ appears in position j in A', then $\mathsf{SSA}_2[i]$ will take the value from $\mathsf{SSA}'[j]$. Since $A'[j]$ and $\mathsf{SSA}'[j]$ have a length-ℓ common prefix, $\mathsf{SSA}_1[i]$ and $\mathsf{SSA}_2[i]$ must as well.   □

Lemma 6. *The instance SSA_2 of SSA returned at the end of* Parameterized-Algo *(Line 14), contains the suffixes of A sorted lexicographically.*

Proof. We prove this by showing that for any two consecutive positions i and $i+1$, $\mathsf{SSA}_2[i]$ and $\mathsf{SSA}_2[i+1]$ appear in the right order. Let SSA_1 be the instance of SSA after the first invocation of Main-Algo.

We already know that SSA_1 is sorted correctly up to ℓ positions. This means that for any i, if the longest common prefix of $\mathsf{SSA}_1[i]$ and $\mathsf{SSA}_1[i+1]$ is shorter than ℓ, they already appear in the correct order in this array. If neither suffix is overwritten after the second phase, this is also trivially the case for them in SSA_2. Now suppose that exactly one of the two (wlog $\mathsf{SSA}_1[i+1]$) is replaced by some other suffix s while the other remains the same. Let k be the LCP of $\mathsf{SSA}_1[i]$ and $\mathsf{SSA}_1[i+1]$. By Lemma 5, $\mathsf{SSA}_1[i+1]$ and s have a longest common prefix of length at least ℓ. This is longer than k, which is strictly less than ℓ. This means that the $(k+1)$-th letter of s is the same as that of $\mathsf{SSA}_1[i+1]$, which is the first position in which it differs from $\mathsf{SSA}_1[i]$. Thus $\mathsf{SSA}_2[i] = \mathsf{SSA}_1[i]$ and $\mathsf{SSA}_2[i+1] = s$ are sorted correctly relative to one another.

The remaining case is when $\mathsf{SSA}_1[i]$ and $\mathsf{SSA}_1[i+1]$ have a longest common prefix of length ℓ or longer. In this case, both suffixes are added to A' to be re-sorted in the second invocation, and both $\mathsf{SSA}_2[i]$ and $\mathsf{SSA}_2[i+1]$ may take the value of another suffix. The second invocation of the main algorithm sorts all suffixes in A' completely, returning SSA'. The suffixes in SSA' are then re-inserted into SSA_2, in which they will appear in the same order as they did in SSA'. Therefore, no matter which suffixes end up at $\mathsf{SSA}_2[i]$ and $\mathsf{SSA}_2[i+1]$, they also appeared consecutively in SSA' and therefore must be sorted correctly.   □

Lemma 7. *For any two consecutive positions i and $i+1$, $\mathsf{SLCP}[i+1]$, as returned by Algorithm 3, gives the length of the longest common prefix of $\mathsf{SSA}[i]$ and $\mathsf{SSA}[i+1]$.*

Proof. Let SSA_1 and SLCP_1 be the arrays returned by the first invocation of Main-Algo, and SSA_2 and SLCP_2 the arrays produced at the end. By Lemma 4, if $\mathsf{SLCP}_1[i+1] < \ell$, this value is correct. Therefore, the only values that need to be overwritten for SLCP_2 are when $\mathsf{SLCP}_1[i+1] = \ell$. The check at Line 12 ensures this. Of course, when $\mathsf{SLCP}_1[i+1] = \ell$, then both $\mathsf{SSA}_1[i]$ and $\mathsf{SSA}_1[i+1]$ are added to A' in order to be re-sorted in the second invocation. The values at $\mathsf{SSA}_2[i]$ and $\mathsf{SSA}_2[i+1]$ are then replaced by two suffixes that appear consecutively in SSA', say $\mathsf{SSA}'[j]$ and $\mathsf{SSA}'[j+1]$. By the correctness of Main-Algo, the LCP value of these two suffixes is given by $\mathsf{SLCP}'[j+1]$, which is the value that $\mathsf{SLCP}_2[j+1]$ takes.   □

Random Strings. Finally, we show that PARAMETERIZED-ALGO can be trivially amended to work in $\mathcal{O}(n)$ time for any string chosen uniformly at random from Σ^n. In particular, we show the following result.

Theorem 3. *For any string T chosen uniformly at random from Σ^n and any set $T_\mathcal{B}$ of b suffixes of T, SSA and SLCP of $T_\mathcal{B}$ can be computed in $\mathcal{O}(n)$ time using $\mathcal{O}(b)$ space. The output is correct with high probability.*

Proof. We assume $|\Sigma| \geq 2$, otherwise the problem has a trivial solution. Bollobás and Letzter [8, Theorem 4] showed that the maximum length of an LCE on T is at most $2\log_{|\Sigma|} n + \log_{|\Sigma|}\log_{|\Sigma|} n$ with high probability. We bound this from above by $3\log n$ and amend PARAMETERIZED-ALGO as follows:

Case (a): $b\log n < n$. We invoke MAIN-ALGO by setting j_{start} to the smallest integer such that $2^{j_{\text{start}}} \geq 2\lfloor\log n\rfloor$, which gives $\ell = 2\lfloor\log n\rfloor \cdot 2 - 1 = 4\lfloor\log n\rfloor - 1$. After the $\mathcal{O}(n)$-time preprocessing of Lemma 1, computing the KR fingerprints takes $\mathcal{O}(b)\cdot 4(\mathcal{O}(\frac{\log n}{1})+\mathcal{O}(\frac{\log n}{2})+\mathcal{O}(\frac{\log n}{4})+\ldots) = \mathcal{O}(b\log n)$ time. Hashing the fingerprints takes $\mathcal{O}(b)$ time per iteration with high probability, and so $\mathcal{O}(b\log n)$ total time. Merge sort takes $\mathcal{O}(b\log b)$ time. Since $\ell > 3\log n$, all suffixes of $T_\mathcal{B}$ will be fully sorted from the first invocation of MAIN-ALGO. If $b' = \mathcal{O}(b/\log b)$ suffixes are still unsorted after the first invocation, these will be fully sorted in the second invocation of MAIN-ALGO in $\mathcal{O}(n)$ time (Theorem 2). If $b' = \omega(b/\log b)$, we output incorrect arrays. The total time complexity is thus $\mathcal{O}(n + b\log n) = \mathcal{O}(n)$. The total space used is the space used by MAIN-ALGO, which is $\mathcal{O}(b)$.

Case (b): $b\log n \geq n$. Assume that we have $\mathcal{O}(s)$ space to sort the b suffixes; we can do it efficiently using radix sort because it suffices to sort all prefixes of them of length $\mathcal{O}(\log_\sigma n)$ by the Bollobas and Letzter's result, where $\sigma = |\Sigma|$ (otherwise, we output incorrect arrays). The b prefixes are each of length at most $c\log_\sigma n$, for some $c = \mathcal{O}(1)$; so radix sort takes $\mathcal{O}((b + s)(c \cdot \log n/\log \sigma) \cdot (\log \sigma/\log s))$ time, because we have at most $(c\log n/\log \sigma)$ letters in every prefix, and each time we sort b letters, one from each prefix, we use $(\log \sigma/\log s)$ rounds of counting sort. Conveniently, the $\log \sigma$ terms cancel out. Then, because we set $s = b$, and by the fact that we are in the case $b \geq n/\log n$, we have that $\log n/\log s = \mathcal{O}(1)$. The total time complexity is thus $\mathcal{O}(b+s) = \mathcal{O}(b)$. The total space used is $\mathcal{O}(s) = \mathcal{O}(b)$. By comparing adjacent suffixes we compute the SLCP array within the same complexities.

$\square$

References

1. Arbitman, Y., Naor, M., Segev, G.: Backyard cuckoo hashing: Constant worst-case operations with a succinct representation. In: FOCS, pp. 787–796 (2010)
2. Ayad, L.A.K., Loukides, G., Pissis, S.P.: Text indexing for long patterns: anchors are all you need. Proc. VLDB Endow. **16**(9), 2117–2131 (2023)

3. Ben-Nun, S., Golan, S., Kociumaka, T., Kraus, M.: Time-space tradeoffs for finding a long common substring. In: CPM. LIPIcs, vol. 161, pp. 5:1–5:14 (2020)
4. Bender, M.A., Conway, A., Farach-Colton, M., Kuszmaul, W., Tagliavini, G.: Iceberg hashing: optimizing many hash-table criteria at once. J. ACM 70(6) (2023)
5. Bernardini, G., Fici, G., Gawrychowski, P., Pissis, S.P.: Substring complexity in sublinear space. In: ISAAC. LIPIcs, vol. 283, pp. 12:1–12:19 (2023)
6. Bille, P., Fischer, J., Gørtz, I.L., Kopelowitz, T., Sach, B., Vildhøj, H.W.: Sparse text indexing in small space. ACM Trans. Algorithms 12(3), 39:1–39:19 (2016)
7. Birenzwige, O., Golan, S., Porat, E.: Locally consistent parsing for text indexing in small space. In: SODA, pp. 607–626 (2020)
8. Bollobás, B., Letzter, S.: Longest common extension. Eur. J. Comb. 68, 242–248 (2018)
9. Chan, T.M., Munro, J.I., Raman, V.: Selection and sorting in the "restore" model. ACM Trans. Algorithms 14(2), 11:1–11:18 (2018)
10. Christiansen, A.R., Ettienne, M.B., Kociumaka, T., Navarro, G., Prezza, N.: Optimal-time dictionary-compressed indexes. ACM Trans. Algorithms 17(1), 8:1–8:39 (2021)
11. Dietzfelbinger, M., Gil, J., Matias, Y., Pippenger, N.: Polynomial hash functions are reliable. In: Kuich, W. (ed.) ICALP 1992. LNCS, vol. 623, pp. 235–246. Springer, Heidelberg (1992). https://doi.org/10.1007/3-540-55719-9_77
12. Fischer, J., Tomohiro, I., Köppl, D.: Deterministic sparse suffix sorting in the restore model. ACM Trans. Algorithms 16(4), 50:1–50:53 (2020)
13. Franceschini, G., Muthukrishnan, S., Pătraşcu, M.: Radix sorting with no extra space. In: Arge, L., Hoffmann, M., Welzl, E. (eds.) ESA 2007. LNCS, vol. 4698, pp. 194–205. Springer, Heidelberg (2007). https://doi.org/10.1007/978-3-540-75520-3_19
14. Gawrychowski, P., Kociumaka, T.: Sparse suffix tree construction in optimal time and space. In: SODA, pp. 425–439 (2017)
15. Grabowski, S., Raniszewski, M.: Sampled suffix array with minimizers. Softw. Pract. Exp. 47(11), 1755–1771 (2017)
16. Tomohiro, I., Kärkkäinen, J., Kempa, D.: Faster sparse suffix sorting. In: STACS. LIPIcs, vol. 25, pp. 386–396 (2014)
17. Kärkkäinen, J., Sanders, P., Burkhardt, S.: Linear work suffix array construction. J. ACM 53(6), 918–936 (2006)
18. Kärkkäinen, J., Ukkonen, E.: Sparse suffix trees. In: Cai, J.-Y., Wong, C.K. (eds.) COCOON 1996. LNCS, vol. 1090, pp. 219–230. Springer, Heidelberg (1996). https://doi.org/10.1007/3-540-61332-3_155
19. Karp, R.M., Rabin, M.O.: Efficient randomized pattern-matching algorithms. IBM J. Res. Dev. 31(2), 249–260 (1987)
20. Kasai, T., Lee, G., Arimura, H., Arikawa, S., Park, K.: Linear-time longest-common-prefix computation in suffix arrays and its applications. In: Amir, A. (ed.) CPM 2001. LNCS, vol. 2089, pp. 181–192. Springer, Heidelberg (2001). https://doi.org/10.1007/3-540-48194-X_17
21. Katajainen, J., Pasanen, T., Teuhola, J.: Practical in-place mergesort. Nord. J. Comput. 3(1), 27–40 (1996)
22. Loukides, G., Pissis, S.P.: Bidirectional string anchors: a new string sampling mechanism. In: ESA. LIPIcs, vol. 204, pp. 64:1–64:21 (2021)
23. Loukides, G., Pissis, S.P., Sweering, M.: Bidirectional string anchors for improved text indexing and top-K similarity search. IEEE Trans. Knowl. Data Eng. 35(11), 11093–11111 (2023)

24. Navarro, G., Prezza, N.: Universal compressed text indexing. Theor. Comput. Sci. **762**, 41–50 (2019)
25. Paige, R., Tarjan, R.E.: Three partition refinement algorithms. SIAM J. Comput. **16**(6), 973–989 (1987)
26. Prezza, N.: Optimal substring equality queries with applications to sparse text indexing. ACM Trans. Algorithms **17**(1), 7:1–7:23 (2021)
27. Salowe, J.S., Steiger, W.L.: Simplified stable merging tasks. J. Algorithms **8**(4), 557–571 (1987)

Wheeler Maps

Andrej Baláž[1], Travis Gagie[2], Adrián Goga[1(✉)], Simon Heumos[3,4,5],
Gonzalo Navarro[6], Alessia Petescia[1], and Jouni Sirén[7]

[1] Comenius University in Bratislava, Bratislava, Slovakia
adrian.goga@fmph.uniba.sk
[2] CeBiB & Dalhousie University, Nova Scotia, Canada
[3] Quantitative Biology Center (QBiC), University of Tübingen, Tübingen, Germany
[4] Department of Computer Science, University of Tübingen, Tübingen, Germany
[5] M3 Research Center, University Hospital Tübingen, Tübingen, Germany
[6] CeBiB & DCC, University of Chile, Santiago, Chile
[7] University of California Santa Cruz Genomics Institute, Santa Cruz, USA

Abstract. Motivated by challenges in pangenomic read alignment, we
propose a generalization of Wheeler graphs that we call Wheeler maps.
A Wheeler map stores a text $T[1..n]$ and an assignment of tags to the
characters of T such that we can preprocess a pattern $P[1..m]$ and then,
given i and j, quickly return all the distinct tags labeling the first characters of the occurrences of $P[i..j]$ in T. For the applications that most
interest us, characters with long common contexts are likely to have the
same tag, so we consider the number t of runs in the list of tags sorted by
their characters' positions in the Burrows-Wheeler Transform (BWT) of
T. We show how, given a straight-line program with g rules for T, we can
build an $O(g + r + t)$-space Wheeler map, where r is the number of runs
in the BWT of T, with which we can preprocess a pattern $P[1..m]$ in
$O(m \log n)$ time and then return the k distinct tags for $P[i..j]$ in optimal
$O(k)$ time for any given i and j.

1 Introduction

For years, geneticists have been worried about the fact that using a single reference for the human genomes biases scientific studies and medical diagnoses,
undermining the potential of personalized medicine, particularly for people from

Research funded in part by European Union's Horizon 2020 research and innovation
program under Marie Skłodowska-Curie grant agreement No 956229 (ALPACA) and
by grants 1/0463/20 and 1/0538/22 from the Scientific Grant Agency of the Ministry of
Education, Science, Research, and Sport of the Slovak Republic and Slovak Academy of
Sciences (VEGA) and grant APVV-22-0143 from the Slovak Research and Development
Agency. T.G. and G.N. funded in part by Basal Funds FB0001, ANID, Chile. T.G.
funded in part by NSERC RGPIN-07185-2020. S.H. funded in part by the Central
Innovation Programme (ZIM) for SMEs of the Federal Ministry for Economic Affairs
and Energy of Germany. J.S. funded in part by National Human Genome Research
Institute (NHGRI) award R01HG010485.

J. A. Soto and A. Wiese (Eds.): LATIN 2024, LNCS 14578, pp. 178–192, 2024.
https://doi.org/10.1007/978-3-031-55598-5_12

under-represented groups. To address this bias, researchers [18] recently published a pangenome consisting of nearly complete genomes from 47 people from diverse origins and took, according to the *New York Times* [8], "a major step toward a deeper understanding of human biology and personalized medicine for people from a wide range of racial and ethnic backgrounds". Eventually, the plan is to include 350 genomes, but even this many genomes cannot fully capture humanity's genetic diversity. As the *Guardian* [34] put it, "as long as the reference contains only a subset, arguably someone will not make the cut". Ultimately, there will be pressure for a reference of at least thousands of genomes.

One of the primary use of a reference is during read alignment. As a DNA sample passes through a sequencing machine, the machine records the genome in short substrings called *reads*. The length and accuracy of the reads vary depending on the sequencing technology used. Next, software called a read aligner uses an index of a reference to find *seeds*, sections of the reads that exactly match sections in the reference, and uses dynamic programming to extend those seeds to approximate matches of the whole read. These approximate matches form alignments, which are used in many subsequent bioinformatics analyses.

Indexing 47 human genomes is feasible even with standard read aligners such as Bowtie [16] and BWA [17], and even indexing 350 may be possible on supercomputers, but indexing thousands will require new algorithmic insights. The emerging consensus is that we should represent the combined reference sequences as a *pangenome graph* [6] that shows variation between genomes as detours on an otherwise shared path. The necessity of mapping reads to the version of the path that best fits the sample leads to the question of how to index pangenome graphs.

Equi et al. [9] showed that, unless the strong exponential-time hypothesis is false, one cannot index a graph in polynomial time such that pattern matching can run in sub-quadratic time, so several groups have tried constraining pangenome graphs to have a particular structure, such as Wheeler graphs [12], p-sortable graphs [7], elastic degenerate strings [1] or founder block graphs [19]. Unfortunately, merging reference sequences into a graph hides certain variations' tendencies to co-occur, known as *linkage disequilibrium* [30], and creates *chimeric* paths whose labels are not in any of the original sequences. Indexing and using such a graph can result in false-positive matches to these chimeric paØLØths.

The more variations are represented, the noisier the graph becomes and the more possibilities there are for spurious matches. The number of false positives can be reduced by excluding rare variations, but sacrificing inclusivity for the sake of computational convenience goes against the spirit of pangenomics, and the pressure to include more genomes will probably force bioinformaticians to index all the variations. Moreover, excluding variations could be viewed as trading false positives for false negatives. Another approach is to filter out false positives by checking matches against the reference sequences represented as strings, but then the overall query time cannot be bounded in terms of the patterns and the true matches reported. Furthermore, the number of false positives will likely grow as the pangenome does.

Some researchers have eschewed using a pangenome graph altogether and indexed the genomes in the pangenome as a set of strings. This approach allowed them to draw on a rich history of indexing compressible texts: the Burrows-Wheeler Transform [4] (BWT) and FM-indexes [10], for which Burrows, Ferragina and Manzini recently shared the Paris Kanellakis Award and which underpin Bowtie and BWA; RLCSA [20]; the r-index [13], subsampled r-index [5] and r-index-f [29]. Recently, Rossi et al. [32] and Boucher et al. [3] showed how, given a straight-line program with g rules for a text $T[1..n]$, they can build an $O(g+r)$ space index, where r is the number of runs in the BWT of T, with which they can find the maximal exact matches (MEMs) of a given pattern $P[1..m]$ with respect to T in $O(m \log n)$ time and list the occurrences of each MEM in constant time per occurrence. This result means they can index the pangenome compactly with no chance of false positives, find good seeds reasonably quickly, and list the occurrences of those seeds in constant time per occurrence.

The main practical problem with those results is that if there are thousands of genomes in the pangenome, then a MEM can occur thousands of times in those genomes, even if all those occurrences map to only one place in the standard single reference genome. This observation makes extending the seeds and combining the approximate matches of the reads much slower. In this paper, we show how we can combine Rossi et al.'s result with a pangenome graph such that we can still find seeds quickly, with no chance of false positives, but then report their non-chimeric occurrences in the graph in constant time per occurrence. Moreover, we put no constraints on the graph.

A set of genomes can be annotated so that for each character in the genomes, we know at which vertex in a pangenome graph that character occurs. Then, our idea is that if someone gives us a set of genomes and the corresponding annotation, we can store them in a small space so we can later quickly report for each seed its starting positions in the graph. The seeds of a read with respect to the set of genomes can be MEMs, but also f-MEMs [25,35] (maximal substrings that occur at least f times in the genomes) or other kinds of substrings.

We can formalize this problem as follows: we want to store a text $T[1..n]$ and an assignment of tags to the characters of T such that we can preprocess a pattern $P[1..m]$ and then, given i and j, quickly return all the distinct tags labeling the first characters of the occurrences of $P[i..j]$ in T. In a pangenome, characters with long common contexts are more likely to have the same tag, so we consider the number of runs t in the list of tags sorted by their characters' positions in the Burrows-Wheeler Transform (BWT) of T.

Our Contribution. In this paper, we show how, given a straight-line program with g rules for T, we can build an $O(g+r+t)$ space data structure, where r is the number of runs in the BWT of T, with which we can preprocess a pattern P in $O(m \log n)$ time and then return the k distinct tags for $P[i..j]$ in the optimal $O(k)$ time for any given i and j.

We call our data structure a *Wheeler map* since it resembles a Wheeler graph [12] but with less structure. One reason Wheeler graphs were introduced was to provide a model for alignment with a pangenome graph: we start with a string dataset, build a graphical representation, and index that graph; the graphical representation is inherently lossy but, to filter out chimeric matches, we can verify matches against the original dataset. (Even before Wheeler graphs were defined, software for indexing variation graphs [15] used a procedure for making them Wheeler or almost Wheeler, falling back on unwinding the graph and indexing substrings when that procedure failed.) Our idea is to reverse that approach of indexing a graph and then filtering out false positives using the strings. Instead, we index the strings, and then map occurrences onto a graph — but without considering all the occurrences in the strings.

Some researchers (see, e.g., [31] and references therein) argue that having a graph index return matches not found in the original strings is a feature, not a bug, since it allows the index to find matches that can be obtained by recombination. As computer scientists, however, it is not our place to decide what combination of alleles are reasonable and which not, and so we should offer the option of indexing the datasets we are given and nothing else. Indexing the strings means we index all the variations they contain, so we can presumably capture most reasonable combinations by increasing the number of genomes in our dataset. Scaling to larger datasets is thus a solution for us, whereas it is a problem for graphical indexes, which tend to produce more false positives when they include all the variations in large datasets.

From Rossi et al. [32], we know r and g are reasonably small for the datasets in which we are most interested. To check that t is comparable, we computed it for the chromosome-19 component in a Minigraph-Cactus graph based on 90 human haplotypes from the Human Pangenome Reference Consortium [18]. This component was built from 1100 contigs with total length $n = 5{,}070{,}072{,}154$ and t was $208{,}649{,}680$, almost 25 times smaller than n. For comparison, r was $71{,}512{,}609$, just over 70 times smaller than n and not quite 3 times smaller than t.

Roadmap. In Sect. 2 we describe the basic concepts that will be used throughout the rest of the work, together with a preliminary method of computing the tags for the occurrences of a pattern P. In Sect. 3 we show how extended matching statistics can be computed $O(m \log n)$ time without the need for buffering that Rossi et al. [32] used, and extend the method for computing the tag statistics. In Sect. 4 we describe how the tag statistics together with range successor queries on the tag array can be used to get the k distinct tags for the occurrences of $P[i..j]$ in $O(\log^\epsilon t + k)$ time for any $\epsilon > 0$. Using more sophisticated techniques, we improve this time to the optimal $O(k)$ in Sect. 5. We conclude in Sect. 6 with some future work directions.

2   Preliminaries

Our model of computation throughout is the standard word-RAM with $\Theta(\log n)$-bit words. For the sake of brevity, we assume the reader is familiar with suffix arrays (SAs), the Burrows-Wheeler Transform (BWT), FM-indexes, LF-mapping and straight-line programs (SLPs); otherwise, we refer them to appropriate surveys [24,26]. We recall only that $\mathrm{LCP}(S_1, S_2)$ denotes the length of the

#	B	W	tag	BWT	context
1	0		9	T	$AGATACA
2			9	T	$GATACA
3		0	9	T	$GATTACA
4			9	T	$GATTAGA
5	0	0	10	A	$GATTAGAT
6	0	1	9	T	A$GATTAGA
7	1		4	T	ACAT$AGA
8		4	4	T	ACAT$GA
9	4		5	T	ACAT$GAT
10		1	5	T	AGAT$GAT
11			5	T	AGATA$GAT
12	5	8	0	$	AGATACAT
13	1		7	C	AT$AGATA
14			7	C	AT$GATA
15		2	7	C	AT$GATTA
16			7	G	AT$GATTA
17			7	G	ATA$GATTA
18	3		2	G	ATACAT$A
19			2	G	ATACAT$
20	2	2	2	G	ATTACAT$
21			2	G	ATTAGAT$
22			2	G	ATTAGATA$
23	0		6	A	CAT$AGAT
24			6	A	CAT$GAT
25		0	6	A	CAT$GATT
26			6	A	GAT$GATT
27			6	A	GATA$GATT
28	4		1	A	GATACAT$
29			1	$	GATACAT
30	3		1	$	GATTACAT
31			1	$	GATTAGAT
32			1	$	GATTAGATA
33	0		8	A	T$AGATAC
34			8	A	T$GATAC
35	1		8	A	T$GATTAC
36			8	A	T$GATTAG
37			8	A	TA$GATTAG
38	2		3	A	TACAT$AG
39		5	3	A	TACAT$G
40	5		4	T	TACAT$GA
41		2	4	T	TAGAT$GA
42			4	T	TAGATA$GA
43	1		3	A	TTACAT$G
44		3	3	A	TTAGAT$G
45			3	A	TTAGATA$G

(0) —A→ (1) —G→ (2) —A→ (3) —T→ (4) —T→ (5) —A→ (6) —C→ (7) —A→ (8) —T→ (9) —A→ (10), with an A alternative at node (5) and a G alternative below node (7)

Fig. 1. Tables **(top)** for a set of toy genomes GATTACAT$, AGATACAT$, GATACAT$, GATTAGAT$ and GATTAGATA$ and a pangenome graph **(bottom)**. The BWT column shows the characters sorted by their contexts, which are the rest of the genomes (consider to be cyclic) and shown in the right column. The tag array is shown in the column to the left of the BWT, with each entry identifying the source in the pangenome graph of the edge labelled by the first character in the context in the same row (not the BWT character). The leftmost two columns (together called L) show the LCP values for the runs in the tag array, discussed in Sect. 3: column B contains the LCP value *between* each run in the tag array and the preceding run (so the LCP value at the beginning of the run), while column W contains the LCP value *within* each run (the length of the longest prefix common to all the contexts in the run). LCPs extend only up to and not including the terminators $ because they are not searchable. Due to space constraints, the tables are displayed split into three pieces (at the end of runs).

longest common prefix of two strings S_1 and S_2 (which need not be lexicographically consecutive suffixes of a text), and of the bounds for Muthukrishnan's [23] classic document-listing data structure:

Theorem 1 (Muthukrishnan, [23, Thm. 3.1]). *Given an array $A[1..h]$, we can build an $O(h)$-space data structure with which, given i and j, we can return the k distinct elements in $A[i..j]$ in $O(k)$ time.*

In our model, each text suffix $T[i..]$ is labeled with a "tag", which can also be seen as labeling the position i. The tags of T are collected in a so-called "tag array"; see Fig. 1.

Definition 1. *Let $T[1..n]$ be a labeled text, such that the label for the ith position is $T[i]$.lab. The tag array $\mathrm{Tag}[1..n]$ of T is then defined as $\mathrm{Tag}[j] = T[SA[j]]$.lab.*

We say that an occurrence $T[i..i + |P| - 1]$ of a pattern P in T is labeled by the tag that labels $T[i..]$. Consequently, the labels of all the occurrences of P in T are listed in $\mathrm{Tag}[s..e]$, where $SA[s..e]$ is the suffix array interval for P. For example, in Fig. 1 the range for $P = $ A is $SA[6..22]$, and $\mathrm{Tag}[6..22]$ contains the tags 9, 4, 5, 0, 7, and 2. Those are the labels with which P appears in the graph.

For convenience, we first extend the standard definition of matching statistics to include the lexicographic ranks of the suffixes of T starting with the occurrences we consider, and then further extend it to mention the tag array.

Definition 2. *The extended matching statistics of a pattern $P[1..m]$ with respect to a text $T[1..n]$ are an array $\mathrm{XMS}[1..m + 1]$ of (len, pos, rank) triples such that*

- *$\mathrm{XMS}[i]$.len is the length of the longest prefix of $P[i..m]$ that occurs in T,*
- *$\mathrm{XMS}[i]$.pos is the starting position of one occurrence of $P[i..i+\mathrm{XMS}[i].\mathrm{len}-1]$ in T,*
- *$\mathrm{XMS}[i]$.rank is the lexicographic rank of $T[\mathrm{XMS}[i].\mathrm{pos}..n]$ among the suffixes of T.*

We emphasize that we expect the tag array to have long runs of equal consecutive symbols. The following definition considers those runs in the process of matching P in T.

Definition 3. *The tag statistics of a pattern $P[1..m]$ with respect to a text $T[1..n]$ and its tag array $\mathrm{Tag}[1..n]$ are an array $\mathrm{TS}[1..m + 1]$ of (len, pos, rank, run, up, down) sextuples such that $\mathrm{TS}[i]$.len, $\mathrm{TS}[i]$.pos and $\mathrm{TS}[i]$.rank are the same as in the XMS array and*

- *$\mathrm{TS}[i]$.run is the index of the run $\mathrm{Tag}[u..d]$ in the tag array that contains position $\mathrm{TS}[i]$.rank,*
- *$\mathrm{TS}[i]$.up $= \mathrm{LCP}(P[i..m], T[SA[u]..n])$,*
- *$\mathrm{TS}[i]$.down $= \mathrm{LCP}(P[i..m], T[SA[d]..n])$.*

Finally, although we know of no previous work specifically addressing tag arrays, we note a solution that follows directly from the work by Mäkinen et al. [21]:

Theorem 2 (Mäkinen et al., [21, Thm 17.]). *Given a text $T[1..n]$ whose BWT has r runs, we can build an $O(r)$-space data structure called RLBWT such that later, given a pattern $P[1..m]$, we can return the lexicographic range of suffixes of T starting with P in $O(m \log \log n)$ time.*

Corollary 1. *Given a text $T[1..n]$ whose BWT has r runs, and a tag array with t runs, we can build an $O(r + t)$-space data structure such that later, given a pattern $P[1..m]$, we can return the k distinct tags of P's occurrences in T in $O(m \log \log n + k)$ time.*

Proof. We store an $O(r)$-space RLBWT for T, an $O(t)$-space predecessor structure storing where the runs start in Tag, and an $O(t)$-space instance of Muthukrishnan's data structure from Theorem 1 for the array $A[1..t]$ obtained from Tag
by replacing each run by a single copy of the same tag. Given P, we first use the
RLBWT to find the lexicographic range $SA[s..e]$ of suffixes of T starting with P,
in $O(m \log \log n)$ time. We then use predecessor queries to find the range $A[s'..e']$
of the tag run indices overlapping $Tag[s..e]$, in $O(\log \log n)$ time. Finally, we use
Muthukrishnan's data structure to report the distinct tags in $A[s'..e']$, in $O(k)$
time. $\square$

Our main concern with Corollary 1 is that if we want the distinct tags for
a set of substrings of P that can overlap—such as the maximal exact matches
(MEMs) of P with respect to T—and we apply this corollary to each one, then
we can use $\Omega(m^2)$ total time even when the number of tags we return is small.
Our plan is then to preprocess P in a first stage, so that in a second stage we
can more quickly answer (many) questions about substrings of the form $P[i..j]$.

3   Computing Tag Statistics

We rely on results about straight-line programs (SLPs), which we can encapsu-
late in the following lemma.

Lemma 1. *Given an SLP with g rules for $T[1..n]$, in $O(n \log n)$ expected time
we can build an $O(g)$-space data structure with which we can preprocess any
pattern $P[1..m]$ in $O(m)$ time such that later, given i, j and q, we can return
$LCP(P[i..j], T[q..n])$ in $O(\log n)$ time and with no chance of error as long as
$P[i..j]$ occurs somewhere in T.*

Proof. Bille et al. [2] showed how to build, in $O(n \log n)$ expected time, a Karp-
Rabin hash function with no collisions between substrings of T. If $S = S' \cdot S''$
and we have the hashes of two of those strings, we can compute the hash of the
third in constant time, as soon as we store some precomputed values that can
also be maintained in constant time (see, e.g., [27]).
    If necessary, we use Ganardi et al.'s [14] construction to balance the SLP such
that it has $O(g)$ rules and height $O(\log n)$. We then label each symbol x in the
SLP with the length and hash of x's expansion. This takes $O(g)$ time because
we compute in constant time the hash of the left-hand side of a rule from those
of the right-hand side.
    When P arrives, we compute the hashes of its suffixes in $O(m)$ total time.
The hash of any $P[i..j]$ can then be computed in constant time from the hashes
of $P[i..m]$ and $P[j+1..m]$.
    Given i, j and q, we descend to the qth leaf of the parse tree in $O(\log n)$ time.
We then re-ascend toward the root in $O(\log n)$ time, keeping track of the length
and hash of $T[q..e]$, where e is the index of the rightmost leaf in the subtree of
the node we are currently visiting.

When we reach a node such that $T[q..e]$ is either longer than $P[i..j]$ or the hash of $T[q..e]$ does not match the hash of the corresponding prefix of $P[i..j]$, we re-descend in $O(\log n)$ time. At each step in the re-descent, we go left if $T[q..e]$ is either longer than $P[i..j]$ or the hash of $T[q..e]$ does not match the hash of the corresponding prefix of $P[i..j]$, where e is now the index of the rightmost leaf in the subtree of the left child. Otherwise, we go right.

We then find $\mathrm{LCP}(P[i..j], T[q..n])$ in $O(\log n)$ time. As long as $P[i..j]$ occurs somewhere in T, no hash of a prefix of $P[i..j]$ collides with the hash of a different substring of T, so we have no chance of error. □

We now show how to preprocess the tag array. Let $U[1..t]$ and $D[1..t]$ be the arrays such that $U[q]$ and $D[q]$ are the indices of the first and last tags, respectively, in the qth run in the tag array. Let $W[1..t]$ be the array with

$$W[q] = \min_{U[q]+1 \leq p \leq D[q]} \{\mathrm{LCP}(T[\mathrm{SA}[p-1]..n], T[\mathrm{SA}[p]..n])\}$$
$$= \mathrm{LCP}(T[\mathrm{SA}[U[q]]..n], T[\mathrm{SA}[D[q]]..n])$$

for $1 \leq q \leq t$, and let $B[1..t-1]$ be the array with

$$B[q] = \mathrm{LCP}(T[\mathrm{SA}[D[q]]..n], T[\mathrm{SA}[U[q+1]]..n])$$

for $1 \leq q \leq t-1$—so $W[q]$ is the LCP computed *within* run q and $B[q]$ is the LCP computed *between* runs q and $q+1$. Finally, let

$$L[0..2t] = 0, W[1], B[1], W[2], B[2], \ldots, W[t-1], B[t-1], W[t], 0.$$

From now on we will only use L, and we will not refer to U, D, W or B again. We recall that Fig. 1 shows the L array on an example text.

We now describe our preprocessing of the pattern. Our results in this section can be viewed as mainly extending Rossi et al.'s [32] work on computing (extended) matching statistics to computing tag statistics:

Theorem 3 (cf. [32]). *Given an SLP with g rules for a text $T[1..n]$ whose BWT has r runs, we can build an $O(g+r)$-space data structure such that later, given a pattern $P[1..m]$, we can compute the extended matching statistics XMS of P with respect to T in $O(m \log n)$ time.*

Proof. We apply Lemma 1 to the SLP to obtain an $O(g)$-space LCP data structure with $O(\log n)$ query time. We also store $\mathrm{SA}[u]$ and $\mathrm{SA}[d]$, for each run $\mathrm{BWT}[u..d]$, in an $O(r)$-space data structure supporting predecessor and successor queries on the keys u and d. Finally, we use the $O(r)$-space RLBWT of Theorem 2, which can also compute any $\mathrm{BWT}[j]$ and $\mathrm{LF}[j]$. These functions and the predecessor queries can run in $O(\log \log n)$ time, but $O(\log n)$ time is enough for our purposes.

As usual, for technical convenience we add to T a special symbol $T[n+1] = \$$ that is lexicographically smaller than all the other symbols in T (and in potential patterns P). This implies $\mathrm{BWT}[1] = \$$. For a start, then, considering

$P[m + 1..m] = \epsilon$, we set $\text{XMS}[m + 1].\text{len} = 0$, $\text{XMS}[m + 1].\text{rank} = 1$ and $\text{XMS}[m + 1].\text{pos} = n + 1$.

Now, suppose we have already computed the suffix $\text{XMS}[i + 1..m + 1]$ of the extended matching statistics and want to compute $\text{XMS}[i]$. If $\text{BWT}[\text{XMS}[i + 1].\text{rank}] = P[i]$ then

$$\text{XMS}[i].\text{len} = \text{XMS}[i + 1].\text{len} + 1,$$
$$\text{XMS}[i].\text{pos} = \text{XMS}[i + 1].\text{pos} - 1,$$
$$\text{XMS}[i].\text{rank} = \text{LF}[\text{XMS}[i + 1].\text{rank}].$$

Otherwise, let $\text{BWT}[u]$ and $\text{BWT}[d]$ be the occurrences of $P[i]$ immediately preceding and following $\text{BWT}[\text{XMS}[i + 1].\text{rank}]$. We find u and d with predecessor/successor queries.

By the definition of the BWT, at least one of $T[\text{SA}[u]..n]$ and $T[\text{SA}[d]..n]$ has the longest common prefix with $P[i + 1..m]$ of any suffix of T preceded by a copy of $P[i]$. Since $\text{BWT}[u]$ is the last character in a run and $\text{BWT}[d]$ is the first character in a run, we have $\text{SA}[u]$ and $\text{SA}[d]$ stored. Therefore, we can compute

$$\ell_u = \text{LCP}(P[i + 1..i + \text{XMS}[i + 1].\text{len} - 1], T[\text{SA}[u]..n]),$$
$$\ell_d = \text{LCP}(P[i + 1..i + \text{XMS}[i + 1].\text{len} - 1], T[\text{SA}[d]..n]),$$

in $O(\log n)$ time, since $P[i+1..i+\text{XMS}[i+1].\text{len} - 1]$ occurs in T, with no chance of error.

If $\ell_u \geq \ell_d$ then

$$\text{XMS}[i].\text{len} = \ell_u + 1,$$
$$\text{XMS}[i].\text{pos} = \text{SA}[u] - 1,$$
$$\text{XMS}[i].\text{rank} = \text{LF}[u],$$

and, symmetrically, if $\ell_u < \ell_d$ then

$$\text{XMS}[i].\text{len} = \ell_d + 1,$$
$$\text{XMS}[i].\text{pos} = \text{SA}[d] - 1,$$
$$\text{XMS}[i].\text{rank} = \text{LF}[d].$$

$\square$

Corollary 2. *Suppose we are given an SLP with g rules for a text $T[1..n]$ whose BWT has r runs, and a tag array for T with t runs. Then we can build an $O(g + r + t)$-space data structure such that later, given a pattern $P[1..m]$, we can compute the tag statistics of P with respect to T in $O(m \log n)$ time.*

Proof. We store an $O(t)$-space predecessor data structure on the starting positions of the runs in Tag. For each run $\text{Tag}[u..d]$, we also store $\text{SA}[u]$ and $\text{SA}[d]$. Given P, we start by applying Theorem 3 to compute the extended matching statistics $\text{XMS}[1..m + 1]$ of P with respect to T in $O(m \log n)$ time. For

$1 \leq i \leq m + 1$, we then set

$$TS[i].len = XMS[i].len,$$
$$TS[i].pos = XMS[i].pos,$$
$$TS[i].rank = XMS[i].rank,$$

and $TS[i]$.run to the index of the run $Tag[u..d]$ in the tag array containing position $TS[i]$.rank (computed with a predecessor query). Further, we use the LCP data structure to compute

$$TS[i].up = LCP(P[i..m], T[SA[u]..n]),$$
$$TS[i].down = LCP(P[i..m], T[SA[d]..n]).$$

This also takes a total of $O(m \log n)$ time.   □

4   Using Tag Statistics

Once we have the tag statistics of P with respect to T, we no longer need Lemma 1, or even the SA samples or BWT, to find out which tags label the occurrences of any $P[i..j]$. We use Muthukrishnan's document-listing data structure in the same way as in the proof of Corollary 1: once we know which runs in the tag array overlap the BWT interval for $P[i..j]$, we use Muthukrishnan's structure to list the k distinct tags in $O(k)$ time. In this section we explain how we find which runs in the tag array overlap the BWT interval for $P[i..j]$, without computing the interval itself (which we do not know how to do quickly in $O(g + r + t)$ space).

Lemma 2. *Suppose we are given a text $T[1..n]$ and a tag array for T with t runs. Then, for any constant $\epsilon > 0$, we can build an $O(g+t)$-space data structure such that later, given the tag statistics of a pattern $P[1..m]$ with respect to T and i and j, we can find which runs in the tag array overlap the BWT interval for $P[i..j]$ in $O(\log^\epsilon t)$ time.*

Proof. We store $O(t)$-space range-predecessor/successor data structures over L with $O(\log^\epsilon t)$ query time [28] (we call them collectively range-successor queries at times). With these data structures and given values ℓ and q, we can find the largest position of a value less than ℓ in $L[0..2q - 2]$ and the smallest position of a value less than ℓ in $L[2q..2t]$ in $O(\log^\epsilon t)$ time. We note that ϵ can be made arbitrarily small for the cost of a larger constant multiplying the space consumption.

Given the tag statistics $TS[1..m + 1]$ of P with respect to T and i and j, we can check that $P[i..j]$ occurs in T at all by verifying that $TS[i].len \geq j - i + 1$. Assuming it does, we can look up the index $q = TS[i].run$ of the run in the tag array containing $Tag[TS[i].rank]$ and we can check in constant time whether

$$TS[i].up \geq j - i + 1,$$
$$TS[i].down \geq j - i + 1.$$

If $TS[i].up < j-i+1$ then $L[2q-1] < j-i+1$ (note $L[2q-1]$ is the LCP within run q) and run q is the first in the tag array to overlap the BWT interval for $P[i..j]$. Otherwise, we use a range-predecessor query to find the largest position in $L[0..2q-2]$ with value less than $j-i+1$. This tells us the first run in the tag array to overlap the BWT interval for $P[i..j]$: If the range-predecessor query returns p, then the index of this first run is $1 + \lfloor p/2 \rfloor$; the run is covered completely if p is even and partially if p is odd.

Symmetrically, if $TS[i].down < j-i+1$ then $L[2q-1] < j-i+1$ and run q is the last one in the tag array to overlap the BWT interval for $P[i..j]$. Otherwise, we use a range-successor query to find the smallest position in $L[2q..2t]$ of a value less than $j-i+1$, which tells us the last run in the tag array to overlap the BWT interval for $P[i..j]$. If the range-successor query returns p, then the index of this last run is $\lceil p/2 \rceil$, and it is covered completely iff p is even.

Notice we never compute the BWT interval for $P[i..j]$.   □

Corollary 3. *Suppose we are given an SLP with g rules for a text $T[1..n]$ whose BWT has r runs, and a tag array for T with t runs. Then, for any constant $\epsilon > 0$, we can build an $O(g + r + t)$-space data structure with which we can preprocess any pattern $P[1..m]$ in $O(m \log n)$ time such that later, given i and j, we can return the k distinct tags labeling occurrences of $P[i..j]$ in T in $O(\log^\epsilon t + k)$ time.*

Proof. We store instances of the data structures from (i) Corollary 2, (ii) Lemma 2, and (iii) Corollary 1. Given P, we use the data structures (i) to compute the tag statistics of P with respect to T in $O(m \log n)$ time. Given i and j, we use the data structures (ii) to find the indices s and e of the runs in Tag that are contained in or overlap the BWT range of $P[i..j]$, in time $O(\log^\epsilon t)$. Finally, using the array $A[1..t]$ (iii) we run Muthukrishnan's algorithm on $A[s..e]$ to find the k distinct tags labeling occurrences of $P[i..j]$ in T, in $O(k)$ time.   □

5   Optimal-Time Tag Reporting

The time in Corollary 3 for reporting the k distinct tags labeling occurrences of $P[i..j]$ in T—that is, $O(\log^\epsilon t + k)$—is optimal if $k \in \Omega(\log^\epsilon t)$. We do not know k in advance, however, and if we always want optimal reporting time we cannot afford range-successor queries right away.

We start with an important property of the ranges we find in L in the proof of Corollary 3.

Lemma 3. *Let q, q' be positions in L with respective thresholds ℓ, ℓ', from which the predecessor/successor queries result in ranges $[u, d], [u', d']$. Then $[u, d], [u', d']$ can be equal, disjoint or nested, but cannot overlap.*

Proof. Consider $L[u..d]$ ($L[u'..d']$), which is as large as possible around q (q') not containing any values less than ℓ (ℓ'), and that $u < u' \le d < 2t$. It follows that $\ell \le L[u'-1] < \ell'$, therefore, since $L[d+1] < \ell < \ell'$, it must be $d' \le d$, so $L[u'..d']$ is contained in $L[u..d]$. The case $u \le d' < d$ is analogous.   □

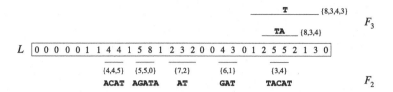

Fig. 2. The array L of Fig. 1 and the sets of segments forming F_3 (above) and F_2 (below). The larger range of F_3 contains the smaller, and thus they represent the same set of tags.

Consider the distinct ranges we can find in L such that the corresponding range in Tag (including both contained and overlapped runs) contains k' distinct tags, for some k'. If two of these ranges in L are nested, then their corresponding ranges in Tag contain exactly the same k' distinct tags—possibly with different multiplicities, but that does not concern us here. Let $F_{k'}$ be the $O(t)$-bit balanced-parentheses representation [22] of these distinct ranges in L, where every range is an ancestor of those it contains. With $O(t)$ further bits, we can find in $O(1)$ time the lowest node of $F_{k'}$ that contains any given entry $L[q]$ [33, Sec. 4.1]. Figure 2 gives an example.

While querying the data structure from Lemma 2, if we somehow guess correctly that our range-successor queries will return a range in L whose corresponding range in Tag contains exactly k' distinct tags, then we can replace those range-successor queries by the constant-time method described above to find the corresponding node in $F_{k'}$.

This node may correspond to a range nested strictly inside the one we would obtain from the range-successor queries but, as we noted above, that makes no difference to our final answer. In fact, the node of $F_{k'}$ we find has the smallest range—corresponding to the largest value of $j - i + 1$—we could obtain from our range-successor queries, while still returning k' distinct tags. If we store an $O(t)$-bits range-minimum data structure [11] over L—which we can reuse for all values of k'—then we can find that largest value $j - i + 1$ in constant time, as it is the minimum value of L in the range.

Of course, we cannot assume we will guess correctly the number k' of distinct tags we will eventually return. Instead, we keep an $O(t)$-bits representation $F_{k'}$ for every $k' \leq \lg^\epsilon t$, which takes $O\left(\frac{t \lg^\epsilon t}{\log t}\right) \subset O(t)$ space. We query $F_1, F_2, F_3, \ldots, F_{\lg^\epsilon t}$ in turn, using constant time for each. If, for some $F_{k'}$, the range-minimum data structure returns a value smaller than $j - i + 1$, then we know that $P[i..j]$ is labeled by $k = k' - 1$ distinct tags, so we use the formulas of Sect. 4 to convert the range in L given by F_k to a range $A[s..e]$, and use Muthukrishnan's algorithm (Corollary 1) to return the distinct tags in $A[s..e]$. Otherwise, after we query $F_{\lg^\epsilon t}$, we know that $k > \lg^\epsilon t$, so we can perform the range-successor queries safely as in Sect. 4. In both cases, we use $O(k)$ total time.

Theorem 4. *Suppose we are given an SLP with g rules for a text $T[1..n]$ whose BWT has r runs, and a tag array for T with t runs. Then we can build an*

$O(g + r + t)$-space data structure that can preprocess any pattern $P[1..m]$ in $O(m \log n)$ time such that later, given i and j, it returns the k distinct tags labeling occurrences of $P[i..j]$ in T in optimal $O(k)$ time.

6   Discussion and Future Work

This paper lays out the theoretical basis for Wheeler maps. We have shown how using compressed space, we can preprocess a pattern P such that later, given any i and j, we report the distinct tags labeling the occurrences of $P[i..j]$ in the optimal constant time per tag reported. To the best of our knowledge, Wheeler maps are the first data structure allowing for an efficient tag listing of subpatterns. Further results on prioritizing and constraining the query tag frequencies will be included in the extended version of this article.

As a future work, we plan to address the question of whether mixing Wheeler graphs with Wheeler maps — to allow some kinds of recombinations while excluding others — is useful and viable. Besides pangenomics, we plan to explore the versatile nature of Wheeler maps and look for other applications. We also believe that for certain cases we can prove analytical bounds on the number of runs in the tag array by relating them to the repetitiveness of the input.

We are now investigating our approach experimentally. Together with the full implementations of the data structures described here, we also still need efficient algorithms for extracting tag arrays from pangenome graphs for large genomic datasets, and good compression schemes for those tag arrays. A tag could contain a lot of information, so representing it explicitly for every run of that tag in the tag array might be very wasteful. It is likely more space-efficient to store each distinct tag only once, separated from the tag array by one or more levels of indirection. Once we can build and store Wheeler maps well in practice, we intend to integrate them into current pangenomics pipelines.

References

1. Bernardini, G., Pisanti, N., Pissis, S.P., Rosone, G.: Pattern matching on elastic-degenerate text with errors. In: Fici, G., Sciortino, M., Venturini, R. (eds.) SPIRE 2017. LNCS, vol. 10508, pp. 74–90. Springer, Cham (2017). https://doi.org/10.1007/978-3-319-67428-5_7
2. Bille, P., Gørtz, I.L., Cording, P.H., Sach, B., Vildhøj, H.W., Vind, S.: Fingerprints in compressed strings. J. Comput. Syst. Sci. **86** 171–180 (2017)
3. Boucher, C., et al. PHONI: Streamed matching statistics with multi-genome references. In: Proceedings 31st Data Compression Conference (DCC), pp. 193–202 (2021)
4. Burrows, M., Wheeler, D.: A block-sorting lossless data compression algorithm. In Digital SRC Research Report, Citeseer (1994)
5. Cobas, D., Gagie, T., Navarro, G.: A fast and small subsampled R-Index. In: Proceedings 32nd Annual Symposium on Combinatorial Pattern Matching (CPM), pp. 13:1–13:16 (2021)

6. Computational Pan-Genomics Consortium: Computational pan-genomics: status, promises and challenges. Brief. Bioinform. **19**(1), 118–135 (2018)
7. Cotumaccio, N., Prezza, N.: On indexing and compressing finite automata. In: Proceedings 32nd ACM-SIAM Symposium on Discrete Algorithms (SODA), pp. 2585–2599 (2021)
8. Dolgin, E.: Scientists unveil a more diverse human genome (2023) Accessed 3 Jan 2024
9. Equi, M., Mäkinen, V., Tomescu, AI.: Graphs cannot be indexed in polynomial time for sub-quadratic time string matching, unless SETH fails. In: Proceedings Theory and Practice of Computer Science (SOFSEM), pp. 608–622 (2021)
10. Ferragina, P., Manzini, G.: Indexing compressed text. J. ACM **52**(4), 552–581 (2005)
11. Fischer, J., Heun, V.: Space-efficient preprocessing schemes for range minimum queries on static arrays. SIAM J. Comput. **40**(2), 465–492 (2011)
12. Gagie, T., Manzini, G., Sirén, J.: Wheeler graphs: framework for BWT-based data structures. Theoret. Comput. Sci. **698**, 67–78 (2017)
13. Gagie, T., Navarro, G., Prezza, N.: Optimal-time text indexing in BWT-runs bounded space. In: Proceedings of the 29th Annual ACM-SIAM Symposium on Discrete Algorithms (SODA), pp. 1459–1477 (2018)
14. Ganardi, M., Jeż, A., Lohrey, M.: Balancing straight-line programs. J. ACM **68**(4), 1–40 (2021)
15. Garrison, E., et al.: Variation graph toolkit improves read mapping by representing genetic variation in the reference. Nature Biotechnol. **36**(9), 875–879 (2018)
16. Langmead, B., Salzberg, S.L.: Fast gapped-read alignment with Bowtie 2. Nature Methods, **9**(4), 357–359 (2012)
17. Li, H., Durbin, R.: Fast and accurate short read alignment with Burrows-Wheeler transform. Bioinformatics **25**(14), 1754–1760 (2009)
18. Liao, W.W., et al.: A draft human pangenome reference. Nature **617**(7960), 312–324 (2023)
19. Mäkinen, V., Cazaux, B., Equi, M., Norri, T., Tomescu, A.I.: Linear time construction of indexable founder block graphs. In: 20th International Workshop on Algorithms in Bioinformatics (WABI), pp. 7:1–7:18 (2020)
20. Mäkinen, V., Navarro, G.: Succinct suffix arrays based on run-length encoding. In: Apostolico, A., Crochemore, M., Park, K. (eds.) Combinatorial Pattern Matching: 16th Annual Symposium, CPM 2005, Jeju Island, Korea, June 19-22, 2005. Proceedings, pp. 45–56. Springer Berlin Heidelberg, Berlin, Heidelberg (2005). https://doi.org/10.1007/11496656_5
21. Mäkinen, V., Navarro, G., Sirén, J., Välimäki, N.: Storage and retrieval of highly repetitive sequence collections. J. Comput. Biol. **17**(3), 281–308 (2010)
22. Munro, J.I., Raman, V.: Succinct representation of balanced parentheses and static trees. SIAM J. Comput. **31**(3), 762–776 (2001)
23. Muthukrishnan, S.: Efficient algorithms for document retrieval problems. In: Proceedings of the 13th ACM-SIAM Symposium on Discrete Algorithms (SODA), pp. 657–666, (2002)
24. Navarro, G., Mäkinen, V.: Compressed full-text indexes. ACM Comput. Surv., **39**(1)article 2 (2007)
25. Navarro, G.: Compact data structures: a practical approach. Cambridge University Press (2016)
26. Navarro, G.: Indexing highly repetitive string collections, part II: compressed indexes. ACM Comput. Surv. **54**(2), article 26 (2021)

27. Navarro, G., Prezza, N.: Universal compressed text indexing. Theoret. Comput. Sci. **762**, 41–50 (2019)
28. Nekrich, Y., Navarro, G.: Sorted range reporting. In: Fomin, F.V., Kaski, P. (eds.) SWAT 2012. LNCS, vol. 7357, pp. 271–282. Springer, Heidelberg (2012). https://doi.org/10.1007/978-3-642-31155-0_24
29. Nishimoto, T., Tabei, Y.: Optimal-time queries on BWT-runs compressed indexes. In: Proceedings of the International Colloquium on Automata, Languages, and Programming (ICALP), pp. 101:1–101:15 (2021)
30. Reich, D.E., et al.: Linkage disequilibrium in the human genome. Nature, **411**(6834), 199–204 (2001)
31. Rizzo, N., Cáceres, M., Mäkinen, V.: Finding maximal exact matches in graphs. In: Proceedings of the Workshop on Algorithms in Bioinformatics (WABI), pp. 10:1–10:17 (2023)
32. Rossi, M., Oliva, M., Langmead, B., Gagie, T., Boucher, C.: MONI: a pangenomic index for finding maximal exact matches. J. Comput. Biol. **29**(2), 169–187 (2022)
33. Russo, L.M., Navarro, G., Oliveira, A.L.: Fully-compressed suffix trees. ACM Trans. Algorithm. **7**(4):article 53 (2011)
34. Steinmark, E.M.: The human genome needs updating but how do we make it fair? (2023). Accessed 3 Jan 2024
35. Tatarnikov, I., Farahani, A.S., Kashgouli, S., Gagie, T.: MONI can find k-MEMs. In: Proceedings 34th Annual Symposium on Combinatorial Pattern Matching (CPM), pp. 26:1–26:14 (2023)

Faster Combinatorial k-Clique Algorithms

Amir Abboud, Nick Fischer, and Yarin Shechter$^{(\boxtimes)}$

Weizmann Institute of Science, Rehovot, Israel
{amir.abboud,nick.fischer,yarin.shechter}@weizmann.ac.il

Abstract. Detecting if a graph contains a k-Clique is one of the most fundamental problems in computer science. The asymptotically fastest algorithm runs in time $O(n^{\omega k/3})$, where ω is the exponent of Boolean matrix multiplication. To date, this is the only technique capable of beating the trivial $O(n^k)$ bound by a polynomial factor. Due to this technique's various limitations, much effort has gone into designing "combinatorial" algorithms that improve over exhaustive search via other techniques.

The first contribution of this work is a faster combinatorial algorithm for k-Clique, improving Vassilevska's bound of $O(n^k/\log^{k-1} n)$ by two log factors. Technically, our main result is a new reduction from k-Clique to Triangle detection that exploits the same divide-and-conquer at the core of recent combinatorial algorithms by Chan (SODA'15) and Yu (ICALP'15).

Our second contribution is exploiting combinatorial techniques to improve the state-of-the-art (even of non-combinatorial algorithms) for generalizations of the k-Clique problem. In particular, we give the first $o(n^k)$ algorithm for k-clique in hypergraphs and an $O(n^3/\log^{2.25} n + t)$ algorithm for listing t triangles in a graph.

1 Introduction

One of the most fundamental problems in computer science is k-Clique: given an n-node graph, decide if there are k nodes that form a clique, i.e. that have all the $\binom{k}{2}$ edges between them. Our interest is in the case where $3 \leq k \ll n$ is a small constant. This is the "SAT of parameterized complexity" being the canonical problem of the W[1] class of "fixed parameter intractable" problems, and its basic nature makes it a core task in countless applications where we seek a small sub-structure defined by pairwise relations.

The naïve algorithm checks all subsets of k nodes and runs in $O(k^2\binom{n}{k})$ time, which is $\Theta(n^k)$ for constant k. Whether and how this bound can be beaten (in terms of worst-case asymptotic time complexity) is a quintessential form of the question: *can we beat exhaustive search?*

This work is part of the project CONJEXITY that has received funding from the European Research Council (ERC) under the European Union's Horizon Europe research and innovation programme (grant agreement No. 101078482). The first author is additionally supported by an Alon scholarship and a research grant from the Center for New Scientists at the Weizmann Institute of Science.

J. A. Soto and A. Wiese (Eds.): LATIN 2024, LNCS 14578, pp. 193–206, 2024.
https://doi.org/10.1007/978-3-031-55598-5_13

The asymptotically fastest algorithms gain a speedup by exploiting fast matrix multiplication – one of the most powerful techniques for beating exhaustive search. In particular, for the important special case of $k = 3$, i.e. the *Triangle Detection* problem, the running time is $O(n^\omega)$ where $2 \leq \omega < 2.3719$ [23] is the exponent in the time complexity of multiplying two $n \times n$ binary matrices.[1] For larger $k > 3$, there is a reduction to the $k = 3$ case by Nešetřil and Poljak [31] that produces graphs of size $O(n^{\lceil k/3 \rceil})$.[2] The resulting time bound is $O(n^{\lceil \omega k/3 \rceil})$. Except for improvements for k that is not a multiple of 3 [24], and the developments in fast matrix multiplication algorithms reducing the value of ω over the years, this classical algorithm remains the state-of-the-art.

The one general technique underlying all fast matrix multiplication, starting with Strassen's algorithm [34], is to find some clever formula to exploit cancellations in order to replace multiplications with additions. To date, this is *the only* technique capable of beating exhaustive search by a polynomial n^ε factor for the k-Clique problem. All techniques have their limitations, and so does Strassen's; we defer a detailed discussion on this to the full paper due to space constraints. Consequently, much research has gone into finding "combinatorial algorithms" that beat exhaustive search by other techniques. Existing techniques have only led to polylogarithmic speedups, leading the community to the following conjectures that have become the basis for many conditional lower bounds.

Conjecture 1 (Combinatorial BMM). Combinatorial algorithms cannot solve Triangle Detection in time $O(n^{3-\varepsilon})$ where $\varepsilon > 0$.[3]

A reduction of Vassilevska and Williams [37] shows that this conjecture is *equivalent* to the classical conjecture that combinatorial algorithms cannot solve Boolean Matrix Multiplication (BMM) in truly subcubic time [28,33]. Following their reduction, many conditional lower bounds were based on this conjecture, e.g. [5,16,19,21] (we refer to the survey [36] for a longer list).

Conjecture 2 (Combinatorial k-Clique). Combinatorial algorithms cannot solve k-Clique in time $O(n^{k-\varepsilon})$ for any $k \geq 3$ and $\varepsilon > 0$.

The latter conjecture is stronger than the former, in the sense that faster algorithms for $k = 3$ imply faster algorithms for larger $k > 3$ but the converse is not known. The first use of this conjecture as a basis for conditional lower bounds was by Chan [17] to prove an $n^{k-o(1)}$ lower bound for a problem in computational geometry. Later, Abboud, Backurs, and Vassilevska Williams [2] used it to prove $n^{3-o(1)}$ lower bounds in P. Several other papers have used it since then, e.g. [1,4,9,11–14,20,22,26,29].

---

[1] Simply compute A^2 where A is the adjacency matrix of the graph and check if $A^2[i,j] > 0$ for any $\{i,j\}$ that are an edge.

[2] Each $k/3$-clique becomes a node and edges are defined in a natural way so that a triangle corresponds to a k-clique.

[3] Note the informality in these combinatorial conjectures stemming from the lack of precise definition for "combinatorial" in this context. See full paper for further discussion.

Previous Combinatorial Bounds. The previous bounds for Triangle detection ($k = 3$) fall under three conceptual techniques (see full paper for more details). We will omit $(\log \log n)$ factors in this paragraph.

1. The *Four-Russians technique* [6] from 1970 gives an $O(n^3 / \log^2 n)$ bound, and is used in all later developments.
2. In 2010, Bansal and Williams [7] use *pseudoregular partitions* to shave off an additional $\log^{1/4} n$ factor.
3. In 2014, Chan [18] introduced a simple *divide-and-conquer technique* to get an $O(n^3 / \log^3 n)$ bound, and a year later, Yu [38] optimized this technique to achieve a bound of $O(n^3 / \log^4 n)$.

For $k > 3$ there are two options: (1) we either apply these algorithms inside the aforementioned reduction to Triangle, getting a bound of $O(n^k / \log^4 n)$, or (2) we apply these combinatorial techniques directly to k-Clique. An early work of Vassilevska [35] from 2009 applied the Four-Russians technique directly to get an $O(n^k / \log^{k-1})$ bound. Note that this generalizes the $\log^2$ shaving from the first bullet naturally to all k, and is favorable to the algorithms from option (1) for $k > 5$. Vassilevska's bound remains state-of-the-art, and in this work, we address the challenge of generalizing the other combinatorial techniques to k-Clique.

1.1 Our Results

The first result of this paper is a faster combinatorial algorithm for k-Clique for all $k > 3$ based on a generalization of the divide-and-conquer technique from Chan's and Yu's algorithms for $k = 3$. We use divide-and-conquer to design a *more efficient reduction* from k-Clique to the $k = 3$ case. The main feature of this reduction is that we get an additional log factor shaving each time k increases by one; this should be contrasted with the classical reduction from option(1) above, in which we gain nothing when k grows.

Theorem 1 (Reduction from k-Clique to Triangle). *Let $k \geq 3$, and let a, b be reals such that there is a combinatorial triangle detection algorithm running in time $O(n^3 (\log n)^a (\log \log n)^b)$. Then there is a combinatorial k-clique detection algorithm in time $O(n^k (\log n)^{a-(k-3)} (\log \log n)^{b+k-3})$.*

Combining our reduction with Yu's state-of-the-art combinatorial algorithm for Triangle detection, we improve Vassilevska's bound by two log factors.

Corollary 1 (Faster Combinatorial k-Clique Detection). *There is a k-clique detection algorithm running in time $O(n^k (\log n)^{-(k+1)} (\log \log n)^{k+3})$.*

It may be interesting to note that our reduction can even be combined with the naïve $O(n^3)$ algorithm for Triangle detection, giving a $(\log n)^{k-3}$ shaving for k-Clique *without using the Four-Russians technique*.

Another interesting implication of our reduction is concerning the framework of Bansal and Williams' [7]. Their algorithm can be improved if better dependencies for regularity/triangle removal lemmas are achieved. The best known upper bound on $f(\varepsilon)$ in a triangle removal lemma is of the form $\frac{c}{(\log^*(1/\varepsilon))^\delta}$ for some constants $c > 1$ and $\delta > 0$.[4] Due to this dependency, their first algorithm [7, Theorem 2.1] only shaves a $\log^*(n)$ factor from the running times achieved with the standard Four-Russians technique. However, it is not ruled out that much better dependencies can be achieved that would accelerate their algorithm to the point where, combined with our reduction, a k-clique algorithm with faster running times than Corollary 1 is obtained.[5]

A primary reason to seek combinatorial algorithms for k-Clique is that the techniques may generalize in ways fast matrix multiplication cannot (see full paper for detailed discussion). Our second set of results exhibits this phenomenon by shaving logarithmic factors over state-of-the-art for general (non-combinatorial) algorithms.

One limitation of the $O(n^\omega)$ algorithm for Triangle detection is that it does not solve the *Triangle listing* problem: we cannot specify a parameter t and get all triangles in the graph in time $O(n^\omega + t)$ assuming their number is up to t. Listing triangles in an input graph is not only a natural problem, but it is also connected to the fundamental 3SUM problem (given n numbers, decide if there are three that sum to zero). A reduction from 3SUM [27,32] shows that in order to beat the longstanding $O(n^2/\log^2 n)$ bound over integers [8] it is enough to shave a $\log^{6+\varepsilon} n$ factor for Triangle listing – i.e., achieve a running time of $O(\frac{n^3}{\log^{6+\varepsilon} n} + t)$ for some $\varepsilon > 0$. Although research has seen some results on triangle listing [10], we are not aware of any previous $o(n^3) + O(t)$ time bound for this problem (even with non-combinatorial techniques). Our second result produces such a time bound, showing that the other combinatorial techniques (namely Four-Russians and regularity lemmas) can be exploited. We shave a $\log^{2.25} n$ factor for this problem, generalizing the Bansal-Williams bound for BMM. Note we use the non-standard notation $\widetilde{\widetilde{O}}(n) = n(\log \log n)^{O(1)}$ to suppress polyloglog factors.

Theorem 2 (Faster Triangle Listing). *There is a randomized combinatorial algorithm that lists up to t triangles in a given graph in time $\widetilde{\widetilde{O}}(\frac{n^3}{(\log n)^{2.25}} + t)$, and succeeds with probability $1 - n^{-100}$.*

Another well-known limitation of Strassen-like techniques is that they are ineffective for detecting hypergraph cliques. They fail to give any speedup even for the first generalization in this direction: detecting a 4-clique in a 3-uniform hypergraph (i.e. a hypergraph where each hyper-edge is a set of three nodes). We are not aware of any non-trivial $o(n^4)$ algorithm for this problem (even with non-combinatorial techniques). The conjecture that $O(n^{4-\varepsilon})$ time cannot be achieved

---

[4] Fox achieved some improved dependencies with a new proof of the removal lemma [25], however, it is not clear whether it can be implemented efficiently.

[5] Note that the same cannot be said about their second algorithm [7, Theorem 2.1]; see the lower bound for pseudoregular partitions due to Lovasz and Szegedy [30]).

has been used to prove conditional lower bounds, e.g. [15,29]. Our third result is a $\log^{1.5} n$ factor shaving for this problem. The following theorem provides our general bound and strengthens the result for listing (detection can be obtained by setting $t = 1$).

Theorem 3 (Faster k-Hyperclique Listing). *There is an algorithm for listing up to t k-hypercliques in an r-uniform hypergraph in time*

$$O\left(\frac{n^k}{(\log n)^{\frac{k-1}{r-1}}} + t\right)$$

(assuming a word RAM model with word size $w = \Omega(\log n)$).

Subsequent Work. Shortly after this work, Abboud, Fischer, Kelly, Lovett, and Meka announced a combinatorial algorithm for BMM with $O(\frac{n^3}{2^{(\log n)^\varepsilon}})$ running time [3]. This implies an improvement for k-Clique as well that is stronger than any poly-log speedup and thus improves over Corollary 1 (by using pseudo-regularity techniques rather than divide-and-conquer). Moreover, building on our proof of Theorem 2 the authors present a speedup for triangle listing as well. However, our result for hypergraphs in Theorem 3 remains unbeaten.

1.2 Outline

We start with some preliminaries in Sect. 2. In Sect. 3 we provide our improved combinatorial k-Clique algorithm. In Sects. 4 and 5 we provide the high-level ideas of our improvements for Triangle Listing and k-Hyperclique Detection; due to space constraints we are forced to defer the technical details to the the full paper.

2 Preliminaries

Let $[n] := \{1, \ldots, n\}$. We write $\widetilde{O}(n) = n(\log n)^{O(1)}$ to suppress polylogarithmic factors and use the non-standard notation $\overset{\approx}{O}(n) = n(\log \log n)^{O(1)}$.

Throughout we consider undirected, unweighted graphs. In the k-clique problem, we are given a k-partite graph $(V_1, \ldots, V_k, E)$ and the goal is to determine whether there exist k vertices $v_1 \in V_1, \ldots, v_k \in V_k$ such that there is an edge $(v_i, v_j) \in E$ for every pair $i \neq j$. Note that the assumption that the input graphs are k-partite is without loss of generality, and can be achieved by a trivial transformation of any non-k-partite graph $G = (V, E)$: We create k copies $V_1, \ldots, V_k$ of the vertex set and for every $(u, v) \in E$ we add the edges (u_i, v_j) for every $i \neq j$. Another typical relaxation is that we only design an algorithm that *detect* the presence of k-cliques (without actually returning one). It is easy to transform a detection algorithm into a finding algorithm using binary search without asymptotic overhead.[6]

We additionally define the following notation for a k-partite graph as before: For a vertex v, let $N_i(v) = \{u \in V_i : (v, u) \in E\}$ denote the neighbourhood of v in V_i and $d_i(v) = |N_i(v)|$ denote the degree of v in V_i. Moreover, for a vertex set $V' \subseteq V$ we let $G[V']$ denote the subgraph of G induced by the vertex set V'. Throughout we further let $n = |V_1| + \cdots + |V_k|$ denote the total number of vertices in the graph.

An *r-uniform hypergraph* is a pair (V, E), where V is a vertex set and $E \subseteq \binom{V}{r}$ is a set of *hyperedges*. In the *r-uniform k-hyperclique* problem we need to decide whether in a k-partite hypergraph $(V_1, \ldots, V_k, E)$ there are vertices $v_1 \in V_1, \ldots, v_k \in V_k$ such that all hyperedges on $\{v_1, \ldots, v_k\}$ are present. Similarly, the assumption that the hypergraph is a k-partite is without loss of generality.

We are using the standard word RAM model with word size $w \in \Theta \log(n)$. In this model a random-access machine can perform arithmetic and bitwise operations on w-bit words in constant time.

3   Combinatorial Log-Shaves for k-Clique

In this section we provide our improved algorithmic reduction from k-clique to triangle detection (see Theorem 1). In our core we follow a divide-and-conquer approach for k-clique reminiscent to Chan's algorithm for triangle detection [18] with a simple analysis. We start with the following observation:

Observation 1 (Trivial Reduction from k-Clique to $(k-1)$-Clique). *Let $k \geq 4$, let $f(n)$ be a nondecreasing function, and assume that there is a combinatorial $(k - 1)$-clique detection algorithm running in time $O(n^{k-1}/f(n))$. Then there is a combinatorial k-clique detection algorithm running in time*

$$O\left(\sum_{v \in V_1} \frac{d_2(v) \cdot \ldots \cdot d_k(v)}{f(\min\{d_2(v), \ldots, d_k(v)\})}\right).$$

Proof. The algorithm is simple: For each vertex $v \in V_1$, we construct the subgraph $G_v = G[N_2(v) \cup \cdots \cup N_k(v)]$ consisting of all neighbors of v and test whether G_v contains a $(k - 1)$-clique. Let $n_v = d_2(v) + \cdots + d_k(v)$ denote the number of vertices in G_v. Our intention is to use the efficient $(k - 1)$-clique algorithm—however, simply running the algorithm in time $O(n_v^k/f(n_v))$ is possibly too slow. Instead, we partition each of the $k - 1$ vertex parts in G_v into

---

[6] More specifically, any detection algorithm can be transformed into a finding algorithm with constant running time overhead by using binary search as follows: Arbitrarily split each of the k vertex parts into two halves. Then for each subgraph induced by one of the 2^k combination of halves whether it contains a k-clique. If the detection algorithm succeeds on some combination, we continue on this combination recursively. For any natural running time the recursive overhead becomes a geometric sum and thus is constant.

blocks of size $d_v := \min\{d_2(v), \ldots, d_k(v)\}$ (plus one final block of smaller size, respectively). Then, for each combination of $k - 1$ blocks, we use the efficient $(k - 1)$-clique detection algorithm. It is clear that the algorithm is correct, since we exhaustively test every tuple $(v_1, v_2, \ldots, v_k)$. For the running time, note that testing whether G_v contains a k-clique takes time

$$\left\lceil \frac{d_2(v)}{d_v} \right\rceil \cdots \cdots \left\lceil \frac{d_k(v)}{d_v} \right\rceil \cdot O\left(\frac{(d_v)^{k-1}}{f(d_v)} \right) = O\left(\frac{d_2(v) \cdots \cdots d_k(v)}{f(\min\{d_2(v), \ldots, d_k(v)\})} \right),$$

and thus the total running time is indeed

$$O\left(\sum_{v \in V_1} \frac{d_2(v) \cdots \cdots d_k(v)}{f(\min\{d_2(v), \ldots, d_k(v)\})} \right)$$

(possibly after preprocessing the graph in time $O(n^2)$ to allow for constant-time edge queries. Note that this also covers the cost of constructing G_v for every $v \in V_1$).   □

Before moving to the formal proof of Theorem 1, let us give a simplified high-level description of this algorithmic reduction in the specific case of 4-clique. For a given 4-partite graph (V_1, V_2, V_3, V_4), the core idea is the following: If the degrees in V_1 tend to be small, i.e. if for every $v \in V_1$ we have $d_2(v) \cdot d_3(v) \cdot d_4(v) \leq \alpha \cdot |V_2| \cdot |V_3| \cdot |V_4|$ for some fraction $\alpha \approx \frac{1}{\log n}$, then we can apply Observation 1. Otherwise, there is a *heavy* vertex $v \in V_1$ with $d_2(v) \cdot d_3(v) \cdot d_4(v) > \alpha \cdot |V_2| \cdot |V_3| \cdot |V_4|$. In this case, we will check every triplet of the form $(u, w, z) \in N_2(v) \times N_3(v) \times N_4(v)$. If any of these triplets form a triangle, we have detected a 4-clique. Otherwise, we have learned that no triplet in $N_2(v) \times N_3(v) \times N_4(v)$ is part of a 4-clique. We will therefore recurse in such a way that ensures we never test these triplets again and thereby make sufficient progress.

Proof. Assume that there is a combinatorial triangle detection algorithm which runs in time $O(n^3 (\log n)^a (\log \log n)^b)$. We prove the claim by induction on k. The base case ($k = 3$) is immediate by the assumption there exists a triangle detection algorithm running in time $O(n^3 (\log n)^a (\log \log n)^b)$.

For the inductive step, consider the following recursive algorithm to detect a k-clique in a given k-partite graph $(V_1, \ldots, V_k, E)$. Let D and α be parameters to be determined later and let d be initialized to 0.

KCLIQUEREC$(G = (V_1, \ldots, V_k, E), d)$:

1. If $d = D$, meaning depth D in the recursion is reached, perform exhaustive search. Return YES if a k-clique was detected, otherwise NO.
2. Test whether there is some $v \in V_1$ with $d_2(v) \cdot \ldots \cdot d_k(v) \geq \alpha \cdot |V_2| \cdot \ldots \cdot |V_k|$. If such a vertex exists:
   a. Test whether the subgraph G_v induced by $N_2(v) \cup \cdots \cup N_k(v)$ contains a $(k-1)$-clique by exhaustive search. If it does return YES since this means we've found a k-clique involving v.

b. For $2 \leq i \leq k$, partition V_i into $V_{i,0} = V_i \setminus N_i(v)$ and $V_{i,1} = V_i \cap N_i(v)$. Recursively solve the $2^{k-1} - 1$ subproblems on $(V_1, V_{2,i_2}, \ldots, V_{k,i_k})$ for $(i_2, \ldots, i_k) \in \{0,1\}^{k-1} \setminus \{1^{k-1}\}$, while incrementing the depth.

In other words, for each $(i_2, \ldots, i_k) \in \{0,1\}^{k-1} \setminus \{1^{k-1}\}$, call KCLI-QUEREC$(G[V_1 \cup V_{2,i_2} \cup \cdots \cup V_{k,i_k}], d+1)$.

c. If any of the calls returned YES, return YES. Otherwise, return NO.

3. Solve the instance using Observation 1.

Correctness. As soon as the algorithm reaches recursion depth D, the algorithm will correctly detect a k-clique in step 1. In earlier levels of the recursion, the algorithm first attempts to find a vertex v with $d_2(v) \cdot \ldots \cdot d_k(v) \geq \alpha \cdot |V_2| \cdot \ldots \cdot |V_k|$ in step 2. If this succeeds, we test whether v is involved in a k-clique (and terminate in this case). Otherwise, we recurse on $(V_1, V_{2,i_2}, \ldots, V_{k,i_k})$ for all combinations $(i_2, \ldots, i_k) \in \{0,1\}^{k-1} \setminus \{1^{k-1}\}$. Note that we can indeed ignore the instance $(V_1, V_{2,1}, \ldots, V_{k,1})$ knowing that $(V_{2,1}, \ldots, V_{k,1})$ does not contain a $(k-1)$-clique. If the condition in step 2 is not satisfied, we instead correctly solve the instance by means of Observation 1 (which reduces the problem to an instance of $(k-1)$-clique).

Running Time. Imagine a recursion tree in which every node corresponds to an execution of the algorithm; the root corresponds to the initial call and child nodes correspond to recursive calls. Thus, every node in the tree is either a leaf (indicating that this execution does not spawn recursive calls), or an internal node with fan-out exactly $2^{k-1} - 1$. The *time* at a node is the running time of the respective call of the algorithm (ignoring the cost of further recursive calls). In other words, the *time* at a node is the amount of local work performed in the corresponding call. To bound the total running time of the algorithm, we bound the total time across all nodes in the recursion tree.

We analyze the contributions of all steps individually. Let us introduce some notation first: At a node x in the recursion tree, let $(V_1^x, \ldots, V_k^x)$ denote the instance associated to the respective invocation. We similarly write $d_2^x(v), \ldots, d_k^x(v)$.

Cost of Step 1. Note that at any node x at depth D in the recursion tree, the time is $O(|V_1^x| \cdot \ldots \cdot |V_k^x|)$ since we solve the instance by exhaustive search. Next, observe that for any internal node x in the recursion tree, we have that

$$|V_1^x| \cdot \ldots \cdot |V_k^x| = |V_1^x| \cdot \sum_{i_2, \ldots, i_k \in \{0,1\}^{k-1}} |V_{2,i_2}^x| \cdot \ldots \cdot |V_{k,i_k}^x|$$

$$\geq |V_1^x| \cdot d_2^x(v) \cdot \ldots \cdot d_k^x(v) + \sum_{y \text{ child of } x} |V_1^y| \cdot \ldots \cdot |V_k^y|$$

$$\geq \alpha \cdot |V_1^x| \cdot \ldots \cdot |V_k^x| + \sum_{y \text{ child of } x} |V_1^y| \cdot \ldots \cdot |V_k^y|,$$

and thus

$$\sum_{y \text{ child of } x} |V_1^y| \cdot \ldots \cdot |V_k^y| \leq (1 - \alpha) \cdot |V_1^x| \cdot \ldots \cdot |V_k^x|.$$

It follows by induction that at any depth $d \leq D$ in the recursion tree, we have that

$$\sum_{x \text{ at depth } d} |V_1^x| \cdot \ldots \cdot |V_k^x| \leq (1-\alpha)^d n^k.$$

In particular, the total time of all nodes at depth D is bounded by $O((1-\alpha)^D n^k)$.

Cost of Step 2. Note that the number of nodes in our recursion tree is at most 2^{kD} since the recursion tree has degree $\leq 2^k$ and the recursion depth is capped at D. At each node, the time of step 2a is bounded by $O(n^{k-1})$ and the cost of step 2b is bounded by $O(n^2)$. Therefore, the total time of step 2 across all nodes is bounded by $O(2^{kD} n^{k-1})$.

Cost of Step 3. By induction we have obtained a $(k-1)$-clique algorithm in time $O(n^{k-1}/f(n))$, where $f(n) = (\log n)^{-a+k-4}(\log \log n)^{-b-(k-4)}$. Therefore, by Observation 1 the total time of step 3 across all nodes x in the recursion tree is

$$O\left(\sum_{x \text{ lea}} \sum_{v \in V_1^x} \frac{d_2^x(v) \cdot \ldots \cdot d_k^x(v)}{f(\min\{d_2^x(v), \ldots, d_k^x(v)\})}\right).$$

To bound this quantity, we distinguish two subcases: A pair (x,v) (where x is a leaf in the recursion tree and $v \in V_1^u$) is called *relevant* if $d_2^x(v), \ldots, d_k^x(v) \geq \sqrt{n}$ (where n is the initial number of nodes). On the one hand, it is easy to bound the total cost of all irrelevant pairs by

$$O\left(\sum_{(x,v) \text{ irrelevant}} \frac{d_2^x(v) \cdot \ldots \cdot d_k^x(v)}{f(\min\{d_2^x(v), \ldots, d_k^x(v)\})}\right) \leq O(2^{kD} n^{k-1/2}),$$

since there are at most 2^{kD} nodes in the recursion tree. On the other hand, for any relevant pair (x,v), we have $\min\{d_2^x(v), \ldots, d_k^x(v)\} \geq \sqrt{n}$. Moreover, since we reach step 3 of the algorithm we further know that $d_2^x(v) \cdot \ldots \cdot d_k^x(v) \leq \alpha |V_2^x| \cdot \ldots \cdot |V_k^x|$ (as otherwise the condition in step 2 had triggered). It follows that

$$O\left(\sum_{(x,v) \text{ relevant}} \frac{d_2^x(v) \cdot \ldots \cdot d_k^x(v)}{f(\min\{d_2^x(v), \ldots, d_k^x(v)\})}\right)$$

$$\leq O\left(\sum_{(x,v) \text{ relevant}} \frac{\alpha |V_2^x| \cdot \ldots \cdot |V_k^x|}{f(\sqrt{n})}\right)$$

$$\leq O\left(n^k \cdot \frac{\alpha}{f(\sqrt{n})}\right).$$

Choosing the Parameters. Summing over all contributions computed before, the total running time is bounded by

$$O\left(n^k \cdot (1-\alpha)^D + n^k \cdot \frac{\alpha}{f(\sqrt{n})} + n^{k-1/2} \cdot 2^{kD}\right).$$

We pick $D = \log n/(4k)$ such that the latter term becomes $n^{k-1/4}$. Next, we pick $\alpha = \log((-a + k) \log n)/D = \Theta((\log n)^{-1} \log \log n)$ such that the first term becomes

$$n^k \cdot (1 - \alpha)^D \leq n^k \cdot 2^{-\alpha D} \leq n^k (\log n)^{a-k}.$$

All in all, the total running time is dominated by the second term

$$n^k \cdot \frac{\alpha}{f(\sqrt{n})} \leq O(n^k \cdot \alpha \cdot (\log n)^{a-(k-4)} (\log \log n)^{b+k-4})$$
$$\leq O(n^k (\log n)^{a-(k-3)} (\log \log n)^{b+k-3}),$$

which is as claimed.   □

4   Combinatorial Log-Shaves for Triangle Listing by Weak Regularity

In this section we quickly outline our triangle listing algorithm which is based on Bansal and Williams' BMM algorithm [7]. Our contribution is in reformulating and reanalyzing their algorithm for the purpose of triangle listing achieving Theorem 2. Note that we cannot achieve the running time stated in the theorem by applying state-of-the-art black-box reductions from triangle listing to Boolean matrix multiplication [37].

The two key ingredients are pseudoregularity and the following lemma which applies four russians to sparse graphs (see full paper for discussion on pseudoregularity and proof of the lemma).

Lemma 1 (Sparse Four-Russians). *There is an algorithm which lists up to t triangles in a given graph (V_1, V_2, V_3, E) (with $n = \min\{|V_1|, |V_2|, |V_3|\}$) in time*

$$\tilde{O}\left(\frac{|V_1| \cdot |V_2| \cdot |V_3|}{(\log n)^{100}} + \sum_{v \in V_1} \frac{d_2(v) \cdot d_3(v)}{(\log n)^2} + t\right).$$

Let us give an informal overview of the algorithm. For a given tripartite graph $G = (V_1, V_2, V_3, E)$, we first compute an ε-pseudoregular partition of the bipartite graph $G[V_2 \cup V_3]$. We then distinguish between two types of pieces— pieces with low density (less than $\sqrt{\varepsilon}$) and pieces with high density. Based on this we divide the instance into two triangle listing instances—G_L which only includes edges connecting low density parts between V_2 and V_3 in G, and its complement G_H consisting of edges connecting the high-density parts between V_2 and V_3. In the former case we can benefit from the sparseness (by construction the total number of edges G_L is at most $\sqrt{\varepsilon}n^2$). In the latter case, due to the pseudoregularity, there must be many triangles in G_H. We can thus charge the extra cost of computing with G_H towards the output-size. For complete specification refer to the full paper.

5   Combinatorial Log-Shaves for k-Hyperclique

In this section we give an intuitive description of the algorithm in the simplest case $k = 4$, $r = 3$ (detecting a 4-clique in a 3-uniform hypergraph in faster than $O(n^4)$ time), for complete and general specification refer to the full paper. We are given a 4-partite 3-uniform graph $G = (V_1, V_2, V_3, V_4, E)$ with vertex sets of size n. For each $v \in V_1$, we can define a tri-partite graph $G_v = (V_2, V_3, V_4, E')$ in which we draw an edge between two vertices if and only if they share a hyperedge with v in G. It is easy to check that there is a 4-hyperclique in G if and only if there are vertices v_2, v_3, v_4 that form a triangle in G_v *and* in G (meaning they are a hyperedge in G). The naive search for such a triplet would take $O(n^3)$, and we present an algorithm that accelerates this search:

1. Let $s = \sqrt{c \log n}$ for some small constant $c > 0$, and partition V_2, V_3 and V_4 each into $g = \lceil n/s \rceil$ blocks of size at most s. We let $V_{i,j}$ denote the j'th block in V_i.
2. For every combination $j_2, j_3, j_4 \in [g]$:
   a. Create a lookup table T_{j_2,j_3,j_4} with an entry for every possible tripartite graph on the vertex sets $V_{2,j_2}, V_{3,j_3}, V_{4,j_4}$ (there are $2^{s^2} = n^c$ such graphs).
   b. For every entry corresponding to a graph G' store whether G' has a triangle that is a hyperedge in G.

Note that this preprocessing is fast: We construct $\frac{n^3}{s^3}$ tables, each consisting of n^c entries, and each entry takes $O(s^3)$ time to determine. So, the total pre-processing time is $O(n^{3+c})$. Given these tables we can now search for a 4-clique more efficiently: For each $v \in V_1$ we break G_v into triples of blocks as before, and query T_{j_2,j_3,j_4} for the graphs $G_v[V_{2,j_2} \cup V_{3,j_3} \cup V_{4,j_4}]$, for all j_2, j_3, j_4. If one the answers is positive we have found a hyperclique. Assuming every query is performed in constant time, the running time is determined by the number of queries which is

$$O\left(n \cdot \frac{n^3}{s^3}\right) = O\left(\frac{n^4}{(\log n)^{1.5}}\right).$$

All that is left now is to justify the assumption that every query is performed in constant time. The main question is given $v \in V_1$ and a combination of blocks $V_{2,j_2}, V_{3,j_3}, V_{4,j_4}$, how can we determine the key corresponding to $G_v[V_{2,j_2}, V_{3,j_3}, V_{4,j_4}]$ in T_{j_2,j_3,j_4} in constant time? For this purpose, we define in the proof a *compact representation* of tripartite graphs (on vertex sets of size s) used to index the tables T_{j_2,j_3,j_4}. This compact representation is chosen in such a way which allows to efficiently precompute the compact representations of all such graphs $G_v[V_{2,j_2}, V_{3,j_3}, V_{4,j_4}]$.

References

1. Abboud, A., Backurs, A., Bringmann, K., Künnemann, M.: Fine-grained complexity of analyzing compressed data: quantifying improvements over decompress-and-solve. In: 2017 IEEE 58th Annual Symposium on Foundations of Computer Science (FOCS), pp. 192–203. IEEE (2017)

2. Abboud, A., Backurs, A., Williams, V.V.: If the current clique algorithms are optimal, so is valiant's parser. SIAM J. Comput. **47**(6), 2527–2555 (2018)
3. Abboud, A., Fischer, N., Kelley, Z., Lovett, S., Meka, R.: New graph decompositions and combinatorial boolean matrix multiplication algorithms. CoRR, abs/2311.09095 (2023). arxiv:2311.09095
4. Abboud, A., et al.: Faster algorithms for all-pairs bounded min-cuts. In: Baier, C., Chatzigiannakis, I., Flocchini, P., Leonardi, S., editors, 46th International Colloquium on Automata, Languages, and Programming, ICALP 2019, July 9–12, 2019, Patras, Greece, volume 132 of LIPIcs, pp. 7:1–7:15. Schloss Dagstuhl - Leibniz-Zentrum für Informatik (2019). https://doi.org/10.4230/LIPIcs.ICALP.2019.7
5. Abboud, A., Williams, V.V.:. Popular conjectures imply strong lower bounds for dynamic problems. In: 55th Annual IEEE Symposium on Foundations of Computer Science (FOCS 2014), pp. 434–443. IEEE Computer Society (2014). https://doi.org/10.1109/FOCS.2014.53
6. Arlazarov, V.L.V., Dinitz, Y.A., Kronrod, M.A., Faradzhev, I.: On economical construction of the transitive closure of an oriented graph. In: Akademii Nauk, D., vol. 194, pp. 487–488. Russian Academy of Sciences (1970)
7. Bansal, N., Williams, R.: Regularity lemmas and combinatorial algorithms. Theor. Comput. **8**(1), 69–94 (2012). https://doi.org/10.4086/toc.2012.v008a004
8. Baran, I., Demaine, E.D., Patrascu, M.: Subquadratic algorithms for 3SUM. Algorithmica **50**(4), 584–596 (2008). https://doi.org/10.1007/s00453-007-9036-3
9. Bergamaschi, T., Henzinger, M., Gutenberg, M.P., Williams, V.V., Wein, N.: New techniques and fine-grained hardness for dynamic near-additive spanners. In: Proceedings of the 2021 ACM-SIAM Symposium on Discrete Algorithms (SODA), pp. 1836–1855. SIAM (2021)
10. Björklund, A., Pagh, R., Williams, V.V., Zwick, U.: Listing triangles. In: Esparza, J., Fraigniaud, P., Husfeldt, T., Koutsoupias, E. (eds.) ICALP 2014. LNCS, vol. 8572, pp. 223–234. Springer, Heidelberg (2014). https://doi.org/10.1007/978-3-662-43948-7_19
11. Bringmann, K., Fischer, N., Künnemann, M.: A fine-grained analogue of schaefer's theorem in p: Dichotomy of exists^ k-forall-quantified first-order graph properties. In: 34th Computational Complexity Conference (CCC 2019). Schloss Dagstuhl-Leibniz-Zentrum für Informatik (2019)
12. Bringmann, K., Gawrychowski, P., Mozes, S., Weimann, O.: Tree edit distance cannot be computed in strongly subcubic time (unless apsp can). ACM Trans. Algorithm. (TALG) **16**(4), 1–22 (2020)
13. Bringmann, K., Grønlund, A., Larsen, K.G.. A dichotomy for regular expression membership testing. In: 2017 IEEE 58th Annual Symposium on Foundations of Computer Science (FOCS), pp. 307–318. IEEE (2017)
14. Bringmann, K., Wellnitz, P.: Clique-based lower bounds for parsing tree-adjoining grammars. arXiv preprint arXiv:1803.00804 (2018)
15. Carmeli, N., Kröll, M.: On the enumeration complexity of unions of conjunctive queries. In: Dan Suciu, Sebastian Skritek, and Christoph Koch, editors, Proceedings of the 38th ACM SIGMOD-SIGACT-SIGAI Symposium on Principles of Database Systems, PODS 2019, Amsterdam, The Netherlands, June 30 - July 5, 2019, pp. 134–148. ACM (2019). https://doi.org/10.1145/3294052.3319700
16. Casel, K., Schmid, M.L.: Fine-grained complexity of regular path queries. arXiv preprint arXiv:2101.01945 (2021)
17. Chan, T.M.: A (slightly) faster algorithm for klee's measure problem. In: Proceedings of the Twenty-fourth Annual Symposium on Computational geometry, pp. 94–100 (2008)

18. Chan, T.M.: Speeding up the four Russians algorithm by about one more loga-rithmic factor. In: Piotr Indyk, editor, Proceedings of the Twenty-Sixth Annual ACM-SIAM Symposium on Discrete Algorithms, SODA 2015, San Diego, CA, USA, January 4–6, 2015, pp. 212–217. SIAM (2015). https://doi.org/10.1137/1. 9781611973730.16
19. Chan, T.M., Rahul, S., Xue, J.: Range closest-pair search in higher dimensions. Comput. Geometry **91** 101669 (2020)
20. Chang, Y.J.: Hardness of RNA folding problem with four symbols. In: Grossi, R., Lewenstein, M., editors, 27th Annual Symposium on Combinatorial Pattern Matching, CPM 2016, June 27–29, 2016, Tel Aviv, Israel, volume 54 of LIPIcs, pp. 13:1–13:12. Schloss Dagstuhl - Leibniz-Zentrum für Informatik (2016). https://doi. org/10.4230/LIPIcs.CPM.2016.13
21. Clifford, R., Grønlund, A., Larsen, K.G., Starikovskaya, T.: Upper and lower bounds for dynamic data structures on strings. arXiv preprint arXiv:1802.06545 (2018)
22. Dalirrooyfard, M., Vuong, T.D., Williams, V.V.: Graph pattern detection: Hard-ness for all induced patterns and faster non-induced cycles. In: Proceedings of the 51st Annual ACM SIGACT Symposium on Theory of Computing, pp. 1167–1178 (2019)
23. Duan, R., Wu, H., Zhou, R.: Faster matrix multiplication via asymmetric hashing. In: 64th IEEE Annual Symposium on Foundations of Computer Science (FOCS 2023). IEEE Computer Society, 2023. To appear. https://doi.org/10.48550/arXiv. 2210.10173
24. Eisenbrand, F., Grandoni, F.: On the complexity of fixed parameter clique and dominating set. Theor. Comput. Sci. **326**(1–3), 57–67 (2004). https://doi.org/10. 1016/j.tcs.2004.05.009
25. Fox, J.: A new proof of the graph removal lemma. CoRR, abs/1006.1300 (2010). arxiv:1006.1300
26. Jin, C., Xu, Y.: Tight dynamic problem lower bounds from generalized bmm and omv. In: Proceedings of the 54th Annual ACM SIGACT Symposium on Theory of Computing, pp. 1515–1528 (2022)
27. Kopelowitz, T., Pettie, S., Porat, E.: Higher lower bounds from the 3SUM conjec-ture. In: Krauthgamer, R., editor, 27th Annual ACM-SIAM Symposium on Dis-crete Algorithms (SODA 2016), pp. 1272–1287. SIAM (2016). https://doi.org/10. 1137/1.9781611974331.ch89
28. Lee, L.: Fast context-free grammar parsing requires fast boolean matrix multipli-cation. J. ACM (JACM) **49**(1), 1–15 (2002)
29. Lincoln, A., Williams, V.V., Williams, R.: Tight hardness for shortest cycles and paths in sparse graphs. In: Proceedings of the Twenty-Ninth Annual ACM-SIAM Symposium on Discrete Algorithms, pp. 1236–1252. SIAM (2018)
30. lóVász, L., Lovász, M., Szegedy, B.: Szemerédi's lemma for the analyst. GAFA Geo-metric Funct. Anal. **17** 252–270 (2007). https://api.semanticscholar.org/CorpusID: 15201345
31. Nešetřil, J., Poljak, S.: On the complexity of the subgraph problem. Comment. Math. Univ. Carol. **26**(2), 415–419 (1985)
32. Pătraşcu, M.: Towards polynomial lower bounds for dynamic problems. In: Schul-man, L.J., editor, 42nd Annual ACM Symposium on Theory of Computing (STOC 2010), pp. 603–610. ACM (2010). https://doi.org/10.1145/1806689.1806772
33. Roditty, L., Zwick, U.: On dynamic shortest paths problems. In: Albers, S., Radzik, T. (eds.) ESA 2004. LNCS, vol. 3221, pp. 580–591. Springer, Heidelberg (2004). https://doi.org/10.1007/978-3-540-30140-0_52

34. Strassen, V.: Gaussian elimination is not optimal. Numer. Math. **13**, 354–356 (1969)
35. Vassilevska, V.: Efficient algorithms for clique problems. Inf. Process. Lett. **109**(4), 254–257 (2009). https://doi.org/10.1016/j.ipl.2008.10.014
36. Williams, V.V.: On some fine-grained questions in Algorithms and Complexity, pp. 3447–3487 (2018). https://doi.org/10.1142/9789813272880_0188
37. Williams, V.V., Williams, R.R.: Subcubic equivalences between path, matrix, and triangle problems. J. ACM **65**(5), 27:1–27:38 (2018). https://doi.org/10.1145/3186893
38. Huacheng, Yu.: An improved combinatorial algorithm for boolean matrix multiplication. Inf. Comput. **261**, 240–247 (2018). https://doi.org/10.1016/j.ic.2018.02.006

Approximation and Online Algorithms

On Approximate Colored Path Counting

Younan Gao[1][(⊠)]🆔 and Meng He[2][(⊠)]🆔

[1] Département D'informatique et D'ingénierie, Université du Québec en Outaouais,
Gatineau, Canada
gaoy03@uqo.ca
[2] Faculty of Computer Science, Dalhousie University, Halifax, Canada
mhe@cs.dal.ca

Abstract. Given an ordinal tree T on n nodes in which each node is assigned a color from $\{0, 1, \ldots, C - 1\}$, an approximate colored path counting query asks for an approximation of the number, occ, of distinct colors assigned to nodes in a query path. We first present data structures that can compute a 2-approximate answer, i.e., a number in [occ, 2occ], and achieve three different time/space trade-offs: i) an $O(n)$-word structure with $O(\lg^{\lambda} n)$ query time for any constant $0 < \lambda < 1$, ii) an $O(n \lg \lg n)$-word structure with $O(\lg \lg n)$ query time and iii) an $O(n \lg^{\lambda} n)$-word structure with $O(1)$ query time. The first trade-off beats the $O(\lg n / \lg \lg n)$ query time of the linear-word 2-approximate structure in previous work. We then design an $O(n)$-word structure which can compute in $O(\epsilon^{-2} \lg n)$ time a $(1 \pm \epsilon)$-approximate answer, i.e, a number in $[(1 - \epsilon)\text{occ}, (1 + \epsilon)\text{occ}]$, for any $\epsilon \in (0, 1)$. Previously, when the space cost is $O(n)$ words, the only known solution computes a $(1 \pm \epsilon)$-approximate answer in $O(\epsilon^{-4} \lg^2 n)$ time with success probability no less than $1 - \delta$, where δ is an arbitrary constant in $(0, 1)$; our solution not only has faster query time but also always returns a $(1 \pm \epsilon)$-approximation. When designing $(1 \pm \epsilon)$-approximate solutions, our techniques also yield an $O(n)$-word structure that can answer a colored type-2 path counting query in $O(\text{occ})$ time; this query reports the number of occurrences of each distinct color in a query path. This result improves the best previous linear-word solution in which the query time is $O(\text{occ} \lg \lg n)$.

Keywords: Path queries · Colored path counting · Colored path reporting · Approximate colored path counting

1   Introduction

In tree-structured data, information such as categories can be viewed as colors assigned to tree nodes. One query which can retrieve such information is the *colored path counting query*. It is defined over an ordinal tree[1] T on n nodes, each assigned a color from $\{0, 1, \ldots, C - 1\}$, where $C \leq n$, and it computes the number, occ, of distinct colors assigned to the nodes in any query path in T.

---

[1] This query can be defined over free trees. Following [12], we assume that the input tree is ordinal, so that we can use data structures for ordinal trees directly [13,14].

This work was supported by NSERC of Canada.

© The Author(s), under exclusive license to Springer Nature Switzerland AG 2024
J. A. Soto and A. Wiese (Eds.): LATIN 2024, LNCS 14578, pp. 209–224, 2024.
https://doi.org/10.1007/978-3-031-55598-5_14

When the tree structure is a single path, this query becomes the well-known 1D colored range counting query, for which Nekrich [19] designed an $O(n)$-word solution with $O(\lg \text{occ}/\lg \lg n + 1)$ query time. However, a conditional lower bound gives evidence that the colored path counting query problem is much harder: He and Kazi [12] showed how to multiply two $\sqrt{n} \times \sqrt{n}$ boolean matrices by answering n colored path counting queries over a tree of $O(n)$ nodes. This reduction means, with current knowledge, the total running time of answering n colored path counting queries, including preprocessing, cannot be faster than $n^{\omega/2}$, save for polylogarithmic speedups, where $\omega < 2.37286$ denotes the exponent of matrix multiplication [1]. Furthermore, since the best known combinatorial approach of multiplying two $n \times n$ Boolean matrices under the word RAM model requires $\Theta(n^3/\texttt{polylog}(n))$ time [23], the total time of answering n of these queries cannot be faster than $n^{1.5}$, save for polylogarithmic speedups, using pure combinatorial methods with current knowledge. He and Kazi designed an $O(n)$-word structure with $O(\sqrt{n} \lg \lg C)$ query time and $O(n^{3/2} \lg \lg C)$ preprocessing time under the word RAM model. More recently, Gao and He [10] considered the batched version of this problem and showed how to answer n queries, including preprocessing, in $O(n^{1.40704})$ time.

To achieve faster queries, approximate colored path counting problems have been studied. Two different ways of bounding approximate ratios have been considered [12]: a *c-approximate colored path counting query* computes a number in $[\text{occ}, c \cdot \text{occ}]$, while a $(1 \pm \epsilon)$-*approximate query* returns a number in $[(1 - \epsilon)\text{occ}, (1 + \epsilon)\text{occ}]$ for any $\epsilon \in (0,1)$. In this paper, we study this problem and aim at improving previous results under both approximate measures.

We also note that 1D colored range counting is sometimes called *1D colored type-1 range counting* in the literature [4,11], while *1D colored type-2 range counting* reports the number of occurrences of each distinct color in a query range. We can also generalize the latter to consider tree topology by defining *colored type-2 path counting* over a colored tree, which reports the number of occurrences of each distinct color in a query path. The $O(n)$-word data structure of Durocher et al. [7] can be used to answer a colored type-2 path counting query in $O(\text{occ} \lg \lg n)$ time. This is slower than the $O(\text{occ} + 1)$-time support for 1D colored type-1 range counting over points in rank space [9,11]. Thus, another goal is to close this gap.

Previous Work. By reducing colored path counting to path counting over weighted trees [15] using the chaining approach [11], He and Kazi [12] designed a linear space data structure that supports 2-approximate colored path counting in $O(\lg n/ \lg \lg n)$ time. For $(1 \pm \epsilon)$-approximate colored path counting queries, they showed a sketching data structure that occupies $O(n + \frac{n}{\epsilon^2 t} \lg n)$ words and answers a query in $O(\epsilon^{-2} t \lg n)$ time[2] with success probability no less than $1 - \delta$, where t is an arbitrary integer in $[1, n]$ and δ is an arbitrary constant in $(0,1)$. Setting $t = \lceil \epsilon^{-2} \lg n \rceil$ makes the space cost linear and the query time $O(\epsilon^{-4} \lg^2 n)$.

---

[2] He and Kazi [12] originally stated their result for constant ϵ, but it is easy to generalize their bounds when $\epsilon = o(1)$.

Similar approximate problems can also be defined for colored 1D range counting and colored 2D orthogonal range counting which generalizes the former by preprocessing colored points in 2D to efficiently compute the number of distinct colors assigned to points in an axis-aligned query rectangle. The same conditional lower bound for colored path counting also applies to the latter [17]. In 1D, given n colored points in the rank space, El-Zein et al. [8] designed an encoding data structure that uses $O(n)$-bits of space and supports c-approximate colored counting for any constant $c > 1$ in $O(1)$ time without accessing the given point set. In higher dimensions, Rahul [21] showed that $(1 \pm \epsilon)$-approximate colored range counting can be answered by combining a colored range reporting structure and a c-approximate colored range counting structure. With it, he designed an $O(n \lg n)$-word structure to support $(1 \pm \epsilon)$-approximate colored 2D range counting in $O(\epsilon^{-2} \lg n)$ time.

Regarding 1D colored type-2 range counting, Gupta et al. [11] designed an $O(n)$-word structure with $O(\lg n + \mathsf{occ})$ query time, over a set of n colored points on a real line. Ganguly et al. [9] stated that, by combining the approach of Gupta et al. and some other results [18, 22], the query time can be further improved to $O(1 + \mathsf{occ})$ if all points are in rank space. This query problem can also be generalized to point sets on the plane, and we refer to [4,6] for recent work on 2D colored type-2 range counting. For colored trees, the linear word structures designed by Durocher et al. [7] can answer a colored type-2 path counting query in $O(\mathsf{occ} \lg \lg n)$ time. They did not state this result explicitly, but it is implied by the algorithmic steps stated in the proof of Theorem 6 in their article.

Our Results. Under the word RAM model, we first design 2-approximate colored path counting structures with i) $O(n)$ words of space and $O(\lg^\lambda n)$ query time for any constant $0 < \lambda < 1$, ii) $O(n \lg \lg n)$ words of space and $O(\lg \lg n)$ query time and iii) $O(n \lg^\lambda n)$ words of space and $O(1)$ query time. In all three cases, the preprocessing time is $O(n \lg n)$. Hence the first trade-off beats the $O(\lg n / \lg \lg n)$ query time of the linear-word 2-approximate structure of He and Kazi [12]. We then design an $O(n)$-word $(1 \pm \epsilon)$-approximate colored path counting structure with $O(\epsilon^{-2} \lg n)$ deterministic query time and $O(n^2 \lg C \lg \lg C)$ expected preprocessing time. Compared to the sketching structure by He and Kazi [12] with $O(n)$-word space and $O(\epsilon^{-4} \lg^2 n)$ query time, we not only achieve improvement for query time but also guarantee that the query algorithm always returns a $(1 \pm \epsilon)$-approximation, though the cost of preprocessing is higher. In the new $(1 \pm \epsilon)$-approximate solution, our techniques also lead to a linear-word data structure supporting colored type-2 path counting in $O(\mathsf{occ})$ time. This result improves the solution of Durocher et al. [7] which has $O(\mathsf{occ} \lg \lg n)$ query time. See Table 1 for a comparison of our results to all previous results.

To achieve these results, we develop new techniques. For 2-approximate colored path counting, note that no further improvement can be made using the strategy of He and Kazi for this problem, due to the lower bound on (uncolored) 2D orthogonal range counting [20] (which is a special case of path counting over weighted trees). Instead, we adopt the strategy of Gao and He [10] for batched

Table 1. A summary of our results on approximate colored path counting and colored type-2 path counting, in which space costs are measured in words, ϵ is an arbitrary parameter in $(0,1)$, λ is an arbitrary constant in $(0,1)$, † marks an expected bound, and ‡ marks a solution that returns a $(1 \pm \epsilon)$-approximation with probability no less than $1 - \delta$ for any constant $\delta \in (0,1)$.

	Space	Query	Preprocess	Ref
2-approx	$O(n)$	$O(\frac{\lg n}{\lg \lg n})$	$O(n \frac{\lg n}{\lg \lg n})$	[12]
	$O(n)$	$O(\lg^{\lambda} n)$	$O(n \lg n)$	Thm 1a)
	$O(n \lg \lg n)$	$O(\lg \lg n)$		Thm 1b)
	$O(n \lg^{\lambda} n)$	$O(1)$		Thm 1c)
$(1 \pm \epsilon)$-approx	$O(n + \frac{n \lg n}{\epsilon^2 t})$	$O(\epsilon^{-2} t \lg n)$	$O(\epsilon^{-2} n \lg n)$	[12]‡
	$O(n)$	$O(\epsilon^{-2} \lg n)$	$O(n^2 \lg C \lg \lg C)$†	Thm 2
Type-2	$O(n)$	$O(\text{occ} \lg \lg n)$	$O(n \lg n)$	[7]
		$O(\text{occ})$		Lemma 6

colored path counting which applies centroid decomposition to decompose the tree into a hierarchy of components. A query is answered by locating and querying the component that satisfies these two conditions: This component contains the entire query path, and its centroid is in the path. This means, within each component, we need only support queries for the paths that pass through a fixed node, e.g., the selected centroid, given during the construction. This strategy however incurs $O(n \lg n)$ words of space cost in [10]. To address this, we design a data structure of $O(n)$ bits that answers 2-approximate queries in constant time, provided that a query path must contain a fixed node. To speed up the mapping of the endpoints of a query path to nodes in a specific component in a space-efficient manner, we borrow ideas from the solutions to the *ball inheritance* problem [5] and design data structures with different time-space trade-offs.

With regard to $(1 \pm \epsilon)$-approximate colored path counting queries, we adapt the reduction by Rahul [21] and make it work for trees. When doing so, we design a solution to colored type-2 path counting with optimal query time.

2   Preliminaries

This section introduces the notation and the previous results used in this paper.

Notation. Given an ordinal tree T, we use $|T|$ to represent the number of nodes in T and $\perp$ to represent its root. Each node of T is assigned a color encoded by an integer in $\{0, 1, \cdots, C - 1\}$, where $C \leq n$. We identify each node by its preorder rank, i.e., node x is the x-th node in a preorder traversal (x starts from 0), and $c(x)$ denotes the color of node x. Furthermore, $P_{x,y}$ denotes the path between nodes x and y, and $C(P_{x,y})$ denotes the set of colors that appear in it.

Navigation in Colored Ordinal Trees. To support navigational operations over the input tree, we apply the succinct representations of ordinal trees [13] and labeled trees [14]. The operations and their complexity are summarized in Lemma 1. As in previous work, we refer to a node (resp. ancestor) colored in α as an α-node (resp. α-ancestor).

Lemma 1 ([13,14]). *Let T denote a labeled ordinal tree on n nodes, each of which is assigned a color from $\{0, 1, \ldots, C-1\}$, where $C \leq n$. A data structure occupying $n \lg C + 2n + o(n \lg C)$ bits can be built over T in $O(n)$ time[3] to support:*

- *counting the number, $\mathtt{depth}(x)$, of ancestors of x in $O(1)$ time,*
- *counting the number, $\mathtt{depth}_\alpha(x)$, of α-nodes in $P_{x,\perp}$ in $O(\lg \lg C)$ time,*
- *finding the lowest common ancestor, $\mathtt{lca}(x, y)$, of x and y in $O(1)$ time,*
- *finding the parent node, $\mathtt{parent}(x)$, of non-root node x in $O(1)$ time,*
- *finding x's lowest proper α-ancestor, $\mathtt{parent}_\alpha(x)$, in $O(\lg \lg C)$ time and*
- *finding x's ancestor, $\mathtt{level\_anc}(x, d)$, at depth $\mathtt{depth}(x) - d$ in $O(1)$ time.*

Given a query path $P_{x,y}$ in a tree and a color α, a *colored path emptiness* query determines whether color α appears in $P_{x,y}$. He and Kazi [12] showed how to use $\mathtt{depth}_\alpha$ and $\mathtt{lca}$ to compute the number of appearances of α in $P_{x,y}$ in $O(\lg \lg C)$ time. Via counting the number of appearances of color α in $P_{x,y}$, one can figure out whether or not α appears in $P_{x,y}$. Hence, Lemma 1 can support colored path emptiness query in $O(\lg \lg C)$ time.

Partial Rank. Let $A[0..n-1]$ be a sequence of symbols over alphabet $\{0, 1, \ldots, \sigma - 1\}$. The *partial rank* operation [2], $\mathtt{rank}'(A, i)$, counts the number of elements equal to $A[i]$ in $A[0..i]$.

Lemma 2 ([2]). *Given a sequence $A[0..n-1]$ over alphabet $\{0, 1, \ldots, \sigma - 1\}$, where $\sigma \leq n$, a structure of $O(n \lg \sigma)$ bits can be constructed in $O(n)$ time to support $\mathtt{rank}'$ in $O(1)$ time.*

Tree Extraction [15]. Given a subset, X, of nodes of an ordinal tree T, the extracted tree, T_X, can be constructed by deleting each node $v \notin X$ using the following approach: If v is not the root, let $u = \mathtt{parent}(v)$. We remove v and its incident edges from T and insert its children into the list of children of u, replacing v in this list while preserving these children's original left-to-right order. This means that v's left and right siblings before the deletion will respectively become the left and right siblings of its children after the deletion. If v is the root, then, before we apply the same procedure to delete v, we add a dummy root to T and make it the parent of v, so that T_X will remain a tree.

To map a path $P_{x,y}$ in T to a path in T_X, we use the operation, $\mathtt{decompose}(x, y)$, defined in [12]: If $P_{x,y} \cap X = \emptyset$, it returns $\mathtt{null}$. Otherwise,

---

[3] As mentioned by Gao and He [10], the string representation of Belazzougui *et al.* [2] needs to be used in the framework of He *et al.* [14] to achieve $O(n)$ deterministic preprocessing time, at the cost of slowing down labeled operations from $O(\lg \frac{\lg C}{\lg w})$ in [14] to $O(\lg \lg C)$ in this lemma.

let $\hat{x}$ and $\hat{y}$ denote the nodes in $P_{x,y} \cap X$ that are closest to x and y, respectively (this can be the node x or y itself if it is in X). Then decompose(x, y) returns the nodes x' and y' in T_X that correspond to (i.e., whose original copies are) nodes $\hat{x}$ and $\hat{y}$ in T, respectively.

Lemma 3 ([12, Proposition 9]). *Given a tree T on n nodes and a tree extraction T_X, an $O(n)$-bit structure on top of T and T_X can be constructed in $O(n)$ time to support* decompose *in $O(1)$ time.*

3   2-Approximate Colored Path Counting

For 2-approximate colored path counting, we first consider a special case in which query paths must contain a fixed tree node specified in the preprocessing step (Sect. 3.1). Then we generalize it for arbitrary paths (Sects. 3.2 and 3.3).

3.1   Counting over a Path that Contains a Fixed Node

Fix a node v of T, and we design an $O(n)$-bit encoding data structure that supports 2-approximate colored path counting over any query path containing v. To answer a query, our encoding data structure does not need to access T after preprocessing, provided that the preorder ranks of the endpoints of the query path are known.

Lemma 4. *Let T be a colored tree on n nodes and fix any node v in T. A data structure of $O(n)$ bits can be constructed in $O(n)$ time to support 2-approximate colored path counting over any query path containing v in $O(1)$ time.*

Proof. We associate a binary label $B(u)$ to each node u of T as follows: If $u = v$, then $B(u) = 1$. Otherwise, locate the node, t, in $P_{u,v}$ that is adjacent to u. If color $c(u)$ appears in $P_{t,v}$, then set $B(u) = 0$. If not, set $B(u) = 1$. We discard the original colors of T, treat these labels as node colors and represent T with these labels using Lemma 1. Since there are only two possible labels, this uses $3n + o(n)$ bits.

Let $P_{x,y}$ be a query path containing v. Consider the nodes in the subpath $P_{x,v}$ one by one in the direction from v to x. Observe that, each time we see a node labeled by 1, we encounter a color that has not been seen previously. Therefore, the number of 1-bits assigned to nodes in $P_{x,v}$ is equal to $|C(P_{x,v})|$. Similarly, the number of 1-bits assigned to nodes in $P_{y,v}$ is equal to $|C(P_{y,v})|$. Therefore, the number of 1-bits assigned to nodes in $P_{x,y}$ is a 2-approximation of the precise answer. Following the discussion after Lemma 1, this number can be computed in $O(\lg \lg C) = O(1)$ time using operations lca and depth$_1$, as $C = 2$.

To prove the bound on construction time, it suffices to show that these binary labels can be assigned in $O(n)$ time. This can be done by performing a depth-first traversal of T using v as the starting node. During this traversal, we also update an array $A[0..C - 1]$, and the invariant that we maintain is that, each time we visit a node u, $A[i]$ stores the number of nodes in $P_{v,u}$ that are assigned color i

in the original tree T. The following are the steps: We start the traversal from vertex v, set $A[c(v)] = 1$ and initialize all other entries of A to 0. During the traversal, each time we follow an edge (x, y) with $x \in P_{v,y}$, there are two cases. In the first case, we follow this edge to visit node y. Then this is the first time we visit y. We check if $A[c[y]] = 0$. If it is, then the color of y does not appear in $P_{v,x}$, so we set $B(y) = 1$. Afterwards, we increment $A[c[y]]$ to maintain the invariant. Otherwise, we set $B(y) = 0$ and also increment $A[c[y]]$. In the second case, we follow this edge to visit x. Since x is closer to v than y is, we have visited x before and we will not traverse any paths containing y in the future. In this case, we decrement $A[c[y]]$ to maintain the invariant.   □

3.2   Counting over Arbitrary Paths

To support 2-approximate colored path counting queries over arbitrary paths, we transform the given ordinal tree T on n colored nodes into a binary tree $\tilde{T}$. Our transformation, similar to that used by Chan et al. [3], works as follows: For each node v with degree d, where $d > 2$, we remove the edges between v and its children, $v_1, v_2, \ldots, v_d$, to detach the subtrees rooted at these children from T. For the convenience of the description, let v_0 denote the node v. We then create $d - 2$ dummy nodes, $\tilde{v}_1, \tilde{v}_2, \ldots, \tilde{v}_{d-2}$, each assigned the color of v_0. Next we add edges to reconnect v_0 and its d children with the newly created dummy nodes as follows: For $t = 1, 2, \ldots, d - 2$, make v_t and $\tilde{v}_t$ the left and right children of v_{t-1}, respectively. Afterwards, make v_{d-1} and v_d the left and right children of $\tilde{v}_{d-2}$, respectively. The resulting tree is $\tilde{T}$ which has at most $2n$ nodes. This transformation preserves preorder among the original nodes of T. More importantly, any path $P_{x,y}$ in T and its corresponding path, $\tilde{P}$, in $\tilde{T}$ share the same set of colors. To see this, observe that the lowest common ancestor, z, of nodes x and y in T is the only original node that appears in $P_{x,y}$ but may not necessarily appear in $\tilde{P}$. If z does not appear in $\tilde{P}$, then $\tilde{P}$ must contain a dummy node created for z which is also colored in $c(z)$. On the other hand, any original node in $\tilde{P}$ must also be in $P_{x,y}$ in T, while any dummy node that appears in $\tilde{P}$ must satisfy the condition that the original node it is created for must be in $P_{x,y}$ in T.

After this transformation, for each node in T, we store the preorder rank of its corresponding node in $\tilde{T}$. Then we follow the strategy in [10] to decompose $\tilde{T}$ recursively using *centroid decomposition* [16], but different data structures will be constructed this time. Here a *centroid* of an m-node tree is a node whose removal splits the tree into connected components, each containing at most $m/2$ nodes. It is known that a centroid can be found in $O(m)$ time.

We now give the details of the recursion. At level 0 of the recursion, we call tree $\tilde{T}$ the level-0 component. We find a centroid, u, of $\tilde{T}$ and construct the data structure in Lemma 4 supporting 2-approximate queries over paths containing u. Since $\tilde{T}$ is a binary tree, after removing u, we are left with at most three connected components, and we add u back into the smallest component. In this way, tree $\tilde{T}$ is partitioned into at most three pairwise-disjoint connected components in the level-0 recursion, each of which is a tree on no more than $|\tilde{T}|/2$ nodes. We call each of these three components a *level-1 component* and build the data structure recursively upon each of them. In general, at level i of the recursion, we compute a centroid, v, of each level-i component γ. We then use Lemma 4 to construct a data structure $D(\gamma)$ supporting 2-approximate colored path counting over paths that are entirely contained within γ and also contain v. This component can be partitioned into up to three level-$(i+1)$ components using the approach described above. We call component γ the *parent* of these up to three level-$(i+1)$ components, and each of these up to three level-$(i+1)$ components is a *child* of γ. A component that has a single node is called a *base component* and is not partitioned further. Hence, the recursion has $O(\lg n)$ levels.

Suppose that query path $P_{x,y}$ is contained entirely within a level-t component γ but not in any level-$(t+1)$ components. This means that $P_{x,y}$ contains the centroid of γ. Therefore, we can find a 2-approximate of $|C(P_{x,y})|$ using the data structure $D(\gamma)$, provided that the preorder ranks of x and y in γ are known. To locate component γ and then to compute the preorder rank of x and y in it, we define a *component tree* CT. A component tree is a 3-ary tree in which each node represents a component and the edges represent the parent-child relationship between components. More specifically, a node v at level l of CT represents a level-l component $\mathcal{C}_v$, where l starts from 0. A node v of CT is the parent of another node u iff component $\mathcal{C}_v$ is the parent of component $\mathcal{C}_u$. Among the nodes that share the same parent in CT, the relative order between them does not matter, so we order them arbitrarily. The height of CT is bounded by $O(\lg n)$, and each leaf in it represents a base component. Since each internal node has at least two children, CT has $O(n)$ nodes in total.

At each internal node v of CT, we build an array $\mathtt{SP}(v)$ of length $|\mathcal{C}_v|$, in which $\mathtt{SP}(v)[i]$ is set to be d if the i-th node (in preorder) in $\mathcal{C}_v$ is stored in the d-th child component of v in the next level. Then we represent $\mathtt{SP}(v)$ using Lemma 2 to support $\mathtt{rank}'$. Since v can have at most 3 children, the alphabet size of $\mathtt{SP}(v)$ is constant. Therefore, $\mathtt{SP}(v)$ is represented in $O(|\mathtt{SP}(v)|)$ bits. With these data structures, we can support queries over arbitrary paths and achieve Lemma 5.

Lemma 5. *Let T be an ordinal tree on n nodes in which each node is assigned a color. A data structure of $O(n)$ words can be constructed in $O(n \lg n)$ time to support 2-approximate colored path counting over T in $O(\lg n)$ time.*

Proof. Given a node $i \in \tilde{T}$, let π denote the path from the root of CT to the leaf of CT representing the base component that contains i, and let π_l denote the node in π whose depth is l. First, we show how to locate π_l and to compute the preorder rank of i in component $\mathcal{C}_{\pi_l}$ for $l = 0, 1, 2, \ldots$. The procedure proceeds as follows: We start at the root π_0 of CT. The preorder rank of i in $\mathcal{C}_{\pi_0}$ is i, and π_1 is the $\mathtt{SP}(\pi_0)[i]$-th child of π_0, following the definition of array $\mathtt{SP}(\pi_0)$. In general, given that the preorder rank of i in $\mathcal{C}_{\pi_l}$ is j, one can find node π_{l+1}, which is the $\mathtt{SP}(\pi_l)[j]$-th child of π_l. Since tree extraction preserves preorder, the preorder rank of i in $\mathcal{C}_{\pi_{l+1}}$ is $\mathtt{rank}'(\mathtt{SP}(\pi_l), j) - 1$. Each $\mathtt{rank}'$ query takes constant time, so this procedure uses constant time per level of CT.

Since each node of T stores the preorder rank of its corresponding node in $\tilde{T}$, to answer a query, it is sufficient to compute a 2-approximation of $|C(P_{x,y})|$ for a query path $P_{x,y}$ in $\tilde{T}$. This can be done by performing the top-down traversals of $\mathcal{CT}$ described in the previous paragraph for x and for y simultaneously until we reach the lowest level, l, of $\mathcal{CT}$ such that x and y are contained in the same level-l component γ. This process also gives us the preorder ranks of x and y in γ, which allows us to query $D(\gamma)$ to find a 2-approximate answer. Since $\mathcal{CT}$ has $O(\lg n)$ levels, the query algorithm uses $O(\lg n)$ time.

To analyze the space cost, observe that the total number of nodes in the components at the same level is at most the number of nodes of $\tilde{T}$, which is $2n$. The component tree contains $O(\lg n)$ levels, and all $D(\gamma)$'s and $SP(v)$'s at the same level use $O(n)$ bits, for a total of $O(n \lg n)$ bits, or $O(n)$ words. The $O(n)$-node component tree $\mathcal{CT}$ itself occupies another $O(n)$ words of space. Therefore, the total space cost is $O(n)$ words. The data structure at each level can be constructed in linear time, so the overall construction time is $O(n \lg n)$.   □

3.3 Speeding Up the Query

To further improve the query efficiency in Lemma 5, observe that two procedures introduced before require $O(\lg n)$ time for a query $P_{x,y}$ in $\tilde{T}$: The first locates the lowest component γ in $\mathcal{CT}$ that contains both nodes x and y, and the second computes the preorder ranks of x and y in γ. Previously, both procedures proceed in the same top-down traversal of $\mathcal{CT}$. Now, we perform them separately. For the first procedure, observe that the node representing component γ in $\mathcal{CT}$ must be the lowest common ancestor of the two leaves of $\mathcal{CT}$ representing the base components that contain nodes x and y, respectively. To locate γ in constant time, we can represent $\mathcal{CT}$ using Lemma 1 to support lca in $O(1)$ time and store with each node x of $\tilde{T}$ a pointer to the base component that contains x. This incurs $O(n)$ words of space and $O(n)$ preprocessing time. To improve the second procedure, we model it by defining an operation, $\text{locate}(v, x)$; given a node v of $\mathcal{CT}$ and a tree node x of $\tilde{T}$ that appears in component $\mathcal{C}_v$, $\text{locate}(v, x)$ returns the preorder rank of node x in $\mathcal{C}_v$.

To support $\text{locate}(v, x)$, we borrow ideas from the solution to the ball inheritance problem [5] and achieve various trade-offs. Let π denote the path between the root of $\mathcal{CT}$ and the leaf representing the base component that contains x, and let π_l denote the node in π whose depth is l. Then each component $\mathcal{C}_{\pi_l}$ contains a copy of node x. Array SP's in Sect. 3.2 work as pointers between these copies in components at consecutive levels. The algorithm in the proof of Lemma 5 follows these pointers one by one till we locate x in $\mathcal{C}_v$, which requires $O(\lg n)$ time. To speed it up, we construct *skipping pointers* which allow us to jump over many levels at one time: Suppose that we have computed the preorder rank, i, of node x in component $\mathcal{C}_{\pi_l}$, and we need to locate x in $\mathcal{C}_{\pi_{l+\Delta}}$ for some positive integer Δ. What we can do is to build an array $SP_\Delta(\pi_l)$ with length $|\mathcal{C}_{\pi_l}|$, in which $SP_\Delta(\pi_l)[k]$ is set to d if the k-th node in preorder in $\mathcal{C}_{\pi_l}$ appears in the component represented by the d-th descendant of π_l at depth $l + \Delta$ of $\mathcal{CT}$. If this node is not stored in any level-$(l + \Delta)$ descendant component of π_l (this

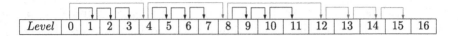

Level	0	1	2	3	4	5	6	7	8	9	10	11	12	13	14	15	16

Fig. 1. An example of the first traversal strategy. In this example, the component tree CT has 16 levels, and $B = 4$. The arrows represent the skipping pointer. By following 6 skipping pointers, one can reach level-15 from the root level.

may happen when CT is not a complete tree), then $\mathrm{SP}_\Delta(\pi_l)[k] = -1$. Since π_l has up to 3^Δ descendants at level $l + \Delta$, we can represent $\mathrm{SP}_\Delta(\pi_l)$ in $O(|\mathcal{C}_{\pi_l}|\Delta)$ bits by Lemma 2 to support $\mathtt{rank'}$. Then $\mathtt{rank'}(\mathrm{SP}_\Delta(\pi_l), i) - 1$ is the preorder rank of node x in $\mathcal{C}_{\pi_{l+\Delta}}$ and can be computed in $O(1)$ time. We regard $\mathrm{SP}_\Delta(\pi_l)$ as an array of skipping pointers that connect the nodes in component $\mathcal{C}_{\pi_l}$ to the nodes in level-$(l + \Delta)$ descendant components of π_l.

If an array of skipping pointers map nodes in a level-l component to nodes in level-$(l+\Delta)$ descendants of this component, then we say that the length of each of these skipping pointers is Δ. Furthermore, based on previous discussions, storing a skipping pointer of length Δ incurs a space cost of $O(\Delta)$ bits. To achieve good time/space trade-offs, we design two strategies to decide what skipping pointers to construct for each level. Henceforth, let $h = O(\lg n)$ denote the height of CT. Let $B \in [2, h]$ be an integer parameter to be chosen later, and let $\tau = \log_B h$; for simplicity, assume that τ is an integer.

In the first strategy, consider level l of the component tree CT. For each integer $i \in [0, \tau - 1]$ such that l is a multiple of B^i but $l + B^i$ is not a multiple of B^{i+1}, we build an array of length B^i skipping pointers for each level-l component. See Fig. 1 for an example. Since at most $\frac{h}{B^i}$ levels of CT have skipping pointers of length B^i, the total space cost of all the skipping pointers in this strategy is $\sum_{i=0}^{\tau-1}(\frac{h}{B^i}) \cdot O(nB^i) = O(n \lg n \log_B \lg n)$ bits, which is $O(n \log_B \lg n)$ words.

To use these skipping pointers to compute $\mathtt{locate}(v, x)$, let $b_{\tau-1}b_{\tau-2}\cdots b_0$ denote the base-B expression of the depth[4], l_v, of node v in CT. That is, each b_i is in $[0, B-1]$ and $l_v = \sum_{i=0}^{\tau-1} b_i B^i$. We then compute $\mathtt{locate}(v, x)$ in τ phases. In phase-1, we start from the root of CT and follow length $B^{\tau-1}$ skipping pointers $b_{\tau-1}$ times. Each time after we follow a skipping pointer to reach a level of CT, we use $\mathtt{level\_anc}$ to locate the ancestor, u, of v at that level. We then follow the skipping pointers in $\mathrm{SP}_{B^{\tau-1}}(u)$ to continue this phase. At the end of phase-1, we have located the ancestor of v at level $b_{\tau-1}B^{\tau-1}$ of CT and computed the preorder rank of node x in the component that this ancestor represents. In phase-2, we start from this ancestor and follow length $B^{\tau-2}$ skipping pointers $b_{\tau-2}$ times, and so on. In general, in phase-p, we follow length $B^{\tau-p}$ skipping pointers $b_{\tau-p}$ times, reach the ancestor of v at level $\sum_{j=1}^{p} b_{\tau-p}B^{\tau-p}$ of CT and compute the preorder rank of node x in the component represented by this ancestor. Thus, we reach v and compute the answer after τ phases. Since we

---

[4] Note that a component tree has $O(\lg n)$ depths and the base-B expression of any depth can be encoded in $O(\log B \times \log_B \lg n) = O(\lg n)$ bits. Storing the base-B expressions of all $O(\lg n)$ depths uses $O(\lg n)$ words of space overall.

follow at most $B - 1$ skipping pointers in each phase, the total running time is $O(B\tau) = O(B \log_B \lg n)$. Setting $B = \lg^\lambda n$ for an arbitrary constant $\lambda \in (0, 1)$ yields a solution with $O(n)$ space and $O(\lg^\lambda n)$-time support for locate.

The second strategy improves the running time of the above process by constructing a different set of skipping pointers so that each phase can be completed by following exactly one skipping pointer. Let l be an arbitrary level of $\mathcal{CT}$. For each integer $i \in [0, \tau - 1]$ such that l is a multiple of B^i, we construct for level-l clusters of skipping pointers of length $B^{i-1}, 2B^{i-1}, \ldots, (B-1)B^{i-1}$. With these skipping pointers, in phase-p, for each $p \in [\tau]$, of the above process, we only need one hop by following a skipping pointer of length $b_{\tau-p}B^{\tau-p}$, decreasing the total query time to $O(\tau) = O(\log_B \lg n)$. The total space cost of these skipping pointers is then at most $\sum_{i=0}^{\tau-1} \frac{h}{B^i} \cdot (B - 1)O(nB^i) = O(nB \lg n \log_B \lg n)$ bits, which is $O(nB \log_B \lg n)$ words. Setting $B = 2$ bounds the space cost by $O(n \lg \lg n)$ and query time by $O(\lg \lg n)$, while setting $B = \lg^\lambda n$ bounds the space cost by $O(n \lg^\lambda n)$ and query time by $O(\lambda^{-1}) = O(1)$ for any constant $0 < \lambda < 1$.

With these three trade-offs for locate, we have the following theorem:

Theorem 1. *Let T be an ordinal tree on n nodes in which each node is assigned a color. A data structure of $s(n)$ words can be constructed in $O(n \lg n)$ time to support 2-approximate colored path counting over T in $q(n)$ time, where a) $s(n) = O(n)$ and $q(n) = O(\lg^\lambda n)$; b) $s(n) = O(n \lg \lg n)$ and $q(n) = O(\lg \lg n)$; or c) $s(n) = O(n \lg^\lambda n)$ and $q(n) = O(1)$ for any constant $0 < \lambda < 1$.*

4 $(1 \pm \epsilon)$-Approximate Colored Path Counting

We first present in Sect. 4.1 a data structure for optimal colored type-2 path counting, which implies the support for colored path reporting. This data structure is further used in our solution to $(1 \pm \epsilon)$-approximate colored path counting. In Sect. 4.2, we perform random sampling of node colors and construct an extracted tree accordingly. We also determine a condition under which this extracted tree can be used to compute a $(1 \pm \epsilon)$-approximate answer with high probability. Then, in Sect. 4.3, we show how to combine the techniques in the previous two subsections to design a solution for the case in which the number of distinct colors in a query path is in $[\kappa/2, 2\kappa]$, where κ is an integer parameter specified in the preprocessing stage. Finally, in Sect. 4.4, we construct a set of data structures from Sect. 4.3, each for a different parameter κ. We then use 2-approximate colored path counting to determine a range that the exact answer to the query must be in, so that we can use an appropriate building block to compute a $(1 \pm \epsilon)$-approximate answer.

4.1   A New Solution to Colored Type-2 Path Counting

Let $P_{x,x'}$ denote a query path such that x' is an ancestor of x. Consider colored type-2 path counting for $P_{x,x'}$. Our strategy can be described as follows: For each color $c \in C(P_{x,x'})$, we locate the lowest node, ℓ_c, in $P_{x,x'}$ whose color is

c, as well as the highest node, h_c, in $P_{x,x'}$ colored in c. Then the frequency of color c in $P_{x,x'}$ is $\mathtt{depth}_c(\ell_c) - \mathtt{depth}_c(h_c) + 1$. If we precompute the value of $\mathtt{depth}_{c(v)}(v)$ for each node v, then, after locating ℓ_c and h_c, we can compute the frequency of color c in $P_{x,x'}$ immediately. The details of finding h_c's and ℓ_c's and extending this method to general query paths are deferred to the full version of this paper. The result is summarized as Lemma 6.

Lemma 6. *Let T be an ordinal tree on n nodes with each node assigned a color from $\{0, 1, \ldots, C-1\}$, where $C \leq n$. An $O(n)$-word data structure can be constructed over T in $O(n \lg n)$ time to support colored type-2 path counting in $O(\mathtt{occ})$ time, where $\mathtt{occ}$ denotes the number of distinct colors in a query path.*

4.2   Random Sampling

We now perform random sampling of node colors and apply tree extraction accordingly. Set $\theta = \frac{6(c_1+3)\lg n}{\epsilon^2 \lg e}$, where e denotes Euler's number and $c_1 \geq 1$ is an arbitrary positive constant. Let $\kappa \in (\theta, n]$ be an integer parameter to be chosen later, and define $\mathtt{M} = \theta/\kappa$. We create a random color set C' by choosing each color that appears in T independently at random with probability $\mathtt{M}$. Then we construct a tree extraction T' from T by removing nodes whose colors are not in C' using the approach described in Sect. 2. All the nodes in T' are assigned their original color in T except for the dummy root; if a dummy root is added, it is uncolored. For each color $c \in \{0, \cdots, C-1\}$, let X_c denote a random variable indicating whether color c is sampled: X_c is set to 1 if c has been sampled and 0 otherwise. Thus, $\Pr[X_c = 1] = \mathtt{M}$. Furthermore, for an arbitrary path $P_{x,y}$ in T, we define a random variable $X_{x,y} = \sum_{c \in C(P_{x,y})} X_c$. Lemma 7 states the conditions under which $X_{x,y}/\mathtt{M}$ is a $(1 \pm \epsilon)$-approximation of $|C(P_{x,y})|$ with high probability; the proof is deferred to the full version of this paper.

Lemma 7. *Consider an arbitrary path $P_{x,y}$ in T. If $\mathtt{occ} \geq \kappa/2$, where $\mathtt{occ}$ denotes $|C(P_{x,y})|$, then $\Pr[(1-\epsilon)\mathtt{occ} \leq \frac{X_{x,y}}{\mathtt{M}} \leq (1+\epsilon)\mathtt{occ}] > 1 - \frac{2}{n^{c_1+3}}$.*

4.3   Approximate Colored Path Counting over Canonical Paths

To use Lemma 7 and also due to other considerations, we call a path in T *canonical* if the number of colors that appear in the path is in $[\kappa/2, 2\kappa]$, for an integer $\lceil \theta \rceil \leq \kappa \leq C/2$ to be decided later. We first solve the $(1 \pm \epsilon)$-approximate problem for canonical paths:

Lemma 8. *Let T be an ordinal tree on n nodes represented by Lemma 1. With success probability more than $1 - \frac{1}{n^{c_1+1}}$, one can construct a data structure in $O(n \lg n)$ worst-case time to answer $(1 \pm \epsilon)$-approximate colored path counting queries over canonical paths in $O(\frac{1}{\epsilon^2} \lg n)$ worst-case time. The space cost is $O(n \cdot \mathtt{M} + n/\lg n)$ words in the expected case (and $O(n)$ words in the worst case).*

Proof. First, we present the data structures. As described in Sect. 4.2, we choose a random color set and construct a tree extraction T' consisting of nodes whose colors are sampled. Tree T' has $O(n \cdot M)$ expected number of nodes but $O(n)$ nodes in the worst case. We represent T' by Lemma 1 in $O(|T'| \lg C)$ bits to support fast navigation. We also construct the data structure of Lemma 3 over T and T' to support decompose in constant time. The data structure uses $O(n)$ bits which is $O(n/\lg n)$ words. Finally, we construct in $O(n \lg n)$ time in the worst case the linear space data structure for colored path reporting queries over T' by applying Lemma 6. The overall space cost is $O(n \cdot M + n/\lg n)$ words in the expected case, and $O(n)$ words in the worst case. The construction time is bounded by $O(n \lg n)$.

To describe the query algorithm, let $P_{x,y}$ be a canonical query path. Since $|C(P_{x,y})| \geq \kappa/2$, $X_{x,y}/M$ is a $(1 \pm \epsilon)$-approximate of $|C(P_{x,y})|$ with probability greater than $1 - \frac{2}{n^{c_1+3}}$ by Lemma 7. Since M is given in preprocessing, we need only compute $X_{x,y}$. Let x' and y' be the nodes of T' returned by decompose(x, y) in $O(1)$ time. If x' and y' are null, no color in $P_{x,y}$ has been sampled, so we set $X_{x,y}$ to be 0. Otherwise, $X_{x,y}$ is either $|C(P_{x',y'})|$ or $|C(P_{x',y'})| - 1$, and we can determine which case it is by performing these steps: If the color of $z = \mathrm{lca}(x, y)$ in T is sampled, then its corresponding node, z', in T' belongs to $P_{x',y'}$. In this case, there is a one-to-one correspondence between the nodes in $P_{x',y'}$ and the nodes in $P_{x,y}$ whose colors are sampled, so $X_{x,y} = |C(P_{x',y'})|$. If the color of z is not sampled, then the node $z'' = \mathrm{lca}(x', y')$ in T' does not correspond to z, but all other nodes in $P_{x',y'}$ correspond to nodes with sampled colors in the query path. Then there are two sub-cases to be considered, depending on whether the color of z'' also happens to appear in $P_{x',y'} \setminus \{z''\}$. If it does, then we have $X_{x,y} = |C(P_{x',y'})|$, and otherwise, $X_{x,y} = |C(P_{x',y'})| - 1$. Navigational operations such as lca can be performed over T and T' in constant time, and whether $c(z'')$ appears in $P_{x',y'} \setminus \{z''\}$ can be tested by two path emptiness queries in $O(\lg \lg C)$ time. Therefore, if we know the value of $|C(P_{x',y'})|$, we can compute $X_{x,y}$ in $O(\lg \lg C)$ extra time.

It remains to show how to compute $|C(P_{x',y'})|$. To do it, observe that if $X_{x,y}/M$ is a $(1 \pm \epsilon)$-approximation, then $X_{x,y} \leq (1+\epsilon) \cdot M \cdot |C(P_{x,y})| \leq (1+\epsilon) \cdot M \cdot 2\kappa$. Since $X_{x,y}$ is at least $|C(P_{x',y'})| - 1$, we have $|C(P_{x',y'})| \leq (1+\epsilon) \cdot M \cdot 2\kappa + 1$. With this, we can apply Lemma 6 to report the distinct colors in $C(P'_{x',y'})$, and instead of reporting all these colors, we stop when the number of reported colors reaches $(1+\epsilon) \cdot M \cdot 2\kappa + 2$. If this happens, we terminate our query algorithm with failure. Otherwise, the number of colors reported is $|C(P_{x',y'})|$. Since $(1+\epsilon) \cdot M \cdot 2\kappa + 2 = O(\epsilon^{-2} \lg n)$, this process uses $O(\epsilon^{-2} \lg n)$ time.

By Lemma 7, for an arbitrary query path, our data structure fails to return a $(1 \pm \epsilon)$-approximation with probability $\Pr[|\frac{X}{M} - \mathrm{occ}| > \epsilon \cdot \mathrm{occ}] < \frac{2}{n^{c_1+3}}$. Since there are $\binom{n}{2}$ different query paths, the probability of constructing a data structure that answers all queries correctly is more than $1 - \binom{n}{2} \cdot \frac{2}{n^{c_1+3}} > 1 - \frac{1}{n^{c_1+1}}$. $\square$

Next, we keep resampling colors and building the structure of Lemma 8 for the sample, until we find a data structure that occupies $O(n \cdot M + \frac{n}{\lg n})$ words in the worst case and can always return $(1 \pm \epsilon)$-approximations for canonical paths.

This process requires $O(n^2 \lg \lg C)$ expected preprocessing time (the analysis is deferred to the full version of this paper). Therefore, we achieve:

Lemma 9. *Let T be an ordinal tree on n nodes represented by Lemma 1. A data structure occupying $O(n \cdot \mathtt{M} + n/\lg n)$ extra words in the worst case can be constructed in $O(n^2 \lg \lg C)$ expected time to support $(1 \pm \epsilon)$-approximate colored path counting over canonical paths in $O(\epsilon^{-2} \lg n)$ worst-case time.*

4.4 Approximate Colored Path Counting over Arbitrary Paths

To solve queries over arbitrary paths, we first represent T by Lemma 1 to support navigational operations. We also construct the data structures of part a) of Theorem 1 to support 2-approximate colored path counting over T. In addition, we build the data structures of Lemma 6 to support colored path reporting. These data structures use $O(n)$ words and can be built in $O(n \lg n)$ time.

Then, for each $i \in [\lceil \lg \theta \rceil, \lceil \lg C \rceil)$, let κ_i be 2^i. We refer to a query path as a *tier-i canonical path* if the number of distinct colors that appear in it is in $[\kappa_i/2, 2\kappa_i]$. For each possible value of i, we apply Lemma 9 to construct a data structure $\mathbb{DS}_i$ to support $(1 \pm \epsilon)$-approximate colored path counting over tier-i canonical paths. Data structure $\mathbb{DS}_i$ uses $O(n\theta/\kappa_i + n/\lg n) = O(n\theta/2^i + n/\lg n)$ words in the worst case and can be constructed in $O(n^2 \lg \lg C)$ expected time. Summing up over all $i \in [\lceil \lg \theta \rceil, \lceil \lg C \rceil)$, the overall space cost of these data structures is $O(n)$ words in the worst case, and they can be constructed in $O(n^2 \lg C \lg \lg C)$ expected time. Theorem 2 summarizes our final result.

Theorem 2. *Let T be an ordinal tree on n nodes with each node assigned a color from $\{0, 1, \dots, C-1\}$, where $C \le n$. A data structure of $O(n)$ words of space in the worst case can be constructed in $O(n^2 \lg C \lg \lg C)$ expected time to support $(1 \pm \epsilon)$-approximate colored path counting in $O(\epsilon^{-2} \lg n)$ worst-case time.*

Proof. It remains to show the query algorithm. Let $P_{x,y}$ denote the query path. We first use the colored path reporting structure to report up to θ distinct colors in $C(P_{x,y})$. If less than θ colors are reported, then we return the exact number of colors, taking $O(\epsilon^{-2} \lg n)$ time. Otherwise, $\mathtt{occ} > \theta$. In this case, we compute a 2-approximate result, $\mathtt{occ}_a$, in $O(\lg^\lambda n)$ time. Then, $\mathtt{occ} \le \mathtt{occ}_a \le 2\mathtt{occ}$. Observe that, for any $i \in [\lceil \lg \theta \rceil, \lceil \lg C \rceil)$, if $\kappa_i \le \mathtt{occ}_a \le 2\kappa_i$, then $\kappa_i/2 \le \mathtt{occ} \le 2\kappa_i$. This allows us to perform a binary search in $O(\lg \lg n)$ time to find the value of i such that $P_{x,y}$ is a tier-i canonical path. Finally, by querying $\mathbb{DS}_i$, we can find a $(1 \pm \epsilon)$-approximation of $\mathtt{occ}$ in $O(\epsilon^{-2} \lg n)$ worst-case time. $\square$

References

1. Alman, J., Williams, V.V.: A refined laser method and faster matrix multiplication. In: 32nd Annual ACM-SIAM Symposium on Discrete Algorithms (2021). https://doi.org/10.1137/1.9781611976465.32

2. Belazzougui, D., Cunial, F., Kärkkäinen, J., Mäkinen, V.: Linear-time string index-ing and analysis in small space. ACM Trans. Algorithms (2020). https://doi.org/10.1145/3381417

3. Chan, T.M., He, M., Munro, J.I., Zhou, G.: Succinct indices for path minimum, with applications. Algorithmica (2016). https://doi.org/10.1007/s00453-016-0170-7

4. Chan, T.M., He, Q., Nekrich, Y.: Further results on colored range searching. In: 36th Annual Symposium on Computational Geometry (2020). https://doi.org/10.4230/LIPICS.SOCG.2020.28

5. Chan, T.M., Larsen, K.G., Pătraşcu, M.: Orthogonal range searching on the RAM, revisited. In: 27th Annual Symposium on Computational Geometry (2011). https://doi.org/10.1145/1998196.1998198

6. Chan, T.M., Nekrich, Y.: Better data structures for colored orthogonal range reporting. In: 31st Annual ACM-SIAM Symposium on Discrete Algorithms (2020). https://doi.org/10.1137/1.9781611975994.38

7. Durocher, S., Shah, R., Skala, M., Thankachan, S.V.: Linear-space data structures for range frequency queries on arrays and trees. Algorithmica (2016). https://doi.org/10.1007/s00453-014-9947-8

8. El-Zein, H., Munro, J.I., Nekrich, Y.: Succinct color searching in one dimension. In: 28th International Symposium on Algorithms and Computation (2017). https://doi.org/10.4230/LIPICS.ISAAC.2017.30

9. Ganguly, A., Munro, J.I., Nekrich, Y., Shah, R., Thankachan, S.V.: Categorical range reporting with frequencies. In: 22nd International Conference on Database Theory (2019). https://doi.org/10.4230/LIPICS.ICDT.2019.9

10. Gao, Y., He, M.: Faster path queries in colored trees via sparse matrix multiplica-tion and min-plus product. In: 30th Annual European Symposium on Algorithms (2022). https://doi.org/10.4230/LIPICS.ESA.2022.59

11. Gupta, P., Janardan, R., Smid, M.: Further results on generalized intersection searching problems: counting, reporting, and dynamization. J. Algorithms (1995). https://doi.org/10.1006/jagm.1995.1038

12. He, M., Kazi, S.: Data structures for categorical path counting queries. Theoret. Comput. Sci. (2022). https://doi.org/10.1016/j.tcs.2022.10.011

13. He, M., Munro, J.I., Rao, S.S.: Succinct ordinal trees based on tree covering. ACM Trans. Algorithms (2007). https://doi.org/10.1007/978-3-540-73420-8-45

14. He, M., Munro, J.I., Zhou, G.: A framework for succinct labeled ordinal trees over large alphabets. Algorithmica (2014). https://doi.org/10.1145/2344422.2344432

15. He, M., Munro, J.I., Zhou, G.: Data structures for path queries. ACM Trans. Algorithms (2016). https://doi.org/10.1145/2905368

16. Jordan, C.: Sur les assemblages de lignes. Journal für die reine und angewandte Mathematik (Crelles Journal) (1869). https://doi.org/10.1515/crll.1869.70.185

17. Kaplan, H., Rubin, N., Sharir, M., Verbin, E.: Efficient colored orthogonal range counting. SIAM J. Comput. (2008). https://doi.org/10.1137/070684483

18. Muthukrishnan, S.: Efficient algorithms for document retrieval problems. In: 13th Annual ACM-SIAM Symposium on Discrete Algorithms (2002). https://doi.org/10.5555/545381.545469

19. Nekrich, Y.: Efficient range searching for categorical and plain data. ACM Trans. Database Syst. (2014). https://doi.org/10.1145/2543924

20. Pătraşcu, M.: Lower bounds for 2-Dimensional range counting. In: 39th Annual ACM Symposium on Theory of Computing (2007). https://doi.org/10.1145/1250790.1250797

21. Rahul, S.: Approximate range counting revisited. J. Comput. Geom. (2021). https://doi.org/10.20382/JOCG.V12I1A3
22. Sadakane, K.: Succinct data structures for flexible text retrieval systems. J. Discrete Algorithms (2007). https://doi.org/10.1016/j.jda.2006.03.011
23. Yu, H.: An improved combinatorial algorithm for Boolean matrix multiplication. Inf. Comput. (2018). https://doi.org/10.1016/j.ic.2018.02.006

Quick-Sort Style Approximation Algorithms for Generalizations of Feedback Vertex Set in Tournaments

Sushmita Gupta[1], Sounak Modak[1(✉)], Saket Saurabh[1,2],
and Sanjay Seetharaman[1]

[1] The Institute of Mathematical Sciences, Chennai, India
{sushmitagupta,sounakm,saket,sanjays}@imsc.res.in
[2] University of Bergen, Bergen, Norway

Abstract. A feedback vertex set (FVS) in a digraph is a subset of vertices whose removal makes the digraph acyclic. In other words, it hits all cycles in the digraph. Lokshtanov et al. [TALG '21] gave a factor 2 randomized approximation algorithm for finding a minimum weight FVS in tournaments. We generalize the result by presenting a factor 2α randomized approximation algorithm for finding a minimum weight FVS in digraphs of independence number α; a generalization of tournaments which are digraphs with independence number 1. Using the same framework, we present a factor 2 randomized approximation algorithm for finding a minimum weight Subset FVS in tournaments: given a vertex subset S in addition to the graph, find a subset of vertices that hits all cycles containing at least one vertex in S. Note that FVS in tournaments is a special case of Subset FVS in tournaments in which $S = V(T)$.

1 Introduction

Quicksort is a randomized divide-and-conquer algorithm for sorting a list of numbers. In this we randomly pick a pivot, partition the rest of the list into two parts, and recursively solve the two parts. The choice of pivot determines the size of the subproblems and consequently the overall running time. This forms a central idea in the polynomial time factor 2 randomized approximation algorithm for finding a minimum weight feedback vertex set in tournaments, given by Lokshtanov et al. [7]. In this paper one of our goals is to find further problems for which this approach can be applied in designing approximation algorithms. Without further ado we formally define the problem studied by Lokshtanov et al. [7], which we generalize in this article.

A tournament $T = (V, E)$ is a digraph in which there is exactly one arc between each pair of vertices (an orientation of a clique). A feedback vertex set is a subset of vertices whose removal makes the digraph acyclic. In the FEEDBACK VERTEX SET IN TOURNAMENTS (FVST) problem, we are given a tournament T and a weight function $w : V(G) \to \mathbb{N}_{\geq 0}$. The goal is to find a minimum weight subset of vertices whose removal makes the digraph acyclic. It is a folklore that

J. A. Soto and A. Wiese (Eds.): LATIN 2024, LNCS 14578, pp. 225–240, 2024.
https://doi.org/10.1007/978-3-031-55598-5_15

a tournament is acyclic if and only if it has no triangles (directed cycles of length 3). This together with the local ratio technique [1] gives a simple factor 3 approximation algorithm for FVST. Cai et al. [2] gave the first improvement over this algorithm and designed a 2.5-approximation algorithm based on the total dual integral system combined with the local ratio technique. Years later, Mnich et al. [8] gave a 7/3 approximation algorithm using the iterative rounding technique. Finally in 2020, Lokshtanov et al. [7] gave a "quicksort style" randomized 2-approximation algorithm. It is optimal (i.e. there is no polynomial time approximation algorithm with a better factor) assuming the Unique Games Conjecture [5].

In this paper we apply this methodology to two generalizations of FVST.

1. **Beyond Tournaments.** The *independence number* of a digraph is the size of a largest independent set in it. In [4], Fradkin and Seymour introduced the class of digraphs of bounded independence number as a generalization of tournaments (which are digraphs with independence number 1). Problems studied in such digraphs include k-EDGE DISJOINT PATHS [4], EDGE ODD CYCLE TRANSVERSAL, and FEEDBACK ARC SET [6]. In the DIRECTED FVS IN GRAPHS WITH BOUNDED INDEPENDENCE NUMBER (DFVS-bIN) problem, we are given a digraph G with independence number α, called α-*bounded digraph*, and a weight function $w : V(G) \to \mathbb{N}_{\geq 0}$. We are interested in finding a minimum weight feedback vertex set in G.

2. **Subset Version of FVST.** A feedback vertex set is equivalently defined as a subset of vertices that hits all cycles in a digraph. In the SUBSET FEEDBACK VERTEX SET IN TOURNAMENTS (S-FVST) problem, we are given as subset of vertices S (often called as *terminal set*) as input, in addition to a tournament T and a weight function $w : V(T) \to \mathbb{N}_{\geq 0}$. We are interested in finding a minimum weight subset of vertices that hits all cycles that contain a vertex in S. Note that FVST is a special case of S-FVST in which $S = V(T)$.

We design non-trivial randomized approximation algorithms for both S-FVST and DFVS-bIN, with the techniques used in [7] as a starting point. Note that neither problem is a special case of the other. In DFVS-bIN, we are generalizing the tournament graph by allowing the independence number to be α and in S-FVST, we are generalizing the obstruction set to be cycles that contain a vertex from S. Thus, the algorithm for one does not seem to apply for the other in a straightforward manner.

1.1   Our Results

Our first result is a randomized 2α-approximation algorithm for DFVS-bIN. Observe that G has a cycle if and only if G has a cycle of length at most $2\alpha + 1$. Due to this fact, we can apply the local ratio technique and obtain a $(2\alpha + 1)$-approximate feedback vertex set in polynomial time. Improving upon this, we present the following.

Theorem 1. DFVS-bIN *admits a randomized* 2α-*approximation algorithm that runs in time* $n^{\mathcal{O}(\alpha^2)}$.

Observe that for $\alpha = 1$, we get a 2-approximation, which is the case of tournaments. Note that the running time is a polynomial for fixed α. A natural question that follows from our result is whether there exists an approximation algorithm that runs in time $f(\alpha)n^{\mathcal{O}(1)}$ for some computable function f (i.e., for a fixed α, the running time is a polynomial in n whose degree does not depend on α).

Next, we present a randomized 2-approximation algorithm for the S-FVST problem. We observe that for any vertex $s \in S$, T has a cycle containing s if and only if T has a triangle containing s. Due to this fact, we can apply the local ratio technique and obtain a 3-approximate subset feedback vertex in polynomial time. Improving upon this, we present the following.

Theorem 2. *S-FVST admits a randomized* 2-*approximation algorithm that runs in time* $n^{\mathcal{O}(1)}$.

The proof of this theorem uses ideas similar to that of Theorem 1, with a twist. In both the results, we carefully exploit the fact that the size of the "obstructions" (to a vertex not being part of a cycle) is bounded.

1.2   Our Methodology

In this section we describe our main ideas in proving Theorems 1 and 2. For simplicity of presentation we present our ideas for the unweighted case. To describe our methods, we first give a very brief outline of the algorithm of Lokshtanov et al. [7]. This algorithm starts with the assumption that the tournament T has a feedback vertex set F of size at most $n/2$ (else returning all the vertices of the tournament is a 2-approximation). Observe that if F is a minimum feedback vertex set of T, then $T - F$ is acyclic and has a unique topological ordering. Next it selects a vertex v uniformly at random. The probability that v belongs to the middle $n/6$ vertices of the topological ordering of $T - F$ is at least $1/6$. This v will act as a pivot. The algorithm first deletes all the vertices that participate in directed triangles with v (of course not v!). Since, it is known that v does not belong to F, this step can be carried out at the cost of factor 2 in approximation. After this step it is guaranteed that there is no directed triangle containing v and hence the problem decomposes into two disjoint subproblems of size βn ($\beta < 1$ a constant): one on the tournament induced by the in-neighbors of v (i.e., $T[N^-(v)]$) and the other on the tournament induced by the out-neighbors of v (i.e., $T[N^+(v)]$). The algorithm recursively solves these problems and combines their solutions to obtain a 2-approximation for T.

To generalize the techniques from tournaments to α-bounded digraphs, our first challenge is to come up with a notion equivalent to the *unique topological ordering of an acyclic tournament*, that was exploited in the algorithm of Lokshtanov et al. [7].

Towards this we first observe that every α-bounded digraph G on n vertices has a vertex with *both in-degree and out-degree (at least)* $(n - 2\alpha)/4\alpha$ (Lemma 4). Thereafter we define an HL-degree ordering (HL stands for high to low) of an

α-bounded digraph G as $\sigma(G) = \langle v_1, \ldots, v_n \rangle$ over the vertices of G by the recursive application of Lemma 4. Thus, for v_1 both $d_G^+(v_1)$ and $d_G^-(v_1)$ are at least $(n - 2\alpha)/4\alpha$; for v_2 both $d_{G-\{v_1\}}^+(v_2)$ and $d_{G-\{v_1\}}^-(v_2)$ at least $(n - 1 - 2\alpha)/4\alpha$. Therefore for any $i \in \{2, \ldots, n\}$, both $d_{G-\{v_1,\ldots,v_{i-1}\}}^+(v_i)$ and $d_{G-\{v_1,\ldots,v_{i-1}\}}^-(v_i)$ are at least $(n - (i-1) - 2\alpha)/4\alpha$.

For our algorithm, given an α-bounded digraph G, we will work with a fixed feedback vertex set F and a fixed HL-degree ordering $\sigma(G - F)$. Analogous to the tournament's analysis, we can assume that $|F| \leq \frac{n}{2\alpha}$ (else the whole vertex set of G is a 2α-approximation). Thereafter, we select a vertex v uniformly at random. We say that v is *good* if v belongs to the first $(1/3)^{rd}$ part of $\sigma(G - F)$. The probability of v being good is $\frac{(n - \frac{n}{2\alpha})/3}{n} = (\frac{1}{3} - \frac{1}{6\alpha})$. This vertex v acts as our pivot. Now, we either delete all the vertices that participate in directed cycles of length at most $2\alpha + 1$ with v or delete all the vertices of a directed cycle that is part of the large cycle (i.e., length more than $2\alpha + 1$) containing v, of length at most 2α (follows from Lemma 5). Suppose that v is good. Then this step can be carried out at the cost of factor 2α in approximation. After this step we are guaranteed that there is no directed cycle containing v, and hence the problem decomposes into two disjoint subproblems of size βn, where $\beta < 1$ is a constant depending on α alone: one on the digraph induced by the vertices reachable from v (denoted by $G[R_{G-P}(v)]$) and the other on the digraph induced by the vertices not reachable from v (denoted by $G[\overline{R}_{G-P}(v)]$) , where P is a subset of vertices in G such that the graph $G - P$ does not contain a cycle that contains vertex v. We recursively solve these problems, and combine their solutions to obtain a 2α-approximation for G.

For the S-FVST problem, given a tournament T and terminal set S, we again work with a fixed subset feedback vertex set F and an HL-degree ordering $\sigma(T[S \setminus F])$. But here observe that we only focus on the vertices of S, that is, we want to see how F interacts with S. First we have observed that hitting the triangles containing vertices in S is equivalent to hitting the cycles that contain the vertices of S (Lemma 11). As before, our algorithm starts with the assumption that $|F| \leq |S|/2$ (else returning all the vertices of S gives a 2-approximation). Next it selects a vertex v uniformly at random from S. The probability that v belongs to the first $(1/3)^{rd}$ of $\sigma(T[S - F])$ is at least $1/6$. This vertex v acts as a pivot. We first delete all the vertices that participate in directed triangles with v. Let the set of vertices which participate in directed triangles with v be denoted by P. Since we know that v does not belong to F, this step can be carried out at the cost of factor 2 in approximation. After this step we are guaranteed that there is no directed triangle containing v, and hence there is no directed cycle containing v. Thereafter the problem decomposes into two disjoint subproblems with strictly smaller set of terminals: one on the tournament induced by $R_{T-P}(v)$ with terminal set $S \cap R_{T-P}(v)$ and the other on the tournament induced by $\overline{R}_{T-P}(v)$ with terminal set $S \cap \overline{R}_{T-P}(v)$. We recursively solve these problems, and combine their solutions to obtain a 2-approximation for T.

2   Preliminaries

In this paper, we deal with simple directed graphs (*digraphs*, in short) containing no parallel edges/arcs. We work in the setting of vertex weighted digraphs: by (G, w) we denote a vertex weighted digraph G with weight function $w : V(G) \to \mathbb{N}_{\geq 0}$. The weight of a subset of vertices is the sum of weights of the vertices in the subset. Note that the setting of unweighted graphs is a special case of weighted graphs. For any induced subgraph H of a vertex weighted graph (G, w), we will assume that w defines a weight function when restricted to $V(H)$. If there is an arc $(u, v) \in E(G)$, then u is an *in-neighbor* of v, and v is an *out-neighbor* of u. The *in-neighborhood* of a vertex x, denoted by $N^-(x) = \{v \mid (v, x) \in E(G)\}$, is the set of in-neighbors of x. The *in-degree* of a vertex x, denoted by $d_G^-(x) = |N^-(x)|$, is the number of in-neighbors of x. The *out-neighborhood* and *out-degree* of a vertex x, denoted by $N^+(x)$ and $d_G^+(x)$ resp., are defined analogously.

We say a vertex u is *reachable* from a vertex v, if there is a directed path which contains both the vertices u and v and vertex v appears before vertex u in the sequence of the vertices which defines the directed path. By $R_G(x)$, we denote the set of all vertices other than x reachable from x in G. By $\overline{R_G}(x)$, we denote the set of all vertices not reachable from x in G. Observe that $(R_G(x), \overline{R_G}(x), \{x\})$ form a partition of $V(G)$.

A *feedback vertex set* (FVS) in G is a subset of vertices $S \subseteq V(G)$ such that $G - S$ is acyclic. Given a family $X = \{S_1, \ldots, S_l\}$ where $S_i \subseteq V(G)$ for each $i \in [l]$, we call $S_a \in X$ *lightest* if for all $i \in [l]$ we have $w(S_a) \leq w(S_i)$. Similarly, we call $S_b \in X$ *heaviest* if for all $i \in [l]$ we have $w(S_i) \leq w(S_b)$. An FVS F_{opt} in G is an *optimal* (also *minimum*) solution of the instance (G, w) if for every other FVS S in G we have $w(S) \geq w(F_{opt})$ i.e. F_{opt} is the lightest among all FVSs in G. An FVS F in G is called a 2α-*approximate solution* of the instance (G, w) if $w(F) \leq 2\alpha \cdot w(F_{opt})$.

An algorithm is a *randomized factor f approximation algorithm* for a problem $\mathcal{P}$ if, for each instance $\mathcal{I}$ of $\mathcal{P}$, with probability at least $1/2$, it returns a solution for $\mathcal{I}$ of weight at most $f \times OPT_{\mathcal{I}}$ where $OPT_{\mathcal{I}}$ denotes the weight of an optimal solution for $\mathcal{I}$.

We will use the following structural results on digraphs with bounded independence number throughout our paper[1].

Lemma 3 ([4]). *Let G be a simple digraph with n vertices and independence number $\alpha \geq 1$. Then there exists a vertex in $V(G)$ with out-degree at least $\frac{n-\alpha}{2\alpha}$. Similarly, there exists a vertex in $V(G)$ with in-degree at least $\frac{n-\alpha}{2\alpha}$.*

Lemma 4 (†). *Let G be a simple digraph with n vertices and independence number $\alpha \geq 1$. Then there exists a vertex in $V(G)$ which has both in-degree and out-degree at least $(n - 2\alpha)/4\alpha$.*

Lemma 5. *Let G be a simple digraph with n vertices and independence number $\alpha \geq 1$. For a vertex $x \in V(G)$, let $C \subseteq V(G)$ denote the shortest cycle containing*

---

[1] Missing proofs (marked with †) are in the full version of the paper.

x. Then either $|C| \leq 2\alpha + 1$ or there exists an induced cycle $C' \subset C$ of length at most 2α that does not contain x (i.e., $C' \subseteq C \setminus \{x\}$).

Proof. Consider the case $|C| > 2\alpha + 1$. Assume for the sake of contradiction that $C' = \langle v_1, \ldots, v_\ell \rangle$ is the shortest induced cycle in C that does not contain x with $\ell \geq 2\alpha + 1$. Additionally, assume without loss of generality that C' is enumerated in such a way that it appears in that order in C and x appears before v_1 and after v_ℓ in C.

There is no arc $xv_j \in E(G)$ such that $v_j \in C' \setminus \{v_1\}$ since such an arc would imply the existence of the cycle $\langle v_j, \ldots, v_\ell, \ldots, x \rangle$ which contradicts the fact that C is the shortest cycle in G containing x. Since C' is an induced cycle, the set $I = \{v_2, v_4, \ldots, v_{2\alpha}\}$ forms an independent set of size α. By the previous argument $I \cup \{x\}$ is an independent set of size $\alpha + 1$ in G, a contradiction. $\qquad \square$

3   DFVS in Graphs of Bounded Independence Number

Throughout this section we assume that G is a digraph on n vertices with independence number α. The following two observations about FVSs in (G, w) which we will use throughout our results in this section, follow from the hereditary property of acyclicity of digraphs.

Observation 1. *Let F be an FVS in (G, w); and let $S \subseteq V(G)$. Then, $F \setminus S$ is an FVS in $(G - S, w)$.*

Observation 2. *Suppose that F is an optimal FVS in (G, w) and that S is a subset of F. Then, $F \setminus S$ is an optimal FVS in $(G - S, w)$, of weight $w(F) - w(S)$.*

The following lemma shows the interaction of an FVS F in G with the subgraphs $G[R_G(x)]$ and $G[\overline{R}_G(x)]$.

Lemma 6 (†). *Suppose that $x \in V(G)$ is a vertex that is not part of any cycle in G, then the following holds: F is an FVS in G if and only if $F \cap R_G(x)$ is an FVS in $G[R_G(x)]$ and $F \cap \overline{R}_G(x)$ is an FVS in $G[\overline{R}_G(x)]$.*

3.1   Technical Overview

In this section we will briefly describe the work flow of our recursive algorithm FindFVS (Algorithm 1) which is formally presented in the following section and its correctness analyzed in Sect. 3.2. For the base case $n \leq 30\alpha$, we compute an optimal FVS in $n^{\mathcal{O}(\alpha)}$ time by checking all subsets of vertices. Let F_{opt} be an optimal solution for (G, w) which is an instance for the DFVS-bIN problem. We consider the following two cases:

- **Case 1 ($|F_{opt}| \geq 2n/3\alpha$):** Let $L \subseteq V(G)$ be a set of $n/6\alpha$ lightest vertices in G and r be the heaviest vertex in L. We define the new weight function $w' : V(G) \setminus L \rightarrow \mathbb{N}_{\geq 0}$ which assigns the weight $w(v) - w(r)$ to each vertex v in $G - L$. Our algorithm recursively finds an FVS F in $(G - L, w')$. Combining F with L (i.e., $F \cup L$), we have a 2α-approximate FVS in (G, w) with probability at least $1/2$ (by Claim 8).

- **Case 2** ($|F_{opt}| < 2n/3\alpha$)**:** In this case, we randomly pick/sample a vertex $x \in V(G)$ and find a subset of vertices P such that in $G - P$ there is no cycle that contains x. Then, since any cycle in $G - P$ is contained completely inside exactly one of $G[R_{G-P}(x)]$ and $G[\overline{R}_{G-P}(x)]$, we obtain an FVS in G by combining P with FVSs in those two subgraphs.

  With probability at least $(n - \frac{2n}{3\alpha})/n = (1 - \frac{2}{3\alpha})$, x is not part of F_{opt}, i.e., $x \in V(G) \setminus F_{opt}$. Moreover, with probability at least $(1 - \frac{2}{3\alpha})/3 = (\frac{1}{3} - \frac{2}{9\alpha})$, x is in the first $1/3^{rd}$ part of $\sigma(G - F_{opt})$. Such a vertex x has in-degree and out-degree at least $\frac{n}{18\alpha} + \frac{1}{4\alpha} - \frac{1}{2}$ (by Claim 9). Consequently, both $|R_{G-P}(x)|$ and $|\overline{R}_{G-P}(x)|$ are at most $n(1 - 1/30\alpha)$ (by Equation (4)).

  We perform the following iterative procedure to compute P, which is initialized as $\emptyset$. Let C be a shortest cycle in $G-P$ that contains x and let $C' \subseteq C$ be the shortest induced cycle in C. As $x \notin F_{opt}$, therefore $F_{opt} \cap (C' \setminus \{x\}) \neq \emptyset$. The crucial point to note here is that $|C' \setminus \{x\}| \leq 2\alpha$ regardless of $|C|$ (by Lemma 5).

  Since F_{opt} is an FVS and $x \notin F_{opt}$, we have $|F_{opt} \cap (C' \setminus \{x\})| \geq 1$. Even though C' may not contain x, *hitting* (i.e., picking vertices from) the cycle C' implies hitting the cycle C that contains x. Now observe that if we are in the unweighted setup of the DFVS-bIN problem, then to hit the cycle C' we can pick all vertices in $C' \setminus \{x\}$ in P. Therefore, we are getting a 2α-approximate solution conditioning on the event that $x \notin F_{opt}$.

  Observe that this strategy fails if we are in the weighted setup of the DFVS-bIN problem. We cannot simply pick all the vertices of $C' \setminus \{x\}$. We resolve this issue by using the *"local ratio"* technique as follows. We find the lightest vertex (say u) in $C' \setminus \{x\}$, add it to P and update the weights of each vertex $v \in C' \setminus \{x\}$ to $w(v) - w(u)$. We repeat the procedure until there is no cycle in $G - P$ which contains vertex x.

  Let w' be the weight function at the end of the procedure. Next we recursively get 2α-approximate FVSs in $(G[R_{G-P}(x)], w')$ (say F_1) and in $(G[\overline{R}_{G-P}(x)], w')$ (say F_2) resp., each with probability at least $1/2$. Thereafter we construct $P \cup F_1 \cup F_2$ which is a 2α-approximate FVS in (G, w) (by Claim 10) with probability at least $1/12 - 1/18\alpha$ (by Equation (3)).

  To boost the success probability of the algorithm to the required lower bound $1/2$, we repeat the random experiment 28α times, i.e. we repeat the procedure of sampling the vertex x, computing the set P, and solving the two recursive subproblems $(G[R_{G-P}(x)], w')$ and $(G[\overline{R}_{G-P}(x)], w')$, 28α times.

But we don't know F_{opt} during the execution of the algorithm (and hence which of the cases we fall into). Thus, we compute $28\alpha + 1$ solutions for the instance (G, w): one solution for the Case 1 and 28α solutions for the Case 2. Thereafter we take the lightest among all the $28\alpha + 1$ solutions. Now to analyse the time complexity of the algorithm, observe that in the case that $|F_{opt}| < 2n/3\alpha$, we are making two recursive calls on subproblems of size at most $n \cdot (1 - 1/30\alpha)$ and repeating the procedure for 28α times; for the case when $|F_{opt}| \geq 2n/3\alpha$, we are making one recursive call to the instance $(G - L, w')$ of size $n \cdot (1 - 1/6\alpha)$. Thus, we get the recurrence $T(n) \leq 2 \cdot 28\alpha T \left(n \left(1 - \frac{1}{30\alpha}\right)\right) + \mathcal{O}(n^5) + T \left(n \left(1 - \frac{1}{6\alpha}\right)\right)$ for the time complexity of the algorithm FindFVS, which solves to $n^{\mathcal{O}(\alpha^2)}$.

3.2 The Algorithm

We compute $(28\alpha + 1)$ FVSs $\{F_i\}_{i=0}^{28\alpha}$ and return the lightest set among them. The algorithm is recursive. Each recursive call is made on a graph with strictly fewer vertices. When $|V(G)| \leq 30\alpha$ we solve the problem by brute force searching over all subsets of vertices.

Definition 7. *We use two "weight update" functions both of which take as input a weight function w and a subset of vertices Q and return a new weight function defined as follows.*

1. *$update1(w, Q)$: Let h be the heaviest vertex in Q. It returns $w' : V(G) \backslash Q \to \mathbb{N}$ where*

$$w'(v) = w(v) - w(h) \text{ for each } v \in V(G) \setminus Q.$$

2. *$update2(w, Q)$: Let ℓ be the lightest vertex in Q. It returns $w' : V(G) \to \mathbb{N}$ where*

$$w'(v) = \begin{cases} w(v) - w(\ell) & \text{if } v \in Q; \\ w(v) & \text{otherwise.} \end{cases}$$

Algorithm 1. FindFVS

Input: a digraph $G = (V, E)$, vertex weights $w : V \to \mathbb{N}_{\geq 0}$
Output: a subset of vertices X
1: **if** $n \leq 30\alpha$ **then**
2:     Iterate over all subsets of $V(G)$, and return a optimal FVS of G, say X
3: **end if**
4: Let L be a set of $n/6\alpha$ lightest vertices in $V(G)$ acc. to w
5: $F_0 := L \cup$ FindFVS $(G - L, \text{update1}(w, L))$
6: **for each** $i \in [28\alpha]$ **do**
7:     Pick x_i uniformly at random from $V(G)$
8:     **if** $\min \left\{ d_G^+(x_i), d_G^-(x_i) \right\} < \frac{n}{18\alpha} + \frac{1}{4\alpha} - \frac{1}{2}$ **then**
9:         $F_i := V(G)$
10:         continue
11:     **end if**
12:     $C_i = \{\}$
13:     $w_i = w$
14:     **while** there is a cycle in $G - C_i$ containing x_i **do**   ▷ Eliminating cycles with x_i
15:         Let C be a shortest cycle in $G - C_i$ containing x_i
16:         Let $C' \subseteq C$ be a shortest induced cycle inside C
17:         Let v be a lightest vertex in $C' \setminus \{x_i\}$ acc. to w_i
18:         $C_i = C_i \cup \{v\}$
19:         $w_i = \text{update2}(w_i, C' \setminus \{x_i\})$
20:     **end while**
21:     $F_i := C_i \cup$ FindFVS$(G[R_{G-C_i}(x_i)], w_i) \cup$ FindFVS$(G[\overline{R}_{G-C_i}(x_i)], w_i)$
22: **end for**
23: Return a lightest set among $F_0, F_1, \ldots, F_{28\alpha}$ acc. to w, say X

Analysis. *Proof of* Theorem 1. We will prove this by induction on n. For the base case we consider $n \leq 30\alpha$; where by iterating over all subsets of vertices, we can, in $\mathcal{O}(2^{30\alpha}n^2)$ time, find a minimum weight FVS in G. Hence, from now on, we will analyze when $n > 30\alpha$.

Let F_{opt} be an optimal FVS in G. If $|F_{opt}| \geq 2n/3\alpha$, then we claim that F_0 satisfies the theorem statement. Note that the algorithm returns X such that $w(X) \leq w(F_i)$ for each $i \in [0, 28\alpha]$, i.e. X is the lightest set among $\{F_0, \ldots, F_{28\alpha}\}$.

Claim 8 (†). *Suppose that $|F_{opt}| \geq 2n/3\alpha$. Then, with probability at least $1/2$, F_0 is a 2α-approximate FVS in (G, w).*

Proof. Let w' denote the weight function returned by update1(w, L), Definition 7. Let F' denote the set returned by the recursive call FindFVS$(G - L, w')$. Let v denote the heaviest vertex in L. By applying the induction hypothesis on $G - L$, we have that with probability at least $1/2$, F' is a 2α-approximate FVS in $(G - L, w')$. Therefore, we have $w(F_0) = w(F' \cup L) \leq 2\alpha \cdot w(F_{opt})$. □

Thus, from now on we assume that $|F_{opt}| < 2n/3\alpha$. Next, we will analyze the probabilistic events in our algorithm, by which we will obtain a lower bound on the algorithm accuracy.

Consider the ordering $\sigma(G - F_{opt}) = \langle v_1, \ldots, v_{n-|F_{opt}|} \rangle$ of vertices in $G - F_{opt}$. For each $i \in [28\alpha]$, we say that the randomly chosen vertex x_i is good if $x_i \notin F_{opt}$ and the position of x_i in $\sigma(G - F_{opt})$ is in the first $(n - |F_{opt}|)/3$ vertices. Let E_i^1 denote the event that x_i is good. Thus, for each i, E_i^1 occurs with probability at least $\left(\frac{(n-|F_{opt}|)}{3}\right)/n \geq \frac{1}{3} - \frac{2}{9\alpha}$. The underlying goal of this definition is to bound the size of the recursive subproblems. That is, if x_i is good for some $i \in [28\alpha]$, then the size of the subproblem in the ith iteration of the for loop is bounded, established via eq. (4). Towards this, we first show the following.

Claim 9 (†). *If x_i is good, then both $d_G^+(x_i)$ and $d_G^-(x_i)$ are at least $\frac{n}{18\alpha} + \frac{1}{4\alpha} - \frac{1}{2}$.*

Thus, if either $d_G^+(x_i)$ or $d_G^-(x_i)$ is strictly less than $\frac{n}{18\alpha} + \frac{1}{4\alpha} - \frac{1}{2}$, we conclude that x_i is not good. Then we set $F_i = V(G)$ and continue to find the next FVS (lines 8–11).

Approximation factor analysis. For a fixed $i \in [28\alpha]$, suppose that x_i is good (i.e., we condition on the event E_i^1). Let G_1 and G_2 denote $G[R_{G-C_i}(x_i)]$ and $G[\overline{R}_{G-C_i}(x_i)]$, resp. Let F_i^+ an F_i^- denote the FVSs returned by the recursive calls FindFVS$(G[R_{G-C_i}(x_i)], w_i)$ and FindFVS$(G[\overline{R}_{G-C_i}(x_i)], w_i)$, resp.

Observe that all cycles in $G - C_i$ are contained completely inside either G_1 or G_2. Thus, $F_i^+ \cup F_i^-$ is an FVS in $G - C_i$ and $F_i = C_i \cup F_i^+ \cup F_i^-$ is an FVS in G. Consequently, each set in $\{F_i\}_{i=0}^{28\alpha}$ is an FVS in G and the algorithm always returns an FVS in G. Moreover, $F_{opt} \cap V(G_1)$ and $F_{opt} \cap V(G_2)$ are FVSs in G_1 and G_2 resp., by Lemma 6. Thus, $F_{opt} \setminus C_i = F_{opt} \cap (V(G_1) \cup V(G_2))$ is an FVS in $G_1 \cup G_2 = G - C_i$, by Observation 1.

Let E_i^2 and E_i^3 denote the events that F_i^+ and F_i^- are 2α-approximate FVS in (G_1, w_i) and (G_2, w_i), resp. By applying the induction hypothesis on G_1 and G_2, we have that each of E_i^2 and E_i^3 happens individually with probability at least $1/2$. Since E_i^2 and E_i^3 are independent, both E_i^2 and E_i^3 happen with probability at least $1/2 \cdot 1/2 = 1/4$.

Suppose that F_i^+ and F_i^- are 2α-approximate FVSs in (G_1, w_i) and (G_2, w_i), resp. Therefore, $F_i^+ \cup F_i^-$ is a 2α-approximate FVS in $(G_1 \cup G_2, w_i)$. Consequently, we have that $w_i(F_i^+ \cup F_i^-) \leq 2\alpha \cdot w_i(F_{opt} \setminus C_i)$, since $F_{opt} \setminus C_i$ is an FVS in $G - C_i$, by Observation 1. From now on we condition on the events E_i^1, E_i^2, and E_i^3 and then prove the following.

Claim 10 (†). *The set $F_i = C_i \cup F_i^+ \cup F_i^-$ is a 2α-approximate FVS in (G, w).*

Proof Since F_i is an FVS in G, it suffices to show that $w(C_i \cup F_i^+ \cup F_i^-) \leq 2\alpha \cdot w(F_{opt})$.

Suppose that C_i contains ℓ vertices at the end of the while loop (lines 14–20). Then, w_i, which was initially w, was updated ℓ times using the method update2. Let $w_i^0 = w$. For each $j \in [\ell]$, let w_i^j be the function w_i after j updates and let v_j denote the jth vertex added to C_i.

We will prove a more general condition, which implies the claim, that for each $j \in [\ell]$

$$w_i^{j-1}(F_i^+ \cup F_i^- \cup \{v_j, \ldots, v_\ell\} \cup \{v_1, \ldots, v_{j-1}\}) \leq 2\alpha \cdot w_i^{j-1}(F_{opt}). \quad (1)$$

Note that for a fixed i and $j = 1$, we have $w_i^0 = w$ and $C_i = \{v_1, \ldots, v_\ell\}$ and so the above condition yields $w(F_i^+ \cup F_i^- \cup C_i) \leq 2\alpha \cdot w(F_{opt})$. Observe that by the definition of update2,

$$w_i^{j-1}(v_k) = 0 \text{ for all } k \in [j-1]. \quad (2)$$

Then, Eq. (1) is equivalent to the following:

$$w_i^{j-1}(F_i^+ \cup F_i^- \cup \{v_j, \ldots, v_\ell\}) \leq 2\alpha \cdot w_i^{j-1}(F_{opt} \setminus \{v_1, \ldots, v_{j-1}\}). \quad \text{by eq.(2)}$$

Our proof will use induction on the value of j, in decreasing order. For the base case $j = \ell + 1$ (in which case $w_i^{j-1} = w_i^\ell = w_i$), we have

$$w_i^\ell(F_i^+ \cup F_i^- \cup \{v_1, \ldots, v_\ell\}) = w_i(F_i^+ \cup F_i^-) \leq 2\alpha \cdot w_i^\ell(F_{opt} \setminus C_i). \quad \text{by eq.(2)}$$

We provide a proof of the inductive case in the full version of the paper. This concludes the proof of the claim. ☐

We will conclude the proof of the theorem by showing that our algorithm succeeds with bounded probability within time $\mathcal{O}(n^{122\alpha^2})$.

Probability Analysis. We have conditioned upon three events: (E_i^1) x_i is good, (E_i^2) F_i^+ is a 2α-approximate FVS in (G_1, w_i), and (E_i^3) F_i^- is a 2α-approximate FVS in (G_2, w_i). For a fixed i, these three events happen with probability at least

$$(1/3 - 2/9\alpha) \cdot 1/2 \cdot 1/2 = 1/12 - 1/18\alpha. \quad (3)$$

The probability that for each $i \in [28\alpha]$ at least one of $\{E_i^1, E_i^2, E_i^3\}$ does not happen is at most $(11/12 + 1/18\alpha)^{28\alpha} \leq 1/2$ because $\alpha \geq 1$. Thus, with probability at least $1/2$ there exists $i \in [28\alpha]$ such that all the three events occur and consequently $F_i^+ \cup F_i^- \cup C_i$ is a 2α-approximate FVS in (G, w).

Running Time Analysis. If $n \leq 30\alpha$, then the algorithm runs in $\mathcal{O}(2^{30\alpha}n^2)$ time. From now on, consider the case $n > 30\alpha$. Each iteration of the while loop (lines 14–20) can be done in $\mathcal{O}(n^3)$ time since finding a shortest cycle C (line 15), a shortest induced cycle $C' \subseteq C$ (line 16), and a lightest vertex in C' (line 17) can all be done in $\mathcal{O}(n^3)$ time.

Since in each iteration, a vertex $v \in G \setminus C_i$ is added to C_i (which was initially empty), the repeat loop is carried out for at most n steps. Therefore, the repeat loop can be done in time $\mathcal{O}(n^4)$. Before we consider recursive calls, we would like to note that finding the $n/6\alpha$ lightest vertices in G (line 4) can be done in $\mathcal{O}(n \log n)$ time and finding a lightest set (line 23) can be done in $\mathcal{O}(n\alpha)$ time.

If x_i is not good, then F_i is set to $V(G)$ and no further recursive calls are made. Recall that by Claim 9, if x_i is good, both $d_G^+(x_i)$ and $d_G^-(x_i)$ are at least $n/18\alpha + 1/4\alpha - 1/2$. Since $G - C_i$ does not contain any cycle that x_i is part of, the number of vertices in $G[R_{G-C_i}(x_i))]$ and $G[\overline{R}_{G-C_i}(x_i)]$ is at most $n - (n/18\alpha + 1/4\alpha - 1/2)$. Upon simplification, we note that

$$n - (n/18\alpha + 1/4\alpha - 1/2) \leq n(1 - 1/18\alpha) + 1/2 \leq n(1 - 1/30\alpha), \qquad (4)$$

where the last inequality follows from the assumption that $n > 30\alpha$.

Thus, the overall running time is given by an application of the Master theorem [3] to the recurrence relation

$$T(n) \leq 2 \cdot 28\alpha T(n(1 - 1/30\alpha)) + 28\alpha\mathcal{O}(n^4) + T(n(1 - 1/6\alpha)) = \mathcal{O}(n^{122\alpha^2}). \quad (5)$$

This concludes the proof of the theorem.

4 Subset FVS in Tournaments

In addition to a tournament T on n vertices and a weight function $w : V(T) \to \mathbb{N}_{\geq 0}$, we are given as input a vertex subset $S \subseteq V(T)$ of size s. We say that $F \subseteq V(T)$ is a *subset feedback vertex set* (SFVS, in short) in T if there is no cycle containing vertices of S in $T - F$. The goal is to find a minimum weight SFVS in T. Observe that if $S = V(T)$, then the problem is a case of DFVS-bIN with $\alpha = 1$.

By (T, S, w), we denote an instance of S-FVST. The following observations follow from the hereditary property of subset-acyclicity.

Observation 3. *Let F be an SFVS in (T, S, w) and let $X \subseteq V(G)$. Then $F \setminus X$ is an SFVS in $(T - X, S \setminus X, w)$.*

Observation 4. *Suppose that F is an optimal SFVS in (T, S, w) and X is a subset of F. Then, $F \setminus X$ is an optimal SFVS in $(T - X, S \setminus X, w)$, of weight $w(F) - w(X)$.*

In our discussions, a *triangle* ($\triangle$, in short) is a directed cycle of length three. The following structural lemma gives us the fact that, hitting all triangles passing through the vertices of S is equivalent to hitting all cycles passing through S. As a consequence, we have that $F \subseteq V(T)$ is an SFVS if and only if in $T - F$ there is no triangle that contains a vertex of S.

Lemma 11 (†). *For a vertex $x \in S$, any shortest cycle containing x is a $\triangle$.*

The following lemma, analogous to Lemma 6, shows the interaction of an SFVS F in (T, S, w) with $T[R_T(x)]$ and $T[\overline{R}_T(x)]$.

Lemma 12 (†). *Suppose that $x \in S$ is a vertex that is not part of any cycle in T, then the following holds: F is an SFVS in (T, S, w) if and only if $F \cap R_T(x)$ is an SFVS in $(T[R_T(x)], S \cap R_T(x), w)$ and $F \cap \overline{R}_T(x)$ is an SFVS in $(T[\overline{R}_T(x)], S \cap \overline{R}_T(x), w)$.*

Next, we present a randomized 2-approximation algorithm for S-FVST that runs in time $n^{\mathcal{O}(1)}$ extending the ideas that we used to solve DFVS-bIN.

The base case of the recursive algorithm is given by $s = |S| \leq 30$. Unlike Algorithm 1, we cannot handle the base case by simply iterating over all subsets of vertices of size at most 30 to find an optimal solution. An SFVS may contain vertices outside S (i.e., in $T - S$). To overcome this, we use the notion of vertex covers. A subset of vertices $B \subseteq V(T)$ is called a *vertex cover* in T if for each arc $(u, v) \in E(T)$ we have $B \cap \{u, v\} \neq \emptyset$. Towards handling the base case, we use the well known fact that a 2-approximate minimum weight vertex cover in a digraph on n vertices can be computed in $\mathcal{O}(n^2)$ time [1] using the subroutine FindVertexCover(T', w') which takes $T' = (V', E')$ as input together with a weight function $w : V(T') \to \mathbb{N}_{\geq 0}$. From now on, by $\binom{S}{\leq 30}$, we denote the subsets of S of size at most 30.

4.1 Technical Overview

Definition 13. *Given a set $S \subseteq V(T)$, we use two "weight update" functions both of which take as input a weight function w and a set of vertices Q and return a new weight function:*

1. *update3(w, Q): Let h be the heaviest vertex in Q. It returns $w' : V(T) \setminus Q \to \mathbb{N}_{\geq 0}$ where*

$$w'(v) = \begin{cases} w(v) - w(h) & \text{if } v \in S \setminus Q; \\ w(v) & \text{otherwise.} \end{cases}$$

2. *update4(w, Q): Let ℓ be the lightest vertex in Q. It returns $w' : V(G) \to \mathbb{N}_{\geq 0}$ where*

$$w'(v) = \begin{cases} w(v) - w(\ell) & \text{if } v \in Q; \\ w(v) & \text{otherwise.} \end{cases}$$

The high level structure of the solution is similar to the one for DFVS-bIN. In this overview, we highlight the key differences from the previous algorithm. Let F_{opt} be an optimal SFVS in T. We consider the following three cases.

- **Case 1** ($|S \cap F_{opt}| \leq 30$): We "guess" this intersection by iterating over all $Q \in \binom{S}{\leq 30}$, i.e., we guess the part of S which is *inside* (say Q) the solution and the part which is *outside* (say O) the solution. After the guessing if we find any $\triangle$ containing only vertices from O, then we cannot extend Q and trivially set S to be the solution, denoted by F_Q. Otherwise, we initially set F_Q to be the vertices which are inside the solution and extend it in two phases. In the first phase, we deal with $\triangle$s where two of its vertices are in O and we add the third vertex of such a $\triangle$ to F_Q. In the second phase, we deal with $\triangle$s where one vertex is in O: we find a 2-approximate weighted vertex cover using FindVertexCover() on the (undirected) graph containing the edges between the end vertices that are not in O, of such $\triangle$s. We show that the extension F_Q corresponding to $Q = S \cap F_{opt}$ is a 2-approximate solution (Claim 14). As a base case of the recursive algorithm, if $s < 30$ then we return the lightest solution in $\{F_Q\}_{Q \in \binom{S}{\leq 30}}$ (say Y).
- **Case 2** ($|S \cap F_{opt}| \geq 2s/3$): Let L be a set of $s/6$ lightest vertices in S and F be the SFVS returned by the recursive call FindSFVS($T - L, S \setminus L$, update3(w, L)). We show that $F_0 = F \cup L$ is a 2-approximate SFVS in (T, S, w) with probability at least $1/2$ (Claim 15).
- **Case 3** ($30 < |S \cap F_{opt}| < 2s/3$): First, we randomly sample a vertex $x \in S$. With probability at least $1/3 - 2/9$, x is in the first $1/3^{rd}$ part of $\sigma(T[S \setminus F_{opt}])$. Such a vertex x has in-degree and out-degree at least $s/18 + 1/4 - 1/2$ in $T[S \setminus F_{opt}]$ (Claim 16). We compute P (and a weight function w'), a set of vertices such that in $T - P$ there is no cycle that contains x. Consequently, both $|S \cap R_{T-P}(x)|$ and $|S \cap \overline{R}_{T-P}|$ are at most $s(1 - 1/30)$. We recursively compute SFVSs F_1 and F_2 in $(T[R_{T-P}(x)], S \cap R_{T-P}(x), w')$ and $(T[\overline{R}_{T-P}], S \cap \overline{R}_{T-P}, w')$ resp. and obtain an 2-approximate SFVS in T: $P \cup F_1 \cup F_2$ (Claim 17). Let $F_1, \ldots, F_{28}$ be the solutions that we get by repeating the random experiment 28 times.

  We show that with probability at least $1/2$, the lightest set among $\{Y, F_0, F_1, \ldots, F_{28}\}$ according to w is a 2-approximate SFVS in T.

For the running time, we show that the number of subproblems is $s^{\mathcal{O}(1)}$ and the time spent at each subproblem is $n^{\mathcal{O}(1)}$. Thus, the overall running time is $n^{\mathcal{O}(1)}$.

4.2 The Algorithm

The full version of the paper contains the pseudocode of the algorithm for SFVS in tournaments.

Analysis. *Proof of* Theorem 2. Let F_{opt} denote an optimal SFVS in T. We will prove by induction on n. For the base case, we consider $|S \cap F_{opt}| \leq 30$ (which subsumes the case $n \leq 30$).

Claim 14 (†). *If $|S \cap F_{opt}| \leq 30$, then Y is a 2-approximate SFVS in (T, S, w).*

From now on, we assume that $|S \cap F_{opt}| > 30$. We will restate the key statements, claims and definitions. Proofs of the following claims are similar to that of the claims that we have proved for the DFVS-bIN problem. The only part of problem which need to be argued is the running time analysis, as in the S-FVST problem the base case is non-trivial.

Now similar to before, we first consider the case $|S \cap F_{opt}| \geq 2s/3$.

Claim 15 (†). *Suppose that $|S \cap F_{opt}| \geq 2s/3$. Then, with probability at least $1/2$, F_0 is a 2-approximate SFVS in (T, S, w).*

From now on, we assume that $|S \cap F_{opt}| < 2s/3$. Consider the ordering $\sigma(T[S \setminus F_{opt}]) = \langle v_1, \ldots, v_{s-|S \cap F_{opt}|} \rangle$ of vertices in $S \setminus F_{opt}$. For each $i \in [28]$, we say that the randomly chosen vertex $x_i \in S$ is *good* if $x_i \notin F_{opt}$ and the position of x_i in $\sigma(T[S \setminus F_{opt}])$ is at most $(s-|S \cap F_{opt}|)/3$. Let E_i^1 denote the event that x_i is good. Thus, for each i, E_i^1 occurs with probability at least $\left(\frac{(s-|S \cap F_{opt}|)}{3}\right)/s \geq \frac{1}{3} - \frac{2}{9}$. Analogous to Claim 9, we have the following.

Claim 16 (†). *If x_i is good, then $\min(d_{T[S]}^+(x_i), d_{T[S]}^-(x_i)) \geq \frac{s}{18} + \frac{1}{4} - \frac{1}{2}$.*

For a fixed $i \in [28]$, suppose that x_i is good (i.e., we condition on the event E_i^1). Similar to Algorithm 1, the set C_i and function w_i are computed as follows.

$C_i = \{\}, w_i = w$
while there is a $\triangle$ in $T - C_i$ that contains x_i **do**   ▷ Eliminating $\triangle$s with x_i
    Let C' be a $\triangle$ in $T - C_i$ containing x_i
    Let v be a lightest vertex in $C' \setminus \{x_i\}$ acc. to w_i
    $C_i = C_i \cup \{v\}$
    $w_i = \mathsf{update4}(w_i, C' \setminus \{x_i\})$
end while

Let $R_i^1 = R_{T-C_i}(x_i)$, $R_i^2 = T[\overline{R}_{T-C_i}(x_i)]$, $T_1 = T[R_i^1]$, and $T_2 = T[R_i^2]$. Observe that all cycles passing through a vertex in S in $T - C_i$ are contained completely inside either T_1 or T_2; otherwise, x_i would form a triangle with an arc from such a cycle (by Lemma 11). Hence, we can deduce that $F_{opt} \setminus C_i = F_{opt} \cap (V(T_1) \cup V(T_2))$. Moreover, since $T_1 \cup T_2 = T - C_i$, we have that $F_{opt} \setminus C_i$ is an SFVS in $(T - C_i, S \setminus C_i, w)$ (by Observation 3).

Let F_i^+ an F_i^- denote the solutions returned by the recursive calls Find-SFVS$(T[R_i^1], S \cap R_i^1, w_i)$ and FindSFVS$(T[R_i^2], S \cap R_i^2, w_i)$, resp. Given that $R_i^1 = R_{T-C_i}(x_i)$ and $R_i^2 = \overline{R}_{T-C_i}(x_i)$, we note that $F_i^+ \cup F_i^-$ is an SFVS in $(T - C_i, S \setminus C_i, w_i)$ (by Lemma 12). Thus, $F_i = C_i \cup F_i^+ \cup F_i^-$ is an SFVS in (T, S, w_i).

Consequently, each set in $\{F_i\}_{i=0}^{28}$ is an SFVS in T. Since each set in $\{F_Q\}_{Q \in \binom{s}{\leq 30}}$ is also an SFVS, the algorithm always returns an SFVS in T. Next, we will analyze the quality of the solution F_i.

Let E_i^2 and E_i^3 denote the events that F_i^+ and F_i^- are 2-approximate SFVSs in $(T_1, S \cap R_i^1, w_i)$ and $(T_2, S \cap R_i^1, w_i)$, resp. From now on, we condition on the events E_i^1, E_i^2, and E_i^3.

By applying the induction hypothesis on T_1 and T_2, we have that each of E_i^2 and E_i^3 happens individually with probability at least $1/2$. As F_i^+ and F_i^- are 2-approximate SFVS in $(T_1, S \cap R_i^1, w_i)$ and $(T_2, S \cap R_i^2, w_i)$, resp., $F_i^+ \cup F_i^-$ is a 2-approximate SFVS in $(T - C_i, S \setminus C_i, w_i)$. Since, $F_{opt} \setminus C_i$ is also an SFVS in $(T - C_i, S \setminus C_i, w_i)$, we can infer that $w_i(F_i^+ \cup F_i^-) \leq 2 \cdot w_i(F_{opt} \setminus C_i)$. The following result is analogous to Claim 10.

Claim 17 (†). *The set $F_i = C_i \cup F_i^+ \cup F_i^-$ is a 2-approximate SFVS in (T, S, w).*

We will conclude the proof of the theorem by showing that our algorithm succeeds with bounded probability in time $n^{\mathcal{O}(1)}$.

Probability Analysis. With probability at least $1/2$ there exists $i \in [28]$ such that all the three events occur and $F_i^+ \cup F_i^- \cup C_i$ is a 2-approximate SFVS in (T, S, w).

Running Time Analysis. If $|S \cap F_{opt}| \leq 30$, then since there are $\mathcal{O}(n^{30})$ subsets of S of size at most 30 and for each subset, its extension to a solution can be computed in time $n^{\mathcal{O}(1)}$, the set $\{F_Q\}_{Q \in \binom{S}{\leq 30}}$ can be computed in time $n^{\mathcal{O}(1)}$. Else for each i, the set C_i and function w_i can be computed in time $\mathcal{O}(n^4)$. Finding the $s/6$ lightest vertices in S can be done in $\mathcal{O}(s \log s)$ time.

If x_i is not good, then F_i is set to S and no further recursive calls are made. By Claim 16, if x_i is good, both $d_{T[S]}^+(x_i)$ and $d_{T[S]}^-(x_i)$ are at least $s/18 + 1/4 - 1/2$. Since $T - C_i$ does not contain any cycle that x_i is part of, the number of vertices in $S \cap R_i^1$ and $S \cap R_i^2$ is at most $s - (s/18 + 1/4 - 1/2) \leq s(1 - 1/18) + 1/2 \leq s(1 - 1/30)$ (since we have assumed that $s > 30$). The total number of recursive subproblems is given by an application of the Master theorem [3] to the recurrence relation $T(s) \leq 2 \cdot 28 T(s(1 - 1/30)) + T(30) + T(s(1 - 1/6)) = \mathcal{O}(s^{122})$.

Thus, the overall running time is $T(s) \cdot$ (time spent at each subproblem) $= T(s) \cdot n^{\mathcal{O}(1)} = n^{\mathcal{O}(1)}$ (since $s \in [n]$). This concludes the proof of the theorem.

Acknowledgement. We thank the anonymous reviewers for their helpful comments and suggestions.

References

1. Bar-Yehuda, R., Bendel, K., Freund, A., Rawitz, D.: Local ratio: a unified framework for approximation algorithms. in memoriam: Shimon even 1935–2004. ACM Comput. Surv. (CSUR) **36**(4), 422–463 (2004)
2. Cai, M.c., Deng, X., Zang, W.: An approximation algorithm for feedback vertex sets in tournaments. SIAM J. Comput. **30**(6), 1993–2007 (2001)

3. Cormen, T.H., Leiserson, C.E., Rivest, R.L., Stein, C.: Introduction to algorithms. MIT press (2022)
4. Fradkin, A., Seymour, P.: Edge-disjoint paths in digraphs with bounded independence number. J. Combinatorial Theory, Series B **110**, 19–46 (2015)
5. Khot, S., Regev, O.: Vertex cover might be hard to approximate to within 2- ε. J. Comput. Syst. Sci. **74**(3), 335–349 (2008)
6. Lochet, W., Lokshtanov, D., Misra, P., Saurabh, S., Sharma, R., Zehavi, M.: Fault tolerant subgraphs with applications in kernelization. In: 11th Innovations in Theoretical Computer Science Conference (ITCS 2020). Schloss Dagstuhl-Leibniz-Zentrum für Informatik (2020)
7. Lokshtanov, D., Misra, P., Mukherjee, J., Panolan, F., Philip, G., Saurabh, S.: 2-approximating feedback vertex set in tournaments. ACM Trans. Algorithms (TALG) **17**(2), 1–14 (2021)
8. Mnich, M., Vassilevska Williams, V., Végh, L.A.: A 7/3-approximation for feedback vertex sets in tournaments. In: 24th Annual European Symposium on Algorithms (ESA 2016). Schloss Dagstuhl-Leibniz-Zentrum fuer Informatik (2016)

Better Algorithms for Online Bin Stretching via Computer Search

Matej Lieskovský$^{(\boxtimes)}$ [ID]

Charles University, Ovocný Trh 560/5, Praha 1 116 36, Czechia
ml@iuuk.mff.cuni.cz

Abstract. ONLINE BIN STRETCHING is a problem closely related to ONLINE BIN PACKING and various scheduling problems. There is extensive history of computer search being used to establish lower bounds for this problem by identifying difficult sets of inputs. We demonstrate a novel approach enabling the use of computer search for finding new algorithms and therefore upper bounds for this problem. This not only leads to improved results for ONLINE BIN STRETCHING, but also shows that computer search can be used to find new algorithms, even for problems that might not appear suitable for this approach.

Keywords: Bin stretching · Multiprocessor scheduling · Bin packing · Online algorithms · Computer search

1 Introduction

ONLINE BIN STRETCHING, introduced by Azar and Regev [1], is a problem somewhat similar to ONLINE BIN PACKING. We are given the number of bins m and then a sequence of items with sizes between 0 and 1 arrives. Each item must be assigned to one of the m bins before processing the next item. The goal is to minimize the total load packed into the largest bin. Importantly, we are assured that the entire input does fit into m bins of size at most 1. We measure the performance of an algorithm by the worst-case load of the largest bin which is known as the stretching factor.

Example: Let us consider the case $m = 2$. The first item that arrives can be packed into the first bin without loss of generality. Suppose that first item is of size $1/3$ and the second item is also of size $1/3$. If we pack the second item into the first bin, two items of size $2/3$ may then arrive, forcing the total load of the largest bin to be at least $4/3$. If we pack the second item into the second bin, an item of size 1 may then arrive, forcing the total load of the largest bin to be at least $4/3$.

In the language of scheduling problems, ONLINE BIN STRETCHING is a variant of $Pm||C_{max}$ where jobs must be assigned in an online manner but there are

Partially supported by GAUK project 234723, and GA ČR project 19-27871X.

J. A. Soto and A. Wiese (Eds.): LATIN 2024, LNCS 14578, pp. 241–253, 2024.
https://doi.org/10.1007/978-3-031-55598-5_16

no release times and we know the optimal makespan of the instance in advance. We will, however, stick to the original terminology of bins and items.

Since lower bounds for these problems typically consist of sets of inputs utilizing a finite number of item sizes, computer search for ONLINE BIN STRETCHING lower bounds has been successful via constraining the adversary to a finite set of possible item sizes. This approach cannot be directly used for computing upper bounds as the resulting algorithm must be able to pack items of arbitrary size, which prevents both an upper bound on the number of distinct item sizes and on the number of items. We demonstrate a novel approach that where we instead limit the decision abilities of the potential algorithms and thus enable computer search for ONLINE BIN STRETCHING upper bounds.

1.1   Previous Results

See Fig. 1 for a graphical representation of the best known results.

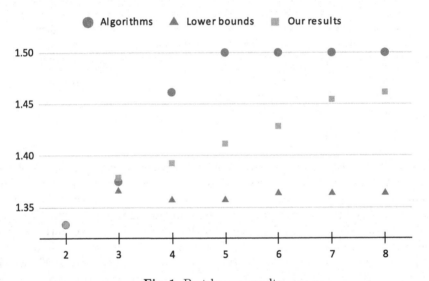

Fig. 1. Best known results

Let us denote the optimal stretching factor for m bins as α_m. Trivially, $\alpha_1 = 1$. Kellerer et al. [7] showed that $\alpha_2 = 4/3 < 1.334$ by demonstrating an algorithm using two bins of size $4/3$ and a set of inputs such that any deterministic algorithm packs at least $4/3$ load into one of the bins for one of the inputs. (See the example earlier.) They also proved that a similar set of inputs exists for three bins and thus $\alpha_3 \geq 4/3 > 1.333$. Azar and Regev [1] generalised the lower bound further, proving that $\alpha_m \geq 4/3 > 1.333$ for all $m \geq 2$. They also presented an algorithm proving $\alpha_m \leq \frac{5m-1}{3m+1}$ for $m \in \{3, 4, \ldots, 21\}$ and $\alpha_m \leq 13/8 = 1.625$ for all m.

The upper bounds have since been improved by multiple algorithms. Kellerer and Kotov [8] proved $\alpha_m \leq 11/7 < 1.572$ for all m, Gabay et al. [5] improved that to $\alpha_m \leq 26/17 < 1.530$ for all m, and the latest results are by Böhm et al. [2] achieving $\alpha_3 \leq 11/8 = 1.375$ and $\alpha_m = 3/2 = 1.5$ for all $m \geq 5$.

Although no better lower bound for general m was found, there are results for small values of m. Gabay et al. [4] introduced the idea of using computer search to find new lower bounds, leveraging the fact that online problems can be viewed as two-player games. This approach was then improved by Böhm and Simon [3] who proved $\alpha_3 \geq 56/41 > 1.365$ and $\alpha_m \geq 19/14 > 1.357$ for $m \in \{4,5,6,7,8\}$. Ongoing efforts to further improve these results have so far yielded $\alpha_m \geq 15/11 > 1.363$ for $m \in \{6,7,8\}$ [9] with more results known to be in preparation.

So far, computer search approaches for lower bounds constrain the possible item sizes to the equally-spaced values $1/k, 2/k, \ldots, 1$ for some granularity k, and then view the problem as a two player game where one player generates items and the other assigns them to the m bins. This game has a finite branching factor $\max(m, k)$ and length $O(mk)$. Unfortunately, placing additional constraints on the item-generating adversary is not permitted in the search for new algorithms.

1.2   Our Results

We modify the computer search approach so that it can be used for finding new algorithms for small values of m. This yields new algorithms, improving upper bounds on α_m for $m \in \{4,5,6,7,8\}$ significantly.

Number of bins	4	5	6	7	8
Previous upper bound	$19/13 < 1.462$	$3/2 = 1.5$	$3/2 = 1.5$	$3/2 = 1.5$	$3/2 = 1.5$
Our upper bound (algorithm)	$39/28$ < 1.393	$31/22$ < 1.410	$20/14$ < 1.429	$16/11$ < 1.455	$19/13$ < 1.462
Time needed (hours)	91	468	23	3.6	235

The computer search was done using a server with an Intel Xeon E5-2630 v3 CPU and 126 GB of RAM. The time complexity grows rapidly with increasing m and granularity, but we do not believe that we have reached any fundamental limit of this method.

For three bins, we found an upper bound of $91/66 < 1.379$ with a run time of 46 h. This is worse than the previous upper bound of $11/8 = 1.375$ by Böhm et al. [2] For $m \geq 9$ we did not find any upper bound better than $3/2 = 1.5$.

2   Reductions Between Games

We view ONLINE BIN STRETCHING as a game. The game is:

- deterministic
- 2-player – we call the players $\mathcal{A}$ (the adversary) and $\mathcal{B}$ (the algorithm)
- sequential – players alternate in taking turns with $\mathcal{A}$ going first
- partisan – $\mathcal{A}$ generates items and $\mathcal{B}$ assigns them to bins
- perfect information
- zero-sum – when the game ends, one player is the winner and the other is the loser

When we talk about winning or losing the game, it will be from the point of view of $\mathcal{B}$. The following definitions are worded carefully since we shall need to apply these methods to games with both unbounded branching and unbounded number of turns.

We shall define a pair of functions, $M_{\mathcal{A}}^G$ and $M_{\mathcal{B}}^G$, which will indicate the possible moves. For example, $M_{\mathcal{B}}^G(p)$ will return the set of positions to which $\mathcal{B}$ can move from position p, always including $lose^G$ (letting $\mathcal{B}$ resign) and possibly including win^G (if $\mathcal{B}$ can claim victory).

We formally define the type of games we are interested in as follows.

Definition 2.1. *Game G is a 7-tuple $(P_{\mathcal{A}}^G, P_{\mathcal{B}}^G, init^G, win^G, lose^G, M_{\mathcal{A}}^G, M_{\mathcal{B}}^G)$ where:*

- $P_{\mathcal{A}}^G$ *is the set of all possible game positions where $\mathcal{A}$ is the next player to move*
- $P_{\mathcal{B}}^G$ *is the set of all possible game positions where $\mathcal{B}$ is the next player to move*
- $init^G \in P_{\mathcal{A}}^G$ *is the initial position*
- win^G *is the winning position where the game ends and $\mathcal{B}$ wins*
- $lose^G$ *is the losing position where the game ends and $\mathcal{B}$ loses*
- $M_{\mathcal{A}}^G : P_{\mathcal{A}}^G \to \mathcal{P}\left(P_{\mathcal{B}}^G \cup \{win^G, lose^G\}\right)$ *indicates all possible moves of $\mathcal{A}$*
- $M_{\mathcal{B}}^G : P_{\mathcal{B}}^G \to \mathcal{P}\left(P_{\mathcal{A}}^G \cup \{win^G, lose^G\}\right)$ *indicates all possible moves of $\mathcal{B}$*

We further require that the following constraints hold:

- $P_{\mathcal{A}}^G, P_{\mathcal{B}}^G, \{win^G\}$ *and $\{lose^G\}$ are disjoint sets*
- $\forall p \in P_{\mathcal{A}}^G : win^G \in M_{\mathcal{A}}^G(p)$ *– $\mathcal{A}$ can resign whenever it is their turn*
- $\forall p \in P_{\mathcal{B}}^G : lose^G \in M_{\mathcal{B}}^G(p)$ *– $\mathcal{B}$ can resign whenever it is their turn*

When talking about positions, we are mostly interested in the case where the next player to move is $\mathcal{A}$. Since $\mathcal{A}$ can make the game of ONLINE BIN STRETCHING last for an unbounded number of turns, and an algorithm that can always pack the next item is satisfactory, we are interested in non-losing rather than winning positions. Therefore, we define the concept of losing positions as follows by building up the set of losing positions iteratively.

Definition 2.2. *Let G be a game. We define*

- $L_0^G = \{lose^G\}$

- $L_{n+1}^G = L_n^G \cup \{a \in P_{\mathcal{A}}^G \mid (\exists b \in M_{\mathcal{A}}^G(a))(\forall c \in M_{\mathcal{B}}^G(b))(c \in L_n^G)\}$
- $L^G = \bigcup_{i=0}^{\infty} L_i^G$. We then call L^G the set of losing positions.

A position is non-losing if it is not losing and game G is non-losing if $init^G$ is a non-losing position.

Our paper relies on the concept of reductions among games. We say that a game G reduces to game H if we can translate the moves of the adversary from G to H and then translate the algorithms response back from H to G in such a way, that if the algorithm avoids losing H, then the same result is achieved in G.

Definition 2.3 (Reduction). *For any two games G and H, we say that $f :$*
$P_{\mathcal{A}}^G \cup P_{\mathcal{B}}^G \cup \{win^G, lose^G\} \to P_{\mathcal{A}}^H \cup P_{\mathcal{B}}^H \cup \{win^H, lose^H\}$ *is a reduction from game*
G to game H if it satisfies the following four conditions:

1. $f(init^G) = init^H$, $f(win^G) = win^H$, and $f(lose^G) = lose^H$
2. $f(P_{\mathcal{A}}^G) \subseteq P_{\mathcal{A}}^H$ and $f(P_{\mathcal{B}}^G) \subseteq P_{\mathcal{B}}^H$
3. $(\forall a \in P_{\mathcal{A}}^G)(\forall b \in M_{\mathcal{A}}^G(a))(f(b) \in M_{\mathcal{A}}^H(f(a)))$
4. $(\forall b \in P_{\mathcal{B}}^G)(\forall c' \in M_{\mathcal{B}}^H(f(b)))(\exists c \in M_{\mathcal{B}}^G(b))(f(c) = c')$

We also say that a reduction is computable if there exists an algorithm that can find

- *$f(b)$ for any $b \in P_{\mathcal{B}}^G$*
- *$c \in M_{\mathcal{B}}^G(b)$ such that $f(c) = c'$ for any $b \in P_{\mathcal{B}}^G$ and any $c' \in P_{\mathcal{A}}^H(f(b))$*

Theorem 2.4. *If f is a reduction from game G to game H, and H is non-losing, then G is non-losing. Furthermore, if the reduction is computable and we are given an algorithm for H, this gives an algorithm for G.*

Proof. In order to prove that $init^G \in L^G$ implies $init^H \in L^H$, we prove by induction on n that $(\forall a \in P_{\mathcal{A}}^G)(\forall n \in \mathbb{N})(a \in L_n^G \implies f(a) \in L_n^H)$.

By Definition 2.2, $L_0^G = \{lose^G\}$ and $L_0^H = \{lose^H\}$. By Condition 1, $f(lose^G) = lose^H$. The basis of the induction thus holds.

For the induction step, suppose that $a \in L_{n+1}^G$. We now need to prove that $f(a) \in L_{n+1}^H$. By definition of L_{n+1}^G, either position a was already contained in L_n^G or there exists a position $b \in M_{\mathcal{A}}^G(a)$ such that $(\forall c \in M_{\mathcal{B}}^G(b))(c \in L_n^G)$. In the first case, $f(a) \in L_n^H$ by induction and $L_n^H \subseteq L_{n+1}^H$ by definition of L_{n+1}^H.

In the latter case, let us take an arbitrary $b \in M_{\mathcal{A}}^G(a)$ such that $(\forall c \in M_{\mathcal{B}}^G(b))(c \in L_n^G)$. We shall now prove that position $f(b)$ in game H has properties similar to those of position b in game G.

Condition 3 gives us $f(b) \in M_{\mathcal{A}}^H(f(a))$. Condition 4 ensures that $(\forall c' \in M_{\mathcal{B}}^H(f(b)))(\exists c \in M_{\mathcal{B}}^G(b))(f(c) = c')$. By induction we already know that $(\forall c \in M_{\mathcal{B}}^G(b))(f(c) \in L_n^H)$, giving us $(\forall c' \in M_{\mathcal{B}}^H(f(b)))(c' \in L_n^H)$ and thus $f(a) \in L_{n+1}^H$.

Now that we know that $a \in L_{n+1}^G$ implies $f(a) \in L_{n+1}^H$, we simply observe that if G is losing, then $init^G$ is in some L_n^G which implies that $init^H$ is in L_n^H and thus H is losing.

The resulting algorithm will, when in position $b \in P_{\mathcal{B}}^G$, find $f(b)$, observe to which position $c' \in P_{\mathcal{A}}^H(f(b))$ the algorithm for H moves from $f(b)$, and move to a position $c \in M_{\mathcal{B}}^G(b)$ such that $f(c) = c'$.

3 Defining the Games Formally

We view ONLINE BIN STRETCHING as a game between two players, $\mathcal{A}$ and $\mathcal{B}$. Let us first introduce a version that represents ONLINE BIN STRETCHING the most directly. For a given number of bins m and target stretching factor α, we call this game REAL GAME(m, α) and it proceeds as follows:

In every round, $\mathcal{A}$ either generates an item or resigns. $\mathcal{B}$ must then place into one of the m bins or resign. We keep track of the sequence s of items generated by $\mathcal{A}$, and their packing t into the bins by $\mathcal{B}$. We restrict $\mathcal{A}$ to sequences that can be packed into m bins of size 1 and $\mathcal{B}$ to packings where no bin exceeds load α, forcing them to resign if they have no other choice.

Definition 3.1 (Real Game). *We define* REAL GAME(m, α) *as the game G with the following components:*

- *each position in $P_\mathcal{A}^G \setminus \{init^G\}$ is an ordered triple (n, s, t) where*
  - *$n \in \mathbb{N}$ is the number of already packed items*
  - *$s \in (0,1]^n$ is a sequence of n items that can be packed into m bins of size at most 1*
  - *$t \in \{1, \ldots, m\}^n$ such that $(\forall i)(\sum_{j:t_j=\alpha} s_j \leq 1)$, which is a packing of those n items into m bins of size at most α*
- *each position in $P_\mathcal{B}^G$ is an ordered triple (n, s, t) where*
  - *$n \in \mathbb{N}$ is the number of already packed items*
  - *$s \in (0,1]^{n+1}$ is a sequence of $n+1$ items that can be packed into m bins of size at most 1*
  - *$t \in \{1, \ldots, m\}^n$ is a packing of the first n items into m bins of size at most α*
- *For any position $(n, s, t) \in P_\mathcal{A}^G \setminus \{init^G\}$ the set $M_\mathcal{A}^G((n,s,t))$ of possible moves by $\mathcal{A}$ contains:*
  - *positions $(n, (s_1, \ldots, s_n, s'), t)$ for all $s' \in (0,1]$ such that the position is in $P_\mathcal{B}^G$*
  - *win^G*
- *The set $M_\mathcal{A}^G(init^G)$ of possible moves by $\mathcal{A}$ contains:*
  - *positions $(0, (s_1), \emptyset)$ for all $s_1 \in (0,1]$*
  - *win^G*
- *For any position $(n, s, t) \in P_\mathcal{B}^G$ the set $M_\mathcal{B}^G((n,s,t))$ of possible moves by $\mathcal{B}$ contains:*
  - *positions $(n+1, s, (t_0, \ldots, t_n, t'))$ for all $t' \in \{1, \ldots, m\}$ such that the position is in $P_\mathcal{A}^G$*
  - *$lose^G$*

Theorem 3.2. *An algorithm for* ONLINE BIN STRETCHING *with m bins and stretching factor α exists if a non-losing* REAL GAME *strategy exists for $\mathcal{B}$.*

Proof. Consider any finite input for ONLINE BIN STRETCHING. If we have $\mathcal{A}$ play according to the input and the algorithm plays according to $\mathcal{B}$, then at the end of input no bin exceeds size α.

There are several main obstacles to implementing a computer search of REAL GAME:

1. $\mathcal{A}$ has an infinite selection of item sizes to pick from.
2. By sending arbitrarily small items, $\mathcal{A}$ can make REAL GAME last arbitrarily many rounds.
3. Computing the maximum size of item $\mathcal{A}$ can generate in a given position requires solving BIN PACKING PROBLEM, which is NP-hard.

In order to avoid these problems, we analyse a simplified version of REAL GAME that we shall call ROUNDED GAME. We then construct a computable reduction from REAL GAME to ROUNDED GAME, proving that a winning strategy for $\mathcal{B}$ in ROUNDED GAME implies an algorithm for ONLINE BIN STRETCHING. We do this by defining ROUNDED GAME such that it corresponds to ONLINE BIN STRETCHING where $\mathcal{B}$ is restricted in its decision-making process and $\mathcal{A}$ is permitted to cheat to a limited extent.

In essence, $\mathcal{A}$ only informs $\mathcal{B}$ about the item sizes and bin loads after rounding them. $\mathcal{A}$ is then allowed to modify the real values afterwards as long as they remain consistent with what $\mathcal{B}$ was told. The branching factor of the game is thus reduced to the granularity of the rounding. This makes it possible to analyse ROUNDED GAME using computer search at the price of a lack of a winning strategy for $\mathcal{B}$ no longer resulting in new lower bounds. Perhaps surprisingly, these modifications do not prevent us from finding new algorithms for ONLINE BIN STRETCHING. We shall now first describe and define ROUNDED GAME, then explain the connection to REAL GAME and finally describe the reduction between them.

An instance of ROUNDED GAME is parameterised by three values—the number of bins m, the granularity level k and the target bin size z. In every round of ROUNDED GAME(m, k, z), $\mathcal{A}$ generates an item of integer size x which $\mathcal{B}$ must then place into one of the m bins. Finally, $\mathcal{A}$ chooses whether the load of that bin increased by x or $x - 1$. We call the latter case an underflow. We keep track of the current loads of the individual bins, not letting any exceed z, and of the multiset of item sizes we have seen so far. $\mathcal{B}$ may claim victory when the total load of the bins exceeds $m(k + 1) - 1$ or, after decreasing the size of each item by 1, the items do not fit into m bins of size $k - 1$. We define these promises formally below as (A) and (B) and will refer to them later. Finally, $\mathcal{B}$ must resign if it cannot claim victory and an item cannot be packed without a bin exceeding load z.

Definition 3.3 (Rounded Game). *We define* ROUNDED GAME(m, k, z) *as the game H with components as follows:*

- *Each position in $P_{\mathcal{A}}^H \setminus \{init^H\}$ is an ordered triple (v, w, u) where*
  - *v is a multiset of integers from $\{1, \ldots, k\}$ representing item sizes*
  - *$w \in \{0, \ldots, z\}^m$ is an m-tuple representing the current loads of the individual bins*
  - *$u \in \{1, \ldots, m\}$ represents the index of the bin the last item was placed into*

- *Each position in P_B^H is an ordered triple (v, w, x) where*
  - *v and w are the same as for P_A^H*
  - *$x \in v$ is the size of the last generated (and not yet packed) item*
- *For any position $(v, w, u) \in P_A^H \setminus \{init^H\}$ the set $M_A^H((v, w, u))$ of possible moves by $\mathcal{A}$ contains:*
  - *positions $(v \cup \{x\}, w', x)$ for all $x \in \{1, \dots, k\}$ such that $w'_u \in \{w_u, w_u - 1\}$ and $w'_i = w_i$ for all $i \neq u$, and the position is in P_B^H*
  - *win^H*
- *The set $M_A^H(init^H)$ of possible moves by $\mathcal{A}$ contains:*
  - *positions $(\{x\}, (0, \dots, 0), x)$ for all $x \in \{1, \dots, k\}$ such that the position is in P_B^H*
  - *win^H*
- *For any position $(v, w, x) \in P_B^H$ the set $M_B^H((v, w, x))$ of possible moves by $\mathcal{B}$ contains:*
  - *positions (v, w', u) for all $u \in \{1, \dots, m\}$ such that $w'_u = w_u + x$ and $w'_i = w_i$ for all $i \neq u$, and the position is in P_A^H*
  - *$lose^H$*
  - *win^H if either of the following two promises is broken:*
    - *(A) the total load of the bins $\sum_{i \in \{1, \dots, m\}} w_i$ is at most $m(k+1) - 1$*
    - *(B) after decreasing the size of each item by 1, the items fit into m bins of size $k - 1$*

We now need to find a reduction from REAL GAME to ROUNDED GAME. Consider how we could modify REAL GAME to make it possible to solve via computer search. Since we cannot restrict the items $\mathcal{A}$ has at their disposal to a finite number of types, we instead restrict information $\mathcal{B}$ has about the game state. We scale the problem to avoid working with fractional values, selecting a granularity $k \in \mathbb{N}$ and target bin size $z \in \mathbb{N}$, which will allow us to analyse REAL GAME with stretching factor $\alpha = z/k$.

We now provide $\mathcal{B}$ with item sizes and bin loads which have been rounded by function $round(x) = \lceil kx \rceil$ instead of their real loads. Due to the uncertainty about the real sizes, placing an item of rounded size x into a bin of rounded load y might result in the new rounded load being merely $x + y - 1$ instead of $x + y$. We call this effect an *underflow*. In order to ensure we do not restrict $\mathcal{A}$, we let $\mathcal{A}$ decide whether that has happened on their following turn.

The game ends with $\mathcal{B}$ losing whenever the rounded load of any bin exceeds z as this corresponds to that bin having a real load higher than z/k. On the other hand, $\mathcal{B}$ gets the option to claim victory when one of the promises defined above is broken, as this implies that $\mathcal{A}$ is cheating in REAL GAME. Breaking promise (A) means that the real total size of all items exceeds m, which is not possible in REAL GAME. Since $round(x) - 1 < kx$, breaking promise (B) means that the real items cannot be packed into m bins of size 1, which is also not possible in REAL GAME.

Definition 3.4. *With the above in mind, we define $round(x) = \lceil kx \rceil$ and construct the reduction f from G to H where G is REAL GAME$(m, z/k)$ and H is ROUNDED GAME(m, k, z) as follows:*

- $f(init^G) = init^H$, $f(win^G) = win^H$, and $f(lose^G) = lose^H$
- For any $(n, s, t) \in P_{\mathcal{A}}^G \setminus \{init^G\}$ we define $f((n, s, t)) = (v, w, u)$ where:
  - v is the multiset $\{round(s_1), \ldots, round(s_n)\}$
  - $u = t_n$
  - $w_u = min(z, round(\sum_{j:t_j=u \wedge j \neq n} s_j) + round(s_n))$
  - $w_i = round(\sum_{j:t_j=i} s_j)$ for $i \neq u$
- For any $(n, s, t) \in P_{\mathcal{B}}^G$ we define $f((n, s, t)) = (v, w, x)$ where:
  - v is the multiset $\{round(s_1), \ldots, round(s_n)\}$
  - $x = round(s_{n+1})$
  - $w_i = round(\sum_{j:t_j=i} s_j)$ for all i

Lemma 3.5. *For any position* $a = (n, (s_1, \ldots, s_n), t) \in P_{\mathcal{A}}^G \setminus \{init^G\}$ *and any successor position* $(n, (s_1, \ldots, s_n, s'), t) \in M_{\mathcal{A}}^G(a) \setminus \{win^G\}$, *let* $u = t_n$, $w_u = min(z, round(\sum_{j:t_j=u \wedge j \neq n} s_j) + round(s_n))$, *and* $w_u' = round(\sum_{j:t_j=u} s_j)$. *Then* $w_u' \in \{w_u, w_u - 1\}$.

Proof. First, observe that $round(\sum_{j:t_j=u \wedge j \neq n} s_j) + round(s_n)$ is equal to either $round(\sum_{j:t_j=u} s_j)$ or $round(\sum_{j:t_j=u} s_j) + 1$. This proves the lemma for all cases where $round(\sum_{j:t_j=u \wedge j \neq n} s_j) + round(s_n) \leq z$. Since $\sum_{j:t_j=u} s_j$ is at most z/k, $round(\sum_{j:t_j=u} s_j)$ must be at most z, and therefore $round(\sum_{j:t_j=u \wedge j \neq n} s_j) + round(s_n)$ is $z + 1$ only if $round(\sum_{j:t_j=u} s_j)$ is z. $\quad$

Theorem 3.6. *The function* f, *as defined by Definition 3.4, is indeed a reduction from* REAL GAME$(m, z/k)$ *to* ROUNDED GAME(m, k, z). *Furthermore, this reduction is computable and any algorithm for* ROUNDED GAME(m, k, z) *thus gives an algorithm for* REAL GAME$(m, z/k)$.

Proof. Let us go over the conditions from Definition 2.3.
Condition 1 obviously holds.
Condition 2:

- All $round(s_i)$ are in $\{1, \ldots, k\}$ as all items in REAL GAME are of size ≤ 1. Thus $\{round(s_1), \ldots, round(s_n)\}$ is a multiset of integers from $\{1, \ldots, k\}$.
- Since $\sum_{j:t_j=u} s_j$ is at most z/k, $min(z, round(\sum_{j:t_j=u \wedge j \neq n} s_j) + round(s_n))$ and $round(\sum_{j:t_j=i} s_j)$ are always in $\{0, \ldots, z\}$.
- t_n is always in $\{1, \ldots, m\}$.

Condition 3:

- Let us consider a move from $a \in P_{\mathcal{A}}^G$ to $b \in M_{\mathcal{A}}^G(a)$.
- The condition obviously holds whenever $b = win^G$.
- If $a = init^G$, then:
  - $f(a)$ is $init^H$.
  - Any $b \in M_{\mathcal{A}}^G(init^G) \setminus \{win^G\}$ is $(0, (s_1), \emptyset)$ for some $s_1 \in (0, 1]$.
  - $f(b)$ is $(\{round(s_1)\}, (0, \ldots, 0), round(s_1))$.
  - Since $round(s_1) \in \{1, \ldots, k\}$, we get $f(b) \in M_{\mathcal{A}}^H(f(a))$.
- If a is some $(n, (s_1, \ldots, s_n), t)$ from the set $P_{\mathcal{A}}^G \setminus \{init^G\}$, then:

- $f(a)$ is $(\{round(s_1), \ldots, round(s_n)\}, w, u)$
- Any position b from $M_{\mathcal{A}}^G(a) \setminus \{win^G\}$ is $(n+1, (s_1, \ldots, s_n, s'), t)$ for some $s' \in (0, 1]$.
- $f(b)$ is $(\{round(s_1), \ldots, round(s_n), round(s')\}, w', round(s'))$ where $w_i' = w_i$ for all $i \neq u$ and, by Lemma 3.5, $w_u' \in \{w_u, w_u - 1\}$.
- Since $round(s') \in \{1, \ldots, k\}$, we get $f(b) \in M_{\mathcal{A}}^H(f(a))$.

Condition 4:

– Let us consider some $b \in P_{\mathcal{B}}^G$ and $c' \in M_{\mathcal{B}}^H(f(b))$.

– The condition obviously holds whenever $c' = lose^G$.

– Since neither of the two promises is ever broken in state $f(b)$, there are no moves from any $f(b)$ to win^G.

– If $b = (n, (s_1, \ldots, s_n, s_{n+1}), ())$, then
  - $f(b) = (\{round(s_1), \ldots, round(s_{n+1})\}, w, round(s_{n+1}))$.
  - Any position $c' \in M_{\mathcal{B}}^H(f(b))$ is either contained in $\{win^G, lose^G\}$ or is $(\{round(s_1), \ldots, round(s_{n+1})\}, w', u)$ for some $u \in \{1, \ldots, m\}$, where:
    * By definition of $f(b)$, $w_i = round(\sum_{j:t_j=i} s_j) \leq z$ for all i.
    * By definition of $M_{\mathcal{B}}^H(f(b))$, $w_i' = w_i \leq z$ for all $i \neq u$ and $w_u' = w_u + round(s_{n+1}) \leq z$.
  - For any $c' \in M_{\mathcal{B}}^H(f(b))$, since $w_i' \leq z$, we know that $\sum_{j:t_j=i} s_j \leq z/k$ for all i and thus $c = (n+1, (s_1, \ldots, s_n, s_{n+1}), (, u))$ is in $M_{\mathcal{B}}^G(b)$.

We also observe that $f(a)$ is easily computable for any position a and $M_{\mathcal{B}}^G(b)$ is easily computable for any $b \in P_{\mathcal{B}}^G$, making the entire reduction computable.

In order to analyse the game using computer search, we employ some further modifications of the game, which will be described in the following section. However, those modifications are much simpler.

4   Computer Search

Let us first describe the simple modifications of the game that we use to make the computer search easier.

ROUNDED GAME still has an infinite number of states due to not placing any limit on the number of items of rounded size 1 as long as almost all of them underflow. To avoid this problem, we do not include items of size 1 in v, thus ensuring that the number of positions is finite. This is obviously a reduction as items of size 1 never affect the set of possible moves. However, this reduction causes the game to include cycles among the states, caused by adding an item of rounded size 1 and then having it underflow.

In order to prevent these cycles, we forbid $\mathcal{A}$ from making the move from any (v, w, u) to $(v, w', 1)$ where $w_u' = w_u - 1$. While this is not a reduction, we can observe that if $\mathcal{A}$ has a winning strategy, then $\mathcal{A}$ also has a winning strategy that never makes such a move. This is because $\mathcal{B}$ can pack the new item of size 1 into bin u, returning the game to position (v, w, u). We are now guaranteed that the sum of w_i is strictly increasing with every pair of successive moves, ensuring that the game takes at most $O(mk)$ moves.

The computer search uses special game positions which occur after $\mathcal{A}$ chooses whether an underflow occurred and before they select the size of the next item. This is because such positions are completely described by the multiset of items and the m-tuple of current loads, minimizing the number of positions to analyse. Instead of forbidding, as per the previous paragraph, the generation of an item of rounded size 1 after an underflow, we can now forbid the underflowing of items of size 1 with very similar reasoning.

We can now describe the basic version of our computer program, which searches for a strategy that would allow $\mathcal{B}$ to win ROUNDED GAME(m, k, z) where m, k and z are, for the purposes of the program, global constants. We have two helper functions:

- CHECK takes a multiset of item sizes and checks whether promise (B) was broken
- ADD takes (w, x, u) and returns a copy of w with w_u increased by x

The program is then a straightforward application of the MINIMAX algorithm first described by von Neumann [11]. Ultimately, if our search returns *True* for the initial game state, the whole ROUNDED GAME is winnable for $\mathcal{B}$, an algorithm for ONLINE BIN STRETCHING corresponding to the winning strategy exists and therefore $\alpha_m \leq s/k$.

5 Optimizations

We observe that permuting the bins does not affect the game significantly. We thus have the helper function ADD sort the bins by their load, decreasing the number of states. This is trivially a reduction. By sorting the bins in decreasing order, the search tries BESTFIT first, further improving overall computation time.

We implemented custom caching of results for SOLVE. Since caching for all possible game states used too much memory, we made use of the fact that, for a given game state, we can often find either a less favourable game state that we managed to win previously or a more favourable game state that we could not win. LRU caching was tried and was clearly inferior when compared to the following custom caching approach.

Definition 5.1 (Order on v). *We define $v_1 \preceq v_2$ to be true if the items from v_1 can fit into bins with sizes corresponding to the items in v_2 after decreasing all rounded sizes in both multisets by 1.*

Theorem 5.2. *If $v_1 \preceq v_2$, then* SOLVE(v_1, w) *implies* SOLVE(v_2, w)

Proof. If $v_1 \preceq v_2$, then $v_1 \cup v'$ breaking promise (B) implies $v_2 \cup v'$ breaking promise (B) for any multiset of items v'. Otherwise we could pack items of v_1 into the spaces occupied by the items of v_2, showing that $v_1 \cup v'$ does not break promise (B) after all.

Algorithm 1. Computer search for a winning strategy for ROUNDED GAME

```
1: function SOLVE(v, w)                          ▷ Return True iff position is winnable.
2:     remaining = m(k + 1) − ∑(wᵢ) − 1          ▷ This ensures promise (A).
3:     if remaining + min(w) < z then return True  ▷ Put all into emptiest bin.
4:     end if
5:
6:     limit = min(remaining, k) ▷ Upper bound on largest item that does not let B
   win.
7:
8:     for item ∈ {1, ..., limit} do               ▷ If there exists an item size...
9:         v' = v ∪ {item}
10:         result = False
11:         for bin ∈ {1, ..., m} do         ▷ ...which cannot be packed into any bin...
12:             w' = ADD(w, item, bin)
13:             w'' = ADD(w, item − 1, bin)
14:             if SOLVE(v', w') ∧ (item == 1 ∨ SOLVE(v', w'')) then
15:                 result = True
16:                 break
17:             end if
18:         end for
19:         if not result and not CHECK(v) then   ▷ ...and B cannot claim victory...
20:             return False                       ▷ ...then declare this position lost.
21:         end if
22:     end for
23:     return True
24: end function
```

For any given w, we now store only the set of minimal v for which SOLVE(v, w) is True and the set of maximal v for which SOLVE(v, w) is False. When evaluating SOLVE(v, w), we query the cache for that w, checking if there is any $v' \preceq v$ for which we know SOLVE(v', w) to be True, or there is any $v'' \succeq v$ for which we know SOLVE(v', w) to be False. This is done after line 4 of the algorithm. If the position needs to be evaluated further, we add v to the relevant cache before returning the result and remove any v' made redundant by it.

The comparisons between v are implemented by generalising the function CHECK. Indeed, the original use of CHECK corresponds to comparing v to m items of rounded size k.

The (now generalised) helper function CHECK is relatively computationally intensive. We preprocess the input to remove some easy cases and then use dynamic programming, which proved faster than using an ILP. Results are cached to avoid repeated computation.

As a final optimisation, the program stores all rounded sizes already decreased by 1 to simplify some formulae and avoid off-by-one errors.

These optimizations are how we achieved our results. The code is available on Github [10].

6 Conclusions

We found new algorithms for ONLINE BIN STRETCHING on 4, 5, 6, 7 and 8 bins. Perhaps surprisingly, these new algorithms were found using discrete computer search, despite the problem inherently using real-valued input. This was possible thanks to a reduction from a game with infinite depth and branching factor to a game with both depth and branching factor being $O(mk)$ where m is the number of bins and k is our chosen granularity. We expect that these techniques will prove to be useful for other similar problems.

Acknowledgements. We would like to thank Martin Böhm for consultation and the reviewers for their valuable feedback.

References

1. Azar, Y., Regev, O.: On-line bin-stretching. Theoret. Comput. Sci. **268**(1), 17–41 (2001)
2. Böhm, M., Sgall, J., van Stee, R., Veselý, P.: A two-phase algorithm for bin stretching with stretching factor 1.5. J. Comb. Optim. **34**, 810–828 (2017)
3. Böhm, M., Simon, B.: Discovering and certifying lower bounds for the online bin stretching problem. arXiv:2001.01125 (2020)
4. Gabay, M., Brauner, N., Kotov, V.: Improved lower bounds for the online bin stretching problem. 4OR **15**, 183–199 (2017)
5. Gabay, M., Brauner, N., Kotov, V.: Semi-online bin stretching with bunch techniques. Theoret. Comput. Sci. **602**, 103–113 (2015)
6. Garey, M. R., Graham, R. L., Ullman J. D.: Worst-case analysis of memory allocation algorithms. In: Proceedings of the Fourth Annual ACM Symposium on Theory of Computing, STOC'72, pp. 143–150 (1972)
7. Kellerer, H., Kotov, V., Speranza, M.G., Tuza, Z.: Semi on-line algorithms for the partition problem. Oper. Res. Lett. **21**(5), 235–242 (1997)
8. Kellerer, H., Kotov, V.: An efficient algorithm for bin stretching. Oper. Res. Lett. **41**(4), 343–346 (2013)
9. Lhomme, A., Romane, O., Catusse, N., Brauner, N.: Online bin stretching lower bounds: Improved search of computational proofs. arXiv:2207.04931 (2022)
10. Lieskovský, M.: BinStretchingAlgorithmicSearch: Computer search for better bounds on the bin stretching problem. Github code repository. https://github.com/MatejLieskovsky/BinStretchingAlgorithmicSearch
11. von Neumann, J.: Zur Theorie der Gesellschaftsspiele. Mathematische Annalen **100**, 295–320 (1928)

Competitive Searching over Terrains

Sarita de Berg[1], Nathan van Beusekom[2](✉), Max van Mulken[2],
Kevin Verbeek[2], and Jules Wulms[2]

[1] Utrecht University, Utrecht, The Netherlands
S.deBerg@uu.nl
[2] TU Eindhoven, Eindhoven, The Netherlands
{n.a.c.v.beusekom,m.j.m.v.mulken,k.a.b.verbeek,j.j.h.m.wulms}@tue.nl

Abstract. We study a variant of the searching problem where the environment consists of a known terrain and the goal is to obtain visibility of an unknown target point on the surface of the terrain. The searcher starts on the surface of the terrain and is allowed to fly above the terrain. The goal is to devise a searching strategy that minimizes the *competitive ratio*, that is, the worst-case ratio between the distance traveled by the searching strategy and the minimum travel distance needed to detect the target. For 1.5D terrains we show that any searching strategy has a competitive ratio of at least $\sqrt{82}$ and we present a nearly-optimal searching strategy that achieves a competitive ratio of $3\sqrt{19/2} \approx \sqrt{82}+0.19$. This strategy extends directly to the case where the searcher has no knowledge of the terrain beforehand. For 2.5D terrains we show that the optimal competitive ratio depends on the maximum slope λ of the terrain, and is hence unbounded in general. Specifically, we provide a lower bound on the competitive ratio of $\Omega(\sqrt{\lambda})$. Finally, we complement the lower bound with a searching strategy based on the maximum slope of the known terrain, which achieves a competitive ratio of $O(\sqrt{\lambda})$.

1 Introduction

The development of autonomous mobile systems has garnered a lot of attention recently. With self-driving cars and autonomous path-finding robots becoming more commonplace, the demand for efficient algorithms to govern the decision-making of these systems has risen as well. A class of problems that naturally arises from these developments is the class of *searching problems*, also known as *searching games*: given an environment, move through the environment to find a target at an unknown location. Many variants of this general problem have been studied in literature, typically differing in the type of search environment, the way the searcher can move through the environment, and the way the target can be detected. In this paper we consider a variant of the problem that is motivated by searching terrains using a flying (autonomous) drone with mounted cameras/sensors, as for example in search-and-rescue operations. Specifically, our environment is defined by a height function $T_d : \mathbb{R}^d \to \mathbb{R}$. For $d = 1$ we refer to the terrain as a 1.5D terrain, and for $d = 2$ we refer to the terrain as a 2.5D

This research was initiated at the 6th Workshop on Applied Geometric Algorithms (AGA 2023), Otterlo, The Netherlands, April 7–21, 2023.

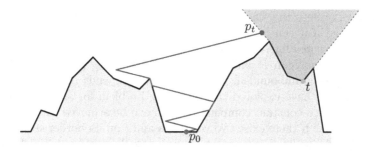

Fig. 1. Our searching strategy starts from p_0 and then follows the blue searching path. When the searcher reaches p_t, it can see the target t. (Color figure online)

terrain. We omit the dimension d from the terrain function T_d when it is clear from the context. The terrain is known to the searcher, and the searcher can fly anywhere above the terrain. The target t is discovered if it can be seen by the searcher along a straight line. The goal is to devise a *searching strategy* (that is, a search path) that finds the (unknown) target t as quickly as possible (see Fig. 1). To the best of our knowledge, this natural variant of the searching problem has not been studied before.

As is common for searching problems, we analyze the quality of the searching strategy using *competitive analysis*. For that we consider the ratio between the travel distance using our searching strategy and the minimum travel distance needed to detect that target. The maximum value of this ratio over all possible environments and all possible target locations is the *competitive ratio* c of the searching strategy. The goal is to find a searching strategy that minimizes c.

Related Work. Searching problems have been studied extensively in the past decades. Here, we mostly restrict ourselves to searching problems with a geometric environment and continuous motion. One of the most fundamental searching problems is that of searching on an infinite line, where the target is detected only when the searcher touches it. The following optimal strategy for this problem was discovered by Beck and Newman [3]: assuming that the distance to the target is at least one, we first move one to the right from the starting point. Next, we move back to the starting point and then move two to the left. We then repeat this process, alternating between moving to the right and left of the starting point, every time doubling the distance from the starting point. This searching strategy has a competitive ratio of 9, which is optimal for this problem.

In subsequent work, researchers have studied searching problems for many other different environments, including lines and grids [1], line arrangements [9], and graphs [7,8]. Other variants include searching on a line when an upper bound on the distance to t is known [6,15], when turns contribute to the cost of the solution [11], or when there are multiple searchers [2].

In settings where the environment is 2-dimensional (or higher), it is not possible to visit every point in the environment, and hence we need to consider different ways of detecting the target. In these settings, the target is often considered detected if it can be seen directly from the searcher's position along a straight, unobstructed line. One example is the problem of finding a target point t inside a simple polygon P with n vertices [21]. For this problem the optimal competitive ratio is unbounded, as it necessarily depends on n [22,25]. As a result, researchers have explored this searching problem for special sub-classes of polygons where a constant competitive ratio can be achieved. A polygon P is considered a *street* if there exist two vertices s and t on its border such that the two boundary chains leading from s to t are mutually weakly visible. Searching for an unknown point in a street can be done with an optimal competitive ratio of $\sqrt{2}$ [18,20]. There is ample further work on searching problems in variants of streets [10,23,26], star-shaped polygons [17,24], or among obstacles [5,19]. For a comprehensive overview of these variants, see [13]. Specifically, López-Ortiz and Schuierer [24] obtain a competitive ratio of 11.51 for star-shaped polygons; note that 1.5D terrains are a special type of unbounded star-shaped polygons.

Other problems strongly related to searching problems are the *exploration problems*, for which the goal is to move through the interior of an unknown environment to gain visibility of its entire interior. Here, the competitive ratio relates the length of the searching strategy to the shortest watchman tour. An unknown simple polygon can be fully explored with competitive ratio 26.5 [16]. For polygons with holes, the competitive ratio is dependent on the number of holes [12], whereas in a rectilinear polygon without holes a competitive ratio as low as $3/2$ can be achieved [14]. Complementary to this problem is the exploration of the *outer* boundary of a simple polygon, where a 23.78 or 26.5 competitive ratio can be achieved for a convex or concave polygon, respectively [27].

Contributions. Given a starting position p_0 on the surface of the terrain (we assume without loss of generality that p_0 is at the origin and that $T_d(p_0) = 0$), the goal is to devise an efficient searching strategy to find an unknown target point t on the surface of the terrain, where the searcher can detect t if it is visible from the searcher's position along a straight unobstructed line (see Fig. 1). In our problem the searcher is not restricted to the surface of the terrain, but it is allowed to move to any position on or above the terrain.

In Sect. 2 we consider the problem for 1.5D terrains. We first prove that any searching strategy for this problem must have a competitive ratio of at least $\sqrt{82}$. We then present a searching strategy with a competitive ratio of $3\sqrt{19/2} \approx \sqrt{82} + 0.19$. Our searching strategy is a combination of the classic searching strategy on an infinite line with additional vertical movement.

In Sect. 3 we consider the problem for 2.5D terrains. We show that no searching strategy can achieve a bounded competitive ratio, as the competitive ratio necessarily depends on the maximum slope, or Lipschitz constant, λ of the terrain function. Specifically, we show that the competitive ratio of any searching strategy is at least $\Omega(\sqrt{\lambda})$. We then present a novel searching strategy that achieves a competitive ratio of $O(\sqrt{\lambda})$, being asymptotically optimal in λ.

In our searching problems we assume that the terrain T_d is known to the searcher. In Sect. 4 we conclude that our strategy for a 1.5D terrain is directly applicable if this is not the case, and discuss to what degree the results for the 2.5D case extend as well. Omitted proofs can be found in the full version [4].

2   Competitive Searching on 1.5D Terrains

In this section we consider our searching problem on a 1.5D terrain, defined by the height function T_1. For 1.5D terrains, the visibility region $Vis(t)$ of a target t can be defined as the set of all points q for which the line segment tq does not properly intersect the terrain. That is, the region that contains all points that can see t. There is a half-line originating from t that needs to be crossed to enter $Vis(t)$ (see Fig. 1). We call this half-line a visibility ray. Thus, the goal of any searching strategy on 1.5D terrain is to enter $Vis(t)$ by crossing a visibility ray r originating at the target point t. We first establish a lower bound on the competitive ratio of any searching strategy.

Theorem 1. *The competitive ratio for searching on 1.5D terrains is at least $\sqrt{82}$.*

Proof. Consider a terrain with pits at integer x-coordinates that have infinite slope, and thus cast (near-)vertical visibility rays. Furthermore, there is a rectangular mountain infinitely far away casting a horizontal visibility ray at some height h (see Fig. 2). Because of the lower bound of 9 on the competitive ratio for searching on a line [3], there must be a pit at distance d such that the distance covered by the optimal strategy for searching on a line is $9d$. We set $h = d$.

   Now consider a search path P for this terrain. If P does not reach height h after covering a horizontal distance of $9d$, then P travels a distance of more than $\sqrt{81d^2 + h^2} = \sqrt{82}d$ before reaching the horizontal ray at height h, resulting in a competitive ratio of more than $\sqrt{82}$. Otherwise, P travels a distance of at least $\sqrt{81d^2 + h^2} = \sqrt{82}d$ before reaching the pit at distance d. Hence the competitive ratio of P is at least $\sqrt{82}$.   □

Searching Strategy. Our strategy is based on the classic searching strategy on an infinite line with additional vertical movement. Specifically, we construct a *projected path* P^* that acts as a guide for the actual *searching path* P. In the description of our strategy we make use of infinitesimally small steps at the start, as this simplifies the description and analysis. We later mention how to avoid this and make our strategy feasible in practice, under the assumption that the length of the shortest path to the target is bounded from below by

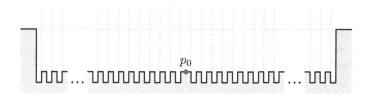

Fig. 2. The lower bound construction for the 1.5D case.

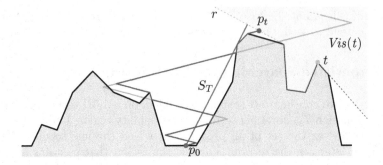

Fig. 3. The strategy P (blue) and the projected path P^* (gray). The goal is seen at p_t. (Color figure online)

some positive constant. Starting from p_0, P^* moves diagonally with slope s to one side, for a horizontal distance of ε, and then moves back to $x = 0$, again with slope s. Subsequently, P^* alternates between moving left and right of p_0, doubling the horizontal distance when moving away from p_0, and always using slope s. As a result, P^* consists of xy-monotone *segments*, and *turning points* where the x-direction swaps. Specifically, P^* is defined by the following functions for $i \in \mathbb{Z}$:

$$h_r^i(x) = s \cdot (2^i + x) \quad \text{for } i \text{ odd and } -2^{i-2} \leq x \leq 2^{i-1},$$
$$h_\ell^i(x) = s \cdot (2^i - x) \quad \text{for } i \text{ even and } -2^{i-1} \leq x \leq 2^{i-2}.$$

We call the line segments h_r^i the *right* segments of P^*, and the line segments h_ℓ^i the *left* segments of P^*. Observe that for two consecutive segments, the values at the ends of the domains coincide, which results in P^* being a connected path. Specifically, we get $h_r^i(2^{i-1}) = h_\ell^{i+1}(2^{(i+1)-2})$ and $h_\ell^{i-1}(-2^{(i-1)-1}) = h_r^i(-2^{i-2})$. The actual search path P follows P^* (see Fig. 3). Whenever P hits the terrain, it follows the terrain upwards until it can continue moving diagonally with slope s again. This diagonal part of P does not coincide with P^*, so once P hits P^*, P starts following P^* again. Observe that P still consists of xy-monotone polygonal chains and turning points, albeit both can differ from P^*. We refer to the monotone chains of P as right and left subpaths of P, when they are monotone in the positive and negative x-direction, respectively. We also refer to a line segment in such a right or left subpath as a left or right segment. We will choose s later to optimize the resulting competitive ratio.

Preliminaries and Definitions. The target t can be seen from any point in $Vis(t)$. We consider all possible visibility rays $r(s_r, d_r)$ that can separate $Vis(t)$ and p_0, where s_r is the slope of r in the positive x-direction, and d_r is the distance between r and p_0.

Let S be the line segment between p_0 and r that is perpendicular to r, so $|S| = d_r$. Furthermore, let S_T be the shortest geodesic path from p_0 to r, taking the terrain T into account. If S does not properly intersect T, then $|S_T| = d_r$. Note that the last line segment of S_T is perpendicular to r. Finally, let $p_t \in P$ be the point where P crosses r to enter $Vis(t)$.

We define $\tau(r)$ as the distance traversed over P until r is crossed, i.e. from p_0 until p_t, and $c(r) = \frac{\tau(r)}{|S_T|}$ as the competitive ratio to cross a ray r. To simplify our proofs, we additionally introduce the following definitions. Let $p_t = (x, y)$, then we define $\tau^*(r) = y\sqrt{1+s^2}/s$. So, $\tau^*(r)$ is the length of a path with slope s up to height y. When P deviates from P^*, the projected path is intersected by T and hence P is steeper than P^*. It follows that $\tau(r) \leq \tau^*(r)$. We use the ratio $c^*(r) = \frac{\tau^*(r)}{|S_T|}$ in our proofs, and analyze the maximum of $c^*(r)$ over all instances to bound the competitive ratio of our searching strategy from above.

For computing the competitive ratio, we only need to consider visibility rays originating from one side. This is due to the symmetric nature of our strategy: we can take any instance with a visibility ray originating left of p_0, and transform it into a case equivalent to having the visibility ray originating from the right of p_0. We achieve this by scaling all distances in P^* by 2 to get the horizontally symmetrical path. To see this observe that for $-2^{i-2} \leq x \leq 2^{i-1}$ and $x' = -x$

$$h_r^i(x) = s \cdot (2^i + x) = 2 \cdot s \cdot (2^{i-1} + x/2) = 2 \cdot h_\ell^{i-1}(x'/2).$$

Additionally, we mirror T horizontally in p_0, hence r is also mirrored with respect to p_0. We thus consider only visibility rays that originate to the right of p_0.

Finally, since T is a height function, the visibility region of any point above the terrain includes the vertical ray cast upwards from that point. Thus, for the visibility ray r that separates $Vis(t)$ from p_0 it holds that $s_r \leq 0$.

Proof Structure. To determine the competitive ratio of our searching strategy, we analyze the competitive ratio in a worst-case instance (T, r) for $c^*(r)$, where T is the terrain and $r = r(s_r, d_r)$ is the visibility ray from the target. To that end we first establish several properties that must hold in some worst-case instance. To exclude various instances from consideration, we can use the following lower bound on the competitive ratio of our strategy. A competitive ratio below this bound would contradict the lower bound for searching on a line [3].

Lemma 1. *The competitive ratio $c^*(r)$ is at least $9\sqrt{1+s^2}$.*

We show that a worst-case instance $(T, r(s_r, d_r))$ has the following properties:

- The ray r is arbitrarily close to a turning point $p_{\hookleftarrow}$ of P when p_t lies on a right subpath (Lemma 4).
- The ray r satisfies $s_r \leq -s$ and p_t lies on a right subpath (Lemma 5).
- If $p_{\hookleftarrow}$ is not a turning point of P^*, and thus a local maximum $\wedge$ of T intersects P^* before $p_{\hookleftarrow}$, then $\wedge$ is at $p_{\hookleftarrow}$ and r is vertical (Lemma 6).
- If $p_{\hookleftarrow}$ coincides with a turning point of P^*, then r is vertical (Lemma 10).

After establishing these properties, the remaining cases can easily be analyzed directly in the proof of Theorem 2.

Close to Turning Point. We first prove two lemmata that help us establish the properties indicated above. Lemma 2 follows from the quotient rule of derivatives.

Lemma 2. *Let $f, g, h : \mathbb{R}_{>0} \to \mathbb{R}_{>0}$ be differentiable functions such that $f(x) = \frac{g(x)}{h(x)}$ and $\frac{\mathrm{d}g}{\mathrm{d}x}, \frac{\mathrm{d}h}{\mathrm{d}x} > 0$ for any $x > 0$. Then, $\frac{\mathrm{d}f}{\mathrm{d}x} < 0$ if and only if $\frac{\mathrm{d}g}{\mathrm{d}x} / \frac{\mathrm{d}h}{\mathrm{d}x} < f(x)$.*

Lemma 3. *Let $(T, r(s_r, d_r))$ be an instance where p_t lies on a right subpath of P with slope s, and let $c^*(d_r) = \frac{\tau^*(r)}{|S_T|}$. If $\frac{1}{9} < s \le 1$, then $\frac{\mathrm{d}c^*}{\mathrm{d}d_r} < 0$.*

Proof. By Lemma 2, to prove $\frac{\mathrm{d}c^*}{\mathrm{d}d_r} < 0$ it is sufficient to show that $\frac{\mathrm{d}\tau^*(r)}{\mathrm{d}d_r} / \frac{\mathrm{d}|S_T|}{\mathrm{d}d_r} < c^*(d_r)$. Since $\frac{\mathrm{d}|S_T|}{\mathrm{d}d_r} \ge 1$, we get that $\frac{\mathrm{d}\tau^*(r)}{\mathrm{d}d_r} / \frac{\mathrm{d}|S_T|}{\mathrm{d}d_r} \le \frac{\mathrm{d}\tau^*(r)}{\mathrm{d}d_r}$. Consider decreasing d_r, i.e., moving the ray r towards the origin. If $s \le 1$, we can bound the ratio between the change in $\tau^*(r)$ and the change in d_r as follows.

$$\frac{\mathrm{d}\tau^*(r)}{\mathrm{d}d_r} \le \frac{\sqrt{s^2 + 1}}{s} < 9\sqrt{1 + s^2} \le c^*(d_r)$$

The second step holds for $s > \frac{1}{9}$, and the final step follows from Lemma 1. $\square$

Thus, as long as p_t lies on a right subpath of P, decreasing d_r increases $c^*(r)$.

Lemma 4. *In a worst-case instance $(T, r(s_r, d_r))$ where p_t lies on a right subpath of P, if $\frac{1}{9} < s \le 1$, then r is infinitesimally close to a turning point of P.*

Proof sketch. Assume for contradiction that r is not infinitesimally close to a turning point. Then either we can move ray r closer to the origin and increase the competitive ratio by Lemma 3, leading to a contradiction, or r reaches a segment of P of slope greater than s. In the latter case, either the competitive ratio is below the lower bound of Lemma 1, or we can alter T without decreasing the competitive ratio, and apply Lemma 3 again. Both cases contradict that $(T, r(s_r, d_r))$ is a worst-case instance. $\square$

Flat Visibility Rays. Next, we deal with all visibility rays $r(s_r, d_r)$ for which $s_r > -s$, which we call *flat* visibility rays. All other visibility rays, which have a slope of at most $-s$, we define as *steep* visibility rays. We show that r is never flat in a worst-case instance, and p_t must then lie on a right subpath of P.

Lemma 5. *In a worst-case instance $(T, r(s_r, d_r))$, if $\frac{2}{9} < s \le 1$ then $s_r \le -s$ and p_t lies on a right subpath of P.*

Proof sketch. Assume for contradiction that $s_r > -s$ or that p_t lies on a left subpath of P. In the former case, p_t can lie on a right or a left subpath. If p_t lies on a right subpath, Lemma 4 implies that in the worst case p_t lies infinitesimally close to a left subpath, so we can apply Lemma 3 until the analysis in the next case applies. If p_t lies on a left subpath, then we can show that the competitive ratio is below the lower bound of Lemma 1, contradicting that (T, r) is a worst-case instance. A similar argument applies when p_t lies on a left subpath and $s_r \le -s$, proving the latter case. $\square$

From now on, we thus consider only steep visibility rays with p_t on a right subpath. Let $p_{\leftarrow}$ denote the turning point infinitesimally close to r. Note that $p_{\leftarrow}$ must lie on the final *left* segment of P^* that is on the search path. We denote this segment by $h_\ell^{\leftarrow}$.

Obstructed Search Path. For steep visibility rays, first consider the case where $p_{\leftarrow}$ is not a turning point of P^*, i.e. T obstructs the right segment before $h_\ell^{\leftarrow}$ (see figure). We call a local maximum of T a peak, denoted by $\wedge$. We prove that a worst-case instance (T, r) has the following three properties.

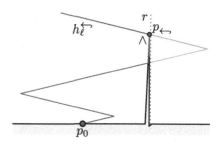

1. If a peak lies on P^* at $p_{\leftarrow}$, then the visibility ray r is vertical (Lemma 7);
2. if the visibility ray r is vertical, then a peak lies on P^* at $p_{\leftarrow}$ (Lemma 8);
3. either a peak lies on P^* at $p_{\leftarrow}$, or the visibility ray r is vertical (Lemma 9).

The following lemma follows directly from the above statements.

Lemma 6. *Let $(T, r(s_r, d_r))$ be a worst-case instance where $p_{\leftarrow}$ is not a turning point of P^*. If $\frac{2}{9} < s < \frac{4}{9}$, then a peak lies on $h_\ell^{\leftarrow}$ at $p_{\leftarrow}$, and $s_r = -\infty$.*

Next we prove Lemmata 7-9, to prove Lemma 6. Let $(T, r(s_r, d_r))$ be a worst-case instance where $p_{\leftarrow}$ is not a turning point of P^* and let $\wedge$ be the last peak on P before $p_{\leftarrow}$. By Lemmata 4 and 5, r is steep and infinitesimally close to $p_{\leftarrow}$.

Lemma 7. *If the peak $\wedge$ lies on left segment $h_\ell^{\leftarrow}$ of P^*, and hence coincides with $p_{\leftarrow}$, then r is vertical.*

Proof. Assume for contradiction that $\wedge$ coincides with $p_{\leftarrow}$ and that r is not vertical. We distinguish between two cases: either the line segment through p_0 perpendicular to r passes above $p_{\leftarrow}$, or not. In the former case, we construct the terrain T' from T by moving $\wedge$ leftwards along $h_\ell^{\leftarrow}$, until we are in the latter case. This does not affect $c^*(r)$. In the latter case, we rotate r around $p_{\leftarrow}$ to become more vertical, resulting in a higher competitive ratio: S_T becomes smaller and $\tau^*(r)$ becomes larger. This contradicts that (T, r) is worst case. $\square$

Lemma 8. *If r is vertical, then the peak $\wedge$ lies on left segment $h_\ell^{\leftarrow}$ of P^*, and hence coincides with $p_{\leftarrow}$, for $s < \frac{4}{9}$.*

Proof. Assume for contradiction that r is vertical and that $\wedge$ does not lie on left segment $h_\ell^{\leftarrow}$. The height value of T at the x-coordinate of $\wedge$ can be increased towards P^*, so that $\wedge$ will lie slightly higher. This changes P, as the turning point $p_{\leftarrow}$ moves to the left by some arbitrarily small distance d (see Fig. 4). By Lemma 3, in the worst case r also moves to the left by distance d. Due to the slope s of P^*, $\wedge$ must have been moved up by a distance of $2ds$. This means $|S_T|$ decreases by at least d, due to r moving to the left and being vertical, and increases by at most $2ds$, due to $\wedge$ moving up: in total $|S_T|$ decreases by at least $d(1 - 2s) > 0$, for $s < 1/2$.

On the other hand, $\tau^*(r)$ also decreases. With r moving d towards $\wedge$, $\tau^*(r)$ decreases by $d\sqrt{1 + s^2}$. We now consider the ratio $\frac{d\tau^*(r)}{d\wedge} / \frac{d|S_T|}{d\wedge} \leq \frac{d\sqrt{1+s^2}}{d(1-2s)}$

between the decrease of $\tau^*(r)$ and $|S_T|$. As this ratio is below the lower bound of $9\sqrt{1+s^2}$ of Lemma 1 for $s < 4/9$, Lemma 2 implies that slightly moving $\wedge$ towards $h_\ell^{\leftharpoondown}$ increases $c^*(r)$, contradicting that (T,r) is a worst-case instance. $\square$

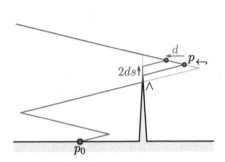

Fig. 4. Moving $\wedge$ upwards causes $p_{\leftharpoondown}$ to move left. If $p_{\leftharpoondown}$ moves a horizontal distance d, then $\wedge$ must have moved $2ds$.

Fig. 5. By moving local maximum $\wedge$ to the green point $\wedge'$, $|S_T|$ (red) strictly decreases, concatenating the green path from $p_0 = q$ to $\wedge'$ and the yellow path. (Color figure online)

Lemma 9. *At least one of the following holds: the peak $\wedge$ lies on left segment $h_\ell^{\leftharpoondown}$ of P^*, and hence coincides with $p_{\leftharpoondown}$, or r is vertical.*

Proof. Assume for contradiction that neither of the two properties holds. For now assume that S_T is routed over $\wedge$ and let $S_\wedge$ be the line perpendicular to r through $\wedge$. We make a case distinction on whether $S_\wedge$ intersects r above $p_{\leftharpoondown}$ or not. We first consider the case where $S_\wedge$ hits r above $p_{\leftharpoondown}$ (see Fig. 5). Let p_{opt} be the point where S_T hits r, and let $p_\wedge$ be the vertex before p_{opt} on the geodesic S_T, coinciding with $\wedge$. Let q be the vertex on S_T before $p_\wedge$ (possibly $q = p_0$, as in Fig. 5). Consider the line segment qp_{opt}. Because $\wedge$ does not lie on $h_\ell^{\leftharpoondown}$, qp_{opt} intersects P between $\wedge$ and $p_{\leftharpoondown}$. Let $p'_\wedge$ be the intersection point, and let p'_{opt} be the point on r hit by the perpendicular on r through $p'_\wedge$. Finally, let $S_T(q)$ be the geodesic S_T from p_0 to q. By the above,

$$|S_T| > |S_T(q)| + |qp_{opt}| > |S_T(q)| + |qp'_\wedge| + |p'_\wedge p'_{opt}|.$$

Consider the terrain T' where, compared to T, $\wedge$ moved rightward along P until it coincides with $p'_\wedge$ (see Fig. 5). For T' we know that $|S_{T'}| = |S_T(q)| + |p'_\wedge q| + |p'_{opt} p'_\wedge| < |S_T|$. Additionally, $\tau^*(r)$ is unaffected. Thus, the ratio $c^*(r)$ strictly increases, contradicting that (T,r) is a worst-case instance.

Notice that, when $\wedge$ is not part of S_T, then S_T hits r above $p_{\leftharpoondown}$. In this case, the above modification to T' does not affect $\tau^*(r)$ and $|S_{T'}| = |S_T|$. Now Lemma 7 applies, contradicting that (T,r) is a worst-case instance.

Finally, consider the case where $S_\wedge$ hits r below or on $p_{\leftharpoondown}$. When we rotate r around $p_{\leftharpoondown}$ to become more vertical, S_T decreases and $\tau^*(r)$ increases.

This results in a strictly higher ratio $c^*(r)$, contradicting that (T,r) is a worst case.   □

Unobstructed Search Path. Next we consider all steep visibility rays in the case that $p_{\leftharpoonup}$ is a turning point of P^*, and show the following.

Lemma 10. *In a worst-case instance* $(T, r(s_r, d_r))$, *where* $p_{\leftharpoonup}$ *is a turning point of* P^*, *if* $\frac{1}{9} < s \le 1$, *then* r *is vertical.*

Proof. Assume for contradiction that r is not vertical. If S_T intersects r below $p_{\leftharpoonup}$, rotating r around $p_{\leftharpoonup}$ to become more vertical results in a strictly higher value $c^*(r)$, as $|S_T|$ becomes smaller and $\tau^*(r)$ becomes larger, contradicting that (T,r) is a worst case. If S_T hits r above $p_{\leftharpoonup}$, then this case is equivalent to having a peak $\wedge$ at exactly $p_{\leftharpoonup}$, because $\wedge$ does not interfere with S_T. By Lemma 7, r is then vertical in the worst-case.   □

Bounding the Competitive Ratio. To finish our analysis, we combine the previous lemmata, and choose s to minimize the competitive ratio across all cases. To obtain a strategy that is feasible in practice, we assume that $|S_T| \ge 1$. That is, we do not use infinitesimally small steps to start in practice. We then adapt our strategy by first moving upwards at most one, up to the final time that P is intersected, and then start following along P. This only shortens the search path, so the competitive ratio holds for this adjusted path as well.

Theorem 2. *Our searching strategy for searching in a 1.5D terrain achieves a competitive ratio of* $3\sqrt{19/2}$ *for* $s = \sqrt{2}/6$.

Proof sketch. We have now constructed two possible worst cases, where either P is obstructed, or P is unobstructed. For the former, careful analysis shows that competitive ratio $c^* \le (8 + d_r)\sqrt{1 + s^2}/\sqrt{d_r^2 + s^2(4 - d_r)^2}$ for $\frac{2}{9} \le s < \frac{4}{9}$. For the latter case, we get $c^* \le 9\sqrt{1 + s^2}$ for $\frac{1}{9} < s \le 1$.

We now choose s such that c^* is minimized. To do so, we observe that $(8 + d_r)\sqrt{1 + s^2}/\sqrt{d_r^2 + s^2(4 - d_r)^2}$ is decreasing in s (when $0 \le d_r \le 1$) and $9 \cdot \sqrt{1 + s^2}$ is increasing in s. We can hence equate the two formulas to find that $s = \sqrt{2}/6$ minimizes c^* over both cases. Then $r(-\infty, 4/13)$ is a worst-case ray, thus $c^* \le 3\sqrt{19/2}$ using $s = \sqrt{2}/6$. As $c \le c^*$, we conclude that our strategy has competitive ratio of at most $3\sqrt{19/2}$.   □

3   Competitive Searching on 2.5D Terrains

In this section we study the searching problem in an environment that is defined by a 2.5D terrain, which is represented by a function T_2. It is easy to see that, without putting additional restrictions on the terrain, achieving a bounded competitive ratio will be impossible: consider a flat terrain with arbitrarily many small pits in the terrain that are arbitrarily steep. Any searching strategy would have to move to the location of each pit in the xy-plane in order to look at the bottom of the pit. As we can place arbitrarily many pits within a small bounded distance from the starting point, and the target may be in any of the pits, the competitive ratio of any searching strategy would always be unbounded. We make this argument more concrete in the lower bound construction below. To restrict the set of 2.5D terrains under consideration, we require that the maximum slope of the terrain, which corresponds to the *Lipschitz constant* λ of T_2, is bounded. A strategy of moving upwards from p_0 results in a competitive ratio of $O(\lambda)$. In the remainder of this section we show that we can achieve a competitive ratio of $O(\sqrt{\lambda})$, which matching the lower bound for 2.5D terrains.

Lower Bound. We first show a lower bound on the competitive ratio for any searching strategy on 2.5D terrains. Since this lower bound is a function of λ, this directly implies that the competitive ratio is unbounded if we do not limit the maximum slope of the terrain.

Theorem 3. *The competitive ratio for searching on 2.5D terrains with maximum slope λ is at least $\Omega(\sqrt{\lambda})$.*

proof. Consider a flat terrain containing a regular grid of $k \times k$ pits formed by a cone of maximum slope λ, where $\delta = 1/k$ is the distance between the centers of two adjacent pits, and $\epsilon > 0$ is the depth

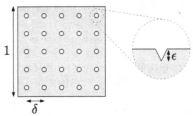

of each pit, where ϵ and k will be chosen later. For convenience we assume that the starting point of the searching problem is exactly a distance δ to the left from the lower-left pit in the grid at height 0. Now consider a searching strategy for this terrain, represented by a path P.

First assume that the maximum height z that P reaches before being able to see the bottom of the last pit satisfies $z > \frac{\sqrt{\lambda}}{4}$. Then P must travel at least a distance $z > \frac{\sqrt{\lambda}}{4}$ before seeing the last pit. The minimum travel distance to see this pit is less than $\sqrt{2}$. Hence, the competitive ratio is at least $\frac{\sqrt{\lambda}}{4\sqrt{2}} = \Omega(\sqrt{\lambda})$.

Now assume that P stays under the height of $z = \frac{\sqrt{\lambda}}{4}$. By construction of the pits, this implies that the searcher must be within a horizontal distance of $(z+\epsilon)/\lambda$ from the center of the pit to see the bottom of the pit (this is the radius of the cone of a pit when extended to height z). As such, after checking one pit, the searcher must travel at least a distance $\delta - 2(z+\epsilon)/\lambda$, which is the distance between two cones at height z, before being able to check another pit. If we choose $\epsilon = \frac{\sqrt{\lambda}}{4}$, then this distance is at least $\delta - \frac{1}{\sqrt{\lambda}}$. The total (horizontal) distance traveled by P before seeing the last pit is then at least $k^2(\delta - \frac{1}{\sqrt{\lambda}}) = k - \frac{k^2}{\sqrt{\lambda}}$, as there are k^2 pits in total. By choosing $k = \frac{\sqrt{\lambda}}{2}$ this total distance is at least $\frac{\sqrt{\lambda}}{2} - \frac{\sqrt{\lambda}}{4} = \frac{\sqrt{\lambda}}{4}$. Since the minimum travel distance to see the last pit is again less than $\sqrt{2}$, the competitive ratio of P is at least $\frac{\sqrt{\lambda}}{4\sqrt{2}} = \Omega(\sqrt{\lambda})$. $\square$

Searching Strategy. We now present a searching strategy for 2.5D terrains with a maximum slope λ. The aim is to match the dependency on λ that is shown in the lower bound. We will use the prior known value of λ to determine our search path P_λ. To simplify the analysis, our searching strategy consists of separate vertical and horizontal movement phases, explained in detail below.

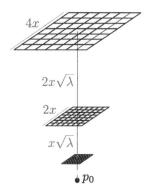

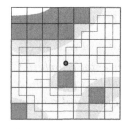

Fig. 6. Three iterations with grids constructed with $2k + 1$ by $2k + 1$ cells. Three iterations with grids constructed with $2k + 1$ by $2k + 1$ cells.

Fig. 7. Connected component $\mathcal{C}$ in red. Gray area is where T is above the grid. (Color figure online)

In the description of our search strategy, we again make use of arbitrarily small steps at the start to simplify the analysis. When a minimum value on the length of the optimal search path is given, all bounds still hold when we simply move upwards up to this value and then continuing on the described search path. Overall, our searching path P_λ works as follows: first, we move vertically up by a distance $\varepsilon\sqrt{\lambda}$, for some arbitrarily small value $\varepsilon > 0$. Next, we construct a square horizontal grid G_ε with total length 2ε centered (horizontally) around the starting point. This grid will consist of $(2k + 1) \times (2k + 1)$ grid cells, where k is chosen large enough such that the side length of a single grid cell is at most $\frac{\varepsilon}{2\sqrt{2\lambda}}$. Specifically, let k be the smallest integer such that $2k + 1 \geq 4\sqrt{2\lambda}$. We perform a horizontal search through this grid, described in detail below, and return to the center of the grid. We then move vertically up again until we are at a height that is $2\varepsilon\sqrt{\lambda}$ above the previous grid. Here we perform a horizontal search on a grid $G_{2\varepsilon}$ with total length 4ε, but where the number of grid cells is still $(2k + 1) \times (2k + 1)$. We then repeat this process, each time doubling the vertical distance between grids and doubling the total length of the grid, but keeping the number of grid cells the same (see Fig. 6). Note that a grid G_x for some $x \geq \varepsilon$ is at height $(2x - \varepsilon)\sqrt{\lambda}$ by construction. Since we assume that ε is arbitrarily small, we will simply say that G_x is at height $2x\sqrt{\lambda}$.

To perform a horizontal search in a grid G_x for some $x > 0$, we first consider the height of the terrain within the grid cells. We say a grid cell σ is *eligible* if at least one point inside σ has a height at most the height of G_x (which is $2x\sqrt{\lambda}$). We consider the connected set of eligible cells $\mathcal{C}$ that includes cell σ_0 containing the starting point (note that σ_0 is always eligible), where two eligible cells are connected if they share a side. To perform the horizontal search in G_x we construct a tour that starts in the center of σ_0, visits all the centers of cells in $\mathcal{C}$, is completely contained within the cells of $\mathcal{C}$, and eventually returns to the center of σ_0 (see Fig. 7). During this horizontal search, the terrain may force the searcher to increase the height, which is allowed. However, the searcher never moves back down, and hence the height will never decrease anywhere on P_λ.

Analysis. We first establish useful properties on the horizontal searches in grids.

Lemma 11. *Let G_x be a horizontal grid used in P_λ for some $x > 0$.*

(1) *The number of grid cells in G_x is $O(\lambda)$.*

    During a horizontal search in G_x:
(2) *The amount of horizontal movement is at most $O(x\sqrt{\lambda})$.*
(3) *The amount of vertical movement is at most $\frac{x\sqrt{\lambda}}{2}$.*

Note that property (3) of Lemma 11 implies that the search path P_λ is indeed valid, as the distance between grids G_x and G_{2x} is $2x\sqrt{\lambda}$, which is greater than $\frac{x\sqrt{\lambda}}{2}$. Thus, it is never necessary to move down again to reach the next grid in P_λ. We can now bound the length of P_λ at a particular height along the path.

Lemma 12. *The length of P_λ up to the point of reaching a horizontal grid G_x is at most $O(x\sqrt{\lambda})$.*

Proof. The total amount of vertical movement in P_λ simply corresponds to the height of G_x, which is $2x\sqrt{\lambda}$ by construction. For the horizontal movement we have to consider the grids $G_{x/2}, G_{x/4}, G_{x/8}, \ldots$, which by Lemma 11 induce a horizontal movement of at most $\sum_{i=1}^{\infty} O(x/2^i\sqrt{\lambda}) = O(x\sqrt{\lambda})$. The stated bound follows from adding the horizontal and vertical movement in P_λ. $\qquad\square$

Next, we use λ to determine when a point on P_λ can see the target t.

Lemma 13. *If p is a point that can see t, then any point p^* in the upwards cone starting at p with slope λ can see t.*

proof. Since the slope is bounded by λ, the upwards cone with slope λ aboveany point that lies above the terrain must be unobstructed. Furthermore, the line segment between p and t is unobstructed. Hence, the upwards wedge with slope λ over the path between p and t is also unobstructed. Since the cone above p is unobstructed and the wedge is unobstructed, the line segment between p^* and t is unobstructed. $\qquad\square$

Theorem 4. *Our strategy for searching in a 2.5D terrain with maximum slope λ achieves a competitive ratio of at most $O(\sqrt{\lambda})$.*

Proof. Let p be the point with the shortest distance to p_0 that can see t, and let $d(p_0, p)$ be the distance from p_0 to p. Furthermore, let $\vee$ be the cone cast upward from p_0 with slope $\frac{\sqrt{\lambda}}{2}$. For our analysis we consider two different cases: (1) p lies inside of $\vee$, or (2) p lies outside of $\vee$.

Case 1: p lies within $\vee$. Let z be the height of p and let r be the horizontal distance from p_0 to p. Since p lies within $\vee$, we know that $z \geq \frac{\sqrt{\lambda}}{2} r$. If we cast a ray directly upwards from p_0, we hit the cone from p with slope λ at height $z^* = z + \lambda r$. By Lemma 13, we see t from that intersection point (or any point directly above it). The next horizontal grid G_x of P_λ is at height $\leq 2z^*$, so $2x\sqrt{\lambda} \leq 2z^*$ or $x \leq \frac{z^*}{\sqrt{\lambda}}$. By Lemma 12 this implies that the searcher travels at most a distance of $O(z^*)$ before seeing t. Since the minimum distance to reach p is at least $z \geq \frac{\sqrt{\lambda}}{2} r$, we get that $z^*/z \leq 1 + 2\sqrt{\lambda}$. Hence, the competitive ratio in this case is at most $O(z^*/z) = O(\sqrt{\lambda})$.

Case 2: p lies below $\vee$. Let again z be the height of p and let r be the horizontal distance from p_0 to p. Since p lies below $\vee$, we know that $z < \frac{\sqrt{\lambda}}{2} r$. Consider the first time that a cell σ directly above p is visited by P_λ during a horizontal search of a grid G_x. Since $x \geq r$, the vertical distance between p and σ is at least $2x\sqrt{\lambda} - z \geq \sqrt{\lambda}(2x - \frac{r}{2}) \geq \frac{3x}{2}\sqrt{\lambda}$. Hence, the upwards cone from p with slope λ intersects the horizontal plane at G_x in a circle with radius $\frac{3x}{2}\sqrt{\lambda}/\lambda = \frac{3x}{2\sqrt{\lambda}}$. Since the side length of σ is at most $\frac{x}{2\sqrt{2\lambda}}$, this circle also contains the center of σ, from which we see t due to Lemma 13. Thus, the target is found at the latest during the horizontal search on G_x. Lemmata 12 and 11 (property 2 and 3) then imply that we travel at most a distance of $O(x\sqrt{\lambda})$ before we find t.

We now consider the distance $d(p_0, p)$. By construction, the horizontal search on grid $G_{x/2}$ did not visit a cell above p. We consider two possible cases. If the grid $G_{x/2}$ does not contain any cell directly above p, then $r > x/2$. In that case $d(p_0, p) > x/2$ and hence we obtain a competitive ratio of $O(x\sqrt{\lambda})/(x/2) = O(\sqrt{\lambda})$. If $G_{x/2}$ does contain a cell σ' directly above p, then σ' was not part of C for $G_{x/2}$. But then, in order to reach the point p from p_0, we must either reach a height of $x\sqrt{\lambda}$ (the height of $G_{x/2}$), or we must leave the horizontal domain of $G_{x/2}$. In both cases the shortest distance from p_0 to p is at least $x/2$ (or even $x\sqrt{\lambda}$ in the first case). Thus, we again obtain a competitive ratio of $O(\sqrt{\lambda})$. $\square$

4   Conclusion

The lower and upper bound for 1.5D terrain might be improved with a more intricate example and more extensive analysis respectively. For our search strategies we assumed that the terrain is given beforehand. However, our searching strategy for 1.5D terrains is affected by the terrain only when obstructed, thus the searching strategy can handle unknown terrains. This does not hold for our strategy on 2.5D terrains. Though we can address terrain on the fly, we crucially use the maximum slope λ to construct our search path. It would be interesting to study whether an efficient strategy exists that does not require λ to be known. Another direction for future research is to extend the result on 2.5D terrains to special types of polyhedral domains, such as star-shaped polyhedra. An important question here is how to redefine the parameter λ for polyhedral domains such that the competitive ratio can be bounded in terms of that parameter.

References

1. Baeza-Yates, R.A., Culberson, J.C., Rawlins, G.J.E.: Searching in the plane. Inf. Comput. **106**(2), 234–252 (1993)
2. Baeza-Yates, R.A., Schott, R.: Parallel searching in the plane. Comput. Geom. **5**, 143–154 (1995)
3. Beck, A., Newman, D.J.: Yet more on the linear search problem. Israel J. Math. **8**(4), 419–429 (1970)
4. de Berg, S., van Beusekom, N., van Mulken, M., Verbeek, K., Wulms, J.: Competitive searching over terrains (2024). https://arxiv.org/abs/2401.01289
5. Blum, A., Raghavan, P., Schieber, B.: Navigating in unfamiliar geometric terrain. SIAM J. Comput. **26**(1), 110–137 (1997)
6. Bose, P., Carufel, J.D., Durocher, S.: Searching on a line: a complete characterization of the optimal solution. Theor. Comput. Sci. **569**, 24–42 (2015)
7. Bose, P., Carufel, J.D., Durocher, S., Taslakian, P.: Competitive online routing on Delaunay triangulations. Int. J. Comput. Geom. Appl. **27**, 241–254 (2017)
8. Bose, P., Morin, P.: Competitive online routing in geometric graphs. Theor. Comput. Sci. **324**(2–3), 273–288 (2004)
9. Bouts, Q.W., Castermans, T., van Goethem, A., van Kreveld, M.J., Meulemans, W.: Competitive searching for a line on a line arrangement. In: Proceedings of the 29th ISAAC, pp. 49:1–49:12 (2018)
10. Datta, A., Hipke, C.A., Schuierer, S.: Competitive searching in polygons—Beyond generalised streets. In: Staples, J., Eades, P., Katoh, N., Moffat, A. (eds.) Algorithms and Computations, pp. 32–41. Springer Berlin Heidelberg, Berlin, Heidelberg (1995). https://doi.org/10.1007/BFb0015406
11. Demaine, E.D., Fekete, S.P., Gal, S.: Online searching with turn cost. Theor. Comput. Sci. **361**(2–3), 342–355 (2006)
12. Deng, X., Kameda, T., Papadimitriou, C.: How to learn an unknown environment. i: the rectilinear case. J. ACM 45(2), 215–245 (1998)
13. Ghosh, S.K., Klein, R.: Online algorithms for searching and exploration in the plane. Comput. Sci. Rev. **4**(4), 189–201 (2010)
14. Hammar, M., Nilsson, B.J., Persson, M.: Competitive exploration of rectilinear polygons. Theor. Comput. Sci. **354**(3), 367–378 (2006)
15. Hipke, C.A., Icking, C., Klein, R., Langetepe, E.: How to find a point on a line within a fixed distance. Discret. Appl. Math. **93**(1), 67–73 (1999)
16. Hoffmann, F., Icking, C., Klein, R., Kriegel, K.: The polygon exploration problem. SIAM J. Comput. **31**(2), 577–600 (2001)
17. Icking, C., Klein, R.: Searching for the kernel of a polygon-a competitive strategy. In: Proceedings of the 11th SoCG, pp. 258–266 (1995)

18. Icking, C., Klein, R., Langetepe, E., Schuierer, S., Semrau, I.: An optimal competitive strategy for walking in streets. SIAM J. Comput. **33**(2), 462–486 (2004)
19. Kalyanasundaram, B., Pruhs, K.: A competitive analysis of algorithms for searching unknown scenes. Comput. Geom. **3**(3), 139–155 (1993)
20. Klein, R.: Walking an unknown street with bounded detour. Comput. Geom. **1**(6), 325–351 (1992)
21. Klein, R.: Algorithmische Geometrie, vol. 80. Springer (1997)
22. Kleinberg, J.M.: On-line search in a simple polygon. In: Proceedings of the 5th SODA, pp. 8–15 (1994)
23. López-Ortiz, A., Schuierer, S.: Generalized streets revisited. In: Diaz, J., Serna, M. (eds.) Algorithms — ESA '96, pp. 546–558. Springer Berlin Heidelberg, Berlin, Heidelberg (1996). https://doi.org/10.1007/3-540-61680-2_81
24. López-Ortiz, A., Schuierer, S.: Searching and on-line recognition of star-shaped polygons. Inf. and Comput. **185**(1), 66–88 (2003)
25. Schuierer, S.: On-line searching in simple polygons. LNCS **1724**, 220–239 (1999)
26. Wei, Q., Tan, X., Ren, Y.: Walking an unknown street with limited sensing. Int. J. Pattern Recognit Artif Intell. **33**(13), 1959042 (2019)
27. Wei, Q., Yao, X., Liu, L., Zhang, Y.: Exploring the outer boundary of a simple polygon. IEICE Trans. Inf. Syst. **104-D**(7), 923–930 (2021)

Computational Geometry

Minimizing the Size of the Uncertainty Regions for Centers of Moving Entities

William Evans and Seyed Ali Tabatabaee[(✉)]

Department of Computer Science, University of British Columbia, Vancouver, Canada
{will,salitaba}@cs.ubc.ca

Abstract. In this paper, we study the problems of computing the 1-center, centroid, and 1-median of objects moving with bounded speed in Euclidean space. We can acquire the exact location of only a constant number of objects (usually one) per unit time, but for every other object, its set of potential locations, called the object's *uncertainty region*, grows subject only to the speed limit. As a result, the center of the objects may be at several possible locations, called the center's uncertainty region. For each of these center problems, we design query strategies to minimize the size of the center's uncertainty region and compare its performance to an optimal query strategy that knows the trajectories of the objects, but must still query to reduce their uncertainty. For the static case of the 1-center problem in $\mathbb{R}^1$, we show an algorithm that queries four objects per unit time and is 1-competitive against the optimal algorithm with one query per unit time. For the general case of the 1-center problem in $\mathbb{R}^1$, the centroid problem in $\mathbb{R}^d$, and the 1-median problem in $\mathbb{R}^1$, we prove that the Round-robin scheduling algorithm is the best possible competitive algorithm. For the center of mass problem in $\mathbb{R}^d$, we provide an $O(\log n)$-competitive algorithm. In addition, for the general case of the 1-center problem in $\mathbb{R}^d$ ($d \geq 2$), we argue that no algorithm can guarantee a bounded competitive ratio against the optimal algorithm.

Keywords: Data in motion · Uncertain inputs · Center problems · Online algorithms

1 Introduction

Many real-world problems, such as controlling air traffic and providing service to cellular phones, involve moving entities. Therefore, analyzing moving objects has become a topic of interest within the area of theoretical computer science. In many problems, the movement of the objects is unpredictable and data processing must be done in real-time. Moreover, obtaining the exact location of an object at any point in time often entails a cost. Hence, knowing the precise location of all objects at any time is impractical. Instead, for every object, a

This work was partially funded by NSERC Discovery Grants and the Institute for Computing, Information and Cognitive Systems (ICICS) at UBC.

region of potential locations that we call its uncertainty region is known. Due to the movement of the objects, the size of the uncertainty region for each object grows over time unless its exact location is acquired. The target is to design cost-effective algorithms for highly accurate analysis of moving objects.

Centers have been used to represent a given point set. They have applications in data clustering and facility location. For a given set of points in the Euclidean space, the three most common centers are the following:

1. **1-center:** Given a set of n points in $\mathbb{R}^d$, the *k-center* is a set of k facility locations such that the farthest distance from an input point to its closest facility, is minimized. For $k = 1$, this is the center of the smallest ball that contains the input set.
2. **centroid:** Given a set of n points in $\mathbb{R}^d$, the *k-centroid*, which is the solution to the *k-means problem*, is a set of k facility locations that minimize the sum of squared distances between each input point and its closest facility. The 1-centroid is called the centroid. Given a set of n weighted points in $\mathbb{R}^d$, the *center of mass* is the average position of all points, weighted according to their masses. When all input points have unit weights, the center of mass is the centroid.
3. **1-median:** Given a set of n points in $\mathbb{R}^d$, the k-median is the set of k facilities such that the average distance from an input point to its closest facility is minimized.

1.1   Model and Definitions

Given a center function and a set of the initial locations of n moving objects ($n \geq 2$) in $\mathbb{R}^d$ where the speed of every object is bounded by v, we consider the problem of finding query strategies (which can query once at the end of each unit of time) for minimizing a measure which is the maximum size of the uncertainty region for the specified center over all query times. We define the size of a region as the maximum pairwise distance between the points of that region. We only care about the uncertainty regions at the times when a query has just been made and one object has an uncertainty region of size 0. Considering that for some instances of the problem the measure is large for every query strategy, we analyze the performance of our algorithms in a competitive framework, where the competitive factor is the ratio of our algorithm's measure to that of the optimal algorithm. We assume that the optimal algorithm knows the trajectories of the objects, however it must still query to keep the uncertainty regions for the moving objects, and hence the uncertainty region of their center, small.

In some parts of this paper, we consider the weighted version of the problem where each object has a positive weight. However, we mostly discuss the unweighted version of the problem; hence, we assume that all objects have unit weights unless stated otherwise. Figure 1 illustrates the weighted version of the centroid problem (the center of mass problem) for a small set of weighted objects with uncertain locations. We also consider a version of the problem where objects have different maximum speeds. But, we assume that all objects have the same maximum speed unless stated otherwise. We define the static case of our problem as a special case where although objects have a maximum speed of v and their uncertainty regions grow accordingly, they are actually static (they do not move). To achieve better and more meaningful competitive ratios, we may allow our algorithms to query more than one object per unit time.

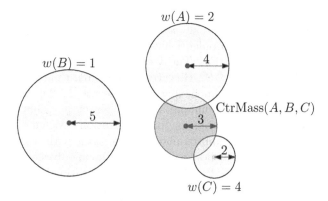

Fig. 1. The uncertainty regions for three weighted objects A, B, and C, along with the potential locations (uncertainty region) of their center of mass.

1.2 Contribution and Organization

In this paper, we study the aforementioned center problems with moving entities in the Euclidean space. We investigate the problem of designing competitive query strategies for minimizing the size of the uncertainty region of these different centers. For the static case of the 1-center problem in $\mathbb{R}^1$, we show that there exists an algorithm that queries a constant number of objects per unit time and is 1-competitive against the optimal algorithm with one query per unit time. For the 1-center problem with moving entities in $\mathbb{R}^1$, we provide an algorithm with the best possible competitive ratio of $O(n)$. For this problem in $\mathbb{R}^d$ ($d \geq 2$), we show that no algorithm can guarantee a bounded competitive ratio against the optimal algorithm. For the centroid problem with moving entities in $\mathbb{R}^d$ ($d \geq 1$), we present an algorithm that is 1-competitive against the optimal algorithm. For the weighted version of the centroid problem (the center of mass problem) in $\mathbb{R}^d$ ($d \geq 1$) where objects have different maximum speeds, we provide an $O(\log n)$-competitive algorithm. For the 1-median problem with moving entities in $\mathbb{R}^1$, we give an algorithm with the best possible competitive ratio of $O(n)$. Several of our algorithms use *Round-robin* scheduling for a subset of the entities, which repeatedly queries the entities in the subset in a fixed order.

The rest of this paper is organized as follows. Section 2 provides background information on the introduced center problems and computing functions with uncertain inputs and moving data. Sections 3, 4, and 5 study the 1-center problem, the centroid problem, and the 1-median problem, respectively, in the presented model. Finally, Sect. 6 concludes the results presented in this paper and lists some open problems.

2   Background

The k-center problem and its special case, the 1-center problem (also known as the minimum enclosing ball problem) have various applications, such as data classification and facility location. The k-center problem is known to be NP-hard in the Euclidean plane [28]. However, polynomial-time 2-approximation algorithms exist for the problem in any metric space [17,20]. The Euclidean 1-center problem always has a unique solution. Exact polynomial-time algorithms [9,30] exist for the problem in fixed dimensions. When the dimension is not fixed, polynomial-time approximation algorithms with a factor of $1 + \epsilon$ exist [31]. The speed of the Euclidean 1-center of moving objects with bounded speed is unbounded in $\mathbb{R}^d$ ($d \geq 2$) [6]. Durocher [13] provided bounded-velocity approximations for the Euclidean k-center of moving objects with bounded speed when $k \leq 2$, and showed that no bounded-velocity approximation is possible for the problem when $k \geq 3$.

The k-means problem has applications in data clustering and facility location. This problem has been proved to be NP-hard in $\mathbb{R}^d$ even for $k = 2$ [12]. Lloyd's method [27] has been widely used to find a local minimum for the objective function of the k-means problem. This method starts with an arbitrary k-clustering of input points and computes the centroids of the clusters. Then the method repeatedly assigns each point to its closest cluster center and recomputes the cluster centers. Constant factor approximation algorithms have been proposed for the Euclidean k-means problem [1,24]. The best known approximation guarantee for this problem is 6.357 [1]. The center of mass is a unique point where the aggregate mass of a set of objects is concentrated. The center of mass problem has applications in various fields including physics, astronomy, and engineering. The speed of the center of mass is the sum of each object's momentum divided by the total weight of all objects. Hence, this speed is bounded by the maximum speed that objects can have.

The k-median problem is another well-studied problem with applications in data clustering and facility location. This problem is known to be NP-hard in the Euclidean plane [28]. Hence, constant factor approximation algorithms have been proposed for this problem [1,8]. The best known approximation ratio for this problem is 2.633 [1]. The 1-median problem, also known as the Fermat-Weber problem [11], is a special case of the k-median problem. The Euclidean 1-median is unique when the input points are not collinear [26] or when the number of input points is odd. However, if the number of input points is even and the points are collinear, then any point that lies on the line segment between the two middle input points is a 1-median of the points. By convention, the Euclidean 1-median of such a set of points is defined as the midpoint of the two middle points in the set. In general, the exact position of the Euclidean 1-median cannot be calculated using radicals over the field of rationals when the number of points is greater than or equal to five [3]. Consequently, $(1 + \epsilon)$-approximate solutions have been designed for this problem [2,10]. The Euclidean 1-median moves discontinuously in $\mathbb{R}^d$ $(d \geq 2)$ [13]. Durocher [13] provided bounded-velocity approximations for the Euclidean 1-median of moving objects with bounded speed, and showed that no bounded-velocity approximation is possible for the Euclidean k-median problem when $k \geq 2$.

Computing functions with uncertain inputs has been the subject of multiple research projects [18,25,29]. Geometric problems have also been studied with uncertainty and methods such as witness algorithms have been proposed to address those problems [7]. Due to the unpredictability of the movements in many problems that involve moving objects, the real-time processing of moving data is closely related to computing with uncertainty. Kahan [22,23] studied the maximum problem, the sorting problem, and some geometric problems with moving data, where the target was to reduce data acquisition costs. Furthermore, competitive query strategies have been developed for problems where the number of queries per unit time is bounded [14,15,32].

3   The 1-Center Problem

In this section, we consider the 1-center problem with moving entities. We first consider a *static* version of the problem where all queries happen to return the object's original location, though we must query to confirm this. Solutions in this case are simpler than in the *general* case, in which the movement of objects may change their importance in calculating the center, however the static case is still challenging.

3.1   The Static Case in $\mathbb{R}^1$

For the static case of the 1-center problem in $\mathbb{R}^1$, we prove the existence of an algorithm that queries four objects per unit time and is 1-competitive against the optimal algorithm with one query per unit time.

We begin by introducing a lemma which we will use to prove the existence of an algorithm with the aforementioned guarantee. This lemma considers a special case of the problem of windows scheduling without migration [5] and entails the notion of pinwheel scheduling [21]. Given a multiset of positive integers $A = \{a_1, ..., a_n\}$, the pinwheel scheduling problem asks for an infinite sequence over $\{1, ..., n\}$ such that any integer $i \in \{1, ..., n\}$ appears at least once in any a_i consecutive entries of the sequence. The density of A is defined as $d(A) = \sum_{i=1}^{n} \frac{1}{a_i}$. A necessary condition for schedulability is $d(A) \leq 1$ [21]. Moreover, $d(A) \leq 0.75$ is a sufficient condition for schedulability [19]. Based on these results, we prove the lemma in the full version of the paper [16].

Lemma 1. *Given a multiset of positive integers A where $d(A) \leq 2$, it is possible to partition A into three subsets, which are themselves multisets, such that each subset has a pinwheel schedule.*

Windows scheduling without migration is a restricted version of the windows scheduling problem [4]. Given a multiset of positive integers $A = \{a_1, ..., a_n\}$ and a positive integer h, the windows scheduling problem asks whether it is possible to schedule n pages on h channels, with at most one page on each channel at any time, such that the time between two consecutive appearances (on any channel) of the i-th page is at most a_i ($1 \le i \le n$). In windows scheduling without migration, all appearances of a page must be on the same channel. Allowing migration would be fine for our use of Lemma 1 in proving the following Theorem 1, even though our proof of the lemma does not make use of this flexibility. It has been shown that windows scheduling is possible on $d(A) + O(\ln(d(A)))$ channels [4]. Nevertheless, Lemma 1 provides a better guarantee for the special case that it considers.

Now, we are ready to prove the main theorem in this subsection.

Theorem 1. *For the static case of the 1-center problem in $\mathbb{R}^1$, there exists an algorithm that queries four objects per unit time and is 1-competitive against the optimal algorithm with one query per unit time.*

Proof. Without loss of generality, we assume that the maximum speed v is unit. We sort the objects based on their positions and let x_i denote the position of the i-th object ($x_1 \le ... \le x_n$). We define $f(x, y) = \lfloor |y - x| \rfloor + 1$. To prevent the uncertainty region of a static object at position x from going beyond a point at position y, we need to acquire the exact location of that object at least once in every $f(x, y)$ queries. We let b denote the smallest positive value such that $\sum_{i=1}^{n} \frac{1}{\min(f(x_i, x_n + b), f(x_i, x_1 - b))} \le 1$. Hence, considering the necessary condition for schedulability [21], no query strategy can ensure that the uncertainty region of each object stays within $[x_1 - b', x_n + b']$ for any value $b' < b$.

The size of the uncertainty region for the 1-center in $\mathbb{R}^1$ is equivalent to the average of the size of the uncertainty region for the maximum and the size of the uncertainty region for the minimum. Given a set of points in $\mathbb{R}^1$, the maximum problem asks for the position of the maximum point and the minimum problem asks for the position of the minimum point in the set. Regardless of the query strategy, at some point in the future, either the uncertainty region for the maximum will include $[x_n, x_n + b]$, or the uncertainty region for the minimum will include $[x_1 - b, x_1]$. If $|x_n - x_1| \le 1$, the union of the uncertainty regions for the maximum and the minimum will always include $[x_1, x_n]$. Otherwise, the union of the two uncertainty regions will always intersect $[x_1, x_1 + 1] \cup [x_n - 1, x_n]$ such that the total size of the intersection is at least 1. Consequently, regardless of the query strategy, the size of the uncertainty region for the 1-center will be at least $\frac{b + \min(|x_n - x_1|, 1)}{2}$ at some point in the future. For the optimal algorithm, this is a lower bound on the maximum size of the uncertainty region for the 1-center.

We now explain how to use four queries per unit time and be 1-competitive against the optimal algorithm with one query per unit time. First, we show how to compute b. We know that $0 \le b \le n$ (because the Round-robin algorithm can ensure that the uncertainty region of each object stays within $[x_1 - n, x_n + n]$).

Furthermore, there exists an index $1 \leq i \leq n$ such that either $\lfloor x_i - x_1 + b \rfloor = x_i - x_1 + b$ or $\lfloor x_n + b - x_i \rfloor = x_n + b - x_i$ (otherwise, there exists a query strategy that can ensure that the uncertainty region of each object stays within $[x_1 - b', x_n + b']$, for some $b' < b$). Therefore, there are $O(n^2)$ possible values for b and we can find b using a binary search. Further, we show that $f(x_i, x_n + b) \leq 2f(x_i, x_n + \frac{b}{2})$. We have

$$f(x_i, x_n + b) = \lfloor x_n + b - x_i \rfloor + 1$$
$$\leq \lfloor x_n + \frac{b}{2} - x_i \rfloor + \lfloor \frac{b}{2} \rfloor + 2$$
$$\leq 2 \lfloor x_n + \frac{b}{2} - x_i \rfloor + 2$$
$$= 2f(x_i, x_n + \frac{b}{2}).$$

Similarly, we have $f(x_i, x_1 - b) \leq 2f(x_i, x_1 - \frac{b}{2})$. Hence, we deduce that

$$\sum_{i=1}^{n} \frac{1}{\min(f(x_i, x_n + \frac{b}{2}), f(x_i, x_1 - \frac{b}{2}))} \leq \sum_{i=1}^{n} \frac{2}{\min(f(x_i, x_n + b), f(x_i, x_1 - b))} \leq 2.$$

Therefore, by Lemma 1, we can use three queries per unit time to maintain the uncertainty regions of all objects within $[x_1 - \frac{b}{2}, x_n + \frac{b}{2}]$. We let the fourth query repeatedly switch between the object at position x_1 and the object at position x_n. This way, the total size of the intersection between $[x_1, x_n]$ and the union of the uncertainty regions for the maximum and the minimum will be at most $\min(|x_n - x_1|, 1)$. Consequently, using this algorithm, the size of the uncertainty region for the 1-center will be at most $\frac{b + \min(|x_n - x_1|, 1)}{2}$ at any point in the future. Hence, the presented algorithm that queries four objects per unit time is 1-competitive against the optimal algorithm with one query per unit time. □

3.2 The General Case in $\mathbb{R}^1$

For the general case of the 1-center problem in $\mathbb{R}^1$, we prove that the Round-robin algorithm for querying objects achieves the best possible competitive ratio against the optimal algorithm.

Theorem 2. *For the 1-center problem in $\mathbb{R}^1$, the Round-robin scheduling algorithm keeps the maximum size of the uncertainty region within $O(vn)$ at any point in the future and achieves the best possible competitive ratio of $O(n)$ against the optimal algorithm.*

Proof. The Round-robin algorithm keeps the size of the uncertainty region of each object within $O(vn)$ because it acquires the exact location of each object once in every n queries. For $1 \leq i \leq n$, let $[s_i, e_i]$ denote the uncertainty region of the i-th object (s_i is the starting point and e_i is the ending point of the uncertainty region). At any time, let m be the index of the object with the highest ending point ($e_m = \max_{1 \leq i \leq n} e_i$). Thus, $[s_m, e_m]$ includes the uncertainty region for the maximum. Consequently, the size of the uncertainty region for the maximum is less than or equal to $|e_m - s_m|$ which is bounded by $O(vn)$. Similarly, the size of the uncertainty region for the minimum is bounded by $O(vn)$ at any point in the future. Therefore, using the Round-robin algorithm, the size of the uncertainty region for the 1-center (which is equivalent to the average of the size of the uncertainty region for the maximum and the size of the uncertainty region for the minimum) will never exceed $O(vn)$.

For the optimal algorithm, at any point after the first query, between the object with the maximum last known exact position and the object with the minimum last known exact position, at least one will have not been queried last; thus, the size of the uncertainty region of that object will be $\Omega(v)$. Consequently, the size of the uncertainty region for at least one of the maximum or minimum will be $\Omega(v)$. Hence, the size of the uncertainty region for the 1-center will be $\Omega(v)$ at any point after the first query. Considering that the Round-robin algorithm keeps the maximum size of the uncertainty region within $O(vn)$, it achieves a competitive ratio of $O(n)$ against the optimal algorithm.

We show that $O(n)$ is the best possible competitive ratio against the optimal algorithm. We consider an example with n objects initially located at the origin. One of those objects moves with a speed of v in the positive direction and another one moves with a speed of v in the negative direction. The rest of the objects do not move. In this example, the optimal algorithm acquires the exact location of each of the two moving objects once in every two queries and maintains the size of the uncertainty region for the 1-center within $O(v)$. However, any algorithm that does not know the future object trajectories may fail to query the two moving objects in its first $n - 2$ queries; hence, the size of the uncertainty region for the 1-center can become $\Omega(vn)$. This is true even if we allow the algorithm to query a constant number of objects (instead of one) per unit time. For this reason, $O(n)$ is the best possible competitive ratio.   □

The Round-robin algorithm keeps the maximum size of the uncertainty region for the 1-center within $O(vn)$, and so does the optimal algorithm. We show that even for the optimal algorithm, the maximum size of the uncertainty region can be $\Omega(vn)$. The proof is presented in the full version of the paper [16].

Proposition 1. *The maximum size of the uncertainty region for the 1-center can be $\Omega(vn)$ for the optimal algorithm.*

We now provide an upper bound on the competitive ratio of the Round-robin algorithm for the 1-center problem where objects have different maximum speeds. The proof is presented in the full version of the paper [16].

Proposition 2. *For the 1-center problem in $\mathbb{R}^1$ where objects have different maximum speeds, the Round-robin scheduling algorithm achieves a competitive ratio of $O(v_M n / v_m)$ against the optimal algorithm, where v_M is the highest maximum speed and v_m is the lowest maximum speed.*

3.3   The General Case in $\mathbb{R}^d$

For the general case of the 1-center problem in $\mathbb{R}^d$ $(d \geq 2)$, we show that the maximum size of the uncertainty region can be unbounded for the optimal algorithm. The proof of the following proposition is inspired by the proof of Theorem 2 of [6] and can be found in the full version of the paper [16].

Proposition 3. *The maximum size of the uncertainty region of the 1-center in $\mathbb{R}^d$ $(d \geq 2)$ can be unbounded for the optimal algorithm.*

Next, we show that it is impossible to find an algorithm that works well compared to the optimal algorithm.

Theorem 3. *The maximum size of the uncertainty region of the 1-center of points in $\mathbb{R}^d$ $(d \geq 2)$ for any algorithm that does not know the trajectories may be an arbitrary factor larger than that obtained by the optimal algorithm.*

Proof. We provide an example for which we prove that no algorithm that does not know the trajectories of the objects can guarantee a bounded competitive ratio against the optimal algorithm. Although we explain the example in $\mathbb{R}^2$ here, we could consider the same example in more than two dimensions, and the proof would be very similar. Let $\mathbf{a} = (0, G)$, $\mathbf{b} = (-2L, 0)$, $\mathbf{c} = (2L, 0)$, $\mathbf{u} = \mathbf{c} - \mathbf{a}$, and $\mathbf{e} = \mathbf{c} + v(n-3)\hat{\mathbf{u}}$, where $n \geq 28$, L is arbitrarily large (much larger than n and v), and G is much larger than L in a way that for $y \geq \frac{G}{3}$, we have $\sqrt{100L^2 + y^2} \leq y + v$. We consider an example with n objects where one object (denoted by A) is always located at $\mathbf{a}$, one object (denoted by B) is always located at $\mathbf{b}$, one object (denoted by E) starts at $\mathbf{e}$ and always moves with a maximum speed of v in the direction of $\hat{\mathbf{u}}$, and $n-3$ objects start at $\mathbf{e}$ and move with a maximum speed of v in the direction of $-\hat{\mathbf{u}}$ until they stop at $\mathbf{c}$.

For the above example, the actual 1-center is always the midpoint of the positions of A and E. Consequently, the x-coordinate of the actual 1-center is always greater than L. Nevertheless, any algorithm that does not know the future object trajectories may fail to query E in its first $n - 3$ queries. Thus, for any such algorithm after the first $n - 3$ queries, $\mathbf{c}$ is a potential location of each of the $n - 2$ objects that were initially at $\mathbf{e}$, $\mathbf{a}$ is a potential location of A, and $\mathbf{b}$ is a potential location of B. Therefore, the 1-center could potentially be located on the line $x = 0$. Consequently, for any algorithm that does not know the future object trajectories, the size of the uncertainty region for the 1-center can become greater than L (which is arbitrarily large), after the $(n - 3)$-th query.

Now, we only need to prove that the optimal algorithm can keep the maximum size of the uncertainty region for the 1-center within a function of n and v for the example mentioned above. We consider an algorithm that queries each of A, B, and E once in every four queries and each other object once in every $4(n - 3)$ queries. Hence, the sizes of the uncertainty regions of A, B, and E will never exceed $6v$. Also, the size of the uncertainty region of each of the other objects will always be $O(vn)$. We prove that using this algorithm, the maximum size of the uncertainty region for the 1-center is at most some function of n and

v for the provided example. This will prove the same for the optimal algorithm which works at least as well as the aforementioned algorithm.

Let us disregard B for now and analyze the uncertainty region for the 1-center of the other objects. For those objects, the radius of the minimum enclosing circle is at most half of the distance between the two farthest points in the uncertainty regions of A and E. This is because the line segment between those points is a diameter of a circle that contains the uncertainty regions of all objects. Therefore, after q queries ($q \geq 0$) made by the algorithm described above, the radius of the minimum enclosing circle is at most $r_q = \frac{1}{2}(3v + |e - a| + qv) = \frac{1}{2}|c - a| + \frac{v}{2}(n + q)$. Considering that the distance between the 1-center and some point in the uncertainty region of A must be at most r_q after the q-th query, the 1-center is within distance $r_q + 3v$ from $\mathbf{a}$. Furthermore, the distance between the 1-center and some point in the uncertainty region of E must be at most r_q after the q-th query. Thus, the distance between the 1-center and $\mathbf{e} + qv\hat{\mathbf{u}}$ is less than or equal to $r_q + 6v$. Consequently, the distance between the 1-center and $\mathbf{a}$ is at least $|e - a| + qv - r_q - 6v = r_q - 9v$ after the q-th query.

The 1-center of a set of points in the Euclidean space is within the convex hull of those points (otherwise, there exists a hyperplane that separates the 1-center from the points, and by moving the 1-center towards that hyperplane, we are reducing its distance to each point). Therefore, the uncertainty region for the 1-center is always within the convex hull of the uncertainty regions of the objects. We deduce that at any time, the uncertainty region for the 1-center of all objects other than B is within a strip-like region inside the convex hull of the uncertainty regions of the objects (see Fig. 2). At any time, the length of this strip is $O(vn)$ because the size of the uncertainty region of each object is $O(vn)$ and the centers of the uncertainty regions of the objects are collinear. Moreover, the width of this strip is at most $r_q + 3v - (r_q - 9v) = 12v$. Hence, the maximum size of the strip-like region is $O(vn)$. Consequently, using the above algorithm, the maximum size of the uncertainty region for the 1-center of all objects other than B is $O(vn)$.

Let $\mathbf{d}$ represent the location of a point within the strip-like region. If the y-coordinate of $\mathbf{d}$ is within $[\frac{G}{3} + 3v, \frac{G}{2} - 4v]$, then the distance between $\mathbf{d}$ and any point in the uncertainty region A is at least $\frac{G}{2} + v$. Also, in this case, the distance between $\mathbf{d}$ and any point in the uncertainty region B is at most $\frac{G}{2}$. Furthermore, it is easy to see that if the y-coordinate of $\mathbf{d}$ is less than $\frac{G}{3} + 3v$, then the distance between $\mathbf{d}$ and any point in the uncertainty region of B is less than the distance between $\mathbf{d}$ and any point in the uncertainty region of A. If $r_q - 9v$ (the minimum distance between $\mathbf{a}$ and any point in the strip-like region) is at least $\frac{G}{2} + 5v$, then the y-coordinate of any point in the strip-like region is at most $\frac{G}{2} - 4v$, which means that the 1-center of all objects other than B is exactly the same as the 1-center of all objects including B. Considering that $n \geq 28$, we have

$$r_q - 9v = \frac{1}{2}|c - a| + \frac{v}{2}(n + q) - 9v \geq \frac{G}{2} + \frac{v}{2}(28 + 0) - 9v = \frac{G}{2} + 5v.$$

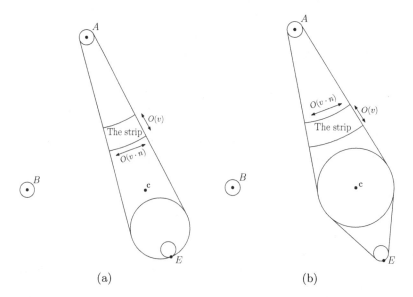

(a)   (b)

Fig. 2. The uncertainty regions of the objects and the strip-like region that contains the 1-center of all objects other than B for the example and the algorithm provided in the proof of Theorem 3 at the time (a) before the $(n-3)$-th query and (b) after the $(5n-15)$-th query.

Hence, using the algorithm described above, the uncertainty region for the 1-center of all objects including B is always within the strip-like region. Thus, the optimal algorithm keeps the maximum size of the uncertainty region for the 1-center within $O(vn)$ for the provided example. □

4 The Centroid Problem

We know that at any time, the uncertainty region of each object is a ball centered at the last known exact position of the object. We introduce a lemma which we will use to obtain the size of the uncertainty region for the center of mass or its special case, the centroid. The proof is provided in the full version [16].

Lemma 2. *Given a set of n weighted objects in $\mathbb{R}^d$ $(d \geq 1)$ with uncertain locations, the uncertainty region for the center of mass is a ball centered at the weighted average of the centers of the objects' uncertainty regions. The radius of this ball is equivalent to the weighted average of the radii of the objects' uncertainty regions.*

Lemma 2 shows that the size of the uncertainty region for the center of mass is equivalent to the weighted average of the sizes of the objects' uncertainty regions. Now, for the centroid problem, we prove that the Round-robin algorithm is 1-competitive against the optimal algorithm. The proof of the following theorem is provided in the full version of the paper [16].

Theorem 4. *For the centroid problem in $\mathbb{R}^d$ $(d \geq 1)$, the maximum size of the uncertainty region for the Round-robin scheduling algorithm is $\Theta(vn)$ and the Round-robin algorithm is 1-competitive against the optimal algorithm.*

For the center of mass problem, the competitive ratio of the Round-robin algorithm is $\Omega(n)$ against the optimal algorithm. This can be easily proved by considering an example where the weight of one of the objects is much higher than the weight of the other objects. Next, we provide an algorithm for the center of mass problem where objects have different maximum speeds and prove that it guarantees a competitive ratio of $O(\log n)$ against the optimal algorithm.

Theorem 5. *For the center of mass problem in $\mathbb{R}^d$ $(d \geq 1)$ where objects have different maximum speeds, there exists an algorithm that achieves a competitive ratio of $O(\log n)$ against the optimal algorithm.*

Proof. We sort the objects in a way that we have $v_1 w_1 \geq ... \geq v_n w_n$. We then partition the objects into $\lfloor \log n \rfloor + 1$ groups. For $1 \leq i \leq n$, we assign the i-th object to the $(\lfloor \log i \rfloor + 1)$-th group. Now, we describe the query strategy. We query objects from different groups in a cyclic way. Also, within each group, we choose the objects to query in a cyclic way. For example, if we have seven objects, we partition the objects into three groups. We assign the first object to the first group, the second and third objects to the second group, and the remaining objects to the third group. Consequently, our algorithm repeatedly follows the sequence $1, 2, 4, 1, 3, 5, 1, 2, 6, 1, 3, 7$ for querying the objects (entries of the sequence indicate the objects that our algorithm queries).

The algorithm described above acquires the exact location of one object from the group that contains the i-th object $(1 \leq i \leq n)$ in every $\lfloor \log n \rfloor + 1$ queries and that group contains at most $2^{\lfloor \log i \rfloor}$ objects. Therefore, the algorithm acquires the exact location of the i-th object at least once in every $(\lfloor \log n \rfloor + 1)2^{\lfloor \log i \rfloor}$ queries. Hence, using this algorithm, the maximum size of the uncertainty region for the center of mass is at most $\frac{1}{W} \sum_{i=1}^{n} 2((\lfloor \log n \rfloor + 1)2^{\lfloor \log i \rfloor} - 1)v_i w_i$ (by Lemma 2), where $W = \sum_{i=1}^{n} w_i$ is the total weight of all objects.

On the other hand, to keep the maximum size of the uncertainty region for the center of mass finite, the optimal algorithm has to query the n-th object at some point. At that time, the size of the uncertainty region for the center of mass will be at least $\frac{1}{W} \sum_{i=1}^{n-1} 2iv_i w_i$ (by Lemma 2). We have

$$\frac{3(\lfloor \log n \rfloor + 1)}{W} \sum_{i=1}^{n-1} 2iv_i w_i \geq \frac{\lfloor \log n \rfloor + 1}{W} \sum_{i=1}^{n} 2iv_i w_i$$

$$\geq \frac{1}{W} \sum_{i=1}^{n} 2((\lfloor \log n \rfloor + 1)i - 1)v_i w_i$$

$$\geq \frac{1}{W} \sum_{i=1}^{n} 2((\lfloor \log n \rfloor + 1)2^{\lfloor \log i \rfloor} - 1)v_i w_i.$$

Therefore, for the center of mass problem, the competitive ratio of the algorithm described above is $O(\log n)$ against the optimal algorithm. $\square$

5   The 1-Median Problem

Given a set of n points in $\mathbb{R}^1$, the 1-median is the $\lceil \frac{n}{2} \rceil$-th smallest point if n is odd and the midpoint of the $\frac{n}{2}$-th and $(\frac{n}{2}+1)$-th smallest points if n is even.

Theorem 6. *For the 1-median problem in $\mathbb{R}^1$, the Round-robin scheduling algorithm keeps the maximum size of the uncertainty region within $O(vn)$ at any point in the future and achieves the best possible competitive ratio of $O(n)$ against the optimal algorithm.*

Proof. (sketch) Let s_k (e_k) be the k-th smallest starting (ending) point of all n objects' uncertainty regions. Even though the region $[s_k, e_k]$ may not be the uncertainty region of any particular object, the k-th smallest object lies within this region, which has size $O(vn)$, since Round-robin keeps the size of every object's uncertainty region within $O(vn)$. The rest of the theorem follows from a lower bound of $\Omega(v)$ on the size of $[s_k, e_k]$ and examples that show $O(n)$ is the best possible competitive ratio. (See the full version of the paper [16]).   □

The Round-robin algorithm keeps the maximum size of the uncertainty region for the 1-median within $O(vn)$. We argue that even for the optimal algorithm, the maximum size of the uncertainty region can be $\Omega(vn)$. The proof is presented in the full version of the paper [16].

Proposition 4. *For the 1-median problem in $\mathbb{R}^1$, the maximum size of the uncertainty region can be $\Omega(vn)$ for the optimal algorithm.*

Next, we provide an upper bound on the competitive ratio of the Round-robin algorithm for the 1-median problem where objects have different maximum speeds. The proof is presented in the full version of the paper [16].

Proposition 5. *For the 1-median problem in $\mathbb{R}^1$ where objects have different maximum speeds, the Round-robin scheduling algorithm achieves a competitive ratio of $O(v_M n/v_m)$ against the optimal algorithm, where v_M is the highest maximum speed and v_m is the lowest maximum speed.*

Now, we consider the 1-median problem in $\mathbb{R}^d$ ($d \geq 2$). The proof of the following proposition is inspired by the proof of Theorem 5.1 of [13] and can be found in the full version of the paper [16].

Proposition 6. *For the 1-median problem in $\mathbb{R}^d$ ($d \geq 2$), the maximum size of the uncertainty region can be unbounded for the optimal algorithm.*

6   Conclusion

We conclude that in the worst case, adjusting query strategies based on the answers to the previous queries and the perceived locations of the objects does not help in achieving better competitive algorithms for any of the center problems discussed in this paper.

Here we list a handful of interesting problems that remain open:

- For the static case of the 1-center problem in $\mathbb{R}^1$, does there exist an algorithm that queries less than four objects per unit time and is 1-competitive against the optimal algorithm with one query per unit time, or an algorithm that queries one object per unit time and achieves a competitive ratio of $o(n)$?
- For the 1-center (1-median) problem in $\mathbb{R}^1$ where objects have different maximum speeds, can we find an algorithm with a competitive ratio of $o(v_M n/v_m)$, where v_M is the highest maximum speed and v_m is the lowest maximum speed?
- For the 1-median problem in $\mathbb{R}^d$ ($d \geq 2$), is there an algorithm that guarantees a bounded competitive ratio against the optimal algorithm?

References

1. Ahmadian, S., Norouzi-Fard, A., Svensson, O., Ward, J.: Better guarantees for k-Means and Euclidean k-Median by primal-dual algorithms. SIAM J. Comput. **49**(4), FOCS17-97 (2019)
2. Bādoiu, M., Har-Peled, S., Indyk, P.: Approximate clustering via core-sets. In: Proceedings of the Thiry-Fourth Annual ACM Symposium on Theory of Computing, pp. 250–257 (2002)
3. Bajaj, C.: The algebraic degree of geometric optimization problems. Discrete Comput. Geom. **3**(2), 177–191 (1988)
4. Bar-Noy, A., Ladner, R.E.: Windows scheduling problems for broadcast systems. SIAM J. Comput. **32**(4), 1091–1113 (2003)
5. Bar-Noy, A., Ladner, R.E., Tamir, T.: Windows scheduling as a restricted version of bin packing. ACM Trans. Algorithms (TALG) **3**(3), 28-es (2007)
6. Bereg, S., Bhattacharya, B., Kirkpatrick, D., Segal, M.: Competitive algorithms for maintaining a mobile center. Mobile Networks Appl. **11**(2), 177–186 (2006)
7. Bruce, R., Hoffmann, M., Krizanc, D., Raman, R.: Efficient update strategies for geometric computing with uncertainty. Theory Comput. Syst. **38**(4), 411–423 (2005)
8. Charikar, M., Guha, S., Tardos, É., Shmoys, D.B.: A constant-factor approximation algorithm for the k-Median problem. In: Proceedings of the Thirty-First Annual ACM Symposium on Theory of Computing, pp. 1–10 (1999)
9. Chrystal, G.: On the problem to construct the minimum circle enclosing n given points in the plane. Proc. Edinb. Math. Soc. **3**(1885), 30–33 (1885)
10. Cohen, M.B., Lee, Y.T., Miller, G., Pachocki, J., Sidford, A.: Geometric median in nearly linear time. In: Proceedings of the Forty-Eighth Annual ACM Symposium on Theory of Computing, pp. 9–21 (2016)
11. Drezner, Z., Hamacher, H.W.: Facility Location: Applications and Theory. Springer, Cham (2004)
12. Drineas, P., Frieze, A., Kannan, R., Vempala, S., Vinay, V.: Clustering large graphs via the singular value decomposition. Mach. Learn. **56**(1), 9–33 (2004)
13. Durocher, S.: Geometric facility location under continuous motion. Ph.D. thesis, University of British Columbia (2006)
14. Evans, W., Kirkpatrick, D., Löffler, M., Staals, F.: Query strategies for minimizing the ply of the potential locations of entities moving with different speeds. In: 30th European Workshop on Computational Geometry (2014)

15. Evans, W., Kirkpatrick, D., Löffler, M., Staals, F.: Competitive query strategies for minimising the ply of the potential locations of moving points. In: Proceedings of the Twenty-Ninth Annual Symposium on Computational Geometry, pp. 155–164 (2013)
16. Evans, W., Tabatabaee, S.A.: Minimizing the size of the uncertainty regions for centers of moving entities. arXiv preprint arXiv:2304.10028v2 (2024)
17. Feder, T., Greene, D.: Optimal algorithms for approximate clustering. In: Proceedings of the Twentieth Annual ACM Symposium on Theory of Computing, pp. 434–444 (1988)
18. Feder, T., Motwani, R., Panigrahy, R., Olston, C., Widom, J.: Computing the median with uncertainty. In: Proceedings of the Thirty-Second Annual ACM Symposium on Theory of Computing, pp. 602–607 (2000)
19. Fishburn, P.C., Lagarias, J.C.: Pinwheel scheduling: achievable densities. Algorithmica **34**(1), 14–38 (2002)
20. Hochbaum, D.S., Shmoys, D.B.: A best possible heuristic for the k-Center problem. Math. Oper. Res. **10**(2), 180–184 (1985)
21. Holte, R., Mok, A., Rosier, L., Tulchinsky, I., Varvel, D.: The pinwheel: a real-time scheduling problem. In: Proceedings of the 22nd Hawaii International Conference of System Science, pp. 693–702 (1989)
22. Kahan, S.: A model for data in motion. In: Proceedings of the Twenty-Third Annual ACM Symposium on Theory of Computing, pp. 265–277 (1991)
23. Kahan, S.H.: Real-time processing of moving data. Ph.D. thesis, University of Washington (1991)
24. Kanungo, T., Mount, D.M., Netanyahu, N.S., Piatko, C.D., Silverman, R., Wu, A.Y.: A local search approximation algorithm for k-Means clustering. In: Proceedings of the Eighteenth Annual Symposium on Computational Geometry, pp. 10–18 (2002)
25. Khanna, S., Tan, W.-C.: On computing functions with uncertainty. In: Proceedings of the Twentieth ACM SIGMOD-SIGACT-SIGART Symposium on Principles of Database Systems, pp. 171–182 (2001)
26. Kupitz, Y., Martini, H.: Geometric aspects of the generalized Fermat-Torricelli problem. Bolyai Soc. Math. Stud. **6**, 55–129 (1997)
27. Lloyd, S.: Least squares quantization in PCM. IEEE Trans. Inf. Theory **28**(2), 129–137 (1982)
28. Megiddo, N., Supowit, K.J.: On the complexity of some common geometric location problems. SIAM J. Comput. **13**(1), 182–196 (1984)
29. Suyadi, S.A.: Computing functions of imprecise inputs using query models. Master's thesis, University of British Columbia (2012)
30. Welzl, E.: Smallest enclosing disks (balls and ellipsoids). In: Maurer, H. (ed.) New Results and New Trends in Computer Science. LNCS, vol. 555, pp. 359–370. Springer, Heidelberg (1991). https://doi.org/10.1007/BFb0038202
31. Yildirim, E.A.: Two algorithms for the minimum enclosing ball problem. SIAM J. Optim. **19**(3), 1368–1391 (2008)
32. Zheng, D.W.: Scheduling queries to moving entities to certify many are distant from a region. Master's thesis, University of British Columbia (2020)

Guarding Polyominoes Under k-Hop Visibility

Omrit Filtser[1], Erik Krohn[2], Bengt J. Nilsson[3], Christian Rieck[4],
and Christiane Schmidt[5(✉)]

[1] Department of Mathematics and Computer Science, The Open University of Israel,
Tel Aviv, Israel
[2] Department of Computer Science, University of Wisconsin - Oshkosh, Oshkosh,
WI, USA
krohne@uwosh.edu
[3] Department of Computer Science and Media Technology, Malmö University,
Malmö, Sweden
bengt.nilsson.TS@mau.se
[4] Department of Computer Science, TU Braunschweig, Braunschweig, Germany
rieck@ibr.cs.tu-bs.de
[5] Department of Science and Technology, Linköping University, Norrköping, Sweden
christiane.schmidt@liu.se

Abstract. We study the ART GALLERY PROBLEM under k-hop visibility
in polyominoes. In this visibility model, two unit squares of a polyomino
can see each other if and only if the shortest path between the respective
vertices in the dual graph of the polyomino has length at most k.

In this paper, we show that the VC dimension of this problem is 3 in
simple polyominoes, and 4 in polyominoes with holes. Furthermore, we
provide a reduction from PLANAR MONOTONE 3SAT, thereby showing
that the problem is NP-complete even in thin polyominoes (i.e., polyominoes that do not a contain a 2×2 block of cells). Complementarily,
we present a linear-time 4-approximation algorithm for simple 2-thin
polyominoes (which do not contain a 3×3 block of cells) for all $k \in \mathbb{N}$.

Keywords: Art Gallery problem · k-hop visibility · polyominoes ·
VC dimension · approximation · k-hop dominating set

1 Introduction

"How many guards are necessary and sufficient to guard an art gallery?" This
question was posed by Victor Klee in 1973, and led to the classic ART GALLERY
PROBLEM: Given a polygon P and an integer ℓ, decide whether there is a guard
set of cardinality ℓ such that every point $p \in P$ is seen by at least one guard,
where a point is seen by a guard if and only if the connecting line segment is
inside the polygon.

Now picture the following situation: A station-based transportation service
(e.g., carsharing) wants to optimize the placement of their service stations.

Due to space constraints, all missing details can be found in the full version [15].

© The Author(s), under exclusive license to Springer Nature Switzerland AG 2024
J. A. Soto and A. Wiese (Eds.): LATIN 2024, LNCS 14578, pp. 288–302, 2024.
https://doi.org/10.1007/978-3-031-55598-5_19

Assume that the demand is given in a granularity of (square) cells, and that customers are willing to walk a certain distance (independent of where they are in the city) to a station. Then, we aim to place as few stations as possible to serve an entire city for a given maximum walking range of k cells. We thus represent the city as a polyomino, potentially with holes, and only walking within the boundary is possible (e.g., holes would represent water bodies or houses, which pedestrians cannot cross).

A *polyomino* P is a connected polygon in the plane formed by joining together $|P| = n$ unit squares (also called *cells*) on the square lattice. The dual graph G_P of a polyomino has a vertex at the center point of each cell of P, and there is an edge between two center points if their respective cells share an edge. Note that G_P is a *grid graph*. A polyomino is *simple* if it has no holes, that is, every inner face of its dual graph has unit area. A polyomino is t-*thin* if it does not contain a block of squares of size $(t + 1) \times (t + 1)$. In particular, a simple polyomino is 1-thin if its dual graph is a tree.

Our real-world example motivates the following type of visibility: a unit square u of a polyomino P is k-*hop visible* to a unit-square guard $v \in P$ if the shortest path from u to v in G_P has length at most k. For $k \geq 2$, this allows a guard to look around corners of the polyomino, as visualized in Fig. 1. The k-*hop-visibility region* of a unit square $u \in P$, is the set of all unit squares that are k-hop-visible from u. Note that this is a subset of the L_1-diamond with diameter $2k$—the maximal k-hop-visibility region.

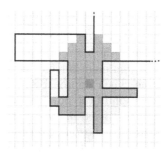

Fig. 1. A unit square in green with its k-hop-visibility region for $k = 6$ (shaded in dark green) within a polyomino—a subset of the diamond shown in light green. (Color figure online)

In this paper, we investigate the MINIMUM k-HOP GUARDING PROBLEM IN POLYOMINOES (MkGP): Given a polyomino P and an integer k, find a minimum-cardinality unit-square guard cover in P under k-hop visibility.

As the dual graph of a polyomino is a grid graph, we analogously formulate the problem as the MINIMUM k-HOP DOMINATING SET PROBLEM IN GRID GRAPHS (MkDSP): Given a grid graph G and an integer k, find a minimum-cardinality subset $D_k \subseteq V(G)$, such that for any vertex $v \in V(G)$ there exists a vertex $u \in D_k$ within hop distance of at most k.

While we formulated the optimization problems, the associated decision problems are defined as expected with an upper bound on the number of guards or dominating vertices.

Our Contributions. For the MkGP in polyominoes, we give the following results.

(1) The VC dimension of the problem is exactly 3 in simple polyominoes, and 4 in polyominoes with holes; see Theorems 3 and 4, respectively.
(2) The decision version of the problem is NP-complete for $k \geq 2$, even in 1-thin polyominoes with holes, see Theorem 5.
(3) A linear-time 4-approximation for simple 2-thin polyominoes, see Theorem 6.

We state our results in terms of the guarding problem. However, they also hold true for the equivalent k-hop dominating set problem in grid graphs.

Related Work. The classic ART GALLERY PROBLEM is NP-hard [24,26], even in the most basic problem variant. Abrahamsen, Adamaszek, and Miltzow [1] recently showed that the AGP is ∃ℝ-complete.

Guarding polyominoes and thin (orthogonal) polygons has been considered for different definitions of visibility. Tomás [28] showed that computing a minimum guard set under the original definition of visibility is NP-hard for point guards and APX-hard for vertex or boundary guards in thin orthogonal polygons; an orthogonal polygon is defined as *thin* if the dual graph of the partition obtained by extending all edges of P through incident reflex vertices is a tree. Biedl and Mehrabi [8] considered guarding thin orthogonal polygons under rectilinear visibility (two points can see each other if the axis-parallel rectangle spanned by these two points is fully contained in the polygon). They showed that the problem is NP-hard in orthogonal polygons with holes, and provided an algorithm that computes a minimum set of guards under rectilinear vision for tree polygons in linear time. Their approach generalizes to polygons with h holes or thickness t (the dual graph of the polygon does not contain an induced $(t+1) \times (t+1)$ grid)—the problem is fixed-parameter tractable in $t+h$. Biedl and Mehrabi [9] extended this study to orthogonal polygons with bounded treewidth under different visibility definitions usually used in orthogonal polygons: rectilinear visibility, staircase visibility (guards can see along an axis-parallel staircase), and limited-turn path visibility (guards can see along axis-parallel paths with at most b bends). Under all these visibility definitions, they showed the guarding problem to be linear-time solvable. For orthogonal polygons, Worman and Keil [30] gave a polynomial time algorithm to compute a minimum guard cover under rectilinear visibility by showing that an underlying graph is perfect.

Biedl et al. [7] proved that determining the guard number of a given simple polyomino with n unit squares is NP-hard even in the all-or-nothing visibility model (a unit square s of the polyomino is visible from a guard g if and only if g sees all points of s under ordinary visibility), and under ordinary visibility. They presented polynomial time algorithms for thin polyominoes, for which the dual is a tree, and for the all-or-nothing model with limited range of visibility. Iwamoto and Kume [20] complemented the NP-hardness results by showing NP-hardness for polyominoes with holes also for rectilinear visibility. Pinciu [27] generalized this to polycubes, and gave simpler proofs for known results and new results for guarding polyhypercubes.

The MINIMUM k-HOP DOMINATING SET PROBLEM is NP-complete in general graphs [3,5]. For trees, Kundu and Majunder [22] showed that the problem can be solved in linear time. Recently, Abu-Affash et al. [2] simplified that algorithm, and provided a linear-time algorithm for cactus graphs. Borradaile and Le [10] presented an exact dynamic programming algorithm that runs in $O((2k+1)^{tw} \cdot n)$ time on graphs with treewidth tw. Demaine et al. [12] considered the (ℓ, k)-center problem on planar and map graphs, i.e., the question whether a graph has at most ℓ many *center* vertices such that every vertex of the graph is within hop distance at most k from some center. They showed that for these graph families, the problem is fixed-parameter tractable by providing an exact $2^{O(k \log k)\sqrt{\text{OPT}}} \cdot \text{poly}(n)$ time algorithm, where OPT is the size of an optimal solution. They also obtained a $(1 + \varepsilon)$-approximation for these families that runs in $k^{O(k/\varepsilon)} \cdot m$ time, where m is the number of edges in the graph.

In the general case where the edges of the graph are weighted, the problem is typically called the ρ-DOMINATING SET PROBLEM. Katsikarelis et al. [21] provided an FPT approximation scheme parameterized by the graphs treewidth tw, or its clique-width cw: In particular, if there exists a ρ-dominating set of size s in a given graph, the approximation scheme computes a $(1+\varepsilon)\rho$-dominating set of size at most s in time $(tw/\varepsilon)^{O(tw)} \cdot \text{poly}(n)$, or $(cw/\varepsilon)^{O(cw)} \cdot \text{poly}(n)$, respectively. Fox-Epstein et al. [16] provided a bicriteria EPTAS for ρ-domination in planar graphs (later improved and generalized by Filtser and Le [14]). Their algorithm runs in $O(n^c)$ time (for some constant c), and returns a $(1+\varepsilon)\rho$-dominating set of size $s \leq (1+\varepsilon)\text{OPT}_\rho$, where OPT_ρ is the size of a minimum ρ-dominating set. Filtser and Le [13] provided a PTAS for ρ-dominating set in H-minor-free graphs, based on local search. Their algorithm runs in $n^{O_H(\varepsilon^{-2})}$ time, and returns a ρ-dominating set of size at most $(1+\varepsilon)\text{OPT}_\rho$.

Meir and Moon [25] showed an upper bound of $\lfloor \frac{n}{k+1} \rfloor$ on the number of vertices in a k-hop dominating set of any tree. This bound holds for general graphs by using any spanning tree. We provide a simple matching lower bound that can be found in the full version.

2 VC Dimension

The *VC dimension* is a measure of complexity of a set system. In our setting, we say that a finite set of (guard) unit squares D in a polyomino P is *shattered* if for any of the $2^{|D|}$ subsets $D_s \subseteq D$ there exists a unit square $u \in P$, such that from u every unit square in D_s but no unit square in $D \setminus D_s$ is k-hop visible (or, by symmetry: from every unit square in D_s the unit square u is k-hop visible, but from no unit square in $D \setminus D_s$ u is k-hop visible). We then say that the unit square u is a *viewpoint*. The VC dimension is the largest d, such that there exists a polyomino P and a set of d unit-square guards D that can be shattered. For detailed definitions, we refer to Haussler and Welzl [19].

In this section, we study the VC dimension of the MkGP in both simple polyominoes and polyominoes with holes, and provide exact values for both cases. The VC dimension has been studied for different guarding problems, e.g., Langetepe and Lehmann [23] showed that the VC dimension of L_1-visibility in a simple polygon is exactly 5, Gibson et al. [17] proved that the VC dimension of visibility on the boundary of a simple polygon is exactly 6. For line visibility in a simple polygon, the best lower bound of 6 is due to Valtr [29], the best upper bound of 14 stems from Gilbers and Klein [18]. Furthermore, given any set system with constant VC dimension, Brönnimann and Goodrich [11] presented a polynomial time $O(\log \text{OPT})$-approximation for SET COVER.

For analyzing the VC dimension, we define the *rest budget* of a unit square $g \in P$ at a unit square $c \in P$ to be $\text{rb}(g, c) = \max\{k - d(g, c), 0\}$, where $d(g, c)$ is the minimum distance between g and c in G_P, and k the respective hop distance. We first state two structural properties which are helpful in several arguments.

Observation 1 (Rest-Budget Observation). *Let P be a polyomino, and let g and u be two unit squares in P such that a shortest path between them contains a unit square c. Then the following holds:*

(1) The unit square g covers u, if and only if u is within distance $\text{rb}(g, c)$ from c.
(2) For any unit square q' with $\text{rb}(g', c) > \text{rb}(g, c)$, if q covers u, then so does g'.

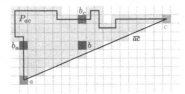

Fig. 2. Location of unit squares used in the proof of Lemma 2.

Lemma 2 (Rest-Budget Lemma). *Let a, c be two unit squares in a simple polyomino P, such that the boundary of P does not cross the line segment $\overline{ac}$ that connects their center points. Let P_{ac} be some path in the dual graph G_P between the center points of a and c, and let b be a unit square whose center point belongs to the area enclosed within $P_{ac} \circ \overline{ac}$. Then, there exists a unit square x on P_{ac} such that $\text{rb}(b, x) \geq \text{rb}(a, x)$ and $\text{rb}(b, x) \geq \text{rb}(c, x)$.*

Proof. Without loss of generality, assume that the center of a is placed on the origin, c lies in the first quadrant, and b is above the line through the centers of a, c; see Fig. 2. If b is above c, then let x be the unit square on P_{ac} directly above b. As P is simple, and the boundary of P does not cross $\overline{ac}$, the area enclosed within $P_{ac} \circ \overline{ac}$ does not contain any boundary piece of P. Thus, the path in G_P from b to x is a straight line segment, and we have $\text{rb}(b, x) \geq \text{rb}(a, x)$

and $\mathrm{rb}(b,x) \geq \mathrm{rb}(c,x)$, as required. Symmetrically, if b is to the left of a, then let x be the unit square on P_{ac} directly to the left of b, and again we have $\mathrm{rb}(b,x) \geq \mathrm{rb}(a,x)$ and $\mathrm{rb}(b,x) \geq \mathrm{rb}(c,x)$, as required.

The only case left is when b lies in the axis-aligned bounding box of a,c. In this case, let b_c (resp. b_a) be the unit square on P_{ac} directly above (resp. to the left of) b. Denote the center point of b by (b_x, b_y), and the center point of c by (c_x, c_y). As b is above the line through a,c, we get that (i) $b_y \geq \frac{c_y}{c_x} b_x$. If both $d(b, b_a) > d(a, b_a)$ and $d(b, b_c) > d(c, b_c)$ hold, then (ii) $b_x > b_y$ and (iii) $c_y - b_y > c_x - b_x$. By (ii) and (iii) we get $c_y > c_x$. On the other hand, by (i) and (ii) we get $b_y > \frac{c_y}{c_x} b_y$, and thus $c_y < c_x$, a contradiction.

We conclude that either $d(a, b_a) \geq (b, b_a)$ or $d(c, b_c) \geq d(b, b_c)$, which means that either $\mathrm{rb}(b, b_a) \geq \mathrm{rb}(a, b_a)$ or $\mathrm{rb}(b, b_c) \geq \mathrm{rb}(c, b_c)$. Furthermore, as b lies above a, we have $d(a, b_c) > d(b, b_c)$, and as b lies also to the left of c, we have $d(c, b_a) > d(b, b_a)$. Therefore, the claim holds for one of b_a or b_c. □

2.1 Simple Polyominoes

In this section, we investigate the VC dimension of k-hop visibility of simple polyominoes. In particular, we show the following.

Theorem 3. *For any $k \in \mathbb{N}$, the VC dimension of k-hop visibility of a simple polyomino is 3.*

Proof. A lower bound construction with $k = 1$ and three guards, indicated by the green 1, the blue 2, and the red 3, is visualized in Fig. 3. All $2^3 = 8$ viewpoints are highlighted, and we denote the guards that see each viewpoint. For larger k, we keep the placement of guards, but the polyomino will be a large rectangle that contains all k-hop-visibility regions. Because of the relative position of the guards they are shattered as before.

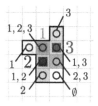

Fig. 3. A lower bound construction for the VC dimension of k-hop visibility of simple polyominoes with $k = 1$. The positions of the guards $1, 2, 3$ are shown as squares in green, blue, and red, respectively. The k-hop-visibility regions are shown in a light shade of those colors. The 8 viewpoints are indicated by circles, and labeled accordingly. (Color figure online)

We now show that four guards cannot be shattered in simple polyominoes. To this end, consider four guards g_1, g_2, g_3, g_4 to be placed in the simple polyomino, and denote the potential viewpoints as v_S with $S \subseteq \{1, 2, 3, 4\}$. For two unit squares $x, y \in P$, we denote by $sp(x, y)$ a shortest path between x and y in P. For $i, j \in \{1, 2, 3, 4\}$, let $P_{i,j} = sp(g_i, v_{i,j}) \circ sp(v_{i,j}, g_j)$. We now distinguish two cases depending on how many of the four guards lie on their convex hull.

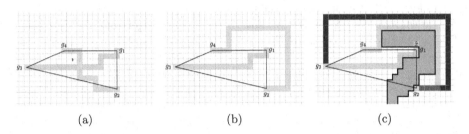

(a)   (b)   (c)

Fig. 4. Case 1 in the proof of Theorem 3, with $P_{1,3}$, $P_{2,4}$ and $P_{2,3}$ shown in orange, pink and turquoise, respectively. (a) $P_{1,3}$ and $P_{2,4}$ cannot cross. (b) $P_{2,4}$ goes around g_1. (c) The boundary of P must pierce $\overline{g_2 g_4}$ and both $P_{2,4}$ and $P_{2,3}$ must go around a square $\hat{x}$ in the exterior of P.

Case 1: The guards lie in convex position. That is, the four center points of their corresponding grid squares are in convex position. Pick any guard and label it g_1 and then label them in clockwise order around the convex hull g_2, g_3 and g_4, see Fig. 4a. Assume, without loss of generality, that g_4 is to the left of g_1 and that g_1 is above the line through g_2 and g_4.

First, we claim that the paths $sp(g_i, v_{1,3})$ and $sp(g_j, v_{2,4})$ cannot cross for $i \in \{1, 3\}, j \in \{2, 4\}$. Indeed, if the paths have unit square x in common (see Fig. 4a), then one of g_i, g_j has a larger rest budget at x (or an equal rest budget). Assume, without loss of generality, that $\mathrm{rb}(g_i, x) \geq \mathrm{rb}(g_j, x)$, then g_i would also cover $v_{2,4}$, which is a contradiction as $i \notin \{2, 4\}$. Therefore, the paths $P_{1,3} = sp(g_1, v_{1,3}) \circ sp(v_{1,3}, g_3)$ and $P_{2,4} = sp(g_2, v_{2,4}) \circ sp(v_{2,4}, g_4)$ cannot cross, and one of them must "go around" a guard in order to avoid a crossing. Without loss of generality, assume that $P_{2,4}$ goes around g_1, that is, g_1 belongs to the area enclosed within $P_{2,4} \circ \overline{g_2 g_4}$; see Fig. 4b. Assume that the boundary of P does not pierce $\overline{g_2 g_4}$. In this case, as P is simple, we get by Lemma 2 that there exists a square x on $P_{2,4}$ such that $\mathrm{rb}(g_1, x) \geq \mathrm{rb}(g_2, x)$ and $\mathrm{rb}(g_1, x) \geq \mathrm{rb}(g_4, x)$. Hence, g_1 covers $v_{2,4}$, a contradiction.

We therefore assume that the boundary of P does pierce $\overline{g_2 g_4}$ (see Fig. 4c), and, hence, there exists a square $\hat{x} \notin P$, which blocks g_1 from reaching the square x on $P_{2,4}$ from Lemma 2. As P is simple, the boundary of P must also cross either $\overline{g_2 g_3}$ or $\overline{g_3 g_4}$ in a way that any path in P between the endpoints of this segment must go around $\hat{x}$. In other words, assume, without loss of generality, that the boundary of P crosses $\overline{g_2 g_3}$. Then there exists a path in the exterior of P connecting $\overline{g_2 g_3}$ and $\hat{x}$, and because P is simple, any path in P from g_2 to g_3 must go around $\hat{x}$ (see Fig. 4c). In particular, the path $P_{2,3} = sp(g_2, v_{2,3}) \circ sp(v_{2,3}, g_3)$ also goes around $\hat{x}$. We get that both $P_{2,3}$

and $P_{2,4}$ go around $\hat{x}$; however, $sp(g_3, v_{2,3})$ and $sp(g_4, v_{2,4})$ cannot intersect. Moreover, consider the region $A_{3,4}^2$ enclosed by $\overline{g_3 g_4} \circ sp(g_3, v_{2,3}) \circ sp(v_{2,3}, v_{2,4}) \circ sp(g_4, v_{2,4})$, and assume that $sp(g_3, v_{2,3})$ is above $sp(g_4, v_{2,4})$ (the other case is argued analogously). As P is simple, the region $A_{3,4}^2$ does not contain any polyomino boundary. Consider the line ℓ through g_4 of slope -1. If g_3 is below ℓ, then for any unit square s to the right of g_4 inside $A_{3,4}^2$, we have $\mathrm{rb}(g_4, s) \geq \mathrm{rb}(g_3, s)$. As g_3 also lies below $\overline{g_2 g_4}$ and to the left of g_4 (and because k-hop-visibility regions are diamond-shaped without boundary), we get that g_4 reaches $v_{2,3}$, a contradiction. On the other hand, if g_3 is above ℓ, consider the region $A_{3,4}^1$ enclosed by $\overline{g_3 g_4} \circ sp(g_3, v_{1,3}) \circ sp(v_{1,3}, v_{1,4}) \circ sp(g_4, v_{1,4})$ and assume that $sp(g_3, v_{1,3})$ is below $sp(g_4, v_{1,4})$. By the same arguments, the region $A_{3,4}^1$ does not contain any polyomino boundary, and for any square s below g_4 inside $A_{3,4}^1$, we have $\mathrm{rb}(g_4, s) \geq \mathrm{rb}(g_3, s)$. In this case, g_4 reaches $v_{1,3}$, a contradiction.

Case 2: Exactly three guards lie on their convex hull. That is, the three center points of their corresponding grid squares are in convex position, and the center point of the fourth guard lie in the convex hull. We label the three guards on the convex hull g_1, g_2, g_3, and g_4 is the guard placed inside the convex hull. We show that the viewpoint $v_{1,2,3}$ is not realizable. Let T be the triangle of grid points that connects the centers of g_1, g_2, and g_3. Consider the three shortest paths connecting g_1, g_2, g_3 to $v_{1,2,3}$. As g_4 lies in T, for any placement of $v_{1,2,3}$, we would get that for some $i, j \in \{1, 2, 3\}$, the area enclosed within $sp(g_i, v_{1,2,3}) \circ sp(v_{1,2,3}, g_j) \circ \overline{g_i g_j}$ contains the center point of g_4. If the boundary of P does not pierce T, then, similar to Case 1, we get by Lemma 2 that g_4 reaches $v_{1,2,3}$, a contradiction. Otherwise, assume that the convex hull of the three guards is pierced by the boundary. Then it is possible to realize the $v_{1,2,3}$ viewpoint. However, similar to the argument in Case 1, the boundary will prevent the realization of a viewpoint of g_4 and one of the other guards (g_4 taking the role of g_1 from Case 1 here).   □

2.2   Polyominoes with Holes

Aronov et al. [4] showed an upper bound of 4 for the VC dimension of hypergraphs of pseudo disks. And while, intuitively, one might suspect that k-hop-visibility regions of unit squares in polyominoes with holes are pseudo disks; that is not the case, as illustrated in Fig. 5a.

Hence, we need to show an upper bound for the VC dimension in this case in another way. In fact, even here, we provide matching upper and lower bounds. These are valid for large enough values of k (e.g., for $k = 1$ we do not gain anything from the holes). In particular, we show the following.

Theorem 4. *For large enough $k \in \mathbb{N}$, the VC dimension of k-hop visibility of a polyomino with holes is 4.*

Proof. A lower bound with $k = 18$ is visualized in Fig. 5c: the four guards are indicated by the green 1, the blue 2, the red 3, and the yellow 4. We highlighted

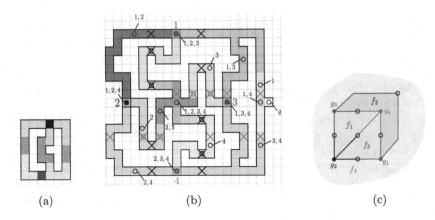

Fig. 5. (a) The k-hop-visibility regions (for $k = 6$) of the two guards intersect more than twice. (b) The lower bound construction for the VC dimension of non-simple polyominoes for $k = 18$. The positions of the four guards $1, 2, 3, 4$ are shown as squares in green, blue, red, and yellow, respectively. Visibility regions are shown in a light shade of those colors. The 16 viewpoints are indicated by circles, and labeled accordingly. The gray and black $\times$'s and boxed $\times$'s indicates where we insert 2 and 1 unit squares, respectively, to increase the value of k by 2. We alternate between using the gray and black markings. (c) The graph G_{4p} with vertices $g_1, \ldots, g_4$ shown in green, blue, red, and yellow, respectively. Gray circles indicate the "pair"-viewpoints, and are not vertices of the graph. (Color figure online)

the $2^4 = 16$ viewpoints, and denoted the guards that see each viewpoint. For larger (even) values of k, we extend the corridors in Fig. 5b at the location marked by "$\times$": We alternate between using the gray and black unit squares. At locations with a simple "$\times$", we insert two unit squares, at locations with a boxed "$\times$", we insert a single unit square. One can verify that by alternating between the gray and black insertions for $k = 18 + 2i$, all viewpoints are realized.

For the upper bound, assume that we can place a set with five unit-square guards g_1, g_2, g_3, g_4, g_5 that can be shattered. We denote viewpoints as v_S with $S \subseteq \{1, 2, 3, 4, 5\}$. Let $P_{i,j}, i \neq j \in \{1, \ldots, 5\}$ denote the shortest path from guard g_i to g_j along which the viewpoint $v_{\{i,j\}}$ is located. In particular, $P_{i,j}$ includes the shortest paths from g_i to $v_{\{i,j\}}$ and from g_j to $v_{\{i,j\}}$ (as this determines the rest budget for both guards at $v_{\{i,j\}}$).

We start with four guards $g_1, \ldots, g_4$. To generate all the "pair" viewpoints, $v_{\{i,j\}}, \{i,j\} \subseteq \{1, 2, 3, 4\}$, we need to embed the graph G_{4p} shown in Fig. 5c where each edge represents a path $P_{i,j}$ (the color of each guard reaches equally far into each edge, e.g., some of the paths reflected in these edges include wiggles).

Of course, in the resulting polyomino, the edges could be embedded in larger blocks of unit squares. However, given the upper bound of 3 on the VC dimension for simple polyominoes, we know that at least one of the four faces $f_1, \ldots, f_4$ (and in fact one of f_1, f_2, f_3) of G_{4p} must contain a hole. A fifth guard g_5 must be located in one of the four faces. Let this be face f_i. As g_i is not incident to f_i, the path from g_5 to g_i, $P_{i,5}$, must intersect one of the other paths represented by the edges in G_{4p}, let this be the path $P_{j,\ell}$. By Observation 1, one of the viewpoints $v_{\{i,5\}}$ and $v_{\{j,\ell\}}$ cannot be realized, as a guard from the other pair will always see such a viewpoint too; a contradiction to our assumption. □

3 NP-Completeness for 1-Thin Polyominoes with Holes

In this section, we note that the decision version of the MkGP is NP-complete, even in 1-thin polyominoes with holes. However, as the dual graph of a 1-thin polyomino without holes is a tree, an optimal solution can be obtained in linear time [2,22].

Theorem 5. *The decision version of the Mk GP is* NP-*complete for $k \geq 2$, even in 1-thin polyominoes with holes.*

Our reduction is from PLANAR MONOTONE 3SAT, which de Berg and Khosravi [6] proved to be NP-complete. Due to space constraints, the proof can be found in the full version [15]. Figure 6 depicts the variable and clause gadgets; the high-level idea is that five guards can be placed in a variable gadget in such a way that one of its exits is already guarded, and a satisfied clause only requires two additional guards, whereas three are needed for an unsatisfied clause. Note that both gadgets scale with k.

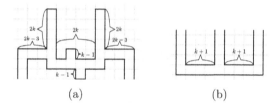

(a)   (b)

Fig. 6. Let $k = 2$: (a) depicts the variable gadget, and (b) the clause gadget.

4 A 4-Approximation for Simple 2-Thin Polyominoes

As already mentioned, there exists a PTAS for k-hop domination in H-minor free graphs [13]. However, the exponent of n in the running time may be infeasible for realistic applications, where n is extremely large. On the other hand, the exact algorithm for graphs with treewidth tw has running time $O((2k+1)^{tw} \cdot n)$ [10], which may be too large if $k = \Omega(n)$, even for small tw (in fact, it is not hard to show that 2-thin polyominoes have constant tw: K_4 is not a minor, hence, we have $tw = 2$ for 2-thin polyominoes).

Therefore, we present a linear-time 4-approximation algorithm for the MkGP in simple 2-thin polyominoes, for any value of $k \in \mathbb{N}$. The running time of our algorithm does not depend on k. The overall idea is to construct a tree T on P, and let T lead us in placing guards in P (inspired by the linear-time algorithm for trees by Abu-Affash et al. [2] for the equivalent problem of k-hop dominating set). In each iteration step, we place 1, 2, or 4 guards and 1 witness. A *witness set* is a set of unit squares W, such that the k-hop-visibility regions of the elements in W are pairwise disjoint. Because the cardinality of a witness set is a lower bound on the cardinality of any guard set, this yields a 4-approximation.

Skeleton Graph Construction. Let P be a simple 2-thin polyomino. A vertex v of a unit square $s \in P$ is called *internal* if it does not lie on the boundary of P. Because P is 2-thin, any square $s \in P$ can have at most 3 internal vertices. Let I be the set of internal vertices of unit squares in P. For any $u, v \in I$, we add the edge $\{u, v\}$ to E_I if one of the following holds:

1. u, v belong to the same unit square and $\|u - v\| = 1$.
2. u, v belong to two different unit squares that share an edge and both vertices of this edge are not internal.
3. u, v belong to the same unit square s and both other vertices belonging to s are not internal.

Because P is a simple 2-thin polyomino, the edges of E_I form a forest T_I on I.

For each unit square $s \in P$ that does not have any internal vertex, place a point b_s in the center of s. We call s a *boundary square*, b_s a *boundary node*, and denote by B the set of all boundary nodes. For any $b_s, b_{s'} \in B$ such that $s, s' \in P$ share an edge, add the edge $\{b_s, b_{s'}\}$ to E_B. Notice that the edges of E_B form a forest T_B on B.

We now connect T_B and T_I. Let s be a boundary square that shares an edge with a non-boundary square s'. Then, b_s must be a leaf in T_B, and s' has at most two internal vertices. If s' has a single internal vertex v, we simply add $\{b_s, v\}$ to E_{con}. Else, s' has two internal vertices v, u, and we add an artificial node $x_{v,u}$ to the set X, and the edges $\{b_s, x_{v,u}\}$, $\{u, x_{v,u}\}$, and $\{v, x_{v,u}\}$ to E_{con}.

Let T be the graph on the vertex set $V = I \cup B \cup X$ and the edge set $E = E_I \cup E_B \cup E_{con}$. Note that, as P is simple, we did not create any cycles when connecting T_I and T_B, thus, T is a tree. Moreover, the maximum degree of a node in T is 4 (for some nodes in X and I).

Associated Squares. Associate with each node $v \in T$ a block $S(v)$ of unit squares from P as follows:

1. For $v = x_{u,v} \in X$, $S(v)$ consists of a two unit squares with the edge $\{u, v\}$.
2. For $v \in I$, $S(v)$ consists of a 2×2 block of unit squares with internal vertex v.
3. For $v = b_s \in B$, $S(v) = \{s\}$.

The Algorithm. As already mentioned, we basically follow the lines of the algorithm of [2] for k-hop dominating sets in trees, with several important changes.

We start by picking an arbitrary node r from T as a root. For a node $u \in T$, denote by T_u the subtree of T rooted at u. Notice that any path between a unit square associated with a node in T_u, and a unit square associated with a node in $T \setminus T_u$, includes a unit square from $S(u)$. For every node $u \in T$, let $h(T_u) = \max_{v \in T_u, s \in S(v)} \min_{s' \in S(u)} d_P(s, s')$, where $d_P(s, s')$ denotes the hop distance between the cells s and s' in P. In other words, $h(T_u)$ is the largest hop distance from a unit square in $\bigcup_{v \in T_u} S(v)$ to its closest unit square from $S(u)$.

For each cell $s' \in \bigcup_{v \in T_u} S(v)$, the minimum distance $\min_{s \in S(u)} d_P(s, s')$ is assumed at a particular unit square s, we denote by $M(s)$ the set of all these cells for which that distance is assumed for s, and set $h_s(T_u) = \max_{s' \in M(s)} d_P(s, s')$. Note that if $h(T_u) = k$, and we pick $S(u)$ for our guard set, then every unit square associated with a node in T_u is guarded.

Initialize an empty set D (for the k-hop-visibility guard set), and compute $h(T_u)$ for every $u \in T$ and $h_s(T_u)$ for every $s \in S(u)$. In addition, for every unit square $s \in P$ set $\mathrm{rb}_D(s) = -1$ (up to a rest budget of 0, s is k-hop visible to the nodes in D). This parameter marks the maximum rest budget of the unit square s over all squares in the guard set D. We run a DFS algorithm starting from r, as follows; let u be the current node in the DFS call.

1. If $h(T_u) = k$, we add $S(u)$ to D, remove T_u from T, and set $\mathrm{rb}_D(s) = k$ for every $s \in S(u)$.
2. Else if, $h(T_u) \geq k - 2$ and $\min_{s \in S(p(u))} d(s, s') > k$ for the parent $p(u)$ of u and $s' \in T_u$ being the unit square that realizes $h(T_u)$, we add $S(u)$ to D, remove T_u from T, and set $\mathrm{rb}_D(s) = k$ for every $s \in S(u)$.
3. Else, for each child v of u with $h(T_v) \geq k - 2$, we run the DFS algorithm on v. Then we update $h(T_u)$ and $h_s(T_u)$, $\mathrm{rb}_D(s)$ for every $s \in S(u)$, according to the values calculated for all children of u.
4. We check if the remaining T_u is already guarded by D, by considering $h_s(T_u)$ and $\mathrm{rb}_D(s)$ for every $s \in S(u)$, where we only consider associated unit squares with negative rest budget.
5. Else, if the new $h(T_u)$ is now exactly k or if the condition from point 2 holds, then again we add $S(u)$ to D, remove T_u from T_r, and set $\mathrm{rb}_D(s) = k$ for every $s \in S(u)$.

If, at the end of the DFS run for r, we have $\mathrm{rb}_D(s) = -1$ for some $s \in S(r)$, then we add $S(r)$ to D. We give an example of our algorithm in Fig. 7.

We show that, after termination of the algorithm, D is a k-hop-visibility guard set for the given polyomino P of size at most $4 \cdot \mathrm{OPT}$ for all $k \in \mathbb{N}$, where OPT is the size of an optimal solution.

Theorem 6. *There is a linear-time 4-approximation for the $Mk\,GP$ in simple 2-thin polyominoes.*

Proof. During the algorithm, we remove a node v from T only if $S(v)$ is covered by cells in D. Since $\bigcup_{v \in T} S(v) = P$, D is a k-hop-visibility guard set for P.

Next, let $u_1, \ldots u_\ell$ be the sequence of nodes of T such $S(u_i)$ was added to the set D during the algorithm. We show that in each T_{u_i} we can find a witness unit square s_i, such that no two witness unit squares $s_i \neq s_j$ have a single unit square in P within hop distance k from both s_i and s_j. This means that any optimal solution has size at least ℓ ($W = \{s_1, \dots, s_\ell\}$ is a witness set with $|W| = \ell$). Since in each step of the algorithm we add at most 4 squares to D, we get a solution of size at most 4ℓ, as required.

We choose s_i to be the unit square from T_{u_i} with maximum distance to its closest unit square from $S(u_i)$—the unit square that realizes $h(T_{u_i})$. We claim that there is no cell in P within hop distance k from both s_i, s_j for any $j < i$.

If $S(u_i)$ was added to D because $T_{u_i} = k$, we had s_i being the node realizing T_{u_i}. Hence, we have $\mathrm{rb}_{D\setminus\{u_i\}}(s_i) = -1$, and thus, the distance from s_i to any $s_j, j < i$ is at least $2k + 1$.

If $S(u_i)$ was added to D because $h(T_{u_i}) \geq k - 2$ and $\min_{s \in S(p(u))} d(s, s_i) > k$ for the parent $p(u)$ of u and $s_i \in T_{u_i}$ being the unit square that realizes $h(T_{u_i})$, we know (because each unit square of the polyomino is an associated unit square of at least one node) that there is a unit square $s'' \in S(u_i)$ with $d(s'', s_i) = k$. Thus, any witness placed after s_i has distance to it of at least $2k + 1$. Moreover, $\mathrm{rb}_{D\setminus\{u_i\}}(s_i) = -1$ and, thus, s_i's distance to any $s_j, j < i$ is at least $2k + 1$.

We initialize $h(T_u)$ for every $u \in T$ and $h_s(T_u)$ for every $s \in S(u)$ with BFS, and we update the values at most once for each square in linear time. $\square$

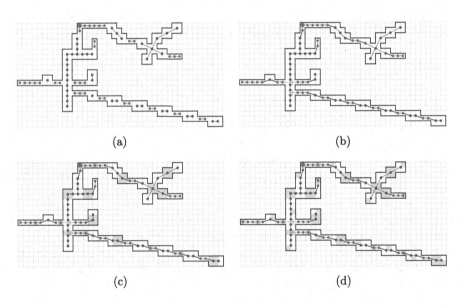

(a)

(b)

(c)

(d)

Fig. 7. Example for our algorithm for $k = 3$: (a) Polyomino P, in black, with the associated vertices: vertices in I in blue, vertices in B in red, the root r is indicated in magenta, the trees T_I and T_B with edges in E_I are shown in blue, edges in E_B are shown in red; (b) connecting T_I and T_B, vertices $x_{v,u}$ and their incident edges are shown in green, all other connecting edges are shown in black; (c) unit squares added to D in light pink; (d) placement of witnesses from the proof of Theorem 6 in turquoise. (Color figure online)

Funding. B. J. N. and C. S. are supported by grants 2021-03810 and 2018-04001 from the Swedish Research Council (Vetenskapsrådet). C. S. was supported by grant 2018-04101 from Sweden's innovation agency VINNOVA.

References

1. Abrahamsen, M., Adamaszek, A., Miltzow, T.: The art gallery problem is ∃ℝ-complete. J. ACM **69**(1), 4:1–4:70 (2022). https://doi.org/10.1145/3486220
2. Abu-Affash, A.K., Carmi, P., Krasin, A.: A linear-time algorithm for minimum k-hop dominating set of a cactus graph. Discret. Appl. Math. **320**, 488–499 (2022). https://doi.org/10.1016/j.dam.2022.06.006
3. Amis, A.D., Prakash, R., Huynh, D.T., Vuong, T.H.: Max-Min d-cluster formation in wireless ad hoc networks. In: Conference on Computer Communications, pp. 32–41 (2000). https://doi.org/10.1109/INFCOM.2000.832171
4. Aronov, B., Donakonda, A., Ezra, E., Pinchasi, R.: On pseudo-disk hypergraphs. Comput. Geom. **92**, 101687 (2021). https://doi.org/10.1016/j.comgeo.2020.101687
5. Basuchowdhuri, P., Majumder, S.: Finding influential nodes in social networks using minimum k-hop dominating set. In: International Conference on Applied Algorithms (ICAA), pp. 137–151 (2014). https://doi.org/10.1007/978-3-319-04126-1_12
6. de Berg, M., Khosravi, A.: Optimal binary space partitions for segments in the plane. Int. J. Comput. Geom. Appl. **22**(03), 187–205 (2012). https://doi.org/10.1142/S0218195912500045
7. Biedl, T.C., Irfan, M.T., Iwerks, J., Kim, J., Mitchell, J.S.B.: Guarding polyominoes. In: Symposium on Computational Geometry (SoCG), pp. 387–396 (2011). https://doi.org/10.1145/1998196.1998261
8. Biedl, T.C., Mehrabi, S.: On r-guarding thin orthogonal polygons. In: International Symposium on Algorithms and Computation (ISAAC), pp. 17:1–17:13 (2016). https://doi.org/10.4230/LIPIcs.ISAAC.2016.17
9. Biedl, T.C., Mehrabi, S.: On orthogonally guarding orthogonal polygons with bounded treewidth. Algorithmica **83**(2), 641–666 (2021). https://doi.org/10.1007/s00453-020-00769-5
10. Borradaile, G., Le, H.: Optimal dynamic program for r-domination problems over tree decompositions. In: International Symposium on Parameterized and Exact Computation (IPEC), pp. 8:1–8:23 (2017). https://doi.org/10.4230/LIPIcs.IPEC.2016.8
11. Brönnimann, H., Goodrich, M.T.: Almost optimal set covers in finite VC-dimension. Discrete Comput. Geom. **14**(4), 463–479 (1995). https://doi.org/10.1007/BF02570718
12. Demaine, E.D., Fomin, F.V., Hajiaghayi, M.T., Thilikos, D.M.: Fixed-parameter algorithms for (k, r)-center in planar graphs and map graphs. ACM Trans. Algorithms **1**(1), 33–47 (2005). https://doi.org/10.1145/1077464.1077468
13. Filtser, A., Le, H.: Clan embeddings into trees, and low treewidth graphs. In: Symposium on Theory of Computing (STOC), pp. 342–355 (2021). https://doi.org/10.1145/3406325.3451043
14. Filtser, A., Le, H.: Low treewidth embeddings of planar and minor-free metrics. In: Symposium on Foundations of Computer Science (FOCS), pp. 1081–1092 (2022). https://doi.org/10.1109/FOCS54457.2022.00105

15. Filtser, O., Krohn, E., Nilsson, B.J., Rieck, C., Schmidt, C.: Guarding polyominoes under k-hop visibility (2023). https://arxiv.org/abs/2308.00334
16. Fox-Epstein, E., Klein, P.N., Schild, A.: Embedding planar graphs into low-treewidth graphs with applications to efficient approximation schemes for metric problems. In: Symposium on Discrete Algorithms (SODA), pp. 1069–1088 (2019). https://doi.org/10.1137/1.9781611975482.66
17. Gibson, M., Krohn, E., Wang, Q.: The VC-dimension of visibility on the boundary of a simple polygon. In: International Symposium on Algorithms and Computation (ISAAC), pp. 541–551 (2015). https://doi.org/10.1007/978-3-662-48971-0_46
18. Gilbers, A., Klein, R.: A new upper bound for the VC-dimension of visibility regions. Comput. Geom. **47**(1), 61–74 (2014). https://doi.org/10.1016/j.comgeo.2013.08.012
19. Haussler, D., Welzl, E.: ε-nets and simplex range queries. Discrete Comput. Geom. **2**(2), 127–151 (1987). https://doi.org/10.1007/BF02187876
20. Iwamoto, C., Kume, T.: Computational complexity of the r-visibility guard set problem for polyominoes. In: Japanese Conference on Discrete and Computational Geometry and Graphs (JCDCGG), pp. 87–95 (2013). https://doi.org/10.1007/978-3-319-13287-7_8
21. Katsikarelis, I., Lampis, M., Paschos, V.T.: Structural parameters, tight bounds, and approximation for (k, r)-center. Discret. Appl. Math. **264**, 90–117 (2019). https://doi.org/10.1016/j.dam.2018.11.002
22. Kundu, S., Majumder, S.: A linear time algorithm for optimal k-hop dominating set of a tree. Inf. Process. Lett. **116**(2), 197–202 (2016). https://doi.org/10.1016/j.ipl.2015.07.014
23. Langetepe, E., Lehmann, S.: Exact VC-dimension for L_1-visibility of points in simple polygons (2017). https://arxiv.org/abs/1705.01723
24. Lee, D., Lin, A.K.: Computational complexity of art gallery problems. IEEE Trans. Inf. Theory **32**(2), 276–282 (1986). https://doi.org/10.1109/TIT.1986.1057165
25. Meir, A., Moon, J.W.: Relations between packing and covering numbers of a tree. Pac. J. Math. **61**(1), 225–233 (1975). https://doi.org/10.2140/pjm.1975.61.225
26. O'Rourke, J., Supowit, K.: Some NP-hard polygon decomposition problems. IEEE Trans. Inf. Theory **29**(2), 181–190 (1983). https://doi.org/10.1109/TIT.1983.1056648
27. Pinciu, V.: Guarding polyominoes, polycubes and polyhypercubes. Electron. Notes Discrete Math. **49**, 159–166 (2015). https://doi.org/10.1016/j.endm.2015.06.024
28. Tomás, A.P.: Guarding thin orthogonal polygons is hard. In: Gasieniec, L., Wolter, F. (eds.) FCT 2013. LNCS, vol. 8070, pp. 305–316. Springer, Heidelberg (2013). https://doi.org/10.1007/978-3-642-40164-0_29
29. Valtr, P.: Guarding galleries where no point sees a small area. Israel J. Math. **104**(1), 1–16 (1998). https://doi.org/10.1007/BF02897056
30. Worman, C., Keil, J.M.: Polygon decomposition and the orthogonal art gallery problem. Int. J. Comput. Geom. Appl. **17**(2), 105–138 (2007). https://doi.org/10.1142/S0218195907002264

Minimum-Width Double-Slabs and Widest Empty Slabs in High Dimensions

Taehoon Ahn[1] , Chaeyoon Chung[1] , Hee-Kap Ahn[2] , Sang Won Bae[3(✉)] ,
Otfried Cheong[4] , and Sang Duk Yoon[5]

[1] Department of Computer Science and Engineering, Pohang University of Science
and Technology, Pohang, Korea
{sloth,chaeyoon17}@postech.ac.kr
[2] Graduate School of Artificial Intelligence, Department of Computer Science and
Engineering, Pohang University of Science and Technology, Pohang, Korea
heekap@postech.ac.kr
[3] Division of Computer Science and Engineering, Kyonggi University, Suwon, Korea
swbae@kgu.ac.kr
[4] SCALGO, Aarhus, Denmark
otfried@scalgo.com
[5] Department of Service and Design Engineering, Sungshin Women's University,
Seoul, Korea
sangduk.yoon@sungshin.ac.kr

Abstract. A *slab* in d-dimensional space $\mathbb{R}^d$ is the set of points enclosed
by two parallel hyperplanes. We consider the problem of finding an optimal pair of parallel slabs, called a *double-slab*, that covers a given set P of
n points in $\mathbb{R}^d$. We address two optimization problems in $\mathbb{R}^d$ for any fixed
dimension $d \geqslant 3$: the *minimum-width double-slab* problem, in which one
wants to minimize the maximum width of the two slabs of the resulting
double-slab, and the *widest empty slab* problem, in which one wants to
maximize the gap between the two slabs. Our results include the first
nontrivial exact algorithms that solve the former problem for $d \geqslant 3$ and
the latter problem for $d \geqslant 4$.

Keywords: computational geometry · hyperplane · slab ·
double-slab · widest empty slab · minimum-width double-slab

1 Introduction

A *slab* in d-dimensional space $\mathbb{R}^d$ is the set of points enclosed by two parallel
hyperplanes. A *double-slab* is a pair of parallel slabs, that is, all four hyperplanes

C. Chung, T. Ahn, and H.-K. Ahn were supported by the Institute of Information
& communications Technology Planning & Evaluation(IITP) grant funded by the
Korea government(MSIT) (No. 2017-0-00905, Software Star Lab (Optimal Data Structure and Algorithmic Applications in Dynamic Geometric Environment)) and (No.
2019-0-01906, Artificial Intelligence Graduate School Program(POSTECH)). T. Ahn,
S.W. Bae, and C. Chung were supported by the National Research Foundation of
Korea(NRF) grant funded by the Korea government(MSIT) (No. RS-2023-00251168).

J. A. Soto and A. Wiese (Eds.): LATIN 2024, LNCS 14578, pp. 303–317, 2024.
https://doi.org/10.1007/978-3-031-55598-5_20

bounding the two slabs are parallel. We consider the problem of finding an optimal double-slab that covers a given set P of n points in $\mathbb{R}^d$, for any fixed dimension $d \geqslant 3$. We consider two optimization problems. In the first problem, which we call the *minimum-width double-slab* problem, we wish to minimize the *width* of the resulting double-slab, where the width of a double-slab is defined as the maximum width of its two slabs. In the second problem, which we call the *widest empty slab* problem, we wish to maximize the *gap* of the double-slab, which is defined as the distance between the two slabs.

The two optimization problems concerning double-slabs extend and/or generalize several fundamental geometric problems, which have been extensively studied mostly in $\mathbb{R}^2$ or $\mathbb{R}^3$, and sometimes in higher dimensions.

The *width* of a point set P in $\mathbb{R}^d$ is defined as the minimum width of a slab that covers P. The width is considered one of the most fundamental extent measures describing a point set, together with the diameter and the radius. The width of a set of n points in the plane can be computed easily in optimal $O(n \log n)$ time [29]. If the convex hull of P is already given, this can be improved to $O(n)$ time using rotating calipers [31]. For $d = 3$, Houle and Toussaint [23] presented an $O(n^2)$-time algorithm, and the first subquadratic-time algorithm was given by Chazelle et al. [13]. The currently best algorithm by Agarwal and Sharir runs in $O(n^{3/2+\epsilon})$ expected time [2]. In higher dimensions, Chan [9] discussed a simple formulation that reduces the width problem to searching a $(d+1)$-dimensional convex polytope, resulting in an $O(n^{\lceil d/2 \rceil})$-time exact algorithm for $d \geqslant 4$. Efficient approximation schemes are also known [1,9].

By definition, computing the width is equivalent to fitting a line for $d = 2$ (a plane for $d = 3$, or in general a hyperplane) to the given point set P. In particular, for $d = 2$, this problem is also known as the *line center* problem. For $k \geqslant 1$, the *k-line center* problem asks to find a set of k lines such that the maximum distance of a point in P to its nearest line center is minimized. Like the k-center problem, this problem is known to be NP-hard when k is part of the input, even in the plane $\mathbb{R}^2$ [28]. In spite of its fundamental position in theory and applications, little is known about exact algorithms for the k-line center problem. Even for $d = 2$, efficient algorithms are known only for $k \leqslant 2$: algorithms with running time $O(n^2 \log^2 n)$ were found in the 1990s [20,25] and there has been no improvement since then. For $d \geqslant 3$, we are not aware of any nontrivial exact algorithm to compute a k-line center for $k \geqslant 2$ in the literature. Agarwal et al. [5] presented an efficient approximation algorithm for any $k \geqslant 1$ and $d \geqslant 2$. Recently, some results on constrained variants of the k-line center problem in the plane $\mathbb{R}^2$ have been published: Bae [6] showed that a *parallel 2-line center* can be computed in $O(n^2)$ time, where the two lines are restricted to be parallel. Das et al. [17] presented an approximation algorithm for orthogonal line centers and Chung et al. [15] studied a variant of parallel k-line centers for $k \geqslant 2$ considering gaps between the induced clusters.

The generalization to higher dimensions is the *k-hyperplane center* problem. Given a set P of n points in $\mathbb{R}^d$, we want to find k hyperplanes that minimize the maximum distance from a point in P to its nearest hyperplane. This is equivalent to finding k slabs of minimum width that cover P. This problem has been studied in the context of the *projective clustering* problem and, except for the variants in $\mathbb{R}^2$ mentioned above, only approximation algorithms are known (see e.g. [4] and the references therein). As discussed in [4] and [5], an exact algorithm with running time $n^{O(dk)}$ can be easily obtained for both the k-line center and the k-hyperplane center problems. To our best knowledge, no better algorithms are known, not even for small constants $k \geqslant 2$ and $d \geqslant 3$.

Our first problem, namely the minimum-width double-slab problem, can be seen as a constrained variant of the 2-hyperplane center problem, in which the two hyperplane centers must be parallel, while it extends and generalizes the fundamental geometric problems mentioned above. We give the first exact algorithm for this problem.

Covering P by a double-slab induces two clusters on P. In some applications, such as in *obnoxious facility location* [16], it is considered more important to achieve a maximum possible separation between the resulting clusters of P. This motivates us to find a double-slab cover of P with the maximum possible gap, or a *widest empty slab* through P. In the planar case $d = 2$, this problem is well known as the *widest empty corridor* problem. Houle and Maciel [22] presented the first $O(n^2)$-time algorithm for the widest empty corridor problem. Many variants of the problem have been studied since, including corridors containing at most k input points [10,24,30], dynamic maintenance of corridors [24,30], and more. Díaz-Báñez et al. [18] studied the widest empty slab problem in $\mathbb{R}^3$, and presented an $O(n^3)$-time algorithm. We are not aware of any known algorithm for the widest empty slab problem in $\mathbb{R}^d$ for any fixed dimension $d \geqslant 4$. We present the first exact algorithm.

We summarize our results as follows:

(1) We solve the minimum-width double-slab problem in $O(n^d)$ time for $d = 3, 4$, and in $O(n^{d+1})$ time for any fixed dimension $d \geqslant 5$.
(2) We solve the widest empty slab problem in $O(n^d \log n)$ time for any fixed dimension $d \geqslant 4$.

We characterize combinatorial properties of optimal solutions to both problems, and find the optimal solution by efficiently enumerating all candidates. We make a heavy use of geometric duality, mapping points in $\mathbb{R}^d$ into non-vertical hyperplanes in $\mathbb{R}^d$, and known algorithms and data structures for hyperplanes and their arrangement in high dimensions.

Due to page limit, most proofs are omitted and will be seen in a full version.

2   Preliminaries

Let $d \geqslant 2$ be an arbitrary fixed dimension. We consider the d-dimensional Euclidean space $\mathbb{R}^d$ with the d coordinate axes, called the $x_1, x_2, \ldots, x_d$-axes. We consider the x_d-axis as the *vertical* direction. We also treat $\mathbb{R}^d$ as a vector space equipped with the standard inner product and the induced Euclidean norm $\|\cdot\|$. Hence, for any two points $p, q \in \mathbb{R}^d$, the length of segment pq is $\|p - q\|$. For any subset $A \subset \mathbb{R}^d$, let $\mathrm{aff}(A)$ be the affine hull of A, that is, the intersection of all affine subspaces containing A.

For any affine subspace A of $\mathbb{R}^d$, we denote by $A^\perp$ the *orthogonal complement* of A in $\mathbb{R}^d$, the vector subspace consisting of all vectors in $\mathbb{R}^d$ that are orthogonal to every vector $a - a_0$ for $a, a_0 \in A$. Note that $\dim(A) + \dim(A^\perp) = d$. Let $\pi_A \colon \mathbb{R}^d \to A$ denote the orthogonal projection onto A. In particular, if $A = \mathbb{R}^{d-1} = \{x_d = 0\}$, then we simply write $\pi = \pi_{\mathbb{R}^{d-1}}$. The projection π drops the x_d-coordinate, so that for any $(a_1, \ldots, a_d) \in \mathbb{R}^d$, we have $\pi(a_1, \ldots, a_{d-1}, a_d) = (a_1, \ldots, a_{d-1}) \in \mathbb{R}^{d-1}$.

Hyperplanes and orientations. We call a k-flat *vertical* if it is parallel to the x_d-axis. A *hyperplane* in $\mathbb{R}^d$ is a $(d-1)$-flat. Any non-vertical hyperplane $h \subset \mathbb{R}^d$ can be seen as the graph of a $(d-1)$-variate linear function $h \colon \mathbb{R}^{d-1} \to \mathbb{R}$: $h \colon x_d = u_1 x_1 + u_2 x_2 + \cdots + u_{d-1} x_{d-1} + b$, for some $u = (u_1, u_2, \ldots, u_{d-1}) \in \mathbb{R}^{d-1}$ and $b \in \mathbb{R}$. We call u the *orientation* of h, and b its *displacement*. Any non-vertical hyperplane h can be uniquely determined by an orientation u and a displacement b in this way, and we have the following.

Observation 1. *Let h be a non-vertical hyperplane in $\mathbb{R}^d$ with orientation $u \in \mathbb{R}^{d-1}$. Then, the vector $(u, -1) \in \mathbb{R}^d$ is a normal of h.*

In particular, hyperplanes with equal orientation are parallel. We can thus identify the space of possible orientations of non-vertical hyperplanes with $\mathbb{R}^{d-1}$. For any orientation $u \in \mathbb{R}^{d-1}$, let $\theta(u)$ be the angle between the vector $(u, -1) \in \mathbb{R}^d$ and the negative direction of the x_d-axis. For any two parallel hyperplanes h_1 and h_2, let $w(h_1, h_2)$ denote the distance between h_1 and h_2. We observe that $w(h_1, h_2) = |b_1 - b_2| \cdot \cos(\theta(u)) = \frac{|b_1 - b_2|}{\|(u, -1)\|}$, where u denotes their common orientation and b_i the displacement of h_i, for $i = 1, 2$.

A *slab* S is the closure of the region between two parallel hyperplanes h_1 and h_2, denoted by $S = (h_1, h_2)$. The *width* of S, denoted by $w(S)$, is $w(h_1, h_2)$. A *double-slab* $D = (S_1, S_2)$ is the union of two disjoint parallel slabs S_1 and S_2. The *width* of D, denoted by $w(D)$, is $\max\{w(S_1), w(S_2)\}$. We say that a slab or double-slab is *vertical* if its defining hyperplanes are vertical. For any non-vertical slab or double-slab, its *orientation* is the orientation of its defining hyperplanes.

Duality. We recall the classic point-to-hyperplane duality transform: for any point $p = (a_1, a_2, \ldots, a_d) \in \mathbb{R}^d$, the dual transformation maps p into its dual hyperplane $p^\star \colon x_d = a_1 x_1 + \cdots + a_{d-1} x_{d-1} - a_d$. Conversely, any non-vertical hyperplane $h \subset \mathbb{R}^d$ is mapped to a point $h^\star \in \mathbb{R}^d$. It is well known that the duality transform $a \mapsto a^\star$ preserves the point-hyperplane incidence relation and the vertical order among points and hyperplanes [8].

Observation 2. *Let $S = (h_1, h_2)$ be a non-vertical slab with orientation u.*

- *The segment $h_1^\star h_2^\star$ is vertical, that is, it is parallel to the x_d-axis in $\mathbb{R}^d$.*
- *It holds that $w(S) = \|h_1^\star - h_2^\star\| \cos(\theta(u))$.*
- *The first $d - 1$ coordinates of $h_1^\star$ and $h_2^\star$ are equal to those of u, that is, $\pi(h_1^\star) = \pi(h_2^\star) = u$.*

– *For any point $p \in \mathbb{R}^d$, $p \in S$ if and only if the segment $h_1^\star h_2^\star$ intersects $p^\star$.*

Therefore, there is a one-to-one correspondence between non-vertical slabs and vertical segments under the duality transform.

Arrangement of Hyperplanes. Let H be a set of n hyperplanes in $\mathbb{R}^d$. Consider the arrangement $\mathcal{A}(H)$ of these n hyperplanes. We introduce some algorithms and data structures for hyperplane arrangements that we will be using.

The arrangement $\mathcal{A}(H)$ consists of $O(n^d)$ faces for $d \geqslant 2$ [21]. The upper and lower envelopes of H, denoted by $\mathcal{U}(H)$ and $\mathcal{L}(H)$, correspond to the convex hull of points that are dual to H, so their complexity is bounded by $O(n^{\lfloor d/2 \rfloor})$. It is well known that $\mathcal{A}(H)$ can be computed in $O(n^d)$ time [21] and the envelopes $\mathcal{U}(H)$ and $\mathcal{L}(H)$ can be computed in $O(n \log n + n^{\lfloor d/2 \rfloor})$ time, using any optimal convex hull algorithm, such as the one by Chazelle [12].

The *zone* $Z(h; H)$ of another hyperplane h in the arrangement of H is the set of all cells in $\mathcal{A}$ intersected by h and their incident faces. Edelsbrunner et al. [19] showed that the complexity of the zone $Z(h; H)$ is $\Theta(n^{d-1})$.

Lemma 1. (de Berg et al. [7]). *For any fixed $d \geqslant 2$, the zone of a hyperplane h in an arrangement $\mathcal{A}(H)$ of n hyperplanes H can be computed in $O(n^{d-1} + n \log n)$ time, without computing the whole arrangement $\mathcal{A}$.*

In our algorithms, it is often required to test the feasibility of a candidate slab. The following query structure for point location in the arrangement will be used for our purpose.

Lemma 2. (Chazelle [11] and Matoušek [26]). *A set H of n hyperplanes in $\mathbb{R}^d$ can be preprocessed in $O(n^d / \log^d n)$ time into a data structure of size $O(n^d / \log^d n)$ that answers the following query in $O(\log n)$ time: Given a query point $q \in \mathbb{R}^d$, locate the face in $\mathcal{A}(H)$ that contains q and count the number of hyperplanes in H above q.*

By the duality transform, this data structure can answer the *half-space counting query* [3]. Thus, given a set P of n points in $\mathbb{R}^d$, by using the query structure in Lemma 2 for the set of n hyperplanes dual to points in P, we can count the number of points in P contained in a query slab $S = (h_1, h_2)$ in $O(\log n)$ time, according to Observation 2.

General Position Assumption. In the following, P will be a set of n input points in $\mathbb{R}^d$. For simplicity, we will assume that P is in general position, meaning that no hyperplane in $\mathbb{R}^d$ contains more than d points of P or, equivalently, that any $d + 1$ points in P are affinely independent. Hence, for any subset $Q \subseteq P$ of $k \leqslant d + 1$ points, its affine hull $\mathrm{aff}(Q)$ is a $(k - 1)$-flat in $\mathbb{R}^d$. This also implies that the orthogonal complement $(\mathrm{aff}(Q))^\perp$ is of dimension $d - k + 1$. Throughout the paper, we often discuss the intersection of two orthogonal complements $V = (\mathrm{aff}(Q_1))^\perp \cap (\mathrm{aff}(Q_2))^\perp$ for two nonempty disjoint subsets $Q_1, Q_2 \subset P$ with $k = |Q_1| + |Q_2| \leqslant d + 1$. We shall call V the (linear) *subspace orthogonal to both* $\mathrm{aff}(Q_1)$ and $\mathrm{aff}(Q_2)$. The general position also implies that the subspace V is always of dimension $d - k + 2$ for any two subsets $Q_1, Q_2 \in P$.

3   Widest Empty Slabs

Let us call a slab S *empty* if it contains no point of P in its interior while separating P into two nonempty subsets. Our goal in this section is to compute an empty slab S of maximum width. The following is an easy observation on empty slabs.

Lemma 3. *Suppose that S is a maximum-width empty slab for P in $\mathbb{R}^d$ for $d \geqslant 3$. Let $h \subset \mathbb{R}^d$ be any hyperplane parallel to the normal of S. Then, $\pi_h(S)$ is a maximum-width empty slab for $\pi_h(P)$ in the $(d-1)$-dimensional space h.*

Lemma 3 allows us to find the widest *vertical* empty slab by projecting the point set to $\mathbb{R}^{d-1}$ and solving the problem there. We can therefore concentrate on finding the widest *non-vertical* empty slab, and in the remainder of this section all slabs will be non-vertical. We represent a non-vertical slab S by a pair (h_1, h_2) of hyperplanes such that h_1 is above h_2.

We prove the following characterization of optimal empty slabs.

Lemma 4. *For any $d \geqslant 2$, if $S = (h_1, h_2)$ is an empty slab of maximum width, then one of the following must hold:*

(i) At least $d+1$ points lie on the boundary of S, that is, $|P \cap (h_1 \cup h_2)| \geqslant d+1$.
(ii) The normal vector to S is parallel to $\mathrm{aff}(P \cap (h_1 \cup h_2))$.

Proof. Let $S = (h_1, h_2)$ be a maximum-width empty slab for P, and let $k := |P \cap (h_1 \cup h_2)|$ be the number of points in P that lie on the boundary of S. Clearly at least one point of P lies on each of h_1 and h_2, so $k \geqslant 2$. Let $u \in \mathbb{R}^{d-1}$ be the orientation of S. Then, $\hat{u} := (u, -1)$ is a normal vector of S by Observation 1.

We will prove by induction on k and d that whenever $k \leqslant d$, the statement in the second case holds, that is, the normal $\hat{u}$ is parallel to $\mathrm{aff}(P \cap (h_1 \cup h_2))$.

Consider first the case $k = 2$. The planar case of the statement was proven by Houle and Maciel [22], see also Janardan and Preparata [24, Theorem 2.1]. Consider now $d > 2$ and assume the claim holds in $\mathbb{R}^{d-1}$ (for $k = 2$). Let $q_1, q_2 \in P$ be the two points such that $q_1 \in h_1$ and $q_2 \in h_2$, and suppose for a contradiction that $\hat{u}$ is not parallel to $\ell := \mathrm{aff}(P \cap (h_1 \cup h_2))$. In this case, ℓ is the line through q_1 and q_2. Let ℓ' be the line parallel to $\hat{u}$ going through q_1. The two lines ℓ and ℓ' make a positive angle $\phi > 0$. We consider a hyperplane h containing both lines ℓ and ℓ'. Since the normal of h is parallel to S, Lemma 3 implies that $\pi_h(S)$ is a maximum-width empty slab for $\pi_h(P)$ in h. The two points q_1 and q_2 lie in h and are still the only two points of P on the boundary of $\pi_h(S)$. The normal of $\pi_h(S)$ is identical to $\hat{u}$ and still makes a positive angle ϕ with ℓ, a contradiction to the optimality of $\pi_h(S)$. See Fig. 1(a). It follows that the claim holds for $k = 2$ in any dimension.

Consider now the general case $d > 2$ and $2 < k \leqslant d$, and assume that the claim holds for all smaller dimensions and smaller values of k. In this case there are at least two points in $P \cap h_1$ or $P \cap h_2$. Without loss of generality, we let $|P \cap h_1| \geqslant 2$. Let f be the $(k-1)$-flat $f = \mathrm{aff}(P \cap (h_1 \cup h_2))$. We pick two points $q_1, q_2 \in P \cap h_1$. Let ℓ be the line through q_1 and q_2, and let h be the hyperplane

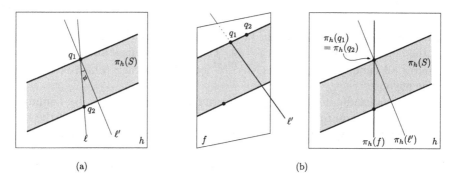

Fig. 1. Illustration to the proof of Lemma 4.

normal to ℓ containing q_1. Since the normal of h is parallel to S, we again apply Lemma 3 to conclude that $\pi_h(S)$ is a maximum-width empty slab for $\pi_h(P)$ in h. Since $\pi_h(q_1) = \pi_h(q_2)$, there are $k-1$ points of $\pi_h(P)$ lying on the boundary of $\pi_h(S)$. By the inductive assumption, the normal $\hat{u}$ of $\pi_h(S)$ is parallel to the $(k-2)$-flat $f' = \mathrm{aff}(\pi_h(P) \cap (\pi_h(h_1) \cup \pi_h(h_2)))$. Since our projection direction lies in f, we have $f' \subset f$, and so $\hat{u}$ is parallel to f, completing the inductive step. See Fig. 1(b) for an illustration when $d = 3$ and $k = 3$.   □

We will now describe our algorithm that computes a maximum-width empty slab for a given set P of n points in $\mathbb{R}^d$. It enumerates all candidate slabs that satisfy one of the two conditions described in Lemma 4. We will show that the number of candidates is bounded by $O(n^d)$ and we will spend $O(\log n)$ time per candidate for the emptiness test.

Let $P^\star$ be the set of n hyperplanes dual to the points P, and let $\mathcal{A} = \mathcal{A}(P^\star)$ be their arrangement. We first build the query structure of Lemma 2 for $\mathcal{A}$. We will separately enumerate all candidates that fall in case (i) and those that fall in case (ii) of Lemma 4.

3.1   Case (i)

Let us call an empty slab S a *candidate slab* if it satisfies the condition of case (i) in Lemma 4. In dual space, Observation 2 immediately implies the following.

Lemma 5. *A candidate slab corresponds to a maximal vertical segment contained in a d-dimensional cell of $\mathcal{A}$, whose endpoints lie in the relative interior[1] of two faces f_1 and f_2 with $\dim(f_1) + \dim(f_2) \leqslant d - 1$.*

We call such a vertical segment a *candidate stick*. Our algorithm collects all candidate sticks from the arrangement $\mathcal{A}$, computes the width of the corresponding slab for each candidate stick, and returns one that maximizes this width.

The candidate sticks are closely related to a vertical decomposition of the arrangement of hyperplanes. Chazelle and Friedman [14] showed that the number

---

[1] Note that the relative interior of a 0-flat (a point) is by definition the point itself.

of candidate sticks is $O(n^d)$ and argued that they can be computed by overlaying the orthogonal projections of all the faces incident to each d-dimensional cell of $\mathcal{A}$ onto the horizontal hyperplane $\{x_d = 0\}$. They also described an algorithm that computes all candidate sticks in $O(n^{\lfloor 3d/2 \rfloor} \log^{\lfloor d/2 \rfloor} n)$ time,[2] which is too slow for our purpose.

Consider a candidate stick s between two faces f_1 and f_2 of $\mathcal{A}$. By Lemma 5, their dimensions sum up to at most $d - 1$, so the smaller dimension of the two, say $\dim(f_2)$, is at most $\lfloor (d-1)/2 \rfloor$. Let $k := \dim(f_2)+1$. Then, we have $\dim(f_1) \leqslant d-k$, and so f_1 lies on (is a subface of) a face f' of $\mathcal{A}$ of dimension $\dim(f') = d-k$. The face f' is a face on the $(d-k)$-flat $f_Q := \bigcap_{q \in Q} q^*$ for some subset $Q \in P$ with $|Q| = k$. Let now $\hat{f}_Q$ be the $(d-k+1)$-flat spanned by f_Q and the vertical direction, and observe that s lies in $\hat{f}_Q$, with one endpoint in f_Q.

Every hyperplane q^* intersects $\hat{f}_Q$ in a $(d-k)$-flat q'. The face f_2 of $\mathcal{A}$ has $\dim(f_2) = k-1$ and therefore lies on the intersection of $d-k+1$ hyperplanes of $\mathcal{A}$. The intersection of these hyperplanes with the $(d-k+1)$-flat $\hat{f}_Q$ is therefore zero-dimensional, that is, a vertex of the arrangement $\mathcal{A}_Q$ of the $(d-k)$-flats q', for $q \in Q$, inside the $(d-k+1)$-flat $\hat{f}_Q$.

To summarize, for our candidate stick s there is a subset $Q \subset P$ of size $|Q| = k$, for some $1 \leqslant k \leqslant \lfloor (d-1)/2 \rfloor + 1 = \lceil d/2 \rceil$, such that one endpoint of s is a vertex v of the arrangement $\mathcal{A}_Q$ in $\hat{f}_Q$ and the other endpoint is the vertical projection of v on the $(d-k)$-flat f_Q. No hyperplane q' crosses the relative interior of s, so the entire segment lies in the zone of f_Q in the arrangement $\mathcal{A}_Q$.

Our algorithm enumerates all possibilities: We consider all non-empty subsets $Q \subset P$ of size at most $\lceil d/2 \rceil$. For each Q, we compute the flat f_Q and its vertical extension $\hat{f}_Q$, and intersect all hyperplanes with $\hat{f}_Q$. We then compute the zone Z of f_Q in the arrangement $\mathcal{A}_Q$, and for each vertex v of Z, generate a vertical segment connecting v with its vertical projection on f_Q. Finally, we test whether the segment actually is a candidate stick by verifying whether its relative interior intersects any hyperplane of $\mathcal{A}$ using the data structure of Lemma 2.

There are $\binom{n}{k} = O(n^k)$ different subsets of k points. Computing the zone of n hyperplanes in the $(d-k+1)$-dimensional space $\hat{f}_Q$ takes time $O(n^{d-k} + n \log n)$ time by Lemma 1. By the zone theorem [19], the zone has at most $O(n^{d-k})$ vertices, so we can test all the resulting vertical segments in time $O(n^{d-k} \log n)$. Summing over all $O(n^k)$ subsets Q of size k results in time $O(n^d \log n + n^{k+1} \log n)$, and summing over k from 1 to $\lceil d/2 \rceil$ results in total time $O(n^d \log n)$.

3.2 Case (ii)

We now turn to the second case. We will enumerate *all* slabs S (not just empty ones) that are determined by a subset Q of at most d points of P in the following way: The intersection of the boundary of S with P is exactly Q, and the normal

---

[2] In fact they construct a vertical decomposition of the entire arrangement.

of S is parallel to aff(Q). We then test whether the slab is actually empty using the data structure of Lemma 2, and return an empty slab maximizing the width.

The following lemma shows that given the boundary points, the unique slab satisfying the condition can be found in constant time.

Lemma 6. *Let $Q_1, Q_2 \subset P$ be two disjoint subsets such that $2 \leqslant |Q_1| + |Q_2| \leqslant d$. There exists a unique orientation u of a slab $S = (h_1, h_2)$ such that $P \cap h_1 = Q_1$, $P \cap h_2 = Q_2$, and the normal of S is parallel to aff$(Q_1 \cup Q_2)$. Moreover, such a unique orientation can be computed in $O(d^3) = O(1)$ time.*

Let $Q_1, Q_2 \subset P$ be a pair of disjoint subsets of P with $2 \leqslant |Q_1 \cup Q_2| \leqslant d$. In constant time we can compute the unique orientation described in Lemma 6, denoted by $u(Q_1, Q_2)$, and construct a slab $S = (h_1, h_2)$ with orientation $u(Q_1, Q_2)$ such that $P \cup h_1 = Q_1$ and $P \cup h_2 = Q_2$. We can then test in $O(\log n)$ time if the relative interior of S is empty using the data structure of Lemma 2. We return the widest slab among all slabs passing the test.

We run the above procedure for all valid pairs (Q_1, Q_2). The number of such pairs is bounded by $\sum_{2 \leqslant k \leqslant d} \binom{n}{k} \cdot 2^k = O(n^d)$, since d is a constant. Hence, the total running time for case (ii) is also bounded by $O(n^d \log n)$.

Theorem 1. *For any constant dimension $d \geqslant 4$, a widest empty slab for a set of n points in $\mathbb{R}^d$ can be computed in $O(n^d \log n)$ time.*

Our algorithm can be implemented faster for $d \leqslant 3$, because the emptiness of a candidate slab can be tested in $O(1)$ amortized time. This results in algorithms with $O(n^2)$ and $O(n^3)$ time for $d = 2$ and $d = 3$, respectively, matching the previously known time bounds [18,22]. The space requirement of our algorithm, however, is larger.

4   Minimum-Width Double-Slabs

In this section, we present algorithms that compute a minimum-width double-slab enclosing P in $\mathbb{R}^d$ for $d \geqslant 3$. Throughout this section, we assume that there exists a minimum-width double-slab enclosing P that is *non-vertical*. As in the previous section, if this is not the case, we can solve the problem by reducing it to an instance in $\mathbb{R}^{d-1}$ after orthogonally projecting P onto $\mathbb{R}^{d-1}$. In addition, we also assume that P consists of at least $d+2$ points, since the problem becomes trivial when $|P| \leqslant d + 1$.

Let $P^\star$ be the set of n hyperplanes dual to points in P, $\mathcal{A} = \mathcal{A}(P^\star)$ be their arrangement $\mathcal{A}$, $\mathcal{L} = \mathcal{L}(P^\star)$ be the lower envelope of $\mathcal{A}$, and $\mathcal{U} = \mathcal{U}(P^\star)$ be the upper envelope of $\mathcal{A}$. We represent any non-vertical double-slab D by a 4-tuple (h_1, h_2, h_3, h_4) of hyperplanes such that h_i is above h_{i+1} in $\mathbb{R}^d$ for each $i = 1, 2, 3$.

It is obvious that there exists a minimum-width double-slab enclosing P such that at least one point in P lies on each of its four hyperplanes. The following lemma describes a characterization of optimal double-slabs for our problem. See also Fig. 2.

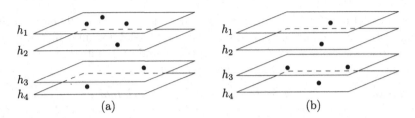

Fig. 2. Two possible configurations of minimum-width double-slabs $D = (h_1, h_2, h_3, h_4)$ in $\mathbb{R}^3$. (a) Case (i) of Lemma 7: There are six points on the four planes. (b) Case (iii): There are five points on the four planes, while we have $w(h_1, h_2) = w(h_3, h_4)$.

Lemma 7. *For any $d \geqslant 2$, if $D = (h_1, h_2, h_3, h_4)$ is a minimum-width double-slab enclosing P such that $|P \cap h_i| \geqslant 1$ for each $i \in \{1, 2, 3, 4\}$. Then, the following conditions must hold:*

- $\sum_i |P \cap h_i| \geqslant d + 2$.
- *Exactly one of the following cases holds:*
   (i) $w(h_1, h_2) > w(h_3, h_4)$ and $|P \cap h_1| + |P \cap h_2| \geqslant d + 1$.
   (ii) $w(h_1, h_2) < w(h_3, h_4)$ and $|P \cap h_3| + |P \cap h_4| \geqslant d + 1$.
   (iii) $w(h_1, h_2) = w(h_3, h_4)$.

Consider any minimum-width double-slab $D = (h_1, h_2, h_3, h_4)$ enclosing P that satisfies the conditions described in Lemma 7. Let $Q_i := P \cap h_i$ for $i = 1, 2, 3, 4$. Since the outer slab $S_{\text{out}} = (h_1, h_4)$ also encloses P, it is obvious that a portion of $\bigcap_{q \in Q_1} q^*$ appears as a face of $\mathcal{L}$ and a portion of $\bigcap_{q \in Q_4} q^*$ appears in $\mathcal{U}$ in the same way. Hence, we have $h_1^\star \in \mathcal{L}$ and $h_4^\star \in \mathcal{U}$. On the other hand, the inner slab $S_{\text{in}} = (h_2, h_3)$ should be empty, so the vertical segment $h_2^\star h_3^\star$ crosses no hyperplane in $P^\star$. Observe also that the dual point $h_i^\star$ for each i lies in the relative interior of a $(d - |Q_i|)$-face of $\mathcal{A}$ that is a portion of the flat $\bigcap_{q \in Q_i} q^\star$.

We will often discuss the overlay $\mathcal{M}$ of the orthogonal projections $\pi(\mathcal{U})$ and $\pi(\mathcal{L})$ of the envelopes $\mathcal{U}$ and $\mathcal{L}$ onto the hyperplane $\{x_d = 0\} = \mathbb{R}^{d-1}$. Each face f of the overlay $\mathcal{M}$ is associated with a pair (Q_1, Q_4) of subsets of P such that: for any orientation $u \in f$, if (h_1, h_4) is the minimum-width slab of orientation u that encloses P, then it always holds that $P \cap h_1 = Q_1$ and $P \cap h_4 = Q_4$.

Lemma 8. *For any fixed $d \geqslant 2$, the overlay $\mathcal{M}$ consists of $O(n^{\lceil d/2 \rceil})$ many faces and can be computed in $O(n^{\lceil d/2 \rceil} + n \log n)$ time.*

In the following, we describe our algorithms for $d \geqslant 3$.

4.1   Three-Dimensional Case

First, we consider the case of $d = 3$. For any point $q \in P$ and any orientation $u \in \mathbb{R}^2$, we consider the double-slab $D_q(u) = (h_1, h_2, h_3, h_4)$ with orientation u such that $P \subset D_q(u)$, $h_1^\star \in \mathcal{L}$, $q \in h_2$ (that is, $h_2^\star \in q'^\star$) for some $q' \in P \setminus \{q\}$, $q \in h_3$ (that is, $h_3^\star \in q^\star$), and $h_4^\star \in \mathcal{U}$. Since $D_q(u)$ encloses P, the point q' on h_2 is determined as the first point in P hit by translating h_3 upwards or, equivalently, as the first hyperplane q'^* in $\mathcal{A}$ hit by the ray emanating downwards from h_3^*. By Lemma 7, there exist some $q \in P$ and $u \in \mathbb{R}^2$ such that $D_q(u)$ is indeed a minimum-width double-slab enclosing P.

Let $w_q(u) := w(D_q(u))$ be its width. Then, we have $w_q(u) = \max\{\|h_1^* - h_2^*\|, \|h_3^* - h_4^*\|\} \cdot \cos(\theta(u))$. Let $v_q(u) := w_q(u)/\cos(\theta(u))$ be the linear part of the right side of the above equation. Observe that the function $v_q \colon \mathbb{R}^2 \to \mathbb{R}$ is piecewise linear, since its values are determined by the vertical distances between faces in $\mathcal{A}$, while $\cos(\theta(u)) = 1/\sqrt{\|u\|^2 + 1}$ is a continuous function over $u \in \mathbb{R}^2$.

Lemma 9. *The number of linear pieces in function v_q is $O(n^2)$ and an explicit description of function v_q can be computed in $O(n^2)$ time.*

By Lemma 9, we also obtain the planar subdivision M_q induced by the breakpoints of v_q such that v_q restricted to any face of M_q (of any dimension) is a linear function. Lemma 7 for $d = 3$ implies the existence of a minimum-width double-slab enclosing P whose orientation lies on a vertex or an edge of M_q. Since the function v_q restricted to each face of M_q (of any dimension) is linear and thus of constant complexity, we can find an optimal double-slab constrained about $q \in P$ in total $O(n^2)$ time. By iterating all $q \in P$, we obtain the following.

Theorem 2. *Given a set P of n points in $\mathbb{R}^3$, a minimum-width double-slab enclosing P can be computed in $O(n^3)$ time.*

4.2 Four-Dimensional Case

We then consider the four-dimensional case, $d = 4$. Consider an optimal solution $D = (h_1, h_2, h_3, h_4)$ that satisfies the conditions of Lemma 7. We distinguish two cases: when (a) $|P \cap h_2| = |P \cap h_3| = 1$ or (b) either $|P \cap h_2| \geqslant 2$ or $|P \cap h_3| \geqslant 2$. Our goal is to find an optimal double-slab that falls into each case.

In the former case (a), we first compute the overlay $\mathcal{M}$ of $\pi(\mathcal{U})$ and $\pi(\mathcal{L})$ by applying Lemma 8. Then, for each face of $\mathcal{M}$, consider its associated pair (Q_1, Q_4) of subsets and a minimum-width double-slab D enclosing P such that $P \cap h_1 = Q_1$, $P \cap h_4 = Q_4$, and it falls into case (a). Since we must have $|Q_1| + |Q_4| \geqslant 4$ by Lemma 7, we ignore those faces of $\mathcal{M}$ whose associated pair (Q_1, Q_4) consists of at most three points. Since our target double-slab contains Q_1 and Q_4 on its outer boundary, its normal should be orthogonal both to $\mathrm{aff}(Q_1)$ and $\mathrm{aff}(Q_4)$. Hence, we consider the subspace $f = (\mathrm{aff}(Q_1))^\perp \cap (\mathrm{aff}(Q_4))^\perp$ orthogonal to both $\mathrm{aff}(Q_1)$ and $\mathrm{aff}(Q_4)$. As discussed in Sect. 2, such a flat f exists and its dimension is always $6 - |Q_1| - |Q_4| \leqslant 2$. Hence, f is a line or a plane in $\mathbb{R}^4$. If f is a line, then we can find our solution in $O(n)$ time by scanning all points in P; if f is a plane, then we project P orthogonally onto f and solve the 2-dimensional problem instance in $O(n^2)$ time [6]. Since $\mathcal{M}$ consists of $O(n^2)$ faces (Lemma 8), we can handle case (a) in a total of $O(n^4)$ time.

Next, we consider the latter case (b). Assume without loss of generality that we have $|Q_3| \geqslant 2$ in an optimal double-slab of this case. We fix a pair of points

$q, q' \in P$ and consider the case where $q, q' \in Q_3$. For the purpose, we consider any hyperplane h in $\mathbb{R}^4$ orthogonal to the line through q and q', and the orthogonal projections of P onto h. Note that q and q' are projected to a common point $\bar{q}$ on h. We then have a problem instance in $\mathbb{R}^3$ with the constraint that $\bar{q}$ should lie on the third plane in the resulting slab. Therefore, we apply Lemma 9 to solve this instance in $O(n^2)$ time. We can thus handle case (b) by iterating all pairs (q, q') of two points in P, taking $O(n^4)$ time.

Theorem 3. *Given a set P of n points in $\mathbb{R}^4$, a minimum-width double-slab enclosing P can be computed in $O(n^4)$ time.*

4.3   Algorithm for Higher Dimension

Finally, we consider higher dimensions for $d \geqslant 5$. The algorithm starts with some preprocessing, including the initialization of the point location structure of Lemma 2 and the query structure by Matoušek and Schwarzkopf [27] for ray shooting for $\mathcal{U}$ and $\mathcal{L}$. The cost of preprocessing is not more than $O(n^d)$ time. In the main part of the algorithm, we handle each of the three cases described in Lemma 7.

Cases (i) and (ii). First, we handle double-slabs $D = (h_1, h_2, h_3, h_4)$ enclosing P of case (i): $w(h_1, h_2) > w(h_3, h_4)$ and $|P \cap h_1| + |P \cap h_2| \geqslant d + 1$. For the purpose, we consider those $D = (h_1, h_2, h_3, h_4)$ with the additional constraint that only one fixed point $q \in P$ lies on h_2.

Lemma 10. *Let $d \geqslant 2$ be any fixed integer and $q \in P$ be any fixed point. A minimum-width double-slab $D = (h_1, h_2, h_3, h_4)$ enclosing P such that $w(h_1, h_2) > w(h_3, h_4)$ and $P \cap h_2 = \{q\}$ can be computed in $O(n^{\lfloor d/2 \rfloor} \log n)$ time, provided the query structure of Lemma 2.*

For each $1 \leqslant k \leqslant d$, consider an optimal double-slab D of this case such that $|P \cap h_2| = k$. Such an optimal solution D can be computed as follows: For each subset $Q_2 \subset P$ with $|Q_2| = k$, let $f := (\text{aff}(Q_2))^{\perp}$ be the orthogonal complement of $\text{aff}(Q_2)$. Note that f is a $(d - k + 1)$-flat, as discussed in Sect. 2. We then orthogonally project P onto f by the projection operator π_f, and find an optimal double-slab $D' = (h_1', h_2', h_3', h_4')$ of case (i) with the additional constraint that $\pi_f(P) \cap h_2' = \pi_f(Q_2)$. This constrained problem can be solved by Lemma 10 if $1 \leqslant k \leqslant d - 1$, since the projection π_f maps all points in Q_2 to a common point in f and thus $\pi_f(Q_2)$ consists of a single point. Since there are $\binom{n}{k}$ different choices of Q_2 and we spend $O(n^{\lfloor (d-k+1)/2 \rfloor} \log n)$ time per each (Lemma 10), it takes $O(n^d \log n)$ time in total over all $1 \leqslant k \leqslant d - 1$.

In case of $k = d$, the dimension of flat f is 1, so it gives us a unique orientation $u \in \mathbb{R}^{d-1}$. Hence, this case can be solved in $O(\log n)$ time as follows: Using the ray shooting queries for $\mathcal{L}$ and $\mathcal{U}$ [27], we locate the points $h_1^{\star}$ and $h_4^{\star}$ on $\mathcal{L}$ and $\mathcal{U}$, respectively, at u in $O(\log n)$ time, and then test if the vertical segment with top endpoint at $h_4^{\star}$ and length $\|h_1^{\star} - h_2^{\star}\|$ crosses the correct number of hyperplanes in

$\mathcal{A}$. This query can be answered in $O(\log n)$ time by Lemma 2. The total running time for case (i) is thus bounded by $O(n^d \log n)$.

Case (ii) can be handled in the same way as above.

Case (iii). Finally, consider double-slabs $D = (h_1, h_2, h_3, h_4)$ enclosing P of case (iii): $w(h_1, h_2) = w(h_3, h_4)$ and $\sum_i |P \cap h_i| \geqslant d + 2$. For any four subsets $Q_1, Q_2, Q_3, Q_4 \subseteq P$, we define $D(Q_1, Q_2, Q_3, Q_4) = (h_1, h_2, h_3, h_4)$ to be a minimum-width double-slab such that $w(h_1, h_2) = w(h_3, h_4)$ and $P \cap h_i = Q_i$ for $i = 1, 2, 3, 4$. Note that $D(Q_1, Q_2, Q_3, Q_4)$ does not have to enclose P and it may be undefined when all the conditions cannot be fulfilled at the same time or there is no minimum width even though there are infinitely many such double-slabs. If $D(Q_1, Q_2, Q_3, Q_4)$ is well defined, then we call it a *candidate double-slab*. Our strategy is to collect all candidate double-slabs over all possible combinations of four subsets and then to test the feasibility of each of them, that is, if it encloses P by Lemma 2.

It turns out that the case of $|Q_2| = |Q_3| = 1$ can be handled efficiently.

Lemma 11. *Let $d \geqslant 2$ be any fixed integer and $q_2, q_3 \in P$ be two fixed points. The number of candidate double-slabs of the form $D(Q_1, \{q_2\}, \{q_3\}, Q_4)$ for some $Q_1, Q_4 \subset P$ is $O(n^{\lfloor d/2 \rfloor})$, and all of them can be computed in $O(n^{\lfloor d/2 \rfloor} + n \log n)$ time. In case of $d \leqslant 3$, given the lower and upper envelopes $\mathcal{L}$ and $\mathcal{U}$, all candidates can be reported in $O(n)$ time.*

Now, we fix $2 \leqslant k \leqslant d$ and any pair (Q_2, Q_3) of nonempty subsets of P with $|Q_2| + |Q_3| = k$. We enumerate all candidate double-slabs of the form $D(Q_1, Q_2, Q_3, Q_4)$ for some $Q_1, Q_4 \subset P$ in the following way. Compute the subspace $f = (\mathrm{aff}(Q_2))^\perp \cap (\mathrm{aff}(Q_3))^\perp$ orthogonal to both $\mathrm{aff}(Q_2)$ and $\mathrm{aff}(Q_3)$. As discussed in Sect. 2, we always have $\dim(f) = d - k + 2$. As done above for case (i), we project points in P orthogonally onto f by projection operator π_f. Observe that all points in Q_2 are projected to a common point $\bar{q}_2 \in f$ and all points in Q_3 are projected to a common point $\bar{q}_3 \in f$. Hence, we can apply Lemma 11 in the $(d-k+2)$-dimensional space f. This results in $O(n^{\lfloor (d-k+2)/2 \rfloor})$ candidate double-slabs, taking the same amount of time if $2 \leqslant k \leqslant d - 2$, or $O(n \log n)$ time if $k = d - 1$ or d. By iterating all $2 \leqslant k \leqslant d$ and all $\binom{n}{k} \cdot (2^k - 2)$ pairs (Q_2, Q_3), we collect a total of $O(n^{d+1})$ candidate double-slabs. We then test each candidate double-slab if it encloses P by point location in $\mathcal{A}$ (Lemma 2) in $O(\log n)$ time per each.

It remains to consider those double-slabs $D = (h_1, h_2, h_3, h_4)$ of case (iii) such that $k = |P \cap h_2| + |P \cap h_3| \geqslant d + 1$. In this case, observe that (h_2, h_3) is an empty slab and its dual segment $h_2^* h_3^*$ is a candidate stick defined in Sect. 3. (See also Lemma 5.) Thus, we run the algorithm described in Sect. 3.1 to compute all $O(n^d)$ candidate sticks and, for each candidate sticks s, we build a corresponding double-slab that encloses P in $O(\log n)$ time by ray shooting queries on $\mathcal{L}$ and $\mathcal{U}$ at $u = \pi(s) \in \mathbb{R}^{d-1}$. So, we can handle this case in $O(n^d \log n)$ time.

The algorithm described above takes $O(n^{d+1} \log n)$ time in total to compute an optimal double-slab of case (iii). The most time-consuming part of the algorithm is the case of $k = d$: for $k = d$, we collect $O(n)$ candidates for each of $O(n^d)$

pairs (Q_2, Q_3) and spend $O(n^{d+1} \log n)$ time, including the feasibility test. The other cases for $2 \leqslant k \leqslant d - 1$ and for $k \geqslant d + 1$ take only $O(n^d \log n)$ time in total.

In order to improve the running time, remark that, when $k = d$, we reduce the problem in $\mathbb{R}^d$ to an instance in $\mathbb{R}^2$. In the following, we show that these 2-dimensional instances can be handled efficiently in a 3-dimensional instance.

Lemma 12. *Let $d = 3$ and $q_2, q_3 \in P$ be two fixed points. A minimum-width double-slab $D = (h_1, h_2, h_3, h_4)$ enclosing P such that $w(h_1, h_2) = w(h_3, h_4)$, $q_2 \in h_2$, $q_3 \in h_3$, and $|P \cap h_2| + |P \cap h_3| = 3$ can be computed in $O(n^2)$ time.*

We exploit Lemma 12 as follows: For each pair (Q_2, Q_3) with $|Q_2| + |Q_3| = d - 1$, we apply Lemma 12. Then, it also handles the pairs (Q'_2, Q_3) and (Q_2, Q'_3) such that $Q'_2 = Q_2 \cup \{q\}$ and $Q'_3 = Q_3 \cup \{q\}$ for any point $q \in P \setminus (Q_2 \cup Q_3)$. Hence, we can now handle those pairs including d points in bundles. Since we spend only $O(n^d \log n)$ time for the cases of $2 \leqslant k \leqslant d - 1$, the final running time is thus improved to $O(n^{d+1})$.

Theorem 4. *Given a set P of n points in $\mathbb{R}^d$ for any fixed $d \geqslant 5$, a minimum-width double-slab enclosing P can be computed in $O(n^{d+1})$ time.*

References

1. Agarwal, P.K., Har-Peled, S., Varadarajan, K.R.: Approximating extent measures of points. J. ACM **51**(4), 606–635 (2004)
2. Agarwal, P.K., Sharir, M.: Efficient randomized algorithms for some geometric optimization problems. Discr. Comput. Geometry **16**(4), 317–337 (1996)
3. Agarwal, P.K.: Range searching. In: Goodman, J., O'Rourke, J., Tóth, C. (eds.) Handbook of Discrete and Computational Geometry, chap. 40, pp. 1057–1092. CRC Press, 3rd edn. (2018)
4. Agarwal, P.K., Procopiuc, C.M.: Approximation algorithms for projective clustering. J. Algorithms **46**(2), 115–139 (2003)
5. Agarwal, P.K., Procopiuc, C.M., Varadarajan, K.R.: Approximation algorithms for a k-line center. Algorithmica **42**(3), 221–230 (2005)
6. Bae, S.W.: Minimum-width double-strip and parallelogram annulus. Theor. Comput. Sci. **833**, 133–146 (2020)
7. de Berg, M., Dobrindt, K., Schwarzkopf, O.: On lazy randomized incremental construction. Discr. Comput. Geometry **14**, 261–286 (1995)
8. de Berg, M., Cheong, O., van Kreveld, M., Overmars, M.: Computational Geometry: Algorithms and Applications. Springer Berlin Heidelberg, Berlin, Heidelberg (2008). https://doi.org/10.1007/978-3-540-77974-2
9. Chan, T.M.: Approximating the diameter, width, smallest enclosing cylinder, and minimum-width annulus. Int. J. Comput. Geom. Appl. **12**(1–2), 67–85 (2002)
10. Chattopadhyay, S., Das, P.: The k-dense corridor problems. Pattern Recogn. Lett. **11**, 463–469 (1990)
11. Chazelle, B.: Cutting hyperplanes for divide-and-conquer. Discr. Comput. Geometry **9**(2), 145–158 (1993)
12. Chazelle, B.: An optimal convex hull algorithm in any fixed dimension. Discr. Comput. Geometry **10**, 377–409 (1993)

13. Chazelle, B., Edelsbrunner, H., Guibas, L., Sharir, M.: Diameter, width, closest line pair and parametric searching. Discr. Comput. Geometry **10**, 183–196 (1993)
14. Chazelle, B., Friedman, J.: Point location among hyperplanes and unidirectional ray-shooting. Comput. Geometry: Theor. Appl. **4**(2), 53–62 (1994)
15. Chung, C., Ahn, T., Bae, S.W., Ahn, H.K.: Parallel line centers with guaranteed separation. In: Proceedings of the 35th Canadian Conference on Computational Geometry (CCCG 2023), pp. 153–160 (2023)
16. Church, R.L., Drezner, Z.: Review of obnoxious facilities location problems. Comput. Oper. Res. **138**, 105468 (2022)
17. Das, A.K., Das, S., Mukherjee, J.: Approximation algorithms for orthogonal line centers. Discret. Appl. Math. **338**, 69–76 (2023)
18. Díaz-Báñez, J.M., López, M.A., Sellarés, J.A.: Locating an obnoxious plane. Eur. J. Oper. Res. **173**(2), 556–564 (2006)
19. Edelsbrunner, H., Seidel, R., Sharir, M.: On the zone theorem for hyperplane arrangements. SIAM J. Comput. **22**(2), 418–429 (1993)
20. Glozman, A., Kedem, K., Shpitalnik, G.: On some geometric selection and optimization problems via sorted matrices. Comput. Geom.: Theory Appl. **11**(1), 17–28 (1998)
21. Halperin, D., Sharir, M.: Arrangements. In: Goodman, J., O'Rourke, J., Tóth, C. (eds.) Handbook of Discrete and Computational Geometry, chap. 28, pp. 723–762. CRC Press, 3rd edn. (2018)
22. Houle, M.E., Maciel, A.: Finding the widest empty corridor through a set of points. Snapshots of computational and discrete geometry, pp. 210–213 (1988)
23. Houle, M., Toussaint, G.: Computing the width of a set. In: Proc. 1st ACM Sympos. Comput. Geom. (SoCG 1985), pp. 1–7 (1985)
24. Janardan, R., Preparata, F.P.: Widest-corridor problems. Nordic J. Comput. **1**(2), 231–245 (1994)
25. Jaromczyk, J., Kowaluk, M.: The two-line center problem from a polar view: a new algorithm and data structure. In: Proc. 4th International Workshop Algorithmic Data Structure (WADS 1995). Lecture Notes Comput. Sci., vol. 955, pp. 13–25 (1995)
26. Matoušek, J.: Range searching with efficient hierarchical cuttings. Discr. Comput. Geometry **10**, 157–182 (1993)
27. Matoušek, J., Schwarzkopf, O.: On ray shooting in convex polytopes. Discr. Comput. Geometry **10**, 215–232 (1993)
28. Megiddo, N., Tamir, A.: On the complexity of locating linear facilities in the plane. Oper. Res. Lett. **1**(5), 194–197 (1982)
29. Preparata, F.P., Shamos, M.I.: Computational Geometry. Springer New York, New York, NY (1985). https://doi.org/10.1007/978-1-4612-1098-6
30. Shin, C.S., Shin, S.Y., Chwa, K.Y.: The widest k-dense corridor problems. Inf. Process. Lett. **68**(1), 25–31 (1998)
31. Toussaint, G.: Solving geometric problems with the rotating calipers. In: Proceedings of the IEEE MELECON (1983)

Complexity Theory

The Complexity Classes of Hamming Distance Recoverable Robust Problems

Christoph Grüne[✉][iD]

Department of Computer Science, RWTH Aachen University, Aachen, Germany
gruene@algo.rwth-aachen.de

Abstract. The well-known complexity class *NP* contains combinatorial problems, whose optimization counterparts are important for many practical settings. In reality, however, uncertainty in the input data is a usual phenomenon, which is typically not covered in *NP* problems.

One concept to model the uncertainty in the input data, is *recoverable robustness*. The instance of the recoverable robust version of a combinatorial problem P is split into a base scenario σ_0 and an uncertainty scenario set S. The task is to calculate a solution $\mathbf{s}_0$ for the base scenario σ_0 and solutions $\mathbf{s}$ for all uncertainty scenarios $\sigma \in \mathsf{S}$ such that $\mathbf{s}_0$ and $\mathbf{s}$ are not too far away from each other according to a distance measure, so $\mathbf{s}_0$ can be easily adapted to $\mathbf{s}$.

We analyze the complexity of *Hamming distance recoverable robust* versions of problems in *NP* for different scenario encodings. The complexity is primarily situated in the lower levels of the polynomial hierarchy. The main contribution of the paper is a gadget reduction framework that reveals that the recoverable robust version of problems in a large class of combinatorial problems is Σ_3^p-complete. We show that this class includes over 20 problems such as Vertex Cover, Independent Set, Hamiltonian Path or Subset Sum. We expect that the number of problems can be easily extended with the help of the gadget reduction framework. Additionally, we expand the results to Σ_{2m+1}^p-completeness for multi-stage recoverable robust problems with $m \in \mathbb{N}$ stages.

Keywords: Computational Complexity · Polynomial Hierarchy · Robust Optimization · Recoverable Robustness · Optimization under Uncertainty

1 Introduction

The concept of *robustness* in the field of optimization problems is a collection of models that consider uncertainties in the input. These uncertainties may for

This work is funded by the Deutsche Forschungsgemeinschaft (DFG, German Research Foundation)–GRK 2236/1. I would like to thank Marc Goerigk, Stefan Lendl and Lasse Wulf for helpful comments and discussions on the paper. Furthermore, I would like to thank the anonymous reviewers for their helpful comments.

J. A. Soto and A. Wiese (Eds.): LATIN 2024, LNCS 14578, pp. 321–335, 2024.
https://doi.org/10.1007/978-3-031-55598-5_21

example arise from faulty or inaccurate sensors or from a lack of knowledge. Robustness measures can model these types of uncertainty that occur in practical optimization instances into an *uncertainty set*. The goal is to find solutions that are stable over all possible *scenarios* in the uncertainty set. That is, these solutions remain good but not necessarily optimal regardless what the uncertainties turn out to be in reality.

One specific robustness concept is *recoverable robustness*, which is a recently introduced concept [15] by Liebchen et al. The input of a recoverable robust version of a problem P is a *base scenario* σ_0, which is an instance of problem P, as well as a set of *uncertainty scenarios* S, whose members are again instances of P. The set of uncertainty scenarios S is the uncertainty set of the problem. We are asked to compute a base solution s_0 to the base scenario σ_0 and to compute recovery solutions s to all members of the uncertainty scenarios $\sigma \in$ S such that s_0 and s are not too far away from each other according to a distance measure. The solution on the base scenario does not directly include the uncertainties but needs to include the potential to adapt the base solution s_0 to solutions s within the given distance between the solutions. Thus, the base solution s_0 may be restricted by these possibly harmful scenarios.

From a worst-case-analysis point of view, we assume that the uncertainty scenarios are chosen by an adversary. The algorithm computes a base solution with the potential to adapt to all scenarios. Then, the adversary chooses the most harmful scenario based on the base solution. Finally, the algorithm computes a recovery solution to adapt to the chosen scenario.

A more general concept is *multi-stage recoverable robustness*, in which not only one set of uncertainty scenarios is provided but m sets of scenarios. This concept was introduced by Cicerone et al. [8]. The *m-stage recoverable robust problem* asks to solve the recoverable robust problem on the individual sets of scenarios inductively. That is, a base solution s_0 has to be found such that one can recover from s_0 for the first set of scenarios S_1 to a solution s_1 such that one can recover from s_1 for the second set of scenarios S_2 and so forth such that one can recover from s_{m-1} for the m-th set of scenarios S_m to a solution s_m.

Related Work. Recoverable robustness is used in many practical settings such as different optimization areas in air transport [9,11,17] or in railway optimization, for which a survey can be found in [16]. Our focus lies on the complexity of recoverable robust problems. In parallel to this paper, Goerigk et al. [12] analyzed the Hamming distance recoverable robust Independent Set, TSP and Vertex Cover. Hamming distance means that at most k elements may be added to or deleted from the base solution in total to obtain a recovery solution. They showed the Σ_3^p-hardness of the variant with discrete budgeted uncertainty over the costs of the elements. To the best of the author's knowledge, this is the only contribution investigating the complexity within the polynomial hierarchy beyond NP-hardness. All other contributions study primarily algorithms and analyze the problems only on their NP-hardness or their approximability, where different distance measures between the solutions are of interest. The concept of *k-dist recoverable robustness*, allowing at most k new elements in recovery solutions, was introduced in [3] but was also used in [14]. Besides the k-dist measures, there are also measures which limit the number of deleted elements [5] or exchanged [7] elements. Furthermore, combinations of these distance measures are analyzed as well in the literature [6]. Further usages of Hamming distance recoverable robustness can be found in [10]. Among the studied recoverable robust problems is Knapsack, which is *NP*-hard for different distance measures between the solutions [4–6]. Recoverable robust versions of problems that are in *PTIME* are shown to be *NP*-complete as well such as Shortest Path, which is *NP*-hard for k-dist [3], or Matching [10]. Furthermore, the recoverable robust Single Machine Scheduling problem is 2-approximable [2] and the recoverable robust TSP is 4-approximable [7]. Moreover, a recoverable robust version of Spanning Tree [14] is shown to be in *PTIME*.

For the complexity analysis, we introduce a gadget reduction framework. Different gadget reduction concepts were studied for example by Agrawal et al. [1], who defined gadget reductions under AC^0 for NP-completeness mapping one bit of the input of one problem to a bounded number of bits in the other problem. A further form of gadget reduction was introduced by Trevisan et al. [19], who formalized constraints of a linear program to be a gadget in the reduction between linear programs.

Contribution. We study Hamming distance recoverable robust problems with different forms of elemental uncertainty. That is, it is uncertain whether an element (e.g. a vertex or object) is included in a scenario or not. This form of uncertainty is different to cost uncertainty, where all elements are present in all scenarios but the costs of the elements are uncertain. We show that recoverable robust versions of typical *NP*-complete combinatorial problems with *xor*-dependencies or Γ-set scenarios are Σ_3^p-complete and the corresponding multi-stage recoverable robust versions are Σ_{2m+1}^p-complete, where $m \in \mathbb{N}$ is the number of stages.

We do this by defining a gadget reduction framework, which uses a specific definition of combinatorial problems. These problems are defined over *combinatorial elements*, which are defined over a *universe U*, and *nested relations $R(U)$* over that universe. We show that this framework is able to "upgrade" many already exsiting NP-hardness reductions by applying it to over 20 well-known problems. Thus, we expect that the results are easily extendable beyond those problems.

In order to explain the idea of these *universe gadget reductions*, consider a problem A for which we want to reduce to another problem B. A gadget reduction creates a gadget for each combinatorial element from U_A and $R_A(U_A)$ to simulate the behavior of this element in B. This gadget consists of universe elements from U_B and/or relation elements $R_B(U_B)$ and is disjoint from all other gadgets. That is, no element from B is in two gadgets at the same time. Additionally, we demand that if we remove a combinatorial element in A, we are able to remove the corresponding gadget in B without invalidating the correctness of the reduction.

This form of reduction preserves the scenarios structurally independent of the underlying encoding. Thus, this gadget reduction framework allows for reductions between Hamming distance recoverable robust problems. Indeed these properties are already achieved by typical polynomial reductions (or slight modifications) of it.

Paper Summary. In Sect. 2, we define necessary complexity theoretical concepts. In Sect. 3, we build a framework for combinatorial decision problems to define Hamming distance recoverable robust problems. Then, we consider typical problems, which are in *NP* or *NP*-complete, and analyze their complexity for polynomially computable scenario encodings in Sect. 4. Section 5 consists of the complexity analysis of succinctly encoded scenarios as well as multi-stage recoverable robustness. At last in Sect. 6, we establish a whole class of Hamming distance recoverable robust problems by using our combinatorial decision problem framework and by introducing universe gadget reductions. With Sect. 7, we conclude the paper. The full version of this paper can be found on arXiv [13].

2   Preliminaries

We define a *language* L as a subset of $\{0, 1\}^*$. The class Σ_p^k contains all languages L such that there is a Turing machine V (the "verifier") and polynomial p such that for all $x \in \{0, 1\}^*$, it holds $x \in L$ iff

$$\exists y_1 \in \{0, 1\}^{p(|x|)} \forall y_2 \in \{0, 1\}^{p(|x|)} \ldots Q y_k \in \{0, 1\}^{p(|x|)} \ V(x, y_1, y_2, \ldots, y_k) = 1,$$

where $Q = \exists$, if k odd, and $Q = \forall$, else. This family of classes is part of the *polynomial-time hierarchy* defined by Stockmeyer [18]. A *many-one reduction* (or *Karp reduction*) from some language L_1 to some language L_2 is defined as a function $f : \{0, 1\}^* \to \{0, 1\}^*$ such that $x \in L_1$ iff $f(x) \in L_2$ for all $x \in \{0, 1\}^*$. A language L_1 is Σ_p^k-hard if all languages $L_2 \in \Sigma_p^k$ can be reduced to L_1 via a polynomial time many-one reduction. A problem is Σ_p^k-complete if it is contained in Σ_p^k and Σ_p^k-hard. The canonical complete problems for Σ_p^k are $\exists_1 \forall_2 \ldots \exists_k \text{CNF-SAT}$, for odd k, and $\exists_1 \forall_2 \ldots \forall_k \text{DNF-SAT}$, for even k.

3   Combinatorial Problem Framework

In theoretical computer science, problems are defined as languages, which consist of all YES-instances of the problem. The instances are encoded as words from $\{0, 1\}^*$. For combinatorial problems, we may assume that an instance contains a universe $U = \{1, \ldots, n\}$, which consists of the encoding atoms of the instance. Furthermore, an instance includes (nested) relations between these atoms. To encode the relations, the atoms are used together with a delimiter symbol.

One example of such a problem is the problem UNDIRECTED s-t-CONNECTIVITY (USTCON). Its input is an undirected graph $G = (V, E)$ together with two vertices $s, t \in V$. The corresponding instance is then encoded by the vertices $V = U$ as universe and three relations $s, t \subseteq V$ and $E \subseteq V \times V$. The instance is a YES-instance iff there is path from s to t in G. Another example is the problem

VERTEX COVER. Again, the vertices $V = U$ are the universe and $E \subseteq V \times V$ is a relation. The instance is a YES-instance iff there is a small vertex cover in G.

In mathematical optimization, a problem is often defined over its feasible solutions F together with a cost function c. The goal is then to find a solution that achieves the minimum (resp. maximum) of all feasible solutions. Oftentimes, an additional ground set of combinatorial elements X is given. For simplicity, the feasible solutions are then combinations of that ground set, that is $F(X) \subseteq 2^X$. We apply this to USTCON by interpreting the edges as the ground set $X = E$ and all paths $F(X) \subseteq 2^E$ from s to t as the feasible solutions. For VERTEX COVER, we define the vertices as ground set $X = V$ and the feasible solutions $F(X) \subseteq 2^V$ are all small vertex covers in the graph. For simplicity, we ignore cost or weight functions and ask for the mere existence of a solution (here: a path, a small vertex cover).

While this is not a general definition, many typical combinatorial problems can be defined this way such as INDEPENDENT SET (an independent set is a subset of vertices), HAMILTONIAN PATH (a Hamiltonian path is a subset of edges), SUBSET SUM (a solution for subset sum is a subset of numbers).

We distinguish the natural encoding universe U from the solution ground set X over which the solutions are defined. Thereby, we reach a larger class of problems. In VERTEX COVER, the encoding universe $U = V$ is the same as the solution ground set $X = V$, because a vertex cover is a set of vertices and a graph is a set of vertices which are in relation via edges. In contrast, the instances of USTCON are still graphs while the solutions are subsets of edges. Thus for USTCON, the solution ground set and the universe do not coincide.

We begin with the definition of nested relations in order to define the instances of combinatorial problems. With these nested relations, we are able to define all possible associations of universe elements as well as between universe elements and relational elements. Thus in a graph $G = (V, E)$, we are not only able to for example encode edges $E \subseteq V \times V$ but also an incidence relation $I \subseteq V \times E$ or the neighborhood relation $N \subseteq V^{\leq |V|}$.

Definition 1 (Nested Relations). *Let U be a set. Then $\mathcal{R}(U)$ is the set of nested relations over U defined by the smallest set fulfilling:*

$$U \in \mathcal{R}(U) \tag{1}$$

$$A \in \mathcal{R}(U), \qquad \text{if } A \subseteq B \text{ for some } B \in \mathcal{R}(U) \tag{2}$$

$$\bigtimes_i A_i \in \mathcal{R}(U), \qquad \text{if for all } i, \ A_i \in \mathcal{R}(U) \tag{3}$$

We denote the set of relation elements that include $r \in A \in \mathcal{R}(U)$ by $R(r)$.

With access to all nested relations over the universe, we are able to define not only a variety of problems but we are also able to meaningfully define gadget reductions between problems. The solution ground set $X = R$ is then a subset of relational elements of one (nested) relation $R \in \mathcal{R}(U)$ over the gadget reduction universe U. Thus the solutions are of the form $F(R) \subseteq 2^R$.

Definition 2 (Combinatorial Decision Problem). *A combinatorial deci-sion problem P_A is a set of tuples $(U_A, R_A, F_A(R_A))$ with the set of universe elements U_A, relations $R_A \in \mathcal{R}(U_A)^r, r \in \mathbb{N}$, and the set of feasible solutions $F_A(R_A) \subseteq 2^{R_A^i}$ for some $1 \le i \le r$. We assume that $R_A^1 = U_A$. We call R_A the instance of the problem and R_A is a YES-instance if and only if $F_A(R_A) \neq \emptyset$. We use an index set I_A to easily address the members of the tuple R_A.*

For simplicity, we may omit the problem in the index of U_A, R_A and $F_A(R_A)$ as well as the dependence of the feasible solutions $F(R)$ on the relations R and write F. For a better understanding, we again use USTCON as an example.

Example 1 (Undirected s-t-Connectivity Problem). The *input* of USTCON is a graph $G = (V, E)$ and two vertices $s, t \in V$. A *feasible solution* is a path from s to t in G. This translates to the following tuple (U, R, F). The universe U consists of the vertices V. The relations in R are the edges E and the vertices s and t, that is, $R = (V, E, s, t)$. The feasible solutions are all s-t-paths $p \in F \subseteq 2^E$ in G defined as subsets of edges.

Observe that for combinatorial problems, the encoding of the input and the solutions depends only on the universe of elements. Thus, the universe elements in U build the atoms of the problem. The (nested) relations R model the relations between these atoms. The feasible solutions F model all possible combinations of solution elements that are feasible.

3.1 Scenarios for Robust Problems

Before we are able to define recoverable robust problems, we need to define scenarios. Scenarios are a central concept in robust optimization, which model the uncertainty. A Hamming distance recoverable robust problem P_A^{HDRR} is based on a combinatorial problem P_A. We then define a scenario as follows.

Definition 3 (Scenarios). *A scenario of the problem P_A^{HDRR} is a problem instance $(U_A, R_A, F_A(R_A))$ of the problem P_A.*

Encoding of Scenarios. For scenarios, we use explicit encodings, implicit encod-ings or succinct encodings. We consider elemental uncertainty, for which it is uncertain whether a combinatorial element is part of a scenario or not. Thus, all of these encodings are based on combinatorial elements of an instance, which include the universe and all relation elements. This is different to uncertainty over the costs of elements, where the underlying combinatorial elements remain the same for all scenarios. If a combinatorial element is not part of a scenario, then all relation elements that include this combinatorial element are discarded as well in the scenario. For example, if a vertex v in a graph problem is dis-carded, then all edges incident to v are discarded, too. We denote this removal of combinatorial elements with $U \setminus \{r\}$ and $R \setminus R(r)$, whereby the removal of r removes all relation elements $R(r)$ that contain r. We call the elements that are part of the current scenario the *active* elements, otherwise we call the elements *inactive*.

First, we will use explicit encodings by providing the complete instance encoding over the base problem P_A. Additionally, we use implicit encodings by providing a set of all elements that are different from base scenario σ_0. Furthermore, we address succinct encodings of scenarios as well. These encodings usually encode an exponential number of scenarios in polynomial space. The well-known Γ-scenarios fall into this last category as well as later defined *xor*-dependencies, which use logical operators between the elements to encode which element is active, i.e. part of a scenario.

3.2   Hamming Distance Recoverable Robust Problems

Now, we define Hamming distance recoverable robust problems. For this, we need a definition of the Hamming distance over a set.

Definition 4 (Hamming Distance of Sets). *Let A, B be two sets. Then, we define the Hamming distance $H(A, B)$ of set A and B to be*

$$H(A, B) := |A \triangle B| = |\{x \mid \text{either } x \in A \text{ or } x \in B\}|$$

Intuitively, a Hamming distance recoverable robust problem P_A^{HDRR} is based on a normal combinatorial decision problem P_A, e.g. UstCon. We distinguish the *base scenario* from *recovery scenarios*. The base scenario σ_0 is the instance on which the first solution s_0 has to be computed. The recovery scenarios $\sigma \in S$ are the scenarios for which the solution s, that has to be adapted from s_0, have to be computed. All scenarios of a problem may share universe elements or relation elements. In conclusion, we not only have to find a solution for one instance, but for one base scenario σ_0 and for all recovery scenarios in S. That is, we can recover from every possible scenario with a new solution to the problem. The recovery solutions, nonetheless, may have a Hamming distance of at most κ to the solution of the base scenario. We always define the Hamming distance over the solution ground set X between the solutions from $F(X) \subseteq 2^X$. Formally, we obtain the following definition.

Definition 5 (Hamming Distance Recoverable Robust Problem). *A Hamming distance recoverable robust problem P_A^{HDRR} is a combinatorial problem based on a combinatorial problem P_A. P_A^{HDRR} is defined as a set of tuples $(U, R, F(R))$ with*

$U = U_0 \cup \bigcup_{\sigma \in S} U_\sigma$ *is the universe. The universe is the union over all universe elements that occur in the scenarios.*

$R = (R_0, (R_\sigma)_{\sigma \in S}) = ((U_0, R_0^2, \dots, R_0^r), (U_\sigma, R_\sigma^2, \dots, R_\sigma^r)_{\sigma \in S})$ *are the relations. The relations are separate for each scenario.*

$F(R) = \{(s_0, (s_\sigma)_{\sigma \in S}) \in F_0(R_0) \times (F_\sigma(R_\sigma))_{\sigma \in S} \mid H(s_0, s_\sigma) \leq \kappa \text{ for all } \sigma \in S\})$ *are the feasible solutions. The Hamming distance $H(s, s')$ is defined over the elements in the solutions s, s'.*

The feasible solutions are not subsets of some relation R but consist of tuples including the solution for each scenario in F, which also adhere to the Hamming distance. In general, we assume that the bound on the Hamming distance κ is part of the input. Observe that the specifications are no restriction because every decision problem can be formulated as one base scenario and no recovery scenarios, that is $S = \emptyset$. On the other hand, the base problem P_A is a restriction of P_A^{HDRR} by setting $S = \emptyset$. Furthermore, the base scenario is defined by $\sigma_0 = (U_0, R_0, F_0)$ and all uncertainty scenarios $\sigma \in S$ are defined by $\sigma = (U_\sigma, R_\sigma, F_\sigma)$. Again, we provide an example for a better understanding of the definition and again, we use the UstCon problem.

Example 2 (Hamming Distance Recoverable Robust UstCon). Let $G = (V, E)$ be a graph, $s, t \in V$ and $\kappa \in \mathbb{N}$. UstCon^{HDRR} is a Hamming distance recoverable robust problem with feasible solutions $F \subseteq 2^E$. Thus, the Hamming distance is defined over the edges. The start and end vertices s and t remain the same for all scenarios. The input R contains the following: Each scenario $\sigma \in S$ encodes the set of active vertices V_σ and edges E_σ. The feasible solutions F consists of all s-t-paths $(p_0, p_{\sigma \in S}) \in 2^{E_{\sigma_0}} \times 2^{E_{\sigma \in S}}$ such that $H(p_0, p_\sigma) \leq \kappa$, for all $\sigma \in S$. In other words, the question is

$$\exists p_0 \in 2^{E_{\sigma_0}} : \forall \sigma \in S : \exists p_\sigma \in 2^{E_\sigma} : p_0 \in F_0, \; p_\sigma \in F_\sigma \text{ and } H(p, p_\sigma) \leq \kappa.$$

4 Recoverable Robust Problems with Polynomially Computable Scenario Encodings

We now consider problems with polynomially computable scenario encodings. A scenario encoding is polynomially computable if the set of scenarios is transformable into a set of explicitly encoded instances in polynomial time. (Consequently, the number of scenarios is bounded by a polynomial.)

Lemma 1. *Let $P_A \in NP$. Then $P_A^{HDRR} \in NP$ if the set of scenarios S of P_A^{HDRR} is polynomially computable.*

Besides general polynomially computable scenarios, we may consider the popular concept of Γ-scenarios. These consist of all scenarios that deviate in at most Γ many elements from the base instance corresponding to a set of activatable elements. If Γ is constant, we may use Lemma 1 to obtain the following result.

Corollary 1. *Let $P_A \in NP$. Then $P_A^{HDRR} \in NP$ if the set of scenarios S of P_A^{HDRR} consists of all possible Γ-scenarios for a constant Γ.*

Theorem 1 follows directly from by Lemma 1 by reusing the original reduction to P_A and setting the scenario set $S = \emptyset$.

Theorem 1. *Let P_A be an NP-complete problem. Then, P_A^{HDRR} is NP-complete if the set of scenarios S of P_A^{HDRR} is polynomially computable.*

4.1 Reduction for Undirected S-t-Connectivity

Theorem 2. *There is a deterministic logarithmic space computable reduction from* 3-SATISFIABILITY *to* USTCON^{HDRR} *with one base and one recovery scenario. In short,* 3-SATISFIABILITY $\leq_L$ USTCON^{HDRR}

5 Recoverable Robust Problems and the Polynomial Hierarchy

In this section, we investigate the connection between multi-stage Hamming distance recoverable robust problems and the polynomial hierarchy. For this, we introduce two succinct encodings: xor-dependencies and Γ-set scenarios. We first prove that the Hamming distance recoverable robust version of problems, which are in *NP*, are in Σ_3^p for both encodings. Then, we prove the hardness of the Hamming distance recoverable robust 3-SATISFIABILITY for both encodings.

Definition 6 (Hamming Distance Recoverable Robust 3-Satisfiability). *The problem* 3-SATISFIABILITY^{HDRR} *with Hamming distance over the literals L is defined as follows.*

Input: *Literals L, clauses C, base scenario $\sigma_0 \subseteq L$, recovery scenarios $S \subseteq 2^L$, $\kappa \in \mathbb{N}$*
Question: *Are there solutions $s_0 \subseteq \sigma_0$ and $s_\sigma \subseteq \sigma$ for all $\sigma \in S$ such that $H(s_0, s_\sigma) \leq \kappa$ for all $\sigma \in S$ and setting s_0 and s_σ to true, all corresponding formulae of clauses $C|_{\sigma_0}$ and $C|_\sigma$ are satisfied?*

At last, we extend these results to the multistage recoverable robustness case by showing the Σ_{2m+1}^p-completeness of the Hamming distance recoverable robust 3-SATISFIABILITY with m recovery stages. We begin with the xor-dependency scenarios.

Definition 7 (xor-Dependency Scenarios). *Let $S = (E', \{(E_{1,1}, E_{1,2}), \ldots, (E_{n,1}, E_{n,2})\})$ be the scenario-encoding, whereby E' and $E_{i,j}$ are pairwise disjoint sets of combinatorial elements for all i, j. Then the corresponding scenario set S includes all σ of the form $\sigma = E' \cup E_1 \cup \ldots \cup E_n$ with either $(E_i = E_{i,1})$ or $(E_i = E_{i,2})$ for all $i = 1, \ldots, n$ and E' is a fixed set of combinatorial elements, which are activated (resp. deactivated) in all recovery scenarios iff the element is inactive (resp. active) in the base scenario.*

Observe that with a linear sized encoding, exponentially many scenarios may be encoded. We study this combinatorial explosion with the result that it introduces more complexity for Hamming distance recoverable robust problems. For this, we use 3-SATISFIABILITY as base problem and show the Σ_3^p-hardness of 3-SATISFIABILITY^{HDRR} with a linear number of xor-dependencies. Furthermore, we show also that if $P_A \in NP$, then P_A^{HDRR} with a linear number of xor-dependencies is in Σ_3^p.

Theorem 3. *If $P_A \in NP$, then P_A^{HDRR} with xor-dependencies is in Σ_3^p.*

Theorem 4. *3-SATISFIABILITY^{HDRR} with xor-dependency scenarios is Σ_3^p-hard.*

While the other parts of the paper are developed independent from Goerigk et al. [12], the results for Γ-set scenarios are built upon it. The results based on xor-dependencies are adaptable to the Γ-set scenarios as described in this section. For the Γ-set scenarios, we use the definition over sets instead of elements as in Γ-scenarios, which is defined as follows.

Definition 8 (Γ-set Scenarios). *Let $S = (E', \{E_1, E_2, \dots E_n\})$ be the scenario-encoding, whereby E' and E_i are pairwise disjoint sets of combinatorial elements for all i. Then, the corresponding scenario set S includes all σ of the form $\sigma = E' \cup \bigcup_{E \in \mathcal{E}} E$ with $\mathcal{E} \subseteq \{E_1, E_2, \dots, E_n\}$, $|\mathcal{E}| \leq \Gamma$, and E' is a fixed set of combinatorial elements, which are activated (resp. deactivated) in all recovery scenarios iff the element is inactive (resp. active) in the base scenario.*

Again, with a linear sized encoding, exponentially many scenarios may be encoded. We show the Σ_3^p-hardness of 3-SATISFIABILITY^{HDRR} with Γ-set scenarios. A proof on the so-called ROBUST ADJUSTABLE SAT was already conducted by Goerigk et al. [12]. This version of 3-SATISFIABILITY uses uncertainties over the costs instead of the elements as in Σ_3^p-hardness of 3-SATISFIABILITY^{HDRR} with Γ-set scenarios. Thus, the proof is not analogous as it is different in technicalities, nevertheless, we reuse their basic idea of introducing the s-variables for our proof. Furthermore, we show also that if $P_A \in NP$, then P_A^{HDRR} with Γ-set scenarios is in Σ_3^p.

Theorem 5. *If $P_A \in NP$, then P_A^{HDRR} with Γ-set scenarios is in Σ_3^p.*

Theorem 6. *3-SATISFIABILITY^{HDRR} with Γ-set scenarios is Σ_3^p-hard.*

5.1   Multi-stage Recoverable Robustness

In *multi-stage recoverable robustness*, the uncertainty is not only modeled by one set of scenarios but multiple sets that are connected inductively.

Definition 9 (Multi-stage Recoverable Robust Problem). *A multi-stage recoverable robust problem with m recoveries $P_A^{m\text{-}HDRR}$ is inductively defined as*

$$P_A^{m\text{-}HDRR} := P_A \qquad\qquad \text{for } m = 0,$$
$$P_A^{m\text{-}HDRR} := (P_A^{(m-1)\text{-}HDRR})^{HDRR} \qquad\qquad \text{for } m > 1.$$

The complexity results naturally extend to the multiple recoverable robustness concept. We make use of the inductive nature of the definition by proving the following theorems by induction. For this, we reuse Theorems 3 to 6 as induction base.

Theorem 7. *3-SATISFIABILITY$^{m\text{-}HDRR}$ with xor-dependency scenarios is Σ_{2m+1}^p-complete. 3-SATISFIABILITY$^{m\text{-}HDRR}$ with Γ-set scenarios is Σ_{2m+1}^p-complete.*

6 Classes of Recoverable Robust Problems

We have shown that 3-SATISFIABILITY^{HDRR} is the canonical Σ_3^p-complete Hamming distance recoverable robust problem. The goal is to "upgrade" the existing reductions on the NP-level to reduce the corresponding Hamming distance recoverable robust problems to each other. If we are additionally able to guarantee transitivity, we are also able to easily achieve complexity results for a large class of problems. Essentially, the reduction between Hamming distance recoverable robust problems needs to preserve the structure of the scenarios. For this, consider problems P_A and P_B. We need to achieve that a combinatorial element e_A in P_A is active if and only if the combinatorial elements E_B, to which e_A is mapped in P_B, are active. Then, we can use this one-to-many correspondence to (de)activate the corresponding elements in the instance of P_B.

Many of the properties from above are already constituted by the informal concept of gadget reductions. Gadget reductions describe that each part of the problem P_A is mapped to a specified part of the problem P_B that inherits the behavior in problem P_A. We adjust this concept to combinatorial elements, that is universe elements and relation elements, for our purpose. The goal is that a gadget is a subset of combinatorial elements in P_B for every combinatorial element in P_A. Thereby, we preserve the (in)activeness of elements in a scenario. We call reductions that fulfill this property *modular* in the sense that all gadgets are easily (de)activatable. Furthermore, the solution size, which is the number of universe elements in a solution, has to adapt accordingly while being easy to compute in order to define the Hamming distance in the reduction correctly. We approach this later by demanding that the solution size of every gadget has to be a constant.

6.1 Universe Gadget Reduction

Let P_A be a combinatorial decision problem with instance tuples (U_A, R_A, F_A) and P_B a combinatorial decision problem with instance tuples (U_B, R_B, F_B). A Universe Gadget Reduction $f_{\preceq}$ that many-one-reduces P_A to P_B is composed of a (possibly empty) constant gadget Y_{const}, which is the same for every instance, and of the independent mappings: $f_{R_A^i, R_B^j} : R_A^i \to 2^{R_B^j}$ for all $(i, j) \in I_A \times I_B$. We, then, call the substructure

$$Y_x = f_{\preceq}(x) = \bigcup_{(i,j)\, \in\, I_A \times I_B} f_{R_A^i, R_B^j}(x)$$

the gadget for the specific universe element or relation element $x \in \bigcup_i R_A^i$. Additionally, we denote the set of all gadgets by $\Upsilon(R_A) = \{Y_r \mid r \in R_A\} \cup \{Y_{const}\}$ for the instance R_A. The mappings must fulfill the following properties.

1. Pre-image uniqueness: Let $y \in R_B^j$ for some $j \in I_B$, then either $y \in Y_{const}$ or there is exactly one $(i, j) \in I_A \times I_B$ and exactly one $x \in R_A^i$ such that $y \in f_{R_A^i, R_B^j}(x)$.

2. Modularity: If a combinatorial element $r \in R_A^i$ from (U_A, R_A, F_A) is removed to form a new instance (U_A', R_A', F_A'), the removal of the gadget of r in (U_B, R_B, F_B) induces a correct reduction instance (U_B', R_B', F_B'). A removal of $r \in R_A^i$ corresponds to the substitution by a (possibly empty) removal gadget Y_r^{rem} in P_B:

$$f_{\preceq}(R_A \setminus R(r)) = (R_B \setminus f_{\preceq}(R(r))) \cup Y_r^{rem}.$$

If the removal gadget is empty for all combinatorial elements, we call the modularity *strong*, otherwise *weak*. We substitute the gadgets Y_x, for $x \in R(r)$, with the removal gadget Y_r^{rem} in $\Upsilon(R_A)$ correspondingly.

This definition of a gadget reduction for combinatorial decision problems ensures that the gadgets are uniquely relatable to the generating combinatorial elements and are modular. Note that only combinatorial elements from P_A can be removed such that the new instance P_A' is a validly encoded instance. That is, combinatorial elements cannot be removed in general as this may void the validity of the instance, e.g. in USTCON the universe elements s and t cannot be deleted.

Additionally, the solution size has to adapt to the modularity of the gadgets in the universe gadget reduction. That is, if a combinatorial element in P_A is removed such that the corresponding gadgets in P_B is removed, the solution size of the instance of P_B is well-defined. For the sake of simplicity and because we later only use reductions originating from 3-SATISFIABILITY, we define this property only for 3-SATISFIABILITY.

Solution Size. In order to correctly define the Hamming distance κ for a reduction from a problem P_A^{HDRR} to P_B^{HDRR} based on a universe gadget reduction from P_A to P_B, we need to find a solution size function. We demand that each gadget $Y \in \Upsilon$ has a constant *local solution size*, which is defined by the universe gadget reduction. A Yes-instance has a solution size, which is defined by the sum of all local solution sizes defined as follows.

Definition 10 (3-Satisfiability-Reduction Solution Size Function). *Let P_B be a problem such that a universe gadget reduction f from 3-SATISFIABILITY to P_B exists. Let (L, C) be a 3-SATISFIABILITY-instance. The gadgets have a local solution size of $size(Y)$ for each $Y \in \Upsilon(L, C)$. The function*

$$size_f : 3\text{-SAT} \to \mathbb{N} : (L, C) \mapsto \sum_{Y \in \Upsilon(L,C)} size(Y)$$

describes the target solution size over universe elements of $f((L, C)) = R_B$ for R_B to be a YES-instance of P_B.

In the following, we only consider universe gadget reductions that have such a solution size function, which is additionally polynomial time computable. While this is necessary, it is not a serious restriction as we see later. Most of the reductions inherently have this property.

6.2   Properties of Universe Gadget Reductions

The definitions of universe gadget reductions and its solutions size function imply the following three properties, which are specifically desired as illustrated before.

Lemma 2. *A universe gadget reduction is total and one-to-many. The inverse to a universe gadget reduction is many-to-one.*

Thus by definition, it is ensured that each element $y \in Y_{const} \cup \bigcup_j R_B^j$ of P_B is left unique and thus belongs to exactly one gadget.

Lemma 3. *Polynomial Universe gadget reductions are transitive. That is, if there are problems P_A, P_B, P_C with $P_A \preceq_p^{UGR} P_B$ and $P_B \preceq_p^{UGR} P_C$, then $P_A \preceq_p^{UGR} P_C$.*

Furthermore, the solution size function adheres to the modularity of the universe gadget reduction.

Lemma 4. *The solution size function is modular. In other words, let (L, C) and (L', C') be instances of 3-SATISFIABILITY with $L' \subseteq L$ and $C' \subseteq C$. Furthermore, let f be a universe gadget reduction from 3-SATISFIABILITY to P_B such that $f(L', C')$ results from $f(L, C)$ by removing the corresponding gadgets. Then,*

$$size_f(L', C') = \sum_{Y \in \Upsilon(L', C')} size(Y).$$

Now, we present a general reduction from 3-SATISFIABILITY based on the structure that a universe gadget reduction provides. If there is a polynomial time universe gadget reduction f from 3-SATISFIABILITY to P_B with a corresponding polynomial time solution size function $size_f$, then there is a polynomial time reduction for the Hamming distance recoverable robust version of P_B with Hamming distance over the universe elements, transforming the scenarios accordingly.

Theorem 8. *If 3-SATISFIABILITY $\preceq_p^{UGR} P_B$ and there is a corresponding solution size function, then 3-SATISFIABILITY$^{m\text{-}HDRR} \preceq_p P_B^{m\text{-}HDRR}$, where the Hamming distance is defined over the universe elements and the scenario encodings are either xor-dependency scenarios or Γ-set scenarios.*

With these structural properties in mind, we can construct a whole set of Hamming distance recoverable robust problems. Note that the transitivity of the universe gadget reduction can be used to deduce further reductions.

6.3   Gadget Reductions for Various Combinatorial Decision Problems

In this section, we examine various but not all problems that are universe gadget reducible from 3-SATISFIABILITY. The reductions are all well-known results or modifications of well-known results. We adapt these results to the universe gadget reduction framework to indicate that Theorem 8 is a general statement. We prove the following theorem by showing that a universe gadget reduction from 3-SATISFIABILITY exists for all the problems. For this, we use the transitivity of the reductions as illustrated in Fig. 1.

Theorem 9. *The m-Hamming distance recoverable robust version of the following problems are NP-complete with polynomially computable scenarios and Σ_{2m+1}^p-complete with xor-dependency scenarios or Γ-set-scenarios:* VERTEX COVER, DOMINATING SET, FEEDBACK ARC SET, FEEDBACK VERTEX SET, HITTING SET, INDEPENDENT SET, CLIQUE, SUBSET SUM, KNAPSACK, PARTITION, TWO MACHINE SCHEDULING, (UN)DIRECTED HAMILTONIAN CYCLE, (UN)DIRECTED HAMILTONIAN PATH, TRAVELING SALESMAN, 3-DIMENSIONAL MATCHING, EXACT COVER BY 3-SETS, K-DISJOINT DIRECTED PATH $(k \geq 2)$.

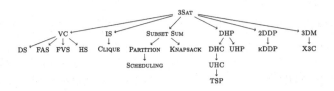

Fig. 1. The tree of gadget reductions for all considered problems.

7   Conclusion

We have defined Hamming distance recoverable robust problems with elemental uncertainty and applied this concept to various well-known problems in *NP*. Further, we have defined universe gadget reductions to build a framework for a large class of Hamming distance recoverable robust problems. The complexity results are that the Hamming distance recoverable robust versions of *NP*-complete problems remain *NP*-complete if the scenarios are polynomially computable and that the *NP*-complete problems are Σ_3^p-complete for *xor*-dependency scenarios and Γ-set scenarios if 3-SATISFIABILITY is universe gadget reducible to them and a corresponding solution size function exists. Furthermore, multi-stage problems with m stages result in Σ_{2m+1}^p-completeness if the encoding of scenarios are *xor*-dependency scenarios or Γ-set scenarios.

Remaining interesting questions are whether there is a (light-weight) reduction framework for other adversarial problems or robustness concepts, for example for interdiction problems or two-stage adjustable problems, to derive completeness for higher levels in the polynomial hierarchy than *NP*. Furthermore, it is of interest whether this concept is adaptable to problems with cost uncertainty and for other distance measures. A more special question is, which succinct encodings also result in Σ_3^p-completeness or if there are succinct encodings which result in the *NP*-completeness of the problem.

References

1. Agrawal, M., Allender, E., Impagliazzo, R., Pitassi, T., Rudich, S.: Reducing the complexity of reductions. Comput. Complex. **10**(2), 117–138 (2001)
2. Bold, M., Goerigk, M.: Investigating the recoverable robust single machine scheduling problem under interval uncertainty. Discret. Appl. Math. **313**, 99–114 (2022)
3. Büsing, C.: Recoverable robust shortest path problems. Networks **59**(1), 181–189 (2012)
4. Büsing, C., Goderbauer, S., Koster, A.M.C.A., Kutschka, M.: Formulations and algorithms for the recoverable Γ-robust knapsack problem. EURO J. Comput. Optim. **7**(1), 15–45 (2019)
5. Büsing, C., Koster, A.M.C.A., Kutschka, M.: Recoverable robust knapsacks: Γ-scenarios. In: Pahl, J., Reiners, T., Voß, S. (eds.) INOC 2011. LNCS, vol. 6701, pp. 583–588. Springer, Heidelberg (2011). https://doi.org/10.1007/978-3-642-21527-8_65
6. Büsing, C., Koster, A.M.C.A., Kutschka, M.: Recoverable robust knapsacks: the discrete scenario case. Optim. Lett. **5**(3), 379–392 (2011)
7. Chassein, A.B., Goerigk, M.: On the recoverable robust traveling salesman problem. Optim. Lett. **10**(7), 1479–1492 (2016)
8. Cicerone, S., Stefano, G.D., Schachtebeck, M., Schöbel, A.: Multi-stage recovery robustness for optimization problems: a new concept for planning under disturbances. Inf. Sci. **190**, 107–126 (2012)
9. Dijk, B., Santos, B.F., Pita, J.P.: The recoverable robust stand allocation problem: a GRU airport case study. OR Spectr. **41**(3), 615–639 (2019)
10. Dourado, M.C., Meierling, D., Penso, L.D., Rautenbach, D., Protti, F., de Almeida, A.R.: Robust recoverable perfect matchings. Networks **66**(3), 210–213 (2015)
11. Froyland, G., Maher, S.J., Wu, C.: The recoverable robust tail assignment problem. Transp. Sci. **48**(3), 351–372 (2014)
12. Goerigk, M., Lendl, S., Wulf, L.: On the complexity of robust multi-stage problems in the polynomial hierarchy. CoRR abs/ arXiv: 2209.01011 (2022)
13. Grüne, C.: The complexity classes of hamming distance recoverable robust problems. CoRR abs/ arXiv: 2209.06939 (2022)
14. Hradovich, M., Kasperski, A., Zielinski, P.: Recoverable robust spanning tree problem under interval uncertainty representations. J. Comb. Optim. **34**(2), 554–573 (2017)
15. Liebchen, C., Lübbecke, M., Möhring, R., Stiller, S.: The concept of recoverable robustness, linear programming recovery, and railway applications. In: Ahuja, R.K., Möhring, R.H., Zaroliagis, C.D. (eds.) Robust and Online Large-Scale Optimization. LNCS, vol. 5868, pp. 1–27. Springer, Heidelberg (2009). https://doi.org/10.1007/978-3-642-05465-5_1
16. Lusby, R.M., Larsen, J., Bull, S.: A survey on robustness in railway planning. Eur. J. Oper. Res. **266**(1), 1–15 (2018)
17. Maher, S.J., Desaulniers, G., Soumis, F.: Recoverable robust single day aircraft maintenance routing problem. Comput. Oper. Res. **51**, 130–145 (2014)
18. Stockmeyer, L.J.: The polynomial-time hierarchy. Theor. Comput. Sci. **3**(1), 1–22 (1976)
19. Trevisan, L., Sorkin, G.B., Sudan, M., Williamson, D.P.: Gadgets, approximation, and linear programming (extended abstract). In: 37th Annual Symposium on Foundations of Computer Science, FOCS 1996, Burlington, Vermont, USA, 14–16 October, 1996, pp. 617–626. IEEE Computer Society (1996)

Geometric Thickness of Multigraphs is ∃ℝ-Complete

Henry Förster[1]([✉])[iD], Philipp Kindermann[2][iD], Tilmann Miltzow[3][iD],
Irene Parada[4][iD], Soeren Terziadis[5][iD], and Birgit Vogtenhuber[6][iD]

[1] Wilhelm-Schickard-Institut für Informatik, Universität Tübingen,
Tübingen, Germany
henry.foerster@uni-tuebingen.de
[2] FB IV - Computer Science, Trier University, Trier, Germany
[3] Department of Information and Computing Sciences, Utrecht University,
Utrecht, Netherlands
[4] Departament de Matemàtiques, Universitat Politècnica de Catalunya,
Barcelona, Spain
[5] Institute of Logic and Computation, Technische Universität Wien , Vienna, Austria
[6] Institute of Software Technology, Graz University of Technology, Graz, Austria

Abstract. We say that a (multi)graph $G = (V, E)$ has geometric thickness t if there exists a straight-line drawing $\varphi : V \to \mathbb{R}^2$ and a t-coloring of its edges where no two edges sharing a point in their relative interior have the same color. The GEOMETRIC THICKNESS problem asks whether a given multigraph has geometric thickness at most t. In this paper, we settle the computational complexity of GEOMETRIC THICKNESS by showing that it is ∃ℝ-complete already for thickness 57. Moreover, our reduction shows that the problem is ∃ℝ-complete for 8280-planar graphs, where a graph is k-planar if it admits a topological drawing with at most k crossings per edge. In this paper we answer previous questions on geometric thickness and on other related problems, in particular that simultaneous graph embeddings of 58 edge-disjoint graphs and pseudo-segment stretchability with chromatic number 57 are ∃ℝ-complete.

1 Introduction

The thickness of a graph G is the minimum number of planar subgraphs whose union is G. It is an old concept; it was formally introduced by Tutte [24] in 1963, but the concept of biplanarity (i.e., geometric thickness 2) had already appeared before, most relevantly, in connection with two open problems: First, Ringel's Earth-Moon problem on the chromatic number of biplanar graphs [20] and second, a question by Selfridge, formulated by Harary, asking whether K_9 is biplanar [16]. In 1983, Mansfield [18] showed that deciding whether a graph is biplanar is NP-complete.

In this article we study the geometric or straight-line version of thickness, which requires that all planar subgraphs are embedded simultaneously with

© The Author(s), under exclusive license to Springer Nature Switzerland AG 2024
J. A. Soto and A. Wiese (Eds.): LATIN 2024, LNCS 14578, pp. 336–349, 2024.
https://doi.org/10.1007/978-3-031-55598-5_22

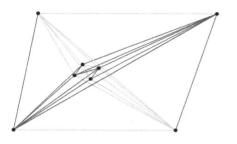 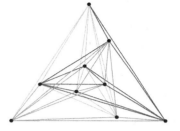

Fig. 1. The (geometric) thickness of K_8 is 2 and of K_9 is 3.

straight-line edges. More precisely, a multigraph $G = (V, E)$ has *geometric thickness* t if there exists a straight-line drawing $\varphi : V \to \mathbb{R}^2$ of G and a t-coloring of all the edges such that no two edges of the same color share a point other than a common endpoint. Figure 1 shows an illustration. Note that by definition, two edges connecting the same two endpoints must be assigned distinct colors in the t-coloring.

The concept of geometric thickness was first introduced by Kainen in 1973 [17] as *real linear thickness*. It was later named *geometric thickness* in Dillencourt, Eppstein, and Hirschberg's study [5] on the geometric thickness of complete and complete bipartite graphs which sparked an increased interest in the topic. Dillencourt et al. already asked about the computational complexity of GEOMETRIC THICKNESS, the problem of deciding whether given a graph G and a value t, G has geometric thickness at most t. Durocher, Gethner, and Mondal [10] partially answered this question by showing that GEOMETRIC THICKNESS is NP-hard even for geometric thickness 2.

Figure 1 shows straight-line drawings of K_8 and K_9 decomposed into two and three plane subgraphs, respectively; these bounds are tight for both the thickness and the geometric thickness. For $n > 10$ the thickness of K_n is $\lfloor \frac{n+2}{6} \rfloor$ [1,2] while the geometric thickness, for which no closed formula is known, is lower bounded by $\frac{n}{5.646}$ [5]. In general, the geometric thickness of a graph is not bounded by any function of its thickness [11].

Motivation. The GEOMETRIC THICKNESS problem consists of a combination of two computationally hard problems: splitting the edges into color classes and positioning the vertices. The first problem, where we are given the straight-line drawing and the goal is to decompose it into the minimum number of plane subgraphs, corresponds to a graph coloring problem for the corresponding segment intersection graph of the drawing. The 2022 Computational Geometry Challenge focused on this problem [14].

The second problem, when we are given the color classes and the goal is to position the vertices such that no two edges of the same color class intersect in their relative interior, is the SIMULTANEOUS GRAPH EMBEDDING problem, which we will formally define later. Both SIMULTANEOUS GRAPH EMBEDDING and GEOMETRIC THICKNESS connect to applications.

In network visualization, in particular of infrastructure, social, and transportation networks, one often has to deal with intersecting systems of connections belonging to different subnetworks. To represent them simultaneously, different visual variables such as colors are used to indicate edge classes. Drawing the edges with straight-line segments and removing/minimizing same-class crossings is often desirable for readability.

In some settings, the vertex positions can be freely chosen, while the edge classes are given. An example would be visualizing a system of communication channels (phone, email, messenger services) between a set of persons. This case corresponds to the SIMULTANEOUS GRAPH EMBEDDING problem. In other settings, both the positions and the edge classes can be freely chosen, which corresponds to the GEOMETRIC THICKNESS problem. It appears in applications such as VLSI design, where a circuit using uninsulated wires requires crossing wires to be placed on different layers.

Since its introduction as a natural measure of approximate planarity [5], geometric thickness keeps receiving attention in Computational Geometry. However, some fundamental questions remain open, including determining the geometric thickness of K_n and the complexity of GEOMETRIC THICKNESS. Moreover, the computational methods currently available are not able to provide straight-line drawings of low geometric thickness for graphs with more than a few vertices. Our results show that, even for small constant values of geometric thickness, computing such drawings is, under widely-believed computational complexity assumptions, harder than any NP-complete problem.

1.1   Results and Related Problems

In this section we state all our results, starting with our main result. To this end, we first need to introduce the complexity class $\exists\mathbb{R}$. The class $\exists\mathbb{R}$ can be defined as the set of problems that are at most as difficult as finding a real root of a multivariate polynomial with integer coefficients. A problem in $\exists\mathbb{R}$ is $\exists\mathbb{R}$-*complete* if it is as difficult as this problem. We give a more detailed introduction to $\exists\mathbb{R}$ in an extended version of this paper [15].

Theorem 1. GEOMETRIC THICKNESS *is $\exists\mathbb{R}$-complete for multigraphs already for any fixed thickness $t \geq 57$.*

Assuming $\exists\mathbb{R} \neq$ NP, Theorem 1 shows that GEOMETRIC THICKNESS is even more difficult than any problem in NP. This implies that even SAT solvers should not be able to solve GEOMETRIC THICKNESS in full generality. Even more, while a planar graph on n vertices can be drawn on an $n \times n$ grid, there are graphs with geometric thickness ≥ 57 that will need more than exponentially large integer coordinates for any drawing.

We prove Theorem 1 in Sect. 2 via a reduction from the problem PSEUDO-SEGMENT STRETCHABILITY, which we discuss next.

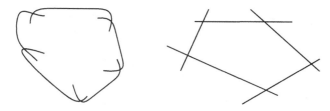

Fig. 2. A stretchable (right) and a non-stretchable (left) pseudo-segment arrangement with the same intersection pattern but different cyclic order around the intersections.

Pseudo-Segment Stretchability. Schaefer showed that the problem PSEUDO-SEGMENT STRETCHABILITY (see below for a definition) is ∃ℝ-complete [23]. We closely inspected the proof by Schaefer and observed some small extra properties that we use for our bound in Theorem 1. To state them we first introduce the corresponding definitions.

A *pseudo-segment arrangement* in $\mathbb{R}^2$ is an arrangement of Jordan arcs such that any two arcs intersect at most once. (A *Jordan arc* is a continuous injective curve from a closed interval to the plane.) A pseudo-segment arrangement $\mathcal{A}$ is called *stretchable* if there exists a segment arrangement $\mathcal{S}$ such that $\mathcal{A}$ and $\mathcal{S}$ are isomorphic. The curves can be encoded using a planar graph, as it completely determines the isomorphism type which describes the order in which each pseudo-segment intersects all other pseudo-segments. However, in contrast to (bi-infinite) pseudolines, this information is not enough. For example the pseudo-segment arrangement in Fig. 2 (left) cannot be stretched, although there exists a segment arrangement with the same intersection pattern (right). Additionally, we require the cyclic order around each intersection point to be maintained.

PSEUDO-SEGMENT STRETCHABILITY asks whether a given pseudo-segment arrangement is stretchable. Schaefer [23] showed that PSEUDO-SEGMENT STRETCHABILITY is ∃ℝ-hard, even if each pseudo-segment intersects at most 72 other pseudo-segments. We consider the intersection graph G of all the pseudo-segments, and we say a pseudo-segment arrangement has *chromatic number* χ if and only if the underlying intersection graph G has this chromatic number. We prove the following based on and strengthening the results of Schaefer [23] in an extended version [15].

Corollary 1. *PSEUDO-SEGMENT STRETCHABILITY is ∃ℝ-hard even for pseudo-segment arrangements of chromatic number 57.*

Simultaneous Graph Embedding. Our main result happens to also resolve a question by Schaefer about simultaneous graph embeddings in the sunflower setting [23]. Before stating the result, we need further definitions. Given simple graphs $G_1, \ldots, G_k$ on the same vertex set V, we say $\varphi : V \to \mathbb{R}^2$ is a *simultaneous graph embedding*, if the straight-line drawing of each G_i, $i = 1, \ldots, k$, on this vertex set is crossing-free. We say $G_1, \ldots, G_k$ form a *sunflower* if every edge is either in all graphs or in exactly one of the graphs. An edge that is present in all graphs is *in the center* of the sunflower or a *public* edge. Edges that are only in one graph are *private* edges or *belonging to a petal* of the sunflower. If all edges are private, we say that $G_1, \ldots, G_k$ form an *empty* sunflower. We define SIMULTANEOUS GRAPH EMBEDDING as the algorithmic problem with k graphs as input that asks if there exists a simultaneous graph embedding.

Theorem 2. *SIMULTANEOUS GRAPH EMBEDDING is* $\exists\mathbb{R}$-*complete already with* $k = 58$ *and* $G_1, \ldots, G_k$ *forming an empty sunflower.*

The proof can be found in Sect. 3. In a series of papers it was shown that SIMULTANEOUS GRAPH EMBEDDING is $\exists\mathbb{R}$-complete [4,13,22,23], already for $k = 240$. We contribute to this line of research by lowering the bound to 58 and by restricting it to families of graphs that form an empty sunflower. Lowering the number of graphs from 240 to 58 is not so significant as 240 did not appear to be tight bound before. The significance lies in the fact that we show $\exists\mathbb{R}$-hardness also for the case that the input graphs form an empty sunflower, which answers a question of Schaefer [23].

Relation Between Geometric Thickness and k-Planarity. In a k-*planar* drawing of a graph each edge can be crossed at most k times. A k-planar graph is a graph admitting such a drawing. For the class of 1-planar graphs, 2 is a tight upper bound for both the thickness and the geometric thickness. The upper bound for the thickness follows immediately from the degree of the intersection graph. The upper bound of 2 for the geometric thickness is non-trivial since not every 1-planar graph has a straight-line 1-planar drawing and was only recently shown [3]. This in particular means that the (geometric) thickness of 1-planar graphs can be decided in polynomial time. In contrast, we show at the end of Sect. 2 that determining the geometric thickness of 8280-planar multigraphs is $\exists\mathbb{R}$-complete.

Corollary 2. GEOMETRIC THICKNESS *is* $\exists\mathbb{R}$-*complete for* 8280-*planar multigraphs.*

Brandenburg [3] explicitly asked about the geometric thickness of k-planar graphs. A combination of previous results [7] shows that *simple* k-planar graphs have bounded geometric thickness: A queue layout of a graph G is a total vertex ordering and a partition of the edges into *queues* such that no queue contains two nested edges ad, bc such that $a < b < c < d$. The *queue number* $\mathrm{qn}(G)$ of G is the minimum number of queues in such a queue layout. First, geometric thickness is known to be bounded by queue number: any graph G with queue number $\mathrm{qn}(G)$ has geometric thickness $\theta(G) \leq 8\,\mathrm{qn}(G)^3$ [9, Corollary 9]. Second, it was recently shown that any k-planar graph G has bounded queue number [6,8], more precisely $\mathrm{qn}(G) \leq 2 \cdot 490^{k+2}$. Together, this yields:

Theorem 3. *Any k-planar graph has geometric thickness at most* $64 \cdot 490^{3(k+2)}$.

2 Geometric Thickness is ∃ℝ-Complete

In this section, we first show ∃ℝ-membership. Then we show ∃ℝ-hardness, splitting the proof into construction, completeness, and soundness.

∃ℝ-membership. To prove that GEOMETRIC THICKNESS is in ∃ℝ, we follow the characterization provided by Erickson, Hoog, and Miltzow [12]. According to this characterization, a problem P is in ∃ℝ if and only if there exists a real verification algorithm A for P that runs in polynomial time on the real RAM. Moreover, for every yes-instance I of P, there should be a polynomial-size witness w for which $A(I, w)$ returns "yes". For every no-instance I and any witness w, $A(I, w)$ should return "no".

We now describe a real verification algorithm A for GEOMETRIC THICKNESS. Given an instance $I = (G, t)$, the witness w consists of the coordinates of the vertices of G and a t-coloring of the edges $\chi : E \rightarrow [t]$.

A then verifies that there are no monochromatic crossings in the induced drawing by examining every pair of edges e and e' that cross. Furthermore, A checks that all edges with higher multiplicity receive distinct colors. This verification can be done on a real RAM in $O(|E|^2)$ time.

Hence, A runs in polynomial time on the real RAM, confirming that GEOMETRIC THICKNESS is in ∃ℝ.

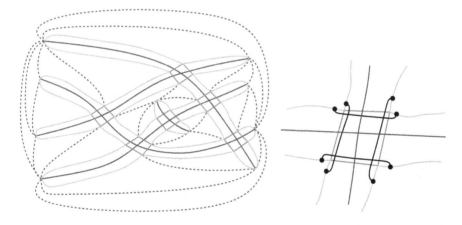

Fig. 3. Left: The pseudo-segment arrangement is drawn in red. The tunnel paths are drawn in green. The crossing boxes are drawn in blue and the blockers are not drawn. The connectors are drawn in purple (dotted). Right: Depiction for adding the blockers.

Construction. We reduce PSEUDO-SEGMENT STRETCHABILITY to GEOMETRIC THICKNESS. For an illustration of this reduction, we refer to Fig. 3. Let $\mathcal{A}$ be a given pseudo-segment arrangement with n pseudo-segments and let $t \geq 57$ be the geometric thickness we are aiming for. We construct a graph G as follows.

- For each pseudo-segment in $\mathcal{A}$, we add one *long edge* (just a single edge) with multiplicity 1.
- For each crossing, we add a *crossing box* (4-cycle) with multiplicity $t - 1$.
- We connect the endpoints of each long edge with the corresponding crossing boxes by *tunnel boundaries* (paths of length n) Tunnel boundaries also connect consecutive crossing boxes.
- We add four *blockers* (edges with multiplicity 1) to each crossing box, s.t. they connect corresponding tunnel paths around each crossing box and two blockers cross each pair of opposite crossing box edges.
- We add *connectors* (paths of length n) of multiplicity t between the endpoints of the long edges and the crossing boxes. First, we add a cycle of them between the endpoints of the long edges on the outer face. Then, we "triangulate" the faces that are bounded by tunnel boundaries and connectors by adding connectors such that each such face is incident to three vertices of degree larger than 2. In other words, each such face becomes a triangle after contracting edges incident to degree-2 vertices. Furthermore, we add the paths in such a way that no triangle contains both endpoints of a long segment.

This finishes the construction. Clearly, it can be done in polynomial time. Note that the illustrations of the graph G in Fig. 3 are meant to help to understand the definition of G. However, we do not know (yet) that the graph G needs to be embedded in the way depicted in the figures.

Completeness. Let S be an arrangement of (straight-line segments) such that S is isomorphic to $\mathcal{A}$. We need to show that there is a drawing of G such that we can color it with t colors avoiding monochromatic crossings.

We start by adding all the edges described in the construction. Note that all added parts are very long, thus, it is easy to see that they offer enough flexibility to realize each edge as a straight-line segment. Next we need to describe a coloring of all edges. The tunnel edges have multiplicity t, but they cross no other edges so this is fine. We know that we can color the long edges with t colors due to Corollary 1. Each crossing box edge has multiplicity $t - 1$. We use the $t - 1$ colors different from the color of the long edge crossing it. Each crossing box has four edges. The two opposite edges receive the same set of colors. The blockers receive the colors of the corresponding long edge that crosses the same crossing box edges. It can be checked that no two edges that cross or overlap have the same color.

Soundness. We now argue that if G has geometric thickness t then $\mathcal{A}$ is stretchable. Let Γ be a straight-line drawing of G with a t-coloring of E.

Frame. Let the *frame* H of G be the subgraph of G that consists only of the crossing boxes, tunnel boundaries, and connectors. Since all edges in H have multiplicity at least $t - 1$, no two edges may cross, so H has to be drawn planar. As a first step, we will argue that the frame of G has a unique combinatorial embedding. It is well-known that a planar graph has a unique combinatorial embedding if and only if it is the subdivision of a planar 3-connected graph [19].

Lemma 1. *The frame H of G has a unique combinatorial embedding.*

Proof. First, consider the *contracted frame* H^* obtained as follows; see Fig. 4a. First, contract each tunnel boundary and each connector to a single edge. Then, contract each crossing box to a single vertex and remove multi-edges that appear by the contraction. By construction of the connectors, the resulting graph H^* is a planar triangulated graph where each vertex v^* corresponds to the endpoint of a long edge or to a crossing box in H, and each edge e^* corresponds to either a connector or to two boundary paths in H. As H^* is a planar triangulated graph, it has a unique embedding.

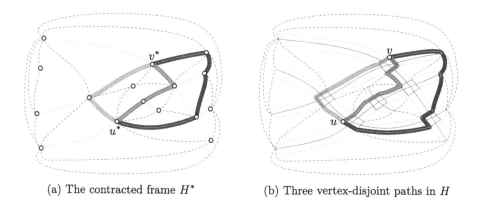

(a) The contracted frame H^* (b) Three vertex-disjoint paths in H

Fig. 4. Illustration for the proof of Lemma 1.

We now argue that H also has a unique embedding by proving that it is the subdivision of a planar 3-connected graph. Obviously, H has no vertices of degree 1. Hence, we only have to prove that there are three vertex-disjoint simple paths between any two vertices of degree at least 3. Let u, v be two such vertices of H.

First, assume that u and v belong to the same crossing box. If u and v are adjacent, then they share two common faces. One of the two faces is the interior of the crossing box, the other face is bounded by two tunnel boundaries and the endpoint of a long edge or an edge of a different crossing box. We obtain the three paths from the edge (u, v) and from following the boundaries of the common faces. If u and v are not adjacent, we find two paths following the edges of the crossing box. For the third path, let w^* be the corresponding vertex in H^*, let e_u and e_v be tunnel paths incident to u and v, respectively, and let e_u^* and e_v^* be the edges of H^* that correspond to e_u and e_v, respectively. Since H^* is 3-connected, there exists a cycle $C^* = \langle w^*, e_0^*, w_1^*, e_1^*, \ldots, e_{k-1}^*, w_k^*, e_k^*, w^* \rangle$ in H^* through $e_u^* = e_0^*$ and $e_v^* = e_k^*$. We find a path P from u to v in H by replacing each edge e_i^* of C^* by the corresponding tunnel path or connector in H, and each vertex w_k^* by the corresponding vertex or a path through the corresponding crossing box.

Now, assume that u and v do not belong to the same crossing box. Let u^* and v^* be the vertices of H^* corresponding to u and v, respectively. Let P_1^*, P_2^*, P_3^* be three vertex-disjoint paths between u^* and v^* in H^*. Similar to the previous case, we aim to find three paths P_1, P_2, P_3 by replacing the edges of P_1^*, P_2^*, P_3^* by corresponding tunnel paths or connectors and interior vertices by paths through corresponding crossing boxes, if necessary. Since the paths are vertex-disjoint, we do not visit any crossing boxes more than once, except those of u and v. In fact, if u is a vertex of a crossing box, it might happen that the tunnel paths that correspond to the first edges of P_1, P_2, P_3 do not start in u, but in different vertices of the crossing box of u.

To get rid of this problem, we now prove a slightly stronger version of Menger's theorem. Let G be a 3-connected graph with vertices u and v and edges $e_u = (u, u')$ and $e_v = (v, v')$. Then there exist three interior-vertex-disjoint paths P_1, P_2, P_3 from u to v such that $e_u, e_v \in P_1 \cup P_2 \cup P_3$. First, find three interior-vertex-disjoint paths P_1', P_2', P_3' from u to v. If these paths contain e_u and e_v, we are done. Otherwise, assume that they do not contain e_u. By 3-connectivity, there are three vertex-disjoint paths from u' to v, so at least one of them does not visit u. Let Q be this path. If Q is interior-vertex-disjoint from two of P_1', P_2', P_3', say P_2' and P_3', then $P_1 = \langle u, e_u, Q \rangle, P_2', P_3'$ are three interior-vertex-disjoint paths from u to v with $e_u \in P_1$. Otherwise, follow Q until it reaches an interior vertex of P_1', P_2', P_3' for the first time, say vertex w on P_1'. Create a path P_1 from u to v by following e_u, then Q until reaching w, then P_1' until reaching v. Then P_1, P_2', P_3' are three interior-vertex-disjoint paths from u to v with $e_u \in P_1$. We can force e_v to be part of one of the paths analogously.

Hence, we can assume that at least one of P_1, P_2, P_3 starts with a tunnel path at u and at least one of P_1, P_2, P_3 ends with a tunnel path at v. For the other two paths at u, we can reach the endpoint of the first tunnel path by following the crossing box, if necessary. An analogous argument works for the last edges to reach v.   □

Tunnels. In Fig. 5, we illustrate the regions in the plane that we refer to as *middle tunnel segments*, *end tunnel segments*, and *crossing boxes*. We believe that those notions are very easy to understand for the reader from the figure and thus we avoid a formal definition. (End or middle) tunnel segments incident to the same crossing box are called *consecutive*.

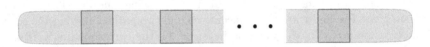

Fig. 5. Each tunnel consists of two end tunnel segments and various middle tunnel segments, separated by crossing boxes.

We say that a tunnel segment has *color c*, if it contains a long segment with color c. We will show that each tunnel segment has exactly one color.

Claim 1. *Each end tunnel segment contains at least one long segment.*

Consider any long edge ℓ and the frame H' without the blockers, and any embedding of $H' \cup \{\ell\}$. Since the frame has a unique combinatorial embedding and a long edge cannot cross a tunnel boundary or connector (as they have multiplicity t), this embedding must be plane and coincide with the unique combinatorial embedding of H on H'. The only way to add ℓ into the embedding of H' is through its end tunnel segment.

Claim 2. *Each tunnel segment has at most one color c and its bounding crossing box edges have all the remaining colors.*

Note that each tunnel segment can only be entered or left by a long segment using the multiedges of the crossing box. As those have multiplicity $t - 1$, it follows that the edges of the crossing box are colored with all colors except the color of the long segment. Now, we see that each tunnel segment is completely surrounded by edges of all but one colors.

Fig. 6. Consecutive tunnel segments must have the same color.

Claim 3. *Consecutive tunnel segments have the same color.*

See Fig. 6, for an illustration. Denote by S, T the two consecutive tunnel segments and by $C = [t]$ the set of all colors. Say tunnel S has color c. We show that tunnel T has the same color. Due to the unique embedding there is a blocker that intersects both tunnels. As the blocker is in a tunnel with color c, the bounding crossing box edges have colors $C \setminus c$. Thus, the blocker must have color c as well. Hence, also T must have color c as all bounding crossing box edges must have colors $C \setminus c$ (unless they are uncrossed). This proves the claim and also immediately the following:

Claim 4. *All tunnel segments of a tunnel have the same color.*

We are now ready to show our central claim:

Claim 5. *Each long segment stays in its respective tunnel.*

First, we note that all the blockers of a tunnel have the same color as the long edge. Furthermore, due to the combinatorial embedding, the tunnel paths together with the blockers form a cycle surrounding the long segment. It remains to show that the blockers cannot leave the tunnel, to show that the long segments cannot leave the tunnel. To this end, notice that a blocker cannot cross a crossing box edge twice, because they are both line segments and two line segments can cross at most once.

Claim 6. *The arrangement S formed by the long segments is combinatorially equivalent to the arrangement A of pseudo-segments.*

This follows from the tunnels crossing combinatorially as in arrangement A and the long segments staying within their respective tunnels.

Claim 6 concludes the proof of Theorem 1. Notice that in our construction, for every crossing, we add $2t$ more crossings ($2(t-1)$ from the parallel crossing box edges and two from the blockers). Since the pseudo-segment graph from Schaefer's construction has maximum degree 72 [21], the graph in our reduction is 8280-planar, which proves Corollary 2.

3   Sunflower Simultaneous Graph Embedding

This section is devoted to proving Theorem 2 by slightly modifying our construction in the proof of Theorem 1. We reduce PSEUDO-SEGMENT STRETCHABILITY to SIMULTANEOUS GRAPH EMBEDDING with the additional restriction that the input graphs of the SIMULTANEOUS GRAPH EMBEDDING instance form an empty sunflower. To this end, let A denote a pseudo-segment arrangement with n pseudo-segments. Further, let c be the chromatic number of the pseudo-segment intersection graph induced by A and let $\chi : A \to [c]$ denote a corresponding c-coloring.

We construct an instance of SIMULTANEOUS GRAPH EMBEDDING consisting of $k := c + 1$ simple graphs $H, G_1, \ldots, G_c$ on a shared vertex set as follows. The graph H contains exactly the edges belonging to the frame graph as defined above, i.e., the crossing boxes, the tunnel boundaries and the connectors. Moreover, for $i \in [c]$, the graph G_i contains the long edges corresponding to each pseudo-segment $S \in A$ for which $\chi(S) = i$ and a 1-subdivision of the edges of H aside from the crossing box edges bounding tunnel segments corresponding to pseudo-segments $S \in A$ for which $\chi(S) = i$. More precisely, in the 1-subdivision of the subgraph of H belonging to G_i, each edge $e = (u, v)$ of H is replaced by a path $(u, x_i(e), v)$ of length 2 where $x_i(e)$ does not belong to H, i.e., $x_i(e)$ is an isolated vertex in all graphs except for G_i. Since χ is a proper c-coloring, the graphs $G_1, \ldots, G_c$ do not share any long edges, while their 1-subdivisions of H are edge-disjoint by construction. As also H is edge-disjoint from any of the 1-subdivisions, we observe that $H, G_1, \ldots, G_c$ form an empty sunflower. Note that the construction here does not require any blockers.

It remains to discuss that A is stretchable if and only if $H, G_1, \ldots, G_c$ admit a simultaneous geometric embedding. First, completeness can be easily shown following the argumentation in the corresponding paragraph in Sect. 2. In particular, we need to discuss how to place the subdivision vertices of edges of H. Namely, for an edge $e = (u, v)$ of H, we can place all subdivision vertices $x_i(e)$ arbitrarily close to the straight-line segment representing (u, v). Completeness now immediately follows by observing that G_i contains no subdivisions of crossing-box edges that bound tunnel segments corresponding to a pseudo-segment $S \in A$ for which $\chi(S) = i$, i.e., the tunnel of a segment $S \in A$ with $\chi(S) = i$ is a single face in G_i minus the long edge representing S.

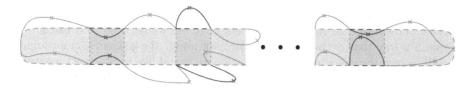

Fig. 7. In our construction for proving Theorem 2, subdivided tunnels do not contain subdivisions of crossing box edges shared by consecutive segments. While subdivided tunnels may cover not all of the tunnel and also additional parts of the plane, each curve connecting between both end segments traverses all segments in order.

Finally, we show soundness. Let φ be a simultaneous geometric embedding of $H, G_1, \ldots, G_c$. For a pseudo-segment $S \in \mathcal{A}$, we define the *subdivided tunnel* as the 1-subdivision of the *outer cycle* of its tunnel belonging to $G_{\chi(S)}$; see Fig. 7. Note that $G_{\chi(S)}$ does not contain a subdivision of the crossing box edges shared by consecutive segments of a subdivided tunnel; see dotted edges in Fig. 7. We now prove the equivalent of Claim 5:

Claim 7. *Each long segment traverses all segments and crossing boxes of its respective tunnel in order.*

First note that by Lemma 1, we know that H has a unique combinatorial embedding. Now consider a pseudo-segment S with $\chi(S) = i$. Since G_i contains subdivisions of all tunnel boundaries and all connectors, the corresponding long edge ℓ must be drawn completely inside tunnels. By construction, the endpoints of ℓ are contained only on the boundary of the subdivided tunnel corresponding to S, i.e., it must start and end inside its respective subdivided tunnel. Moreover, by construction, the entire subdivided tunnel belongs to G_i. Thus, ℓ cannot enter other tunnels in between. While the subdivided tunnel may be covering a superset of the tunnel, ℓ must still traverse all of its segments as the crossing box edges between consecutive segments still separate subdivided segments.

The fact that each long edge still traverses each crossing box in order implies immediately Lemma 6 in our reduction to SIMULTANEOUS GRAPH EMBEDDING and the theorem follows.

4 Conclusion and Discussion

In this paper, we have resolved the algorithmic complexity of GEOMETRIC THICKNESS for multigraphs. Our aim is to resolve the complexity also for simple graphs and in particular for geometric thickness two. It is noteworthy that our result for SIMULTANEOUS GRAPH EMBEDDING provides additional evidence for GEOMETRIC THICKNESS being ∃ℝ-hard for simple graphs as one can interpret SIMULTANEOUS GRAPH EMBEDDING as the subproblem that arises once the coloring has been fixed. Further, the union of simple graphs forming an empty sunflower is again a simple graph.

The main followup problem of our work is adapting our reduction for GEO-METRIC THICKNESS to simple graphs. For that, we would need to construct a graph in a way that we have some control bot about the geometric embedding as well as how the edges are colored. Ideally, we would like to construct a graph in a way that any coloring realizing its geometric thickness leads to the vertices being connected in all colors.

Question 1. Given $t \in \mathbb{N}$, does there always exist a graph with geometric thickness t such that any t-colored drawing of G realizing its geometric thickness is connected in all t colors?

Such a connected construction seems elusive and might not even be possible. However, we remark that connectivity in every color is not necessary in the geometric setting that we are considering.

Another intriguing question related to our reduction is the following:

Question 2. Is PSEUDO-SEGMENT STRETCHABILITY $\exists\mathbb{R}$-hard even for pseudo-segment arrangements of chromatic number 2?

In this direction, we showed that some modifications of Schaefer's reduction lower the current best bound on the chromatic number for $\exists\mathbb{R}$-hardness from 73 to 57 in an extended version [15]. We are convinced that we can further reduce the bound on the chromatic number to at most 48.

We remark that, using our techniques, positive answers to the above questions have the following implications:

(i) Assuming a positive answer to Question 2, our proof in Sect. 3 implies that SIMULTANEOUS GRAPH EMBEDDING is $\exists\mathbb{R}$-complete for three graphs forming an empty sunflower.

(ii) Assuming positive answers to Questions 1 and 2, it holds that GEOMETRIC THICKNESS is $\exists\mathbb{R}$-complete already for *simple* input graphs and thickness 2; see also the extended version of our paper [15].

References

1. Alekseev, V.B., Gončakov, V.S.: The thickness of an arbitrary complete graph. Math. USSR-Sbornik **30**(2), 187 (1976)
2. Beineke, L.W., Harary, F.: The thickness of the complete graph. Canadian J. Math. **17**, 850–859 (1965)
3. Brandenburg, F.J.: Straight-line drawings of 1-planar graphs. arXiv preprint arXiv:2109.01692 (2021)
4. Cardinal, J., Kusters, V.: The complexity of simultaneous geometric graph embedding. J. Graph Algor. Appl. **19**(1), 259–272 (2015)
5. Dillencourt, M.B., Eppstein, D., Hirschberg, D.S.: Geometric thickness of complete graphs. J. Graph Algor. Appl. **4**(3), 5–17 (2000)
6. Dujmovic, V., Joret, G., Micek, P., Morin, P., Ueckerdt, T., Wood, D.R.: Planar graphs have bounded queue-number. J. ACM **67**(4), 22:1–22:38 (2020)
7. Dujmovic, V., Morin, P.: Personal communication (2022)

8. Dujmovic, V., Morin, P., Wood, D.R.: Graph product structure for non-minor-closed classes. J. Comb. Theory, Ser. B **162**, 34–67 (2023)
9. Dujmovic, V., Pór, A., Wood, D.R.: Track layouts of graphs. Discret. Math. Theor. Comput. Sci. **6**(2), 497–522 (2004)
10. Durocher, S., Gethner, E., Mondal, D.: Thickness and colorability of geometric graphs. Comput. Geom. **56**, 1–18 (2016)
11. Eppstein, D.: Separating thickness from geometric thickness. In: Towards a Theory of Geometric Graphs, Contemporary Mathematics, vol. 342, pp. 75–86. American Mathematical Society (2004)
12. Erickson, J., van der Hoog, I., Miltzow, T.: Smoothing the gap between NP and ER. In: Proceedings 61st IEEE Symposium on Foundations of Computer Science (FOCS), pp. 1022–1033. ACM (2020)
13. Estrella-Balderrama, A., Gassner, E., Jünger, M., Percan, M., Schaefer, M., Schulz, M.: Simultaneous geometric graph embeddings. In: Hong, S.-H., Nishizeki, T., Quan, W. (eds.) Graph Drawing. LNCS, vol. 4875, pp. 280–290. Springer, Heidelberg (2008). https://doi.org/10.1007/978-3-540-77537-9_28
14. Fekete, S., Keldenich, P., Krupke, D., Schirra, S.: CG:SHOP 2022. https://cgshop.ibr.cs.tu-bs.de/competition/cg-shop-2022
15. Förster, H., Kindermann, P., Miltzow, T., Parada, I., Terziadis, S., Vogtenhuber, B.: Geometric thickness of multigraphs is ∃R-complete. arXiv preprint arXiv:2312.05010 (2023)
16. Harary, F.: Research problem. Bull. Am. Math. Soc. **67**, 542 (1961)
17. Kainen, P.C.: Thickness and coarseness of graphs. Abh. Math. Semin. Univ. Hambg. **39**, 88–95 (1973)
18. Mansfield, A.: Determining the thickness of graphs is NP-hard. Math. Proc. Camb. Philos. Soc. **93**(1), 9–23 (1983)
19. Nishizeki, T., Chiba, N.: Planar Graphs: Theory and Algorithms. Elsevier (1988)
20. Ringel, G.: Färbungsprobleme auf Flächen und Graphen, Mathematische Monographien [Mathematical Monographs], vol. 2. VEB Deutscher Verlag der Wissenschaften, Berlin (1959)
21. Schaefer, M.: Complexity of some geometric and topological problems. In: Eppstein, D., Gansner, E.R. (eds.) Graph Drawing. LNCS, vol. 5849, pp. 334–344. Springer, Heidelberg (2010). https://doi.org/10.1007/978-3-642-11805-0_32
22. Schaefer, M.: Complexity of geometric k-planarity for fixed k. J. Graph Algor. Appl. **25**(1), 29–41 (2021)
23. Schaefer, M.: On the complexity of some geometric problems with fixed parameters. J. Graph Algor. Appl. **25**(1), 195–218 (2021)
24. Tutte, W.T.: The thickness of a graph. Indagat. Math. (Proc.) **66**, 567–577 (1963)

Author Index

.

Printed in the United States
by Baker & Taylor Publisher Services